Transporters as Drug Carriers

Edited by
Gerhard Ecker and Peter Chiba

Methods and Principles in Medicinal Chemistry

Edited by R. Mannhold, H. Kubinyi, G. Folkers
Editorial Board
H. Timmerman, J. Vacca, H. van de Waterbeemd, T. Wieland

Previous Volumes of this Series:

Faller, Bernard / Urban, Laszlo (Eds.)

Hit and Lead Profiling

Identification and Optimization of Drug-like Molecules

2009
ISBN: 978-3-527-32331-9
Vol. 43

Sippl, Wolfgang / Jung, Manfred (Eds.)

Epigenetic Targets in Drug Discovery

2009
ISBN: 978-3-527-32355-5
Vol. 42

Todeschini, Roberto / Consonni, Viviana

Molecular Descriptors for Chemoinformatics

Volume I: Alphabetical Listing /
Volume II: Appendices, References

2009
ISBN: 978-3-527-31852-0
Vol. 41

van de Waterbeemd, Han / Testa, Bernard (Eds.)

Drug Bioavailability

Estimation of Solubility, Permeability, Absorption and Bioavailability

2., vollständig überarbeitete Auflage
2008
ISBN: 978-3-527-32051-6
Vol. 40

Ottow, Eckhard / Weinmann, Hilmar (Eds.)

Nuclear Receptors as Drug Targets

2008
ISBN: 978-3-527-31872-8
Vol. 39

Vaz, Roy J. / Klabunde, Thomas (Eds.)

Antitargets

Prediction and Prevention of Drug Side Effects

2008
ISBN: 978-3-527-31821-6
Vol. 38

Mannhold, Raimund (Eds.)

Molecular Drug Properties

Measurement and Prediction

2007
ISBN: 978-3-527-31755-4
Vol. 37

Wanner, Klaus / Höfner, Georg (Eds.)

Mass Spectrometry in Medicinal Chemistry

Applications in Drug Discovery

2007
ISBN: 978-3-527-31456-0
Vol. 36

Transporters as Drug Carriers

Structure, Function, Substrates

Edited by
Gerhard Ecker and Peter Chiba

WILEY-VCH

WILEY-VCH Verlag GmbH & Co. KGaA

Series Editors

Prof. Dr. Raimund Mannhold
Molecular Drug Research Group
Heinrich-Heine-Universität
Universitätsstrasse 1
40225 Düsseldorf
Germany
mannhold@uni-duesseldorf.de

Prof. Dr. Hugo Kubinyi
Donnersbergstrasse 9
67256 Weisenheim am Sand
Germany
kubinyi@t-online.de

Prof. Dr. Gerd Folkers
Collegium Helveticum
STW/ETH Zurich
8092 Zurich
Switzerland
folkers@collegium.ethz.ch

Volume Editors

Prof. Dr. Gerhard Ecker
University of Vienna
Department of Medicinal Chemistry
Althanstraße 14
1090 Wien
Austria
gerhard.f.ecker@univie.ac.at

Prof. Dr. Peter Chiba
Institut für Medizinische Chemie
Medizinische Universität Wien
Währinger Straße 10
1090 Wien
Austria
peter.chiba@meduniwien.at

All books published by Wiley-VCH are carefully produced. Nevertheless, authors, editors, and publisher do not warrant the information contained in these books, including this book, to be free of errors. Readers are advised to keep in mind that statements, data, illustrations, procedural details or other items may inadvertently be inaccurate.

Library of Congress Card No.: applied for

British Library Cataloguing-in-Publication Data
A catalogue record for this book is available from the British Library.

Bibliographic information published by the Deutsche Nationalbibliothek
The Deutsche Nationalbibliothek lists this publication in the Deutsche Nationalbibliografie; detailed bibliographic data are available on the Internet at http://dnb.d-nb.de.

© 2009 WILEY-VCH Verlag GmbH & Co. KGaA, Weinheim

All rights reserved (including those of translation into other languages). No part of this book may be reproduced in any form – by photoprinting, microfilm, or any other means – nor transmitted or translated into a machine language without written permission from the publishers. Registered names, trademarks, etc. used in this book, even when not specifically marked as such, are not to be considered unprotected by law.

Cover   Grafik-Design Schulz, Fußgönheim
Typesetting   Thomson Digital, Noida, India
Printing and Binding   Strauss GmbH, Mörlenbach

Printed in the Federal Republic of Germany
Printed on acid-free paper

ISBN: 978-3-527-31661-8

Contents

List of Contributors *XIII*
Preface *XVII*
A Personal Foreword *XIX*

Part One: Human Transporter Families – Structure, Function, Physiology *1*

1 **The ABC Transporters: Structural Insights into Drug Transport** *3*
Robert C. Ford, Alhaji B. Kamis, Ian D. Kerr, and Richard Callaghan
1.1 ABC Proteins – Structure and Function *3*
1.1.1 ABC Proteins *3*
1.1.2 Predicted Topology of ABC Proteins *4*
1.1.3 Nucleotide Binding Domains *4*
1.1.3.1 Conserved Motifs of NBDs *5*
1.1.4 Transmembrane Domains *5*
1.1.5 Mechanisms of Transport *5*
1.1.6 Energy for Translocation *6*
1.1.7 Coupling of ATP Hydrolysis to Transport *7*
1.2 Structures of ABC Transporters *7*
1.2.1 Tertiary Structure *7*
1.2.2 Quaternary Structure of ABC Proteins *11*
1.3 Multidrug Resistance and ABC Transporters *15*
1.3.1 P-Glycoprotein *15*
1.3.1.1 Historical Background *15*
1.3.1.2 The Role of P-gp in Drug Resistance *16*
1.3.1.3 Tissue Distribution and Physiological Roles *17*
1.3.2 Conformational Changes in the Mechanism of P-gp *17*
1.3.3 Comparison of Sav1866 and P-gp Structures *19*
1.3.4 Drug Binding Sites in P-Glycoprotein *21*
1.3.5 Structural Interpretation of Drug Binding *25*
1.3.6 Inhibitors of P-gp *27*
1.3.7 What Properties Are Shared by Drugs that Interact with P-Glycoprotein? *29*

1.3.8	Postscript: Further X-ray Crystallographic Studies and a Structure for the Nucleotide-Free State of P-Glycoprotein *30*
1.4	Summary *31*
	References *31*

2 Biochemistry, Physiology, and Pharmacology of Nucleoside and Nucleobase Transporters *49*

Marçal Pastor-Anglada, Míriam Molina-Arcas, Pedro Cano-Soldado, and Francisco Javier Casado

2.1	Nucleoside and Nucleobase Transporters *49*
2.1.1	Equilibrative Nucleoside Transporters *50*
2.1.2	Concentrative Nucleoside Transporters *51*
2.2	ENT and CNT Tissue Distribution, Regulation, and Physiological Roles *53*
2.2.1	ENT Tissue Distribution and Regulation *54*
2.2.2	CNT Tissue Distribution and Regulation *56*
2.2.2.1	CNTs in Absorptive Epithelia *57*
2.2.2.2	CNTs in Liver Parenchymal Cells *57*
2.2.2.3	CNTs in Immune System Cells *58*
2.2.2.4	CNTs in CNS *59*
2.2.2.5	CNTs in Other Specialized Tissues *60*
2.2.3	NTs as "Transceptors" *60*
2.3	Nucleoside- and Nucleobase-Derived Drug Transport into Cells *61*
2.3.1	Transport of Anticancer Drugs *61*
2.3.2	Transport of Antiviral Drugs *63*
2.4	Drug Transport and Responsiveness to Treatment *65*
2.4.1	Analysis of the Role of NTs in Sensitivity to Nucleoside Anticancer Drugs in Cultured Cell Models *65*
2.4.2	Studies Linking NT Function to Drug Sensitivity and Clinical Outcome in Cancer Patients *67*
2.5	Future Perspectives *69*
	References *70*

3 Organic Anion Transporting Polypeptides (Oatps/OATPs) *81*

Mine Yarim and Meric Koksal

3.1	Introduction *82*
3.2	Nomenclature and Classification *82*
3.3	Tissue Distribution, Structure, and Functions *83*
3.4	Substrate Spectrum *85*
3.5	Members of the Rodent Oatp Family *93*
3.5.1	Oatp1a1 *93*
3.5.2	Oatp1a3-v1/v2 *93*
3.5.3	Oatp1a4 *93*
3.5.4	Oatp1a5 *94*
3.5.5	Oatp1a6 *94*

3.5.6	Oatp1b2	*94*
3.5.7	Oatp1c1	*95*
3.5.8	Oatp2a1	*95*
3.5.9	Oatp2b1	*95*
3.5.10	Oatp3a1	*95*
3.5.11	Oatp4a1	*96*
3.5.12	Oatp4c1	*96*
3.5.13	Oatp6b1/Oatp6c1	*96*
3.5.14	PGT-2	*96*
3.5.15	TST-1 and TST-2	*97*
3.6	Members of Nonmammalian Oatp Family	*97*
3.7	Members of Human OATP Family	*97*
3.7.1	OATP1A2	*97*
3.7.2	OATP1B1	*98*
3.7.3	OATP1B3	*98*
3.7.4	OATP1C1	*99*
3.7.5	OATP2A1	*99*
3.7.6	OATP2B1	*99*
3.7.7	OATP3A1	*100*
3.7.8	OATP4A1	*100*
3.7.9	OATP4C1	*100*
3.7.10	OATP5A1	*100*
3.7.11	OATP6A1	*100*
3.8	Drug Disposition and Drug–Drug Interactions	*101*
3.9	Computational Approaches	*102*
3.10	Conclusions	*103*
	References	*104*

4 CNS-Transporters as Drug Targets *113*
Klaus Gundertofte

4.1	Introduction	*113*
4.2	Structure of Transporters	*113*
4.3	Monoamine Transporters	*114*
4.4	Transporters for Amino Acids	*115*
4.5	Nonneurotransmitter Transporters	*116*
4.6	Concluding Remarks	*116*
	References	*117*

Part Two: Drug Transport in Microorganisms and Fungi *119*

5 Bacterial Multidrug Transporters: Molecular and Clinical Aspects *121*
Olga Lomovskaya, Helen I. Zgurskaya, and Keith Bostian

5.1	Introduction	*121*
5.1.1	The Multiple Antibiotic Resistance Problem	*121*

5.1.2	The Superbugs	*121*
5.1.3	The Multidrug Resistance Transporters	*122*
5.2	Diversity of Bacterial MDR Efflux Systems	*123*
5.2.1	ABC Transporters	*124*
5.2.2	MFS Transporters	*126*
5.2.3	MATE MDR Transporters	*127*
5.2.4	SMR MDR Transporters	*128*
5.2.5	RND MDR Transporters	*129*
5.2.6	Diversity on a Theme	*133*
5.3	Accessory Proteins from Gram-Negative Bacteria	*134*
5.4	Efflux and Antibiotic Resistance	*136*
5.4.1	Gram-Positive Efflux Resistance	*137*
5.4.2	Gram-Negative Efflux Resistance	*139*
5.4.3	Inhibiting Gram-Negative RND Pumps	*141*
5.5	The Search for Efflux Inhibitors	*143*
5.6	Challenges and Perspectives	*144*
	References	*146*

6 **Membrane Transporters in Pleiotropic Drug Resistance and Stress Response in Yeast and Fungal Pathogens** *159*
Tobias Schwarzmüller, Cornelia Klein, Martin Valachovic, Walter Glaser, and Karl Kuchler

6.1	Introduction	*159*
6.2	ABC Protein Genes in *S. cerevisiae*	*162*
6.2.1	The PDR Family	*163*
6.2.2	The MRP/CFTR Family	*167*
6.3	Orchestrating Pleiotropic Drug Resistance: The PDR Network	*168*
6.4	ABC Drug Transporters of Human Fungal Pathogens	*171*
6.5	Physiological Roles of Drug Transporting ABC Proteins – Search for Substrates	*176*
6.6	Conclusions and Perspectives	*177*
	References	*178*

Part Three: Structure Activity Relationship Studies on ABC Transporter *195*

7 **QSAR Studies on ABC Transporter – How to Deal with Polyspecificity** *197*
Gerhard F. Ecker

7.1	The Problem of Polyspecificity/Promiscuity	*197*
7.2	QSAR Approaches to Design Inhibitors of P-glycoprotein (ABCB1)	*197*
7.3	Other ABC Transporter	*206*
7.3.1	ABCG2 (Breast Cancer Resistance Protein, MXR)	*206*
7.3.2	ABCC1 and ABCC2 (Multidrug Resistance-Related Proteins 1 and 2)	*207*

7.3.3	ABCB11 (Bile Salt Export Pump)	208
7.4	Novel Methods	209
7.5	Structural Basis for Polyspecificity	210
7.6	Conclusions and Outlook	210
	References	211
8	**Drug Transporter Pharmacophores**	**215**
	Sean Ekins	
8.1	Introduction	215
8.2	Database Searching with Transporter Pharmacophores	217
8.3	Summary	220
	References	221

Part Four: Transporters and ADME 229

9	**Biological Membranes and Drug Transport**	**231**
	Gert Fricker	
9.1	Biological Membranes	231
9.1.1	Lipid Bilayer	231
9.1.2	Membrane Proteins	233
9.1.3	Membrane Carbohydrates	233
9.2	Membrane Transport	233
9.2.1	Mechanisms of Transport	233
9.2.2	Transport Across Lipid Membranes	233
9.2.3	Protein-Coupled Membrane Transport	235
9.2.4	Kinetics of Carrier-Mediated Transport Processes	236
9.2.5	Ion Gradient-Dependent Transport Processes	237
9.2.6	Cytotic Mechanisms: Transport of Macromolecules	238
9.2.7	Export Proteins	238
9.3	Pharmacokinetic-Relevant Membrane Barriers	239
9.3.1	Intestinal Drug Absorption	239
9.3.2	Liver	244
9.3.3	Kidney	246
9.3.4	Blood–Brain Barrier and Choroid Plexus	251
	References	254
10	**Transport at the Blood–Brain Barrier**	**263**
	Winfried Neuhaus and Christian R. Noe	
10.1	The Blood–Brain Barrier	263
10.2	Transport Mechanisms Across the Blood–Brain Barrier	267
10.3	The Physical Barrier: Paracellular Transport and Its Characterization	270
10.4	The Efflux Barrier: Transport Proteins at the Blood–Brain Barrier	272
10.5	Transporter of the SLC Transporter Family	274

10.5.1	Monocarboxylate Transporters 274
10.5.2	Organic Ion Transporters and Transporting Peptides 276
10.6	Transporter of the ABC Transporter Family 278
10.6.1	P-Glycoprotein 278
10.6.2	Multidrug Resistance-Associated Proteins 282
10.6.3	Breast Cancer Resistance Protein 284
10.7	The Metabolic Barrier: Enzymes at the Blood–Brain Barrier 285
10.8	How to Overcome the Blood–Brain Barrier 286
	References 286

11	**Bile Canalicular Transporters** 299
	Dirk R. de Waart and Ronald P.J. Oude Elferink
11.1	Introduction 300
11.2	ABCC2 301
11.3	ABCG2 304
11.4	ABCB1 305
11.5	ABCB4 306
11.6	ABCB11 307
11.7	ABCG5 and ABCG8 308
11.8	Canalicular Transporters as Targets for Drug Delivery 309
	References 313

12	**Interplay of Drug Metabolizing Enzymes and ABC Transporter** 325
	Walter Jäger
12.1	Combined Role of Cytochrome P450 3A and ABCB1 325
12.2	Combined Role of Cytochrome P450 3A and OATPs 329
12.3	Combined Role of UDP-Glucuronosyltransferases and ABCC2 332
12.4	Biopharmaceutical Classification System 337
	References 342

13	**ABC Transporters – From Targets to Antitargets?** 349
	Gerhard F. Ecker and Peter Chiba
13.1	Introduction 349
13.2	ABC Transporters as Targets 349
13.2.1	P-Glycoprotein (ABCB1) 350
13.2.2	Other ABC Transporter as Drug Targets 353
13.2.2.1	ABCG2 (Breast Cancer Resistance Protein, MXR) 353
13.2.2.2	ABCC1 and ABCC2 (Multidrug Resistance Proteins 1 and 2) 353
13.3	P-Glycoprotein – An Antitarget? 354
13.3.1	Gastrointestinal Absorption 354
13.3.2	Brain Uptake 355
13.4	Predicting Substrate Properties for P-Glycoprotein 355
13.4.1	Data Sets 356
13.4.2	Classification Models 356
13.4.3	Pharmacophore Models 358

13.4.4	Nonlinear Methods	*358*
13.5	Conclusions	*358*
	References	*359*

Part Five: A Systems View of Drug Transport *363*

14 A Systems Biology View of Drug Transporters *365*
Sean Ekins and Dana L. Abramovitz
14.1 Introduction *365*
14.2 Regulation of Transporters *367*
14.3 Network Analysis of Transporters and Ligands *368*
14.4 Transporter Network Examples *369*
14.4.1 MetaCore *369*
14.4.2 Ingenuity Pathways Analysis *372*
14.5 Transporter Gene Expression Data *376*
14.6 Summary and Future Perspectives *377*
References *379*

15 Drug Transporters in Health and Disease *387*
Barbara Bennani-Baïti and Christian R. Noe
15.1 Introduction *387*
15.2 General Mechanisms of Drug Transporter Expression *388*
15.3 Neurological Disorders *389*
15.3.1 Epilepsy *390*
15.3.1.1 Seizure Frequency and RE *394*
15.3.1.2 The Role of Drug Transporters in RE Development *395*
15.3.1.3 Overcoming RE *396*
15.3.2 MDR1 Expression and Neurodegenerative Disorders *397*
15.4 Inflammation *398*
15.4.1 Inflammation – The Acute-Phase Response *398*
15.4.2 BBB and Inflammation *399*
15.4.3 Kidney and Inflammation *400*
15.4.4 Liver and Inflammation *401*
15.4.5 Intestine and Inflammation *404*
References *407*

Index *417*

List of Contributors

Dana L. Abramovitz
Ingenuity Systems, Inc.
1700 Seaport Blvd., Third Floor
Redwood City, CA 94063
USA

Barbara Bennani-Baïti
University of Vienna
Pharmacy Center
Department of Medicinal Chemistry
Althanstraße 14
1090 Vienna
Austria

Keith Bostian
Mpex Pharmaceuticals, Inc.
11535 Sorrento Valley Road
San Diego, CA 92121
USA

Richard Callaghan
University of Oxford
John Radcliffe Hospital
Nuffield Department of Clinical
Laboratory Sciences
Oxford OX3 9DU
UK

Pedro Cano-Soldado
University of Barcelona (IBUB)
Department of Biochemistry and
Molecular Biology
Diagonal 645
08028 Barcelona
Spain

and

CIBER EHD
Barcelona
Spain

Francisco Javier Casado
University of Barcelona (IBUB)
Department of Biochemistry and
Molecular Biology
Diagonal 645
08028 Barcelona
Spain

and

CIBER EHD
Barcelona
Spain

Peter Chiba
Medical University of Vienna
Institute of Medical Chemistry
Waehringerstrasse 10
1090 Vienna
Austria

Transporters as Drug Carriers: Structure, Function, Substrates. Edited by Gerhard Ecker and Peter Chiba
Copyright © 2009 WILEY-VCH Verlag GmbH & Co. KGaA, Weinheim
ISBN: 978-3-527-31661-8

Dirk R. de Waart
Academic Medical Center
AMC Liver Center
Meibergdreef 71
1105 BK Amsterdam
The Netherlands

Gerhard F. Ecker
University of Vienna
Department of Medicinal Chemistry
Althanstraße 14
1090 Wien
Austria

Sean Ekins
Collaborations in Chemistry
601 Runnymede Avenue
Jenkintown, PA 19046
USA

Robert C. Ford
University of Manchester
Faculty of Life Sciences
Manchester Interdisciplinary Biocentre
Manchester M1 7DN
UK

Gert Fricker
Institute of Pharmacy and Molecular Biotechnology
Im Neuenheimer Feld 366
69120 Heidelberg
Germany

Walter Glaser
Medical University Vienna
Max F. Perutz Laboratories
Christian Doppler Laboratory for Infection Biology
Dr. Bohr-Gasse 9/2
1030 Vienna
Austria

Klaus Gundertofte
H. Lundbeck A/S
Medicinal Chemistry Research
Ottiliavej 9
2500 Valby-Copenhagen
Denmark

Walter Jäger
University of Vienna
Department of Clinical Pharmacy and Diagnostics
1090 Vienna
Austria

Alhaji B. Kamis
University of Manchester
Faculty of Life Sciences
Manchester Interdisciplinary Biocentre
Manchester M1 7DN
UK

Ian D. Kerr
University of Nottingham
Queen's Medical Centre
School of Biomedical Sciences
Centre for Biochemistry and Cell Biology
Nottingham NG7 2UH
UK

Cornelia Klein
Medical University Vienna
Max F. Perutz Laboratories
Christian Doppler Laboratory for Infection Biology
Dr. Bohr-Gasse 9/2
1030 Vienna
Austria

List of Contributors

Meric Koksal
Yeditepe University
Faculty of Pharmacy
Department of Pharmaceutical
Chemistry
Kayisdagi, Kadikoy, Istanbul 34755
Turkey

Karl Kuchler
Medical University Vienna
Max F. Perutz Laboratories
Christian Doppler Laboratory for
Infection Biology
Dr. Bohr-Gasse 9/2
1030 Vienna
Austria

Olga Lomovskaya
Mpex Pharmaceuticals, Inc.
11535 Sorrento Valley Road
San Diego, CA 92121
USA

Míriam Molina-Arcas
University of Barcelona (IBUB)
Department of Biochemistry and
Molecular Biology
Diagonal 645
08028 Barcelona
Spain

and

CIBER EHD
Barcelona
Spain

Winfried Neuhaus
Pharma Con GmbH
Riglergasse 4/5
1180 Vienna
Austria

Christian R. Noe
University of Vienna
Department of Medicinal Chemistry
Althanstraße 14
1090 Vienna
Austria

Ronald P.J. Oude Elferink
Academic Medical Center
AMC Liver Center
Meibergdreef 71
1105 BK Amsterdam
The Netherlands

Maral Pastor-Anglada
University of Barcelona (IBUB)
Department of Biochemistry and
Molecular Biology
Diagonal 645
08028 Barcelona
Spain

and

CIBER EHD
Barcelona
Spain

Tobias Schwarzmüller
Medical University Vienna
Max F. Perutz Laboratories
Christian Doppler Laboratory for
Infection Biology
Dr. Bohr-Gasse 9/2
1030 Vienna
Austria

Martin Valachovic
Medical University Vienna
Max F. Perutz Laboratories
Christian Doppler Laboratory for
Infection Biology
Dr. Bohr-Gasse 9/2
1030 Vienna
Austria

Mine Yarim
Yeditepe University
Faculty of Pharmacy
Department of Pharmaceutical
Chemistry
Kayisdagi, Kadikoy, Istanbul 34755
Turkey

Helen I. Zgurskaya
University of Oklahoma
Department of Chemistry and
Biochemistry
620 Parrington Oval, Room 208
Norman, OK 73019
USA

Preface

Although the phenomenon of multidrug resistance of bacteria was observed more than fifty years ago, it took 20 years until the first drug transporter, P-glycoprotein, was discovered as the responsible cellular factor for the outward transport of xenobiotics of different chemical structure. Another ten years later, experimental results on different tumor cell lines indicated that P-glycoprotein also occurs in advanced cancers and plays a major role in contributing to the non-response to chemotherapy.

The broad role of transporters in drug absorption, distribution and elimination, as well as in drug-drug interactions and (multi)drug resistance has only been recognized in recent decades. For almost a century it seemed clear that lipophilicity governs drug absorption and distribution. It was well accepted that drugs that mimic endogenous substrates, like amino acid, sugar and nucleoside analogs, use transporters to cross cell membranes but it was considered to be limited to such compound classes. However, in recent years more and more transporters were discovered and with this increasing number also more and more cases of active drug transport were observed. This fact even generated the speculation that in drug absorption active transport is rather the rule than the exception, another extreme hypothesis. A definite answer to this open question cannot be given at the very moment but it is interesting to watch the engaged discussion on the pros and cons.

In addition to their role in drug absorption, distribution and elimination, transporters are also responsible for certain drug-drug interactions. Drugs like verapamil, propafenone and quinidine are P-glycoprotein inhibitors; co-medication of these drugs with other active agents, normally eliminated by P-glycoprotein, may generate serious side effects. Non-sedating H_1 antihistaminics cross the blood-brain barrier like the classical, sedating antihistaminics but active efflux avoids their interaction with central histamine receptors. The opiate loperamide is a selective antidiarrhoic agent; however, if P-glycoprotein is inhibited by quinidine, loperamide exerts the typical central effects of all other opiates. Some drugs and even "harmless" agents, such as St. Johns Wort or grapefruit juice, induce the expression of drug-metabolizing cytochromes and of drug transporters, leading to other drug-drug interactions.

The book by Gerhard Ecker and Peter Chiba collects and evaluates the available evidence for further research in this hot area. For this purpose, the editors assembled a team of experienced scientists to discuss the important role of drug transporters in detail. We are very grateful to all authors for their excellent contributions, as well as to Frank Weinreich and Waltraud Wüst for their ongoing engagement for our series Methods and Principles in Medicinal Chemistry, in which this book will be another highlight.

April 2009

Raimund Mannhold, Düsseldorf
Hugo Kubinyi, Weisenheim am Sand
Gerd Folkers, Zurich

A Personal Foreword

An Introduction to the Medicinal Chemistry of Drug Transport

Membrane transporters are encoded by numerous gene families, comprising in total 883 genes encoding a broad variety of transporters. This so-called transportome performs important functions for the cell, such as providing nutrients, protecting the cell from xenotoxins, and establishing electrochemical gradients across membranes. Numerous disorders caused by mutations in transporter and channel genes underscore the physiological relevance of transport proteins. These include the Dubin–Johnson syndrome (ABCC2), sitosterolemia (ABCG5/G8), and Tangier disease (ABCA1), to mention a few. Membrane transporters also play a key role in ADME, affecting absorption, distribution, and elimination of drugs. Recently, this has also been recognized by regulatory authorities and the FDA published guidance for how to deal with transport processes. In addition, some ABC (ATP-binding cassette) transporters such as the multiple drug resistance transporter ABCB1 (P-glycoprotein) mediate energy-dependent efflux of drugs and thereby significantly contribute to the development of drug resistance. With respect to the latter, one should also note that numerous transporters have been identified in bacteria, fungi, and plasmodia and are responsible for resistance against chemotherapeutic agents in these organisms.

Given the large number of transport proteins and potential substrates, only a very small percentage of the possible pharmacological interactions among them have been studied so far. Those interactions may be particularly important in the chemotherapy of cancer, for drug–drug and drug–nutrient interactions, and for the bioavailability and brain permeation of drug candidates. Classical examples include the interaction of cyclosporin A with several statins and also the well-known multifactorial interaction of grape fruit juice: naringine is blocking OATP1A2, hesperidin is blocking ABCB1, and bergamottin is interacting with CYP3A4. This leads to a reduction in the plasma concentration of, for example, fexofenadine, talinolol, and celiprolol. However, drug–drug interactions at transport proteins might also be used in a beneficial way. During the Second World War, the so-called wartime tactic was applied by coadministering penicillin with probenecid. Blocking of hOAT

Transporters as Drug Carriers: Structure, Function, Substrates. Edited by Gerhard Ecker and Peter Chiba
Copyright © 2009 WILEY-VCH Verlag GmbH & Co. KGaA, Weinheim
ISBN: 978-3-527-31661-8

transporters in the kidney by probenecid enabled a significant reduction in the penicillin dosing and thus treatment of a higher number of patients. An identical approach has recently been proposed for tamiflu.

Nonetheless, a systematic study of the transportome's role in ADE (absorption, distribution, and elimination), chemosensitivity, and chemoresistance is still lacking. Furthermore, from a systems point of view, the situation is even more complex. Nuclear receptors, cytochromes, and transporters form a protein network responsible for elimination and toxification/detoxification of most of the drugs currently used. This network is subject to multifactorial influences, such as induction of the expression of P-glycoprotein by St. John's wort. Last but not the least, there is increasing evidence of considerable species-specific differences. Thus, using quantitative proteomics could demonstrate that in humans the amount of ABCG2 at the blood–brain barrier is twofold the amount of ABCB1, whereas in rodents the ratio is almost 1:1. Another example is the CNS toxicity of ivermectin in collie dogs, which lack functional P-glycoprotein at the blood–brain barrier due to an MDR1-1delta mutation.

This volume of *Methods and Principles in Medicinal Chemistry* features different classes of membrane transport proteins and highlights their importance in the field of medicinal chemistry. Part One highlights the importance of several human transporter families as drug carriers. Special focus is given on the structure and function of P-glycoprotein, the paradigm protein in the field. The recently published X-ray structure of mouse P-gp now also paves the way for structure-guided modeling studies. The chapter on CNS transporters has been kept short as these will be covered in an upcoming volume of this series. Part Two focuses on drug transporters in bacteria and fungi, which are relevant for medicinal chemists. Part Three gives an overview on rational drug design approaches pursued for prediction of interaction of transporters with their ligands. Part Four is devoted to the role of drug transporters at physiological barriers as well as the interplay of transporters and metabolic enzymes. Finally, an overview on systems biology approaches and the role of drug transporters under pathophysiological conditions is given.

We were in the favorite position to win a number of high-profile research scientists to contribute to this effort and share their views and opinions. We would like to thank all authors for their excellent contributions and also for their patience during the editing process. We would also like to express our sincere appreciation to Frank Weinreich, Waltraud Wüst, and the helpful hands at Wiley-VCH for their excellent support in the production of this book. Finally, we also thank Raimund Mannhold, Hugo Kubinyi, and Gerd Folkers for their enthusiasm and continuous efforts to provide the medicinal chemistry community with this outstanding Methods and Principles series of books.

Enjoy reading!

Vienna, Summer 2009 *Peter Chiba* and *Gerhard Ecker*

Part One:
Human Transporter Families – Structure, Function, Physiology

1
The ABC Transporters: Structural Insights into Drug Transport*

Robert C. Ford, Alhaji B. Kamis, Ian D. Kerr, and Richard Callaghan

Abbreviations

ABC	ATP binding cassette
ATP	adenosine triphosphate
CFTR	cystic fibrosis transmembrane conductance regulator
CsA	cyclosporin A
ICL	intracytoplasmic loop
MRP	multidrug resistance protein
NBD	nucleotide binding domain
NCS	noncrystallographic symmetry
P-gp	P-glycoprotein
SUR	sulfonylurea receptor
TMD	transmembrane domain

1.1
ABC Proteins – Structure and Function

1.1.1
ABC Proteins

A superfamily of membrane proteins involved in transport has been designated the ATP binding cassette (ABC) family [1]. About 5% of the entire *Escherichia coli* genome encodes components of ABC transporters [2]. Their existence across all the domains of life is an indication of their wide physiological roles, and in humans they are associated with several diseases [3]. The members of this superfamily bind ATP and typically use the energy from hydrolysis to translocate a wide range of substances (including sugars, amino acids, glycans, sterols, phospholipids, peptides, proteins, toxins, antibiotics, and xenobiotics) across cytoplasmic and organellar mem-

*This chapter is dedicated to the memory of Alhaji Bukar Kamis.

Transporters as Drug Carriers: Structure, Function, Substrates. Edited by Gerhard Ecker and Peter Chiba
Copyright © 2009 WILEY-VCH Verlag GmbH & Co. KGaA, Weinheim
ISBN: 978-3-527-31661-8

branes [4–8]. In addition to the export and import functions, ABC transporters are known to function as receptors and channels and to mediate other physiological phenomena [4, 9]. The transporters appear to be always unidirectional (see Ref. [10]); but in bacteria, they are either importers or exporters while in eukaryotes they are exclusively exporters (so far) [7].

1.1.2
Predicted Topology of ABC Proteins

Topology prediction indicates the existence of transmembrane domains (TMDs) with between 4 and 10 (TM) α-helices and cytosolic ATP binding/hydrolysis domains designated nucleotide binding domains (NBDs). The minimal structural requirement for an active prokaryotic and eukaryotic transporter is thought to consist of two TMDs and NBDs each (Figure 1.1) [11–13]. This minimum requirement can be satisfied by a single polypeptide chain (full transporter) or can be assembled from two equal (homodimeric) or unequal (heterodimeric) polypeptide chains (half transporters). The assembly of one full transporter protein can be from four, three, two, or one separate polypeptide. Some ABC transporters are observed to possess an additional TMD and also soluble regulatory domains associated with the NBDs. Many bacterial importers have an accessory substrate binding protein (SBP).

1.1.3
Nucleotide Binding Domains

ABC transporters use the energy of nucleotide hydrolysis to transport substances across membranes against a concentration gradient [14]. All ABC proteins have

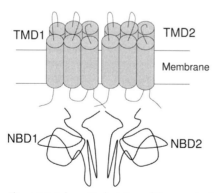

Figure 1.1 Schematic depiction of the organization of ABC proteins. The minimal structural unit appears to consist of two integral transmembrane domains, each with (typically) six helical transmembrane spans, and two extrinsic nucleotide binding domains. At the amino acid sequence level, the nucleotide binding domains are relatively well conserved across the domains of life compared to the transmembrane domains.

two cytoplasmic nucleotide binding domains containing conserved sequence motifs for binding/hydrolysis of ATP; this can be thought of as a standard engine that is bolted onto a specialized translocation pathway – formed by the TMDs [15]. In some prokaryotic NBDs, both are functionally identical and presumably contribute equally to the protein activity, whereas in eukaryotes, the NBDs are often functionally nonequivalent. Deletion of either NBD abolishes transport [14, 16–24], although recently Riordan and colleagues showed that NBD2 was partly dispensable for channel activity in mammalian CFTR [25]. Senior and colleagues proposed that ATP hydrolysis at the two NBDs could take place in an alternate fashion [26], and other workers have argued for functional symmetry [27–29] or asymmetry [30–34].

1.1.3.1 Conserved Motifs of NBDs

ABC transporters translocate many different allocrites (transport substrates), but the primary structures of the NBDs show ~25% sequence identity across the whole superfamily [9]. Marked sequence conservation is observed over five short regions found in the NBDs: (i) the Walker A and (ii) Walker B regions, which are separated by approximately 90–120 amino acids and between which lie the (iii) "signature" motif [2, 7, 9, 35, 36] and (iv) the glutamine loop (Q-loop) [37]. The most C-terminal motif is the histidine loop (H-loop) [2]. The signature, Q-loop, and H-loop seem to be specific to ABC transporters [9], and the function of these is described in more detail below. A recent description of the importance of a highly conserved aromatic residue has led to the naming of an "A-loop" at the N-terminus of the NBD [38].

1.1.4
Transmembrane Domains

The primary function of the TMDs is to provide a pathway through which allocrites can cross the membrane. It is not difficult to imagine how such a pathway could readily be adopted for channel function (e.g., CFTR), but for receptor-type ABC proteins (such as SUR1), the evolutionary step is less obvious. However, for all these functions, the TMDs appear to be the main determinants of specificity [3, 39, 40]. TMDs are much more variable in their amino acid sequences than the NBDs [15, 41]. Hydropathy predictions, which have so far been borne out by later structural studies, imply multiple transmembrane α-helices, typically six per domain (see Section 1.2), but with many exceptions [3]. Eukaryotic ABC proteins in the C class (ABCC subdivision) frequently contain a whole extra TMD with five predicted transmembrane spans [7].

1.1.5
Mechanisms of Transport

It is a general assumption that the TMDs provide the pathway for translocation of allocrites. The crystal structure of a prokaryotic ABC transporter (BtuCD) revealed that the TMDs create an outward-facing cavity at the interface between the two (BtuC) subunits. It was postulated that alternate opening and closing of this cavity to the periplasmic and cytoplasmic sides could allow translocation [15, 42]. The recent structure of a related *Haemophilus influenzae* ABC protein, which shows a similar central cavity but is more open on the inner facing (cytoplasmic) side, has added strength to this hypothesis. However, in neither case was allocrite bound to the protein [43]. On the other hand, a structure with allocrite bound was first published for the bacterial lipid A exporter/flippase MsbA [44]. In this structure, the (two, NCS-related) lipid A molecules were situated on the outside of the protein on the extracellular side, rather than at the interface between the two TMDs. Whether this configuration represents an end point of the translocation pathway or perhaps indicates translocation pathways on the outside rather than the inside of this transporter remains to be established. It was possible that the location of the lipid A molecules may be determined by crystal packing constraints rather than representing the physiological binding sites. However, this and earlier structures of MsbA were removed from the protein data bank (www.rcsb.org), and the papers describing them were retracted [45, 46]. Revised structures for MsbA were later published [47], but in the new manuscript and deposited atomic coordinates, the bound lipid A molecules were not present. Recently, however, the first information on the mechanism and location of allocrite binding has emerged with the publication of the structure of the maltose transporter MalFGK$_2$ [48] with its cognate periplasmic substrate binding protein. These data show that maltose is bound in an outward-facing crevice formed between the two transmembrane proteins (MalF and MalG). This crevice is capped at the top by the SBP, the maltose binding protein, thus forming a cavity in which the allocrite is sequestered. Low- and medium-resolution structures determined for P-gycoprotein (P-gp) and MRP1 also suggested the existence of a central cavity [3, 49–51]. However, findings indicating the existence of two or three allocrite binding sites for P-gp suggest that a single translocation pathway model may be too simplistic [40, 52–56]. That P-gp has a common site for structurally unrelated allocrites may best be explained by allocrite-induced fit [56–58], implying that the translocation pathway may be relatively adaptable. P-gp has also been reported to possess at least one regulatory site [59, 60] in the TMDs.

1.1.6
Energy for Translocation

Early structural studies of NBDs in isolation revealed at least three different possibilities for dimer organization [9, 37, 61–63]. However, accumulating evidence now strongly suggests that the NBDs form a sandwich dimer in which the two ATP molecules are concertedly bound at the NBD interface formed between the Walker A, B, and signature regions [15, 64]. The question remains, especially in eukaryotes, as to the functional equivalence or asymmetry of the NBDs. Several findings suggest that the two NBDs may be functionally equivalent, with one NBD undergoing catalysis at any given point in time (see Section 1.3). However, biochemical and mutational studies have indicated that NBDs of P-gp, MRP1, and CFTR are functionally distinguishable [18, 19, 32, 65–69].

1.1.7
Coupling of ATP Hydrolysis to Transport

The coupling of nucleotide binding/hydrolysis in the NBDs to allocrite transport in the TMDs is envisaged as a coupling of conformational changes in each half of the ABC proteins. Mutational analysis of homodimeric LmrA showed that disruption of NBD inhibited not only the ATPase activity but also the transport function of the protein suggesting interactions between NBDs and TMDs [70–73]. These findings are in agreement with studies of P-gp where ATP binding/hydrolysis and vanadate trapping have profound effects on allocrite binding affinities [40, 74–80]. The switch from high to low affinity for allocrite is probably an integral part of transport/release, and coupled conformational changes could easily bring about such changes in affinity. High-resolution studies of NBD–NBD interactions in the presence or absence of nucleotide suggest small but significant adjustments in the crucial NBD helical regions that interact with the intracellular loops of the TMDs [81]. Low-resolution electron microscopic studies have implicated large conformational changes in the TMDs upon nucleotide binding [50] suggestive of a significant gearing (or amplification) of the nucleotide-induced conformational shifts as they propagate into the TMDs.

It has been observed that ATP binding decreases allocrite binding affinity [76, 77, 82] indicating the precedence of allocrite binding over ATP binding. Similarly, allocrite binding was found to enhance NBD–NBD cross-linking [57, 83, 84] suggesting crosstalk between allocrite binding at the TMDs and the proximity of the NBDs. A model where allocrite binding promotes dimerization of the two NBDs and increased ATP binding is attractive, since it allows allocrite regulation of potentially wasteful ATPase activity [85]. Structurally, one might expect the signal transmission interface between NBDs and TMDs to be highly conserved [15, 57, 83, 84], but so far there is no obvious switch region emerging from the few structures available: For BtuCD and the related HI1470/1, a single loop extends down from the TMDs into the NBDs. This "L-loop" (sometimes also termed the "EAA" loop) in BtuC is close to the BtuD Q-loop, a proposed sensor for γ-phosphate that might change its conformation upon nucleotide binding/hydrolysis [15, 86, 87]. There is some evidence that mutation around the signal transmission interface could perturb assembly as well as the coupling of ATP binding/hydrolysis to allocrite transport [15, 21, 88, 89]. However, in the Sav1866 structure, and as far as we can tell in the MsbA structures, the TMD/NBD interface is much more extensive and involves at least two loops of the TMDs. In the Sav1866 structure, one of the loops is domain swapped with the opposing NBD, inextricably linking the four components of the structure [90].

1.2
Structures of ABC Transporters

1.2.1
Tertiary Structure

Despite technical difficulties, a number of ABC transporters have now been over-expressed and purified sufficiently for 3D crystallization trials [42, 82, 91–94]. The emergence of a wider range of crystallization techniques for membrane proteins has also been a positive development [95, 96]. In addition, electron microscopy approaches have yielded low- to medium-resolution structures.

The first high-resolution structure of the NBD domain of HisP (histidine permease) from *Salmonella typhimurium* was reported in 1998 [62]. In addition, high-resolution structures of RbsA from *E. coli* [97], MalK from *Thermococcus litoralis* [37], MJ0796 from *Methanococcus jannaschii* [87], MJ1267 *Methanococcus jannaschii* [98], Rad50 from *Pyrococcus furiosus* [63], SMC from *Thermotoga maritima* [99], MutS from *E. coli* [100, 101], human Tap1 [102], GlcV from *Sulfolobus solfataricus* [103], HlyB from *E. coli* [104], and human CFTR [105] were determined shortly thereafter. A recent survey of the protein data bank yielded >160 structures, nearly all for NBDs. For such a large data set, structural bioinfomatic studies are now required, but in essence the tertiary structures of the NBDs all have a two-domain organization consisting of catalytic and signaling domains [61]. The catalytic domain houses the Walker A and B motifs and a β-sheet region that interacts with and positions the base and ribose sugar of the nucleotide as well as phosphate moieties [106]. The signaling subdomain contains mostly α-helices that houses the C-loop or ABC signature sequence, and it is this subdomain that has been thought to interact with the TMDs [61, 106].

However, as discussed above, functional insights from these structural studies crucially depend on elucidating the quaternary structure of these proteins [107, 108], and data on this level of structural organization are rare and significantly harder to obtain. Nevertheless, the sheer abundance of ABC proteins has made them particularly attractive for structural proteomics projects, and in many countries (the United Kingdom being a notable exception), funding for structural proteomics of ABC proteins has been forthcoming, leading to significant advances.

The first data for an entire ABC protein emerged from a structural proteomics screen of ~50 *E. coli* ABC proteins [94] (revised in Ward *et al.* [47]). The data for a lipid transporter/flippase termed MsbA revealed a homodimeric cone-shaped protein with base dimensions of 120 Å × 115 Å × 64 Å, with an unusual tilting of the six TMD α-helices to about 30–40° from the normal to the membrane and making monomer–monomer contacts only at the extracellular side [94] (revised in Ward *et al.* [47]). The resulting effect of this tilt is the wide separation of the NBDs by about 50 Å. Being the first crystal structure of an ABC transporter protein, several conclusions were drawn [82, 106]. The structure not only confirmed several previous biochemical findings, including the prediction of six transmembrane α-helices, but also raised numerous questions [82, 106]. Most significant was the observation that the two NBDs were separated by ~50 Å, which contradicted previously observed NBD interactions (fluorescence resonance energy transfer measurements and other cross-linking studies) [9, 109]. Subsequent searches for homologues led to crystal structures of MsbA from *Vibrio cholerae* and *S. typhimurium* (see Figure 1.2), which provided some explanations [44, 93, 94] (all revised in Ward *et al.* [47]). Structural similarities within each MsbA monomer were immediately apparent, but the quaternary structure differed considerably, especially in the orientation of NBDs [93, 94] (both revised in Ward *et al.* [47]). The revised MsbA structures have been corrected for handedness and tracing of the polypeptide chain in the TMDs, but the *E. coli* MsbA structure still has widely separated NBDs.

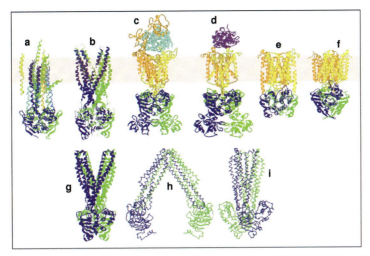

Figure 1.2 Comparison of high-resolution (b–i) and medium-resolution (a) structures of ABC proteins. The predicted transmembrane regions are indicated by the pink boundary for (a)–(f). (a) Medium-resolution model of P-gp with α-helices placed in elongated elements identified in the EM map [110]. NBDs are based on the MJ0796 structure [87] that was manually docked into a bilobed density in the appropriate region of the medium-resolution 3D map. (b) Sav1866 homodimer [90]. (c) MalFGK$_2$ with the periplasmic maltose binding protein (cyan) [48]. (d) ModABC, with the periplasmic binding protein (purple) [111]. (e) BtuCD [42]. (f) Putative metal chelate transporter H10796 [43]. (g–i) MsbA from *S. typhimurium*, *E. coli*, and *V. cholerae*, respectively [47]. Structural similarities are identifiable in the trans-membrane regions of (b) with (g), (c) with (d), and finally (e) with (f).

Another structural protcomics screening of 28 bacterial ABC transporter proteins yielded a second structure (Figure 1.2): the vitamin B12 importer (BtuCD) from *E. coli* [42]. The transporter consists of two subunits BtuC (TMD) and BtuD (NBD) with overall dimensions of $90\,\text{Å} \times 60\,\text{Å} \times 30\,\text{Å}$. Extensive contacts between the subunits [42], and the juxtapositioning within the NBDs of the ABC signature sequence and P-loop (involved in signal transduction and nucleotide binding, respectively), suggested that the BtuCD structure was more consistent with the available biochemical information for ABC proteins. Paradoxically, though, the BtuCD protein with 10 transmembrane α-helices per TMD appears to be very different from eukaryotic ABC proteins and hence has proven less amenable to homology modeling than the MsbA structures and the recently published Sav1866 structure (Figure 1.2) [90]. The interface between the BtuC subunits is formed by the antiparallel packing of two pairs of helices creating a cavity that opens to the periplasmic space and closes on the cytoplasmic side by residues Thr142 and Ser143 in the loops connecting helices 4 and 5 [15, 42, 82]. In BtuC, there is a prominent cytoplasmic loop between helices 6 and 7 folding into two short helices (L1 and L2) making extensive contacts with BtuD (NBD) [15, 42]. Some sequence alignments have suggested architectural conservation especially in helix L2 [15, 82]. For example, this region has been proposed to correspond to the first cytoplasmic loop of drug exporters and to intracellular domain 1 (ICD1) of EC-MsbA [15, 82]. This loop is also often termed the "EAA" loop (see below).

A surprising facet of ABC proteins is the structural plasticity of the transmembrane regions. From the six ABC proteins for which high-resolution structures are available, there are three almost entirely different "folds" for the transmembrane domains (i.e., they are threaded through the membrane on different paths). This is in stark contrast to the NBDs, where structural homology is very clear, and the similarities extend even to the NBD–NBD association in the intact transporter (with the exception of the lower resolution E. coli and V. cholerae MsbA structures) (see Figure 1.2). In one group of exporters, exemplified by Sav1866 and S. typhimurium MsbA (Figure 1.2b and g), two lots of six transmembrane α-helices form the homodimeric ABC protein. Four of the six transmembrane α-helices (TM2–5) are very long (∼70 Å) and have significant helical extensions on the cytoplasmic side, which constitute the so-called intracytoplasmic loops (ICLs). These loops contact the NBDs, presumably act as an interface communicating structural changes in the NBDs to the TMDs, and separate the NBDs from the membrane surface. Intriguingly, the second ICL, joining TM4 and TM5 contacts the opposite NBD to the one contacted by the first ICL. This "crossover" (sometimes termed a domain swap) is mirrored on the extracellular side of the TMDs where the extracellular sides of TM1 and TM2 make contacts with the opposite polypeptide subunit [47, 90].

The second "fold" has been observed with the importers BtuCD and HI1470/1 [42, 43] where a large number (10) of short transmembrane α-helices exist for each TMD. The ICLs for this fold are also very short, with only one loop making significant associations with the underlying NBD. This loop has been termed the "EAA" loop on the basis of this relatively three-residue motif conserved within it. Although there is some "crossover" of contacts in the TMDs, there is no evidence for this in the ICLs. This TMD fold seems (so far) the most distant from the eukaryotic ABC proteins, and hence perhaps the least useful for homology modeling.

The third fold is exemplified by the importers ModBC, MetNI and MalFGK$_2$ [48, 111, 248]. Similar to the second type of fold, the helices are short and there is hardly any ICL region apart from a single loop (similar to the EAA loop described above) that contacts with the underlying NBD. At first sight, the MalFG subunits, with eight and six transmembrane spans, appear quite different from the ModB subunit with six transmembrane spans. (Note: The term "helix" is not used for this group as some of the transmembrane spans are composed only partly of α-helices, and also regions of the subunits apparently dip into the membrane without spanning it.) However, when ModB is aligned with MalF and MalG using the "EAA" loop as an anchoring point, the six ModB transmembrane spans superimpose with reasonable precision on the equivalent regions in the MalF and MalG subunits. Hence, a core in the fold for the TMDs of ModBC and MalFGK$_2$ can be identified. Davidson and colleagues have also noted that the (substrate-lacking) ModBC structure appears to be more open on the cytoplasmic side of the membrane, the reverse being the case for the substrate-containing MalFGK$_2$ structure (Figure 1.3). Whether this reflects conformational changes in the TMDs associated with substrate transport still remains to be established.

Transmission electron microscopy (TEM) studies of ABC proteins have also provided insight into the quaternary structure and also served as a useful tool

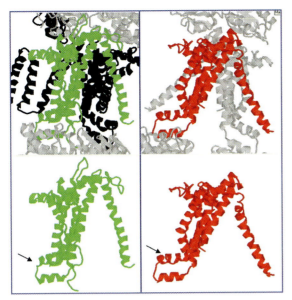

Figure 1.3 Comparison of the transmembrane regions of the MalFGK$_2$ transporter (upper left panel) and the ModBC transporter (upper right). For the former, the MalG subunit is in green, the MalF subunit in black. For the latter, one ModB subunit is highlighted in red. For both transporters, the NBDs (cytoplasmic side) are toward the bottom, while the periplasmic substrate binding subunits are uppermost. The ModBC structure (right) has a more "open" structure on the cytoplasmic side. The lower panels show a simplified representation of the MalG (left) and ModB (right) subunits in the same orientation as in the upper panels, illustrating the similarity in the folds of the two proteins when aligned on the basis of the EAA loop (arrow).

for identifying conformational changes accompanying nucleotide binding (see Figure 1.4).

However, only one of these TEM studies (for P-gp) has so far yielded data at a resolution sufficient for the identification of transmembrane α-helices and other subdomains [51]. Nevertheless, these data were sufficient to confirm the conservation of overall tertiary structure between prokaryotic and eukaryotic ABC proteins that was predicted from the analysis of their primary structure (see Figures 1.2 and 1.5). Indeed, a close examination of the structures suggests that while all display similar NBD folds (Figure 1.2), the transmembrane region of the single eukaryotic example (P-gp) has a closer similarity to Sav1866 and *S. typhimurium* MsbA (revised) than to BtuCD, especially in the spacing of the NBDs from the TMDs by the cytoplasmic helical loop regions.

1.2.2
Quaternary Structure of ABC Proteins

The oligomeric state of detergent-solubilized and purified membrane proteins can depend on a variety of factors such as the detergent used in the purification

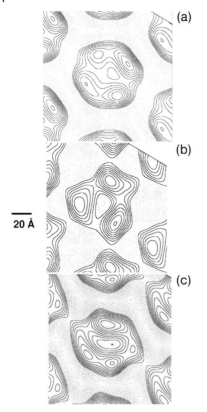

Figure 1.4 Low-resolution projection maps of P-gp obtained from 2D crystals in the nucleotide-free state (a) and in the presence of nucleotides (AMP-PNP) (b) and ADP/vanadate (c). These 2D maps suggest, but do not unequivocally demonstrate, major conformational changes associated with the binding of nucleotides. Major molecular rearrangement between the AMP-PNP-bound state and the nucleotide-free state was later confirmed by generating 3D structures for them. The binding of ADP/vanadate produced only slight changes compared to the nucleotide-free form (a and c), and so far no 3D structure has been generated for that condition. Hence, for the ADP/vanadate trapped P-gp, it is possible that the changes in the projection map are due to differences in crystal packing rather than whole-scale conformational rearrangements in the protein.

and the degree of overexpression of the recombinant protein that may facilitate quaternary interactions [112]. There have been various oligomeric states assigned to different eukaryotic ABC transporters, and no clear consensus is emerging; indeed, entirely different oligomeric forms could exist for the same transporter protein. Bacterial examples of crystallized ABC proteins exist in a form that would be consistent with a single functional transporter – equivalent to a monomer for most eukaryotic ABC proteins where a single polypeptide encodes both TMDs and both NBDs [42, 43, 47, 48, 90, 111]. Similarly, P-gp and CFTR form epitaxial crystals grown in elevated salt and polyethylene glycol, which are composed of monomeric proteins. The only possible exception to this uniformity is the crystalline *E. coli*

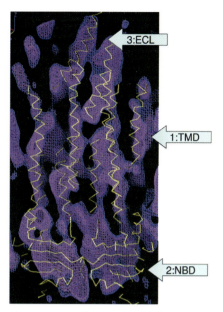

Figure 1.5 Central section through the medium-resolution 3D density map of P-gp in the presence of nucleotide as determined by electron microscopy of 2D crystals. 1: The long cylindrical densities that extend down into the putative NBD region were fitted as simple α-helices. 2: The two heart-shaped lobes of density at the bottom of the map have been fitted using the MJ0796 NBD dimer (with nucleotide bound). Note that individual β-strands cannot be resolved at this resolution (8 Å). 3: Small regions of density extending at the top of the map, presumably extracellular loops and/or glycosylation, have been fitted using short α-helices.

MsbA structure where contacts between the two polypeptides are sufficiently peripheral to suspect that this structure represents a state equivalent to half a functional transporter [47]. However, noncrystalline specimens of eukaryotic and prokaryotic ABC proteins appear to contain both monomeric and dimeric particles [3, 51, 110, 113–117]. Ni-NTA nanogold labeling of particles of these two proteins was also carried out, allowing their C-terminal polyhistidine tags to be localized within the low-resolution structures [113, 117]. This labeling was not able to directly confirm the dimeric nature of the particles, however, since the gold labeling efficiency (<25% of particles were labeled) ensures that few particles are doubly labeled (see Figure 1.6).

Pdr5p, a multidrug exporter in yeast, when examined by EM and single particle analysis, also revealed dimeric (i.e., dimer of homodimer) particles similar to the arrangement observed for P-gp and CFTR [116]. Freeze-fracture electron microscopy of cells expressing high levels of P-gp displayed the presence of large particles and density gradient centrifugation indicated oligomers for the detergent-solubilized protein, although to some extent these, and similar studies with CFTR, may be criticized because of the assumptions made regarding the identity of the particles observed by freeze-fracture and atomic force microscopies. Some oligomeric associations that have been observed are unambiguously the product of the isolation

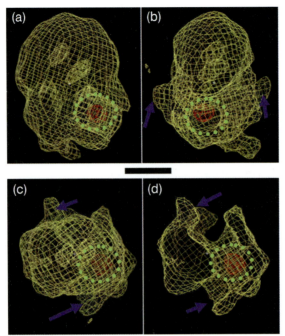

Figure 1.6 Low-resolution structure of human, recombinant, 6-His-tagged P-gp particles labeled with 1.8 nm diameter Ni-NTA nanogold. The structure (yellow netting) shows overall dimensions of 120 Å × 100 Å × 70 Å, sufficient to enclose two P-gp molecules. Panels (a) and (b) show side views orthogonal to each other, while panel (c) shows a view from the top after rotation by 90° around the horizontal axis. Panel (d) is the same view as in panel (c), but with the top half of the structure stripped away. Indicative of a dimeric organization, symmetry-related "arms" (blue arrows) protrude on either side of the structure. The Ni-NTA nanogold (red netting, green circle) breaks the twofold symmetry, probably because the labeling efficiency is only ~25%; hence, few particles are double labeled. The gold sphere is bound on the lower half of the structure, allowing its designation as the NBD region. Scale bar = 40 Å.

procedure, such as the bacterial YvcC protein (now termed BmrA), which after partial reconstitution with lipid forms a 24-membered ring with YvcC proteins forming spokes emanating from the center of the ring [115]. On the other hand, the complex formed between the Kir6.2 potassium channel and the SUR1 ABC protein, which has been observed as an octameric complex (tetramer of heterodimers), seems reasonable given the function of the complex, with four SUR1 proteins arranged around the tetrameric potassium channel [118]. A tetramer of similar external dimensions, but lacking the central potassium channel, has also been observed in EM studies of the isolated multidrug transporter ABCG2^{R482G}, otherwise known as the breast cancer resistance protein [112]. The tetrameric association in both cases appears to be mediated strongly by contacts between adjacent NBDs in separate monomers in the complexes.

The only exception to the general observation of multiple association of purified ABC proteins (so far) is the yeast ABC protein Mdl1 [119], where studies of noncrystalline specimens by EM were interpreted as showing single transporters containing two polypeptides (each containing one TMD and one NBD). However, the overall dimensions of these particles ($\sim$10 nm × 10 nm × 12 nm) suggest that two transporters (four polypeptides) could have been fitted to the 3D density map. A major caveat with all these structural studies is that nonphysiological oligomeric associations could form upon solubilization of the proteins from the membrane and/or during purification in detergent micelles. The availability of structural data for ABC proteins reconstituted back into a lipid bilayer system is very limited at present [50, 115], and clearly there is a need for further structural studies of such reconstituted systems if we are to understand better the true nature of conformational changes and oligomeric associations in ABC proteins.

1.3
Multidrug Resistance and ABC Transporters

Cellular resistance to antibiotic and anticancer drugs is considered to be a major factor for the treatment of infectious diseases or cancers [120, 121]. Sometimes, microorganisms and cancer cells exhibit cross-resistance to a wide variety of functionally and structurally unrelated drugs [121, 122]. This phenomenon is termed multidrug resistance (MDR), a concept introduced into scientific literature in the 1970s [123]. One of the most important mechanisms of multidrug resistance is via active drug export from the cell mediated by membrane proteins [121, 124, 125].

The multidrug transporters are divided into two broad classes, primary and secondary active transporters, although they both mediate the translocation of a range of substances across membranes [121]. Secondary active transporters are predominant in prokaryotic organisms and mediate the drug efflux reaction in a coupled exchange with sodium ions or protons [125, 126]. Primary active transporters use the energy of ATP hydrolysis to translocate compounds and play a key role in eukaryotic drug resistance and to varying degrees in bacteria, fungi, and protozoans [121]. One of the best characterized bacterial proteins, LmrA, shows about 34% amino acid sequence identity with P-glycoprotein, which is probably the best characterized eukaryotic ABC protein [70, 73, 127]. Since the completion of the human genome sequence [128, 129], about 48 human ABC proteins have been identified and as many as 12 might be implicated in drug transport [7, 121].

1.3.1
P-Glycoprotein

1.3.1.1 Historical Background
Biedler and coworkers [123] reported that Chinese hamster cells selected for resistance to actinomycin D also showed cross-resistance to mithramycin, vinblastine, vincristine, puromycin, daunomycin, democolcine, and mitomycin C. Thus, upon selection with a single cytotoxic drug, mammalian cells became simultaneously cross-resistant to a range of drugs with different chemical structures and molecular

mechanisms of actions [123, 130–132], a phenomenon referred to as multidrug resistance [123, 132]. Initially, it was believed that the resistance was due to a membrane alteration that reduced the rate of permeation [133]. However, Juliano and colleagues [133] revealed that drug-resistant Chinese hamster ovary cell membranes contain a cell surface glycoprotein of about 170 kDa that was not observed in drug-sensitive cells. This remarkable observation, achieved well before proteomic analysis became possible, was possible because of the unusually high expression of this protein, which was named P-glycoprotein (where P stands for permeability).

There is a significant body of literature on the biochemistry and pharmacology of P-gp. The recognition that gene amplification and overexpression of the protein could give rise to the multidrug resistance phenotype [134, 135] was followed by the cloning and sequencing of the cDNA that encoded the protein [136–140]. The transfection of the cDNA into cells followed by selection for drug resistance provided an early means of isolating P-gp for functional studies [141]. The deduced amino acid sequence predicted 12 transmembrane α-helices, 2 nucleotide binding domains, and 3 potential glycosylation sites within the 1280 amino acid residues [88, 136]. Biochemical, biophysical, genetic, and microscopic analyses have all been used to investigate the mechanistic behavior of the protein. Cross-linking experiments and pharmacokinetic studies of the proximities of some transmembrane helices have led to speculation on the location of drug binding sites and the number of sites. Some of these studies are described in more detail below.

1.3.1.2 The Role of P-gp in Drug Resistance

One of the major difficulties in cancer chemotherapy is the development of multidrug resistance, a phenomenon of cross-resistance to an array of drugs. Since the discovery of P-gp on the cell surfaces of tumor cells with a wide range of drug-resistant phenotypes [133] and subsequent findings from numerous other laboratories, it has been considered that P-gp when overexpressed in tumor cells can mediate the ATP-dependent extrusion of a variety of drugs, concomitantly reducing intracellular accumulation [141–148].

It is indeed observable that P-gp expression contributes to multidrug resistance. However, establishing a direct and simple relationship between P-gp expression and multidrug resistance is difficult due to differences in populations of tumor cells and methods of measuring P-gp expression [149]. The observation that most tumors contain heterogeneous cell populations with varying degrees of P-gp expression may lead to over/underquantification in various cell populations [150]. Finally, it has been emphasized that only some tumors express P-gp [131].

Although it is apparent that P-gp lowers the concentration of many anticancer agents in tumor cells, the extent of the reduction due to P-gp function alone is often unclear. A recent study investigating the role of P-gp in paclitaxel concentration in tumor cells by comparing the relative importance of extracellular drug concentration, P-gp efflux rate, binding affinity to tubulins/microtubules, and intracellular contents of tubulin proteins indicated that the role of P-gp in multidrug resistance might be less significant compared to other biological factors [131, 151]. The rank order of importance of these factors was reported to be extracellular drug concentration intracellular binding capacity > intracellular binding affinity > P-gp-mediated efflux [131, 151]. It is the conclusion of the above-mentioned authors that the delivery of paclitaxel to tumor cells rather than other mentioned factors determines intracellular drug concentration. This report was supported by a study that showed intravenous administration of radiolabeled daunorubicin to rats bearing bilateral tumors indicated that P-gp accounted only for partial drug resistance [152]. Nevertheless, poor brain penetration of radiolabeled drugs demonstrated a very significant role for P-gp [152, 153]. Despite all these arguments, the general conclusion still remains that P-gp is a major player in multidrug resistance. Therefore, the development of relatively potent but nontoxic P-gp inhibitors could greatly reduce its effect on drug accumulation in cells.

In addition to the significant role P-gp plays in multidrug resistance, multidrug resistance protein 1 (MRP1) is another important ABC transporter protein with similar properties. MRP1 was identified in 1992 as a second drug transporter in humans [154]. It can confer resistance to a variety of drugs when overexpressed in cells. MRP1 has been implicated in the transport of etoposide, teniposide, doxorubicin, vincristine, leukotrienes, glutathione conjugates, glucuronides, and sulfates [155–157]. Like P-gp, MRP1 is thought to provide protection to normal tissues [158, 159]. For other members of the MRP family, such as MRP2, MRP3, MRP4, MRP5, MRP6, ABCC11, and ABCC12, links to multidrug resistance are less well defined [158].

1.3.1.3 Tissue Distribution and Physiological Roles

MDR genes are expressed in normal tissues, prompting researchers to elucidate their physiological roles. Mice lacking P-gp genes (mdr1a and mdr1b) have a subtle phenotype [160] indicating a role for P-gp in physiological defense against xenotoxins. The polarized pattern of P-gp expression in many cells supports this probable role. Thiebaut *et al.* [161] reported the localization of P-gp in the epithelial cells of excretory organs such as the bile canalicular membrane of hepatocytes in the liver, proximal tubules in the kidney, and enterocytes lining the wall of the intestines. The presence of P-gp in the capillary endothelial cells in the brain and testes indicates other roles of significance in biology [162–164] (Sugawara *et al.*, 1990). The expression of P-gp in tissues that partake in steroid hormone biosynthesis is suggestive of its importance in production and secretion of cortisol and other steroids [161, 165–167]. P-gp also occurs in the placental trophoblasts from the first trimester of pregnancy to full term indicating a probable (protective) role in fetal development [167]. Hemopoietic progenitor cells are also shown to contain P-gp [168] where a role in protection against mutagens and teratogens seems likely.

1.3.2
Conformational Changes in the Mechanism of P-gp

It is apparent that if we are to circumvent the unwanted actions of P-gp in cancer chemotherapy, we need to understand the molecular basis of multidrug export. This requires us to understand both the structure and the mechanism of the protein. It is widely accepted that the coupling of ATP binding and hydrolysis at the NBDs to the transport of allocrite in the TMDs could be mediated by conformational changes at different stages in the catalytic cycle. Recent findings seem to highlight the role played by dimerization and dissociation of the two NBDs in bringing about a change in conformation needed for translocation. For P-gp, several experimental approaches have been used to establish the existence of these changes, for instance, the use of differential tryptic digest patterns [169, 170], fluorescence quenching [171, 172], IR spectra [173], and monitoring of changes in accessibility of extracellular antibody (UIC2) epitope [50, 174–176]. The question remains as to what mediates these changes. Recent reports indicate that two molecules of ATP interact at the interface of the Walker A in one NBD and the LSGGQ signature motif in the other NBD and also that allocrites that stimulate or inhibit ATPase activity can cause the above-mentioned sequences to come closer or to move farther apart, respectively [83, 84]. These workers postulated that the LSGGQ sequence conveys the signal of conformational changes from the allocrite binding site to the ATP binding sites. This suggests that the architectural position of the LSGGQ sequence (influenced by allocrite) to Walker A is responsible for influencing the rate of ATPase activity that can be viewed as allocrite-induced conformational crosstalk between NBDs and allocrite binding site [57, 83, 84].

Even though a high-resolution crystal structure was not available for P-gp until recently (see Section 1.3.8), transmission electron microscopy of 2D crystals has yielded low- to medium-resolution 3D structures of P-gp. Conformational rearrangements were indicated in the low-resolution studies where two 3D structures were produced under different conditions – nucleotide-free and nucleotide-bound states [50, 51] (Figure 1.4). Studies of vanadate-trapped P-gp in the presence of ADP were also proposed to give a third conformational state, but no 3D structure has been generated so far for this condition [50]. Changes in the transmembrane region of the protein were particularly apparent. This work implicates major conformational rearrangements (i.e., observable even at ∼20–25 Å resolution) in the transport cycle of P-gp. Higher resolution data for the nucleotide-bound form of P-gp [110] showed that asymmetry in the transmembrane region was mostly caused by different tilts of two of the helices (the remaining 10 helices showed roughly twofold symmetry). Since the nucleotide-free form of P-gp displayed a strong twofold symmetry [50], the authors speculated that the asymmetric tilting of the helices in the nucleotide-bound form could be a result of nucleotide binding at the NBDs. This putative conformational shift could perhaps be significant, since the asymmetry opens up one side of the transmembrane region that may allow access from and to the lipid bilayer. Interestingly, the (ADP-bound) Sav1866 structure also displays similar gaps open to the lipid bilayer on the side of the barrel of transmembrane helices, although in this case, symmetry is retained [90, 177]. Returning to an earlier question, this work also provides a plausible explanation for the evolution of receptor/switch functions of ABC proteins such as SUR1: Presumably, large-scale conformational changes in the TMD regions associated with nucleotide binding or release can be used to induce changes in a transducer protein (such as the Kir6.2 potassium channel).

1.3.3
Comparison of Sav1866 and P-gp Structures

The 3–4 Å resolution structure of the *S. aureus* putative multidrug transporter is the best homologue available for constructing models of eukaryotic ABC proteins such as P-gp [90, 177]. Similarly, the 8 Å resolution map of P-gp 2D crystals represented, until recently, the highest resolution data so far for any eukaryotic ABC protein, and thus a comparison between the two is appropriate. Qualitatively, this has been addressed in the preceding section's discussion of the gaps on the side of the barrel of transmembrane α-helices, but here an overall appraisal of correspondence is attempted. Sequence homology between Sav1866 and P-gp is low except in the conserved regions of the NBDs (see Section 1.3.1). However, the length and spacing of the hydrophobic transmembrane regions relative to the extracellular and intracellular loops are generally well conserved, suggesting that the overall fold of the two proteins may be qualitatively similar in the TMD regions. We therefore fitted the Sav1866 structure into the P-gp density map by hand using the *xfit* program within the *XtalView* software package, initially using the well-conserved NBD regions to guide the process, followed by fine-tuning the rotational and translational operations to give a good fit for the rest of the structure. The results of this exercise are summarized in Figure 1.7.

The views orthogonal to the long molecular axes in panels (a) and (b) confirm that the overall dimensions of Sav1866 fit well to the P-gp map, as expected from the sequence alignment. The distance from the bottom of the NBD (arrow 5) to the long extracellular loop between TM helices 1 and 2 (arrow 3) is ∼130 Å. This latter loop fits into a finger-like protrusion previously identified in the P-gp map as an extracellular domain, but a gap in the density (arrow 4) that may be due to local disorder precluded its assignment as a continuous loop. In contrast, the C2 symmetry-related loop on the other half of the Sav1866 homodimer is a poor fit to the P-gp map (arrows 1 and 2). It is tempting to use this observation to assign this section of the P-gp map to TMD2, which lacks the large extracellular loop of TMD1. As expected, the well-conserved NBD regions of Sav1866 dock into the lower portion of the map with a good overall fit and in a similar position to the MJ0796 dimer that was employed in a previous study [110]. The positions of bound ADP molecules in the Sav1866 structure are indicated by the dashed ellipses. The five-stranded parallel β-sheet, viewed edge-on, running between arrows 5 and 7, appears as a slightly curved slab of density in the P-gp map in panel (a) and has a characteristic heart shape viewed face-on (panel (b), arrow 8). A few discrepancies in this region are observed, however, such as the region of short helices and turns between residues 430–480 that contribute to the signature motif in Sav1866 (arrow 6a). Although this small region does not match density in one half of the map, its C2 symmetry-related equivalent fits well in the other half (see arrows 12a and 12b). Similarly, the region linking TM helix 6 to the start of the NBD in Sav1866 is poorly fit on one side of the structure (arrow 9a), while a good match is found on the other (arrow 9b). These various differences are probably due to the evolutionary divergence from twofold symmetry in P-gp versus the homodimeric Sav1866, although disorder in some regions is an alternative explanation. Presumably,

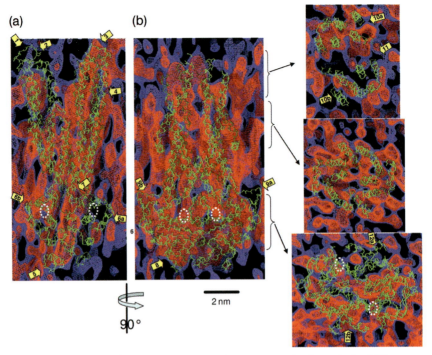

Figure 1.7 Comparison of the Sav1866 structure (green polypeptide Cα trace) and the 8 Å resolution map for P-gp (red and blue netting at 1σ and 1.5σ above the mean density level, respectively). Each panel represents a 25 Å thick slice through the center of the molecule (panels (a) and (b)), or slices perpendicular to the long axis of the molecule, as indicated by the brackets and arrows relative to panel (b). The positions of the two ADP molecules in the Sav1866 structure are indicated by the dashed ellipses. *Note:* One Sav1866 homodimer is shown docked into the center of a P-gp 2D crystal unit cell with a p1 plane group – additional densities around the edges of each panel arise from adjacent unit cells. See main text for description of the numbered arrows.

the presence of a linker region between NBD1 and TMD2 in P-gp also imposes some differences in the organization of the NBD region.

Slices through the map taken perpendicular to the long molecular axes further illustrate the similarities and differences between Sav1866 and the P-gp map. The extracellular transmembrane regions and loops that occupy two separate regions at the top of the Sav1866 structure correspond to two regions of density in the P-gp map (top right panel, arrow 10), which are separated by a low-density region. The correspondence on one side (arrow 10a) is better than that on the other (arrow 10b) where density is weak or lacking at the expected positions of two of the Sav1866 helices. There is a gap on one side of this region (arrow 11) that may allow access to the lipid bilayer in P-gp, as discussed previously [110]. A slice through the cytoplasmic side of the transmembrane region (center right panel) shows a good fit between Sav1866 and the P-gp map. Asymmetry in this region in the P-gp map will arise from additional density that is likely to be due to the longer N-terminal region preceding

TM helix 1 as well as the long linker joining NBD1 with TMD2. The slice through the NBD region (bottom right panel) also shows a good overall fit between the map and the Sav1866 structure. Arrows 12a and 12b indicate regions of asymmetry in the P-gp map corresponding to arrows 6a and 6b in panel (a).

The overall impression from this comparison is that Sav1866 probably represents a reasonably good starting model for P-gp in the presence of AMP-PNP. There are differences in the precise trajectories of transmembrane helical elements, but this is to be expected given the low sequence identity between the two proteins in the transmembrane region (~20%). A model of P-gp based on Sav1866 but using "adjustments" or "constraints" from the EM-derived map would therefore be informative.

1.3.4
Drug Binding Sites in P-Glycoprotein

The involvement of P-gp in the multidrug resistance phenotype has prompted extensive study to uncover the mechanism of this polyspecificity. Two possibilities are most likely: (i) the presence of a single domain to which drugs display a loose association and (ii) multiple drug binding sites, each with a defined specificity.

The former option was proposed initially to take into account the nuances of P-gp-mediated transport. It was suggested [178] that P-gp behaved as a "flippase." The model suggested that the protein extracted drugs directly from the lipid milieu rather than the conventional wisdom of entering the translocation machinery through the cytoplasm. This suggestion was supported by the drug transport studies using acetoxymethyl esters (AM) of fluorescent probes (e.g., calcein-AM) [179]. The AM derivatives were nonfluorescent substrates for P-gp transport and readily converted to the fluorescent, nontransported compound by cytoplasmic esterases. Cells containing P-gp did not display any fluorescence although it rapidly appeared following P-gp inhibition. The results indicated that the drugs were expelled from the cell prior to reaching the cytoplasm. Further support was provided by Raviv et al. who demonstrated that photocross-linking of the membrane localized probe [^{125}I]-INA to P-gp through direct energy transfer from the substrates rhodamine 123 or doxorubicin [180]. Another key premise of the "flippase" model suggested that the "flip-flop" of drugs across the lipid bilayer was a slow process, yet this clearly was not the case for a number of substrates and modulators of P-gp [181, 182]. Finally, it was suggested by this model that drugs interact with the protein on a hydrophobic interface, rather than via a "classical" binding site. As discussed below and in subsequent sections of this chapter, there are clearly defined and specific sites for drug interaction with P-gp. Although two tenets of the flippase model are flawed, the suggestion that drug extraction occurs directly from the lipid bilayer is an accepted and characteristic feature of P-gp-mediated transport.

The quest to determine whether P-gp contains multiple drug binding sites employed a number of distinct approaches. For example, site-directed mutagenesis revealed a multitude of residues that when mutated could alter the pattern of drug resistance conferred by P-gp [183–185]. The effects of mutations on the activity were analyzed by cytotoxicity assays. However, this strategy could not attribute the residues directly to the drug binding event, and the mutations could conceivably interfere at any stage of the translocation process. In contrast, a number of investigators employed a direct approach to the issue of the drug–P-gp interaction. Early studies employing the covalent labeling of P-gp with photolabile derivatives of recognized substrates and subsequent proteolytic cleavage suggested that both halves of the protein contributed to binding [186–189]. However, this does not preclude the possibility that both halves contribute to a single site for drug binding. In addition, the flexibility and high reactivity of photoactivated compounds generate a high degree of nonspecific covalent attachment [190]. The best photoaffinity labeling evidence for the presence of multiple drug recognition sites was obtained by Dey *et al.* [52] while examining the effects of the inhibitor *cis*-flupentixol on [^{125}I]-IAAP binding. P-gp labeled with [^{125}I]-IAAP and subjected to tryptic cleavage produced two polypeptides, both of which contained covalently attached [^{125}I]-IAAP. Addition of *cis*-flupentixol caused an increase in the affinity of [^{125}I]-IAAP binding at the C-terminal site. However, there was no effect of the inhibitor on [^{125}I]-IAAP binding to the N-terminal site. This finding demonstrates the presence of nonidentical sites for drug interaction on P-gp, and that for a subset of drugs, there are overlapping specificities at these sites.

The first account of noncompetitive drug interactions on P-gp was provided by Tamai and Safa using a radioligand binding assay with vinblastine and [^{3}H]-azidopine [191]. The term "competitive inhibition" is often misused in biochemistry and in its purest definition proves interaction at a common site. Consequently, the noncompetitive displacement of [^{3}H]-azidopine binding by vinblastine demonstrated that these two compounds interact at pharmacologically distinct regions. In contrast, the interaction between [^{3}H]-vinblastine and cyclosporin A was competitive [191]. These two investigations demonstrated not only the presence of multiple binding sites but also that these sites could interact with more than one compound. The presence of multiple drug interaction sites was confirmed by a number of groups, and the use of kinetic binding studies indicated that these sites were linked by a negative heterotropic allostery [192–194]. In their most comprehensive study, Martin *et al.* outlined the presence of at least four binding sites [60]. Some of the sites bound transported compounds only, while others were exclusive to modulators. However, this type of classification was achieved with only a small number of compounds compared to the spectrum of drugs recognized by P-gp. Although the interactions described by these investigations do reveal pharmacologically distinct drug binding sites, they do not inform on their precise locations or their spatial proximities.

Residues putatively involved in the drug recognition process have therefore been identified via a combination of direct photolabeling, cysteine scanning mutagenesis, and chemical cross-linking (Table 1.1).

These techniques allow accessibility of residues in P-gp TM helices to be mapped by covalent attachment of substrates. By their nature, these studies are not unbiased. Photolabeling with propafenones identifies accessible methionine residues that are presumed to be at or near the drug binding sites, and cysteine scanning mutagenesis also has an inherent bias. However, these techniques remain the closest we have to definitive identification of the amino acids composing the drug binding sites. A

Table 1.1 Residues implicated in the substrate binding pocket(s) of P-glycoprotein.

TM	Residue	Label	Notes/effect	References
1	65*	MTS-rhodamine	200% Vpl-stimulated ATPase activity after MTS-rhodamine labeling	[55, 56]
		MTS-verapamil	Persistent stimulation after labeling. No further stimulation seen with other drugs suggesting 100% labeling or that the MTS-Vpl blocks other drug binding. CsA and Vpl pretreatment inhibits labeling	[195]
2	118, 125	MTS-verapamil	Persistent stimulation by MTS-Vpl, not inhibited by pretreatment with Vpl	[53, 54, 196]
3	197*	Propafenone	Major photolabeled residue from peptide mass fingerprinting	[197]
4	222*	Dibromobimane	>50% inhibition of ATPase activity. Could be rescued by preincubation with Vpl, Vbl, and Col	[196]
		MTS-verapamil	ATPase activity inhibited by reaction with MTS-Vpl and restored by prior incubation with Vpl	[53, 54]
5	306*	MTS-verapamil	Persistent stimulation with MTS-Vpl. Prior incubation with Vpl inhibits labeling	[198]
	311*	MTS reagents	Reaction of 306C with MTSET and MTSES alters the potency of Vpl stimulation	[199]
		Propafenone	Major photolabeled residue from peptide mass fingerprinting	[197]
6	331	Maleimides	CM accessible all states, BM AMP-PNP	[200]
	333	Maleimides	CM accessible all states, BM Vi-trapped	[53, 54]
	335	Maleimides	CM accessible all states	[201]
	337	Maleimides	CM accessible all states, BM Vi-trapped	
	339	Maleimides	CM accessible all states, BM Vi-trapped and Apo	
	339	Dibromobimane	>50% inhibition of ATPase. Could be rescued by preincubation with Vpl, Vb, and Col	

(Continued)

Table 1.1 (Continued)

TM	Residue	Label	Notes/effect	References
	341	Maleimides	CM accessible all states	
	342[1]	Dibromobimane	>50% inhibition of ATPase. Could be rescued by preincubation with Vpl	[202]
	342*	MTS-verapamil	ATPase activity inhibited by reaction with MTS-Vpl and restored by prior incubation with Vpl	
	343	Maleimides	CM, BM, and FM accessible all states	
7	728*	MTS-verapamil	Persistent stimulation. No further stimulation seen with other drugs. Reduction of this stimulation by cyclosporin A and pretreatment with Vpl	[196]
	725, 729		Persistent stimulation by MTS-Vpl, not inhibited by pretreatment with Vpl	[197]
8	766	MTS-verapamil	Persistent stimulation by MTS-Vpl, not inhibited by pretreatment with Vpl	
	769*	Propafenone	Major photolabeled residue from peptide mass fingerprinting	
9	841, 842	MTS-verapamil	Persistent stimulation by MTS-Vpl, not inhibited by pretreatment with Vpl	[53, 54]
10	868*, 872*	Dibromobimane	>50% inhibition of ATPase activity. Could be rescued by preincubation with Vpl, Vbl, and Col	[196]
11	871	MTS-verapamil	Persistent stimulation by MTS-Vpl, not inhibited by pretreatment with Vpl	[53, 54]
	942*, 945*	Dibromobimane	>50% inhibition of ATPase activity. Could be rescued by preincubation with Vpl, Vbl, and Col	[203]
	951*	Propafenone	Major photolabeled residue from peptide mass fingerprinting	[197]
12	975*[1,2,3], 982[2,3], 985[1]	Dibromobimane	50% inhibition of ATPase activity. Could be rescued by preincubation with Vpl[1], Vbl[2], and Col[3] (verapamil, vinblastine, and colchicine respectively)	[200]
	984*	MTS-verapamil	ATPase activity inhibited by reaction with MTS-Vpl and restored by prior incubation with Vpl	[53, 54]

number of substrates have been used for photolabeling of P-gp to define regions of the protein that play a role in solute binding and transport. Initially, studies with photoactive P-gp ligands were able to ascribe drug binding sites only to the C-terminal ends of the 6th and 12th TM α-helices [189, 204] (Table 1.2).

However, recent advances in trypsin cleavage and subsequent mass spectrometric identification of peptide fragments, coupled with the development of photoactive derivatives of propafenones, have enabled Chiba and colleagues to identify specific amino acids labeled during cross-linking [197]. Quantitative analysis of photolabeling indicates major sites for reactivity within TM3 (methionine 197), TM5 (methionine 311), TM8 (methionine 769), and TM11 (methionine 951) (see Table 1.1 and Ref. [197]). Minor peaks were identified within several other TM α-helices including TM1 and 12 [197] (Table 1.2).

Rather than rely on endogenous residues for cross-linking, the groups of Clarke, and to a lesser extent that of Callaghan, employed directed cysteine mutagenesis and chemical labeling to determine residue accessibility in the TM regions. Both sets of studies relied on the generation of functional cysteine-less versions of P-gp [212, 213]. The ATPase activity of this cysteine-less isoform was stimulated by the cross-linker dibromobimane (Dbbr), and this property was employed by Loo and Clarke to identify amino acids that when mutated to cysteine and then derivatized by Dbbr showed inhibited ATPase activity that could be prevented by prior incubation with another drug substrate (Table 1.1) [31, 196, 200]. Such residues are proposed to be part of the drug binding site for the "protective" drug. Similarly, the synthesis of sulfhydryl reactive rhodamine and verapamil analogues (methanethiosulfonates, MTS), which caused persistent stimulation of ATPase activity, allowed the identification of residues presumed to be involved in rhodamine or verapamil binding [53–56, 195, 198, 199, 202]. In an alternative approach, the Callaghan group introduced cysteine residues in a cysteine-free version of P-gp [213] and investigated their accessibility in distinct states of the catalytic cycle to maleimides of differing physicochemical properties (Table 1.1) [201, 214]. Residue 339 in TM6 appears to be of particular interest as conflicting results have been obtained. C339 (residue 339 mutated to cysteine) is labeled by maleimides with subsequent effect on the ATPase activity of P-gp. However, this is not a consequence of altered drug binding capacity suggesting that 339 is not a direct contributor of the drug binding pocket [214]. In contrast, C339 is labeled by Dbbr with concomitant inhibition of ATPase activity, which is rescued by preincubation with verapamil, vinblastine, and colchicine [200]. This discrepancy is a reflection of the different criteria used to identify a residue involved in drug binding – one of which is an indirect measure of drug binding.

1.3.5
Structural Interpretation of Drug Binding

Are the data in Tables 1.1 and 1.2 of sufficient consistency to predict where drug binding sites might reside on a structural model of P-gp? To answer this question, we have assigned two levels of certainty to a residue being directly implicated in drug binding. In the first category, we place those residues that are heavily labeled in

Table 1.2 Additional localization studies of the drug binding regions on P-glycoprotein.

TM	Residue	Notes/effect	References
1	61, 64	Residues around one-helical turn of TM1 mutated and alteration to the profile of drug resistance determined	[205, 206]
1	68, 69	Few major peaks of labeling with propafenone derivatives	[197]
ECL1	105–111	Few major peaks of labeling with propafenone derivatives	[197]
ICD1	185	Gly to Val at residue 185 of P-gp alters drug resistance and photolabeling profiles	[204]
6	338, 339	P-gp mutants isolated from drug-resistant cell lines. Photolabeling with IAAP that could no longer be inhibited by CsA. A quadruple mutant with residues in TM9 below showed no photolabeling of IAAP	[207, 208]
6-NBD1	358–456	Photolabeling with IAAP; trypsin and chymotrypsin digest and immunoprecipitation	[189]
ICD1	791–796	Secondary peaks of labeling with propafenone derivatives	[197]
9	837, 839	P-gp mutants isolated from drug resistant cell lines. A quadruple mutant together with TM6 mutants above showed no photolabeling of IAAP	[207, 208]
11	948, 949	Secondary peaks of labeling with propafenone derivatives	[197]
	949, 953	Scanning alanine mutagenesis investigating effects at the level of drug transport	[209]
12	969	Secondary peaks of labeling with propafenone derivatives	[197]
	975, 981, 983	Alanine scanning mutagenesis of TM12 and effects on IAAP transport	[210]
11–12	953–1007	Photolabeling of P-gp with a derivative of CsA, combined with purification, digest, and chemical mapping	[211]
12-NBD2	979–1048	Photolabeling with IAAP; trypsin and chymotrypsin digest and immunoprecipitation	[189]

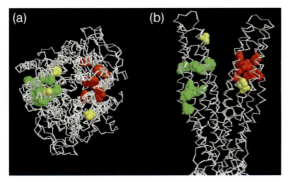

Figure 1.8 Drug binding site localization data in P-gp. Residues (in green, red, and yellow) that are interpreted as contributing to drug binding in P-gp (see Table 1.1 and text) are mapped onto the crystal structure of Sav1866. Residues cluster into two groups: those in green form a slide of residues primarily on the lipid-exposed surface of the molecule (right-hand panel) and those in red form a pocket open to the lumen of the structure. Three residues (yellow) do not fit this pattern.

propafenone photolabeling studies and residues that when altered to cysteine and derivatized result in altered ATPase activity that can be rescued by preincubation with drugs (Table 1.1, indicated by an asterisk in the second column). Other residues listed in Table 1.1 and all those in Table 1.2 are considered to be more indirect measures of drug binding site identification. These former residues have been mapped to their equivalents in Sav1866 [90] and are shown spatially in Figure 1.8.

The residues classified as robust indicators of substrate binding show two obvious clusters on opposite wings of the Sav1866 dimer, with each site primarily contributed by one homologous half. In P-gp this would translate as the first site being exposed to the lumen of the protein, contributed by amino acids at the approximate centers of TM3–6, and a small contribution from TM7 (red, Figure 1.8a). The second site, rather than being a cluster of amino acids, is better portrayed as a slide of residues along the lipid-exposed surface of TM10–12 (green, Figure 1.8b). Whether one of these sites (site 2) represents a hydrophobic interaction site for lipophilic drugs, and the other (site 1) represents a modifying site for hydrophilic ligands, is open to speculation.

1.3.6
Inhibitors of P-gp

The earliest investigations into the resistant phenotype in cancer revealed that cells displayed reduced sensitivity to a large number of anticancer drugs [215, 216]. Moreover, the compounds were often unrelated chemically or functionally. This suggested the presence of multiple mechanisms or perhaps a promiscuous contributing factor. In the case of P-glycoprotein, the latter is certainly true. So far, it has been established that P-gp is capable of interacting with over 200 compounds that may be classed as transported substrates or nontransported modulators [217]. The broad spectrum of resistance and the poor prognosis for patients necessitated strategies to circumvent the actions of P-gp. As a result, a great deal of effort has been directed toward the development of chemical inhibitors of P-gp. During the early 1980s, the emerging broad spectrum of possible inhibitors was greeted with a certain degree of optimism since established, clinically used, compounds could restore some sensitivity to chemotherapy, thereby bypassing the tortuous pathway of preclinical drug development. This use of established drugs for the modulation of P-gp function formed the first generation of inhibitors [218–223]. The strategy was underpinned by numerous biochemical investigations that revealed the successful inhibition of P-gp *in vitro* using a wide range of approaches including cytotoxicity assessment, whole-cell accumulation, and modulation of ATP hydrolysis. The calcium channel blocker verapamil took the most rapid early ascent from such *in vitro* assays to clinical trials [224, 225]. Unfortunately, achieving significant inhibition of P-gp function required plasma concentrations that were considerably higher than those needed for calcium channel blockade. As a result, verapamil treatment was associated with nonspecific toxicity in patients [224, 225]. A similar situation was observed for many first-generation compounds including the immunosuppressant cyclosporin A [226, 227]. In hindsight, the fact that *in vitro* potencies of the drugs rarely reached submicromolar concentrations foreshadowed the lack of effectiveness of their use *in vivo*.

The second-generation P-gp inhibitor emerged directly from the previous generation and involved the use of chemical derivatives of the drugs with a view to eliciting less nonspecific toxicity. For example, the D-isomer of verapamil produced considerably lower calcium channel blockade than the L-isomer, whereas both forms produced equivalent inhibition of P-gp [228]. Therefore, the use of D-verapamil was proposed to raise the tolerated dosage of the drug to levels in the plasma that enabled efficient inhibition of P-gp. Overall, the strategy did achieve lower (but not negligible) levels of toxicity due to D-verapamil; however, there were reports of increased side effects of anticancer drugs [229, 230]. As a result, the increased anticancer drug toxicity required dose reduction of chemotherapy, thereby negating any positive effects generated by the inhibition of P-gp activity. The nonimmunosuppresive cyclosporin A (CsA) derivative PSC833 (valspodar) also offered the potential to reach sufficiently high plasma concentrations to affect P-gp inhibition [231]. Unlike CsA, valspodar displayed little inherent toxicity [232]; however, its addition caused a reduced clearance and metabolism of anticancer drugs [232, 233]. The reason for this effect was the competitive inhibition of anticancer drug metabolism by CYP3A isoforms, thereby prolonging drug residence in the plasma and increasing nonspecific toxicity [232, 233]. However, it was suggested that restricting the bioavailability of valspodar and modification in the dosage of anticancer drugs could overcome these effects and warranted further clinical assessment. Two recent phase III trials reported pharmacokinetic interactions that were overcome; however, the combined treatment with valspodar and anticancer drug did not positively impact patient survival [234].

The problems associated with the first two generations of P-gp inhibitors necessitated an alternative approach to P-gp inhibition. This was achieved through the use of combinatorial chemistry to discover novel classes of lead compounds. The two most notable success stories of this third-generation P-gp inhibitor were GF120918 (elacridar) [235] and XR9576 (tariquidar) [236]. Both drugs were characterized by nanomolar potencies for interaction with P-gp resulting in a high degree of optimism. This was further enhanced by *in vivo* studies demonstrating that the pharmacokinetic interaction between anticancer drugs and these two modulators at the level of CYP3A metabolism was not as significant as with the earlier generations. Inherent toxicities are also less pronounced and both drugs remain in late-stage clinical trials. The only available data from phase III trials has been presented for tariquidar in combination with doxorubicin and taxane containing chemotherapy regimes in breast cancer [237]. The results indicated that only a small subset of patients exhibited benefit with combined administration of anticancer drug and tariquidar. Further clinical trials in a variety of cancer types and chemotherapy regimes are required to elucidate the true worth of tariquidar in the restoration of chemotherapy in resistant cancer.

In summary, over 30 years of research into P-gp inhibitors has generated only a handful of clinically usable compounds. Clearly, there is a pressing need for development of new inhibitors. The use of more rational or directed drug development has not yet been exploited and thus the provision of structural information on the drug binding sites of P-gp will prove instrumental.

1.3.7
What Properties Are Shared by Drugs that Interact with P-Glycoprotein?

A number of attempts have been made to compare physicochemical properties of a large number of compounds capable of interacting with P-gp. This comparative analysis would ideally generate a set of pharmacophoric "rules" to facilitate design of potent inhibitors of P-gp. The earliest attempts [238, 239] failed to reveal any specific criteria; however, they did suggest that substrates and inhibitors shared the following physicochemical properties: (i) planar aromatic rings, (ii) a basic nitrogen atom, and (iii) lipophilicity. Subsequent functional studies [240] classified P-gp modulators and substrates into distinct subsections, but this provided only weak discrimination between interacting drugs.

A comprehensive study by Seelig [241] was the first to produce a specific pharmacophoric pattern for recognition by P-gp and used a strategy involving examination of functional groups capable of hydrogen bonding to P-gp [241]. The strongest interacting compounds contained two or three electron donor groups, and moreover, these groups displayed fixed spatial separation (e.g., type I inhibitors have 2.5 ± 0.3 Å spacing between two e^- donor groups). The data were supported by the high degree of hydrogen bonding donor and acceptor moieties within the transmembrane helices of P-gp. A follow-up study [242] proposed that substrates and inhibitors varied in their propensity to form hydrogen bonds and that this was a key to defining the affinity for interaction. Inhibitors were proposed to display higher affinity due to stronger and more numerous hydrogen bonds with P-gp. This would result in slower dissociation rates for inhibitors, a property that was supported by radioligand binding studies with [^{3}H]-vinblastine (substrate) and [^{3}H]-XR9576 (inhibitor) [60].

During the past 5 years, an increasing number of more sophisticated bioinformatic or modeling approaches have been employed in the quest to generate a detailed map for the pharmacophore of P-gp substrates and inhibitors. The drug–P-gp interaction is a complex one, and as result, these strategies have focused on single classes of compounds. Raad *et al.* used a 3D quantitative structure–activity (QSAR) analysis for natural and synthetic coumarin derivatives [243]. The electrostatic and steric volume factors provided the greatest predictive power in assessing potency of interaction with P-gp. Moreover, a neutral hydrophobic group on the C4 position of the coumarin group greatly affected potency. The nature of the aromatic ring substitution of propafenone derivatives also greatly influenced the affinity of this class of compounds. An e^- donating moiety was a positive factor, whereas a bulky substituent (e.g., diphenylalkamine) weakened the interaction [244]. Labrie *et al.* [245] confined their analysis to derivatives of the potent anthranilamide tariquidar and focused on four distinct chemical regions of the molecule [245]. Once again, the affinity of derivative interaction with P-gp was greatly influenced by steric effects, electrostatic potential, and positioning of the hydrophobic moieties.

Thus, it is clear that two decades of structure–activity analyses have generated only subtle alterations to the originally proposed pharmacophoric elements of drugs interacting with P-gp. What has emerged is that the interaction is clearly a complex one and highly specific at a local level, which is far removed from the earliest suggestions based only on hydrophobicity. In addition, there are subtle, but important, differences between classes of drugs interacting with P-gp as might be expected given that the protein has multiple distinct binding sites. Two recent investigations employing a larger number of compounds have revealed some more specific requirements [245, 246]. Affinity of the drug–P-gp interaction is proportional to the hydrogen bond strength and the specific distance between the hydrogen bonding groups. Substrates may be more hydrophilic than originally proposed, and in fact the K_m for transport does not correlate with log P values. The emerging picture from these investigations is that high-affinity compounds minimally contain (i) two hydrophobic groups separated by 16.5 Å and (ii) two hydrogen bond acceptor groups that are 11.5 Å apart. Fully exploiting these data will require data on the structure of drug binding sites and the molecular basis underlying the "polyspecificity" of P-gp.

1.3.8
Postscript: Further X-Ray Crystallographic Studies and a Structure for the Nucleotide-Free State of P-Glycoprotein

Just prior to the proof stage of this chapter, a report describing the structure of P-gp in the nucleotide-free (apo) state appeared from the group of Chang [247]. Since this happy event merits the short description below, it is also worth updating the chapter with very recent structural insights emerging for other ABC transporters. Two structures for entire bacterial ABC proteins have been added to the growing list: (a) The structure of the methionine importer MetNI, from the laboratory of Rees [248], displays a similar TMD fold to the ModBC and MalFGK2 structures. (b) A second structure for MalFGK2 in the absence of nucleotide and in an apparently inward-facing configuration [249] beautifully confirms the predictions of Davidson about the likely conformational changes in the bacterial importers which were based on comparison of the ModBC and (outward facing) MalFGK2 structures - see Figure 1.3.

Lastly, the P-gp structure: It is the first structure of a eukaryotic ABC protein where the resolution is sufficient to trace most of the (single) polypeptide chain. Moreover, Aller and his co-workers [247] were able to generate two further structures of the nucleotide-free protein with different chiral forms of a cyclic peptide inhibitor bound to the TMD portion of the protein. Although the structural biology group were unable to generate a structure for the nucleotide-bound form of the protein, the existence of a lower resolution EM map (Figure 1.6) and its strong similarity to the nucleotide-bound Sav1866 structure will allow us insight into the conformational changes undergone by P-gp when moving between nucleotide-bound and nucleotide-free states: It seems likely that transmembrane helices 4 and 5 in TMD1, and their equivalents in TMD2 (helices 10 and 11) rotate as pairs around hinge regions in the extracellular loops of P-gp giving a tweezers-like opening and closing of the TMDs. This motion is associated with a separation and coming together of the NBDs, mediated by the intracytoplasmic loops 2 and 4 connecting transmembrane helices 4 to 5 and 10 to 11 respectively. As predicted from the Sav1866 structure, these intracytoplasmic loops cross over to the opposite side of the molecule in P-gp.

1.4
Summary

Progress in the structural description of ABC transporters is beginning to reach the stage where clinically relevant outputs may arise. A combination of low- and high-resolution structural data, alongside molecular homology modeling, is giving us a much clearer picture of transporters such as P-gp. Compared to even 10 years ago, our knowledge has advanced enormously. Given a similar rate of progress, the probability of designing novel inhibitors and drugs with input from structural data seems high over the next decade. Knowledge of the structure of binding sites in P-gp will be vital in the design of new inhibitors of the protein.

References

1 Klein, I., Sarkadi, B., and Varadi, A. (1999) An inventory of the human ABC proteins. *Biochimica et Biophysica Acta*, **1461** (2), 237–262.

2 Linton, K.J. and Higgins, C.F. (1998) The *Escherichia coli* ATP-binding cassette (ABC) proteins. *Molecular Microbiology*, **28** (1), 5–13.

3 Rosenberg, M.F., Mao, Q., Holzenburg, A., Ford, R.C., Deeley, R.G., and Cole, S.P. (2001) The structure of the multidrug resistance protein 1 (MRP1/ABCC1). Crystallization and single-particle analysis. *The Journal of Biological Chemistry*, **276** (19), 16076–16082.

4 Higgins, C.F. (1992) ABC transporters: from microorganisms to man. *Annual Review of Cell Biology*, **8**, 67–113.

5 Childs, S. and Ling, V. (1994) The MDR superfamily of genes and its biological implications, in *Important Advances in Oncology* (eds V.T. DeVita, S. Hellman, and S.A. Rosenberg), Lippincott, Philadelphia, PA, pp. 21–36.

6 Dean, M. and Allikmets, R. (1995) Evolution of ATP-binding cassette

transporter genes. *Current Opinion in Genetics & Development*, **5** (6), 779–785.

7 Dean, M., Rzhetsky, A., and Allikmets, R. (2001) The human ATP-binding cassette (ABC) transporter superfamily. *Genome Research*, **11** (7), 1156–1166.

8 Mutch, D.M., Anderle, P., Fiaux, M., Mansourian, R., Vidal, K., Wahli, W., Williamson, G., and Roberts, M.A. (2004) Regional variations in ABC transporter expression along the mouse intestinal tract. *Physiological Genomics*, **17** (1), 11–20.

9 Kerr, I.D. (2002) Structure and association of ATP-binding cassette transporter nucleotide-binding domains. *Biochimica et Biophysica Acta*, **1561** (1), 47–64.

10 Hosie, A.H., Allaway, D., Jones, M.A., Walshaw, D.L., Johnston, A.W., and Poole, P.S. (2001) Solute-binding protein-dependent ABC transporters are responsible for solute efflux in addition to solute uptake. *Molecular Microbiology*, **40**, 1449–1459.

11 Kast, C., Canfield, V., Levenson, R., and Gros, P. (1995) Membrane topology of P-glycoprotein as determined by epitope insertion: transmembrane organization of the N-terminal domain of mdr3. *Biochemistry*, **34** (13), 4402–4411.

12 Kast, C., Canfield, V., Levenson, R., and Gros, P. (1996) Transmembrane organization of mouse P-glycoprotein determined by epitope insertion and immunofluorescence. *The Journal of Biological Chemistry*, **271** (16), 9240–9248.

13 Tusnady, G.E., Bakos, E., Varadi, A., and Sarkadi, B. (1997) Membrane topology distinguishes a subfamily of the ATP-binding cassette (ABC) transporters. *FEBS Letters*, **402** (1), 1–3.

14 Jha, S., Karnani, N., Dhar, S.K., Mukhopadhayay, K., Shukla, S., Saini, P., Mukhopadhayay, G., and Prasad, R. (2003) Purification and characterization of the N-terminal nucleotide binding domain of an ABC drug transporter of *Candida albicans*: uncommon cysteine 193 of Walker A is critical for ATP hydrolysis. *Biochemistry*, **42** (36), 10822–10832.

15 Locher, K.P. and Borths, E. (2004) ABC transporter architecture and mechanism: implications from the crystal structures of BtuCD and BtuF. *FEBS Letters*, **564** (3), 264–268.

16 Azzaria, M., Schurr, E., and Gros, P. (1989) Discrete mutations introduced in the predicted nucleotide-binding sites of the mdr1 gene abolish its ability to confer multidrug resistance. *Molecular and Cellular Biology*, **9** (12), 5289–5297.

17 Al-Shawi, M.K. and Senior, A.E. (1993) Characterization of the adenosine triphosphatase activity of Chinese hamster P-glycoprotein. *The Journal of Biological Chemistry*, **268** (6), 4197–4206.

18 Gao, M., Cui, H.R., Loe, D.W., Grant, C.E., Almquist, K.C., Cole, S.P., and Deeley, R.G. (2000) Comparison of the functional characteristics of the nucleotide binding domains of multidrug resistance protein 1. *The Journal of Biological Chemistry*, **275** (17), 13098–13108.

19 Hou, Y., Cui, L., Riordan, J.R., and Chang, X. (2000) Allosteric interactions between the two non-equivalent nucleotide binding domains of multidrug resistance protein MRP1. *The Journal of Biological Chemistry*, **275** (27), 20280–20287.

20 Lapinski, P.E., Neubig, R.R., and Raghavan, M. (2001) Walker A lysine mutations of TAP1 and TAP2 interfere with peptide translocation but not peptide binding. *The Journal of Biological Chemistry*, **276** (10), 7526–7533.

21 Saveanu, L., Daniel, S., and van Endert, P.M. (2001) Distinct functions of the ATP binding cassettes of transporters associated with antigen processing: a mutational analysis of Walker A and B sequences. *The Journal of Biological Chemistry*, **276** (25), 22107–22113.

22 Proff, C. and Kolling, R. (2001) Functional asymmetry of the two nucleotide binding domains in the ABC transporter Ste6. *Molecular & General Genetics*, **264** (6), 883–893.

23 Anderson, M.P. and Welsh, M.J. (1992) Regulation by ATP and ADP of CFTR

chloride channels that contain mutant nucleotide-binding domains. *Science*, **257** (5077), 1701–1704.

24 Carson, M.R., Travis, S.M., and Welsh, M.J. (1995) The two nucleotide-binding domains of cystic fibrosis transmembrane conductance regulator (CFTR) have distinct functions in controlling channel activity. *The Journal of Biological Chemistry*, **270** (4), 1711–1717.

25 Cui, L., Aleksandrov, L., Chang, X.B., Hou, Y.X., He, L., Hegedus, T., Gentzsch, M., Aleksandrov, A., Balch, W.E., and Riordan, J.R. (2007) Domain interdependence in the biosynthetic assembly of CFTR. *Journal of Molecular Biology*, **365**, 981–994.

26 Senior, A.E., al-Shawi, M.K., and Urbatsch, I.L. (1995) ATP hydrolysis by multidrug-resistance protein from Chinese hamster ovary cells. *Journal of Bioenergetics and Biomembranes*, **27** (1), 31–36.

27 Szakacs, G., Ozvegy, C., Bakos, E., Sarkadi, B., and Varadi, A. (2001) Role of glycine-534 and glycine-1179 of human multidrug resistance protein (MDR1) in drug-mediated control of ATP hydrolysis. *The Biochemical Journal*, **356** (Pt 1), 71–75.

28 Urbatsch, I.L., Gimi, K., Wilke-Mounts, S., and Senior, A.E. (2000) Conserved Walker A Ser residues in the catalytic sites of P-glycoprotein are critical for catalysis and involved primarily at the transition state step. *The Journal of Biological Chemistry*, **275** (32), 25031–25038.

29 Urbatsch, I.L., Gimi, K., Wilke-Mounts, S., and Senior, A.E. (2000) Investigation of the role of glutamine-471 and glutamine-1114 in the two catalytic sites of P-glycoprotein. *Biochemistry*, **39** (39), 11921–11927.

30 Takada, Y., Yamada, K., Taguchi, Y., Kino, K., Matsuo, M., Tucker, S.J., Komano, T., Amachi, T., and Ueda, K. (1998) Non-equivalent cooperation between the two nucleotide-binding folds of P-glycoprotein. *Biochimica et Biophysica Acta*, **1373** (1), 131–136.

31 Loo, T.W. and Clarke, D.M. (1995) Covalent modification of human P-glycoprotein mutants containing a single cysteine in either nucleotide-binding fold abolishes drug-stimulated ATPase activity. *The Journal of Biological Chemistry*, **270** (39), 22957–22961.

32 Hrycyna, C.A., Ramachandra, M., Germann, U.A., Cheng, P.W., Pastan, I., and Gottesman, M.M. (1999) Both ATP sites of human P-glycoprotein are essential but not symmetric. *Biochemistry*, **38** (42), 13887–13899.

33 Vigano, C., Julien, M., Carrier, I., Gros, P., and Ruysschaert, J.M. (2002) Structural and functional asymmetry of the nucleotide-binding domains of P-glycoprotein investigated by attenuated total reflection Fourier transform infrared spectroscopy. *The Journal of Biological Chemistry*, **277** (7), 5008–5016.

34 Berridge, G., Walker, J.A., Callaghan, R., and Kerr, I.D. (2003) The nucleotide-binding domains of P-glycoprotein. Functional symmetry in the isolated domain demonstrated by *N*-ethylmaleimide labelling. *European Journal of Biochemistry*, **270** (7), 1483–1492.

35 Walker, J.E., Saraste, M., Runswick, M.J., and Gay, N.J. (1982) Distantly related sequences in the alpha- and beta-subunits of ATP synthase, myosin, kinases and other ATP-requiring enzymes and a common nucleotide binding fold. *The EMBO Journal*, **1** (8), 945–951.

36 Young, J. and Holland, I.B. (1999) ABC transporters: bacterial exporters-revisited five years on. *Biochimica et Biophysica Acta*, **1461** (2), 177–200.

37 Diederichs, K., Diez, J., Greller, G., Muller, C., Breed, J., Schnell, C., Vonrhein, C., Boos, W., and Welte, W. (2000) Crystal structure of MalK, the ATPase subunit of the trehalose/maltose ABC transporter of the archaeon *Thermococcus litoralis*. *The EMBO Journal*, **19** (22), 5951–5961.

38 Ambudkar, S.V., Kim, I.W., Xia, D., and Sauna, Z.E. (2006) The A-loop, a novel

conserved aromatic acid subdomain upstream of the Walker A motif in ABC transporters, is critical for ATP binding. *FEBS Letters*, **580**, 1049–1055.

39 Holland, I.B. and Blight, M.A. (1999) ABC-ATPases, adaptable energy generators fuelling transmembrane movement of a variety of molecules in organisms from bacteria to humans. *Journal of Molecular Biology*, **293** (2), 381–399.

40 Martin, C., Berridge, G., Higgins, C.F., Mistry, P., Charlton, P., and Callaghan, R. (2000) Communication between multiple drug binding sites on P-glycoprotein. *Molecular Pharmacology*, **58**, 624–632.

41 Saurin, W., Hofnung, M., and Dassa, E. (1999) Getting in or out: early segregation between importers and exporters in the evolution of ATP-binding cassette (ABC) transporters. *Journal of Molecular Evolution*, **48** (1), 22–41.

42 Locher, K.P., Lee, A.T., and Rees, D.C. (2002) The *E. coli* BtuCD structure: a framework for ABC transporter architecture and mechanism. *Science*, **296** (5570), 1091–1098.

43 Pinkett, H.W., Lee, A.T., Lum, P., Locher, K.P., and Rees, D.C. (2007) An inward-facing conformation of a putative metal-chelate-type ABC transporter. *Science*, **315**, 373–377.

44 Reyes, C.L. and Chang, G. (2005) Structure of the ABC transporter MsbA in complex with ADP·vanadate and lipopolysaccharide. *Science*, **308**, 1028–1031. Retraction in: Chang, G., Roth, C.B., Reyes, C.L., Pornillos, O., Chen, Y.J., and Chen, A.P. (2006) *Science*, **314**, 1875.

45 Chang, G. (2007) Retraction of "Structure of MsbA from *Vibrio cholera*: a multidrug resistance ABC transporter homolog in a closed conformation" [J. Mol. Biol. (2003) 330: 419–430]. *Journal of Molecular Biology*, **369**, 596.

46 Chang, G., Roth, C.B., Reyes, C.L., Pornillos, O., Chen, Y.J., and Chen, A.P. (2006) Retraction. *Science*, **314**, 1875.

47 Ward, A., Reyes, C.L., Yu, J., Roth, C.B., and Chang, G. (2007) Flexibility in the ABC transporter MsbA: alternating access with a twist. *Proceedings of the National Academy of Sciences of the United States of America*, **104**, 19005–19010.

48 Oldham, M.L., Khare, D., Quiocho, F.A., Davidson, A.L., and Chen, J. (2007) Crystal structure of a catalytic intermediate of the maltose transporter. *Nature*, **450**, 515–521.

49 Rosenberg, M.F., Callaghan, R., Ford, R.C., and Higgins, C.F. (1997) Structure of the multidrug resistance P-glycoprotein to 2.5 nm resolution determined by electron microscopy and image analysis. *The Journal of Biological Chemistry*, **272** (16), 10685–10694.

50 Rosenberg, M.F., Velarde, G., Ford, R.C., Martin, C., Berridge, G., Kerr, I.D., Callaghan, R., Schmidlin, A., Wooding, C., Linton, K.J., and Higgins, C.F. (2001) Repacking of the transmembrane domains of P-glycoprotein during the transport ATPase cycle. *The EMBO Journal*, **20** (20), 5615–5625.

51 Rosenberg, M.F., Kamis, A.B., Aleksandrov, L.A., Ford, R.C., and Riordan, J.R. (2004) Purification and crystallization of the cystic fibrosis transmembrane conductance regulator (CFTR). *The Journal of Biological Chemistry*, **279**, 39051–39057.

52 Dey, S., Ramachandra, M., Pastan, I., Gottesman, M.M., and Ambudkar, S.V. (1997) Evidence for two nonidentical drug-interaction sites in the human P-glycoprotein. *Proceedings of the National Academy of Sciences of the United States of America*, **94** (20), 10594–10599.

53 Loo, T.W. and Clarke, D.M. (2001) Defining the drug-binding site in the human multidrug resistance P-glycoprotein using a methanethiosulfonate analog of verapamil, MTS-verapamil. *The Journal of Biological Chemistry*, **276**, 14972–14979.

54 Loo, T.W. and Clarke, D.M. (2001) Determining the dimensions of the drug-

binding domain of human P-glycoprotein using thiol cross-linking compounds as molecular rulers. *The Journal of Biological Chemistry*, **276**, 36877–36880.

55 Loo, T.W. and Clarke, D.M. (2002) Vanadate trapping of nucleotide at the ATP-binding sites of human multidrug resistance P-glycoprotein exposes different residues to the drug-binding site. *Proceedings of the National Academy of Sciences of the United States of America*, **99** (6), 3511–3516.

56 Loo, T.W. and Clarke, D.M. (2002) Location of the rhodamine-binding site in the human multidrug resistance P-glycoprotein. *The Journal of Biological Chemistry*, **277** (46), 44332–44338.

57 Loo, T.W., Bartlett, M.C., and Clarke, D.M. (2003) Substrate-induced conformational changes in the transmembrane segments of human P-glycoprotein. Direct evidence for the substrate-induced fit mechanism for drug binding. *The Journal of Biological Chemistry*, **278** (16), 13603–13606.

58 Loo, T.W., Bartlett, M.C., and Clarke, D.M. (2003) Simultaneous binding of two different drugs in the binding pocket of the human multidrug resistance P-glycoprotein. *The Journal of Biological Chemistry*, **278** (41), 39706–39710.

59 Shapiro, A.B., Fox, K., Lam, P., and Ling, V. (1999) Stimulation of P-glycoprotein-mediated drug transport by prazosin and progesterone. Evidence for a third drug-binding site. *European Journal of Biochemistry*, **259**, 841–850.

60 Martin, C., Berridge, G., Mistry, P., Higgins, C., Charlton, P., and Callaghan, R. (1999) The molecular interaction of the high affinity reversal agent XR9576 with P-glycoprotein. *British Journal of Pharmacology*, **128** (2), 403–411.

61 Schmitt, L. and Tampe, R. (2002) Structure and mechanism of ABC transporters. *Current Opinion in Structural Biology*, **12** (6), 754–760.

62 Hung, L.W., Wang, I.X., Nikaido, K., Liu, P.Q., Ames, G.F., and Kim, S.H. (1998) Crystal structure of the ATP-binding subunit of an ABC transporter. *Nature*, **396** (6712), 703–707.

63 Hopfner, K.P., Karcher, A., Shin, D.S., Craig, L., Arthur, L.M., Carney, J.P., and Tainer, J.A. (2000) Structural biology of Rad50 ATPase: ATP-driven conformational control in DNA double-strand break repair and the ABC-ATPase superfamily. *Cell*, **101** (7), 789–800.

64 Gadsby, D.C., Vergani, P., and Csanady, L. (2006) The ABC protein turned chloride channel whose failure causes cystic fibrosis. *Nature*, **440**, 477–483.

65 Moody, J.E., Millen, L., Binns, D., Hunt, J.F., and Thomas, P.J. (2002) Cooperative, ATP-dependent association of the nucleotide binding cassettes during the catalytic cycle of ATP-binding cassette transporters. *The Journal of Biological Chemistry*, **277** (24), 21111–21114.

66 Qian, Y.M., Qiu, W., Gao, M., Westlake, C.J., Cole, S.P., and Deeley, R.G. (2001) Characterization of binding of leukotriene C4 by human multidrug resistance protein 1: evidence of differential interactions with NH_2- and COOH-proximal halves of the protein. *The Journal of Biological Chemistry*, **276** (42), 38636–38644.

67 Hou, Y.X., Cui, L., Riordan, J.R., and Chang, X.B. (2002) ATP binding to the first nucleotide-binding domain of multidrug resistance protein MRP1 increases binding and hydrolysis of ATP and trapping of ADP at the second domain. *The Journal of Biological Chemistry*, **277** (7), 5110–5119.

68 Dousmanis, A.G., Nairn, A.C., and Gadsby, D.C. (2002) Distinct Mg^{2+}-dependent steps rate limit opening and closing of a single CFTR Cl^- channel. *The Journal of General Physiology*, **119** (6), 545–559.

69 Aleksandrov, L., Aleksandrov, A.A., Chang, X.B., and Riordan, J.R. (2002) The first nucleotide binding domain of cystic fibrosis transmembrane conductance regulator is a site of stable nucleotide interaction, whereas the second is a site of

70 van Veen, H.W., Margolles, A., Muller, M., Higgins, C.F., and Konings, W.N. (2000) The homodimeric ATP-binding cassette transporter LmrA mediates multidrug transport by an alternating two-site (two-cylinder engine) mechanism. *The EMBO Journal*, **19** (11), 2503–2514.

71 van Veen, H.W., Margolles, A., Putman, M., Sakamoto, K., and Konings, W.N. (1999) Multidrug resistance in lactic acid bacteria: molecular mechanisms and clinical relevance. *Antonie van Leeuwenhoek*, **76** (1–4), 347–352.

72 van Veen, H.W., Putman, M., Margolles, A., Sakamoto, K., and Konings, W.N. (1999) Structure–function analysis of multidrug transporters in *Lactococcus lactis*. *Biochimica et Biophysica Acta*, **1461** (2), 201–206.

73 van Veen, H.W. and Konings, W.N. (1996) Multidrug resistance mediated by a bacterial homolog of the human multidrug transporter MDR1. *Proceedings of the National Academy of Sciences of the United States of America*, **93** (20), 10668–10672.

74 Shepard, R.L., Winter, M.A., Hsaio, S.C., Pearce, H.L., Beck, W.T., and Dantzig, A.H. (1998) Effect of modulators on the ATPase activity and vanadate nucleotide trapping of human P-glycoprotein. *Biochemical Pharmacology*, **56** (6), 719–727.

75 Sauna, Z.E. and Ambudkar, S.V. (2000) Evidence for a requirement for ATP hydrolysis at two distinct steps during a single turnover of the catalytic cycle of human P-glycoprotein. *Proceedings of the National Academy of Sciences of the United States of America*, **97** (6), 2515–2520.

76 Kerr, K.M., Sauna, Z.E., and Ambudkar, S.V. (2001) Correlation between steady-state ATP hydrolysis and vanadate-induced ADP trapping in human P-glycoprotein. Evidence for ADP release as the rate-limiting step in the catalytic cycle and its modulation by substrates. *The Journal of Biological Chemistry*, **276** (12), 8657–8664.

77 Sauna, Z.E. and Ambudkar, S.V. (2001) Characterization of the catalytic cycle of ATP hydrolysis by human P-glycoprotein. The two ATP hydrolysis events in a single catalytic cycle are kinetically similar but affect different functional outcomes. *The Journal of Biological Chemistry*, **276** (15), 11653–11661.

78 Sauna, Z.E., Smith, M.M., Muller, M., and Ambudkar, S.V. (2001) Functionally similar vanadate-induced 8-azidoadenosine 5′-[alpha-^{32}P] diphosphate-trapped transition state intermediates of human P-glycoprotein are generated in the absence and presence of ATP hydrolysis. *The Journal of Biological Chemistry*, **276** (24), 21199–21208.

79 Sauna, Z.E., Smith, M.M., Muller, M., and Ambudkar, S.V. (2001) Evidence for the vectorial nature of drug (substrate)-stimulated ATP hydrolysis by human P-glycoprotein. *The Journal of Biological Chemistry*, **276** (36), 33301–33304.

80 Martin, C., Higgins, C.F., and Callaghan, R. (2001) The vinblastine binding site adopts high- and low-affinity conformations during a transport cycle of P-glycoprotein. *Biochemistry*, **40** (51), 15733–15742.

81 Davidson, A.L. and Chen, J. (2004) ATP-binding cassette transporters in bacteria. *Annual Review of Biochemistry*, **73**, 241–268.

82 Borges-Walmsley, M.I., McKeegan, K.S., and Walmsley, A.R. (2003) Structure and function of efflux pumps that confer resistance to drugs. *The Biochemical Journal*, **376** (Pt 2), 313–338.

83 Loo, T.W., Bartlett, M.C., and Clarke, D.M. (2002) The "LSGGQ" motif in each nucleotide-binding domain of human P-glycoprotein is adjacent to the opposing Walker A sequence. *The Journal of Biological Chemistry*, **277** (44), 41303–41306.

84 Loo, T.W., Bartlett, M.C., and Clarke, D.M. (2003) Drug binding in human P-glycoprotein causes conformational

changes in both nucleotide-binding domains. *The Journal of Biological Chemistry*, **278** (3), 1575–1578.
85 Callaghan, R., Ford, R.C., and Kerr, I.D. (2006) The translocation mechanism of P-glycoprotein. *FEBS Letters*, **580**, 1056–1063.
86 Schneider, E. and Hunke, S. (1998) ATP-binding-cassette (ABC) transport systems: functional and structural aspects of the ATP-hydrolyzing subunits/domains. *FEMS Microbiology Reviews*, **22** (1), 1–20.
87 Yuan, Y.R., Blecker, S., Martsinkevich, O., Millen, L., Thomas, P.J., and Hunt, J.F. (2001) The crystal structure of the MJ0796 ATP-binding cassette. Implications for the structural consequences of ATP hydrolysis in the active site of an ABC transporter. *The Journal of Biological Chemistry*, **276** (34), 32313–32321.
88 Chen, H.L., Gabrilovich, D., Tampe, R., Girgis, K.R., Nadaf, S., and Carbone, D.P. (1996) A functionally defective allele of TAP1 results in loss of MHC class I antigen presentation in a human lung cancer. *Nature Genetics*, **13** (2), 210–213.
89 Riordan, J.R. (1993) The cystic fibrosis transmembrane conductance regulator. *Annual Review of Physiology*, **55**, 609–630.
90 Dawson, R.J. and Locher, K.P. (2006) Structure of a bacterial multidrug ABC transporter. *Nature*, **443**, 180–185.
91 Murakami, S., Nakashima, R., Yamashita, E., and Yamaguchi, A. (2002) Crystal structure of bacterial multidrug efflux transporter AcrB. *Nature*, **419** (6907), 587–593.
92 Koronakis, V., Sharff, A., Koronakis, E., Luisi, B., and Hughes, C. (2000) Crystal structure of the bacterial membrane protein TolC central to multidrug efflux and protein export. *Nature*, **405** (6789), 914–919.
93 Chang, G. (2003) Structure of MsbA from *Vibrio cholera*: a multidrug resistance ABC transporter homolog in a closed conformation. *Journal of Molecular Biology*, **330** (2), 419–430.

94 Chang, G. and Roth, C.B. (2001) Structure of MsbA from *E. coli*: a homolog of the multidrug resistance ATP binding cassette (ABC) transporters. *Science*, **293** (5536), 1793–1800.
95 Faham, S. and Bowie, J.U. (2002) Bicelle crystallization: a new method for crystallizing membrane proteins yields a monomeric bacteriorhodopsin structure. *Journal of Molecular Biology*, **316** (1), 1–6.
96 Nollert, P., Navarro, J., and Landau, E.M. (2002) Crystallization of membrane proteins in cubo. *Methods in Enzymology*, **343**, 183–199.
97 Armstrong, S., Tabernero, L., Zhang, H., Hermodsen, M., and Stauffacher, C. (1998) Powering the ABC transporters: the 2.5 Å crystal structure of the ABC domain of RbsA. *Pediatric Pulmonology*, **17**, 91–92.
98 Karpowich, N., Martsinkevich, O., Millen, L., Yuan, Y.R., Dai, P.L., MacVey, K., Thomas, P.J., and Hunt, J.F. (2001) Crystal structures of the MJ1267 ATP binding cassette reveal an induced-fit effect at the ATPase active site of an ABC transporter. *Structure*, **9** (7), 571–586.
99 Lowe, J., Cordell, S.C., and van den Ent, F. (2001) Crystal structure of the SMC head domain: an ABC ATPase with 900 residues antiparallel coiled-coil inserted. *Journal of Molecular Biology*, **306** (1), 25–35.
100 Lamers, M.H., Perrakis, A., Enzlin, J.H., Winterwerp, H.H., de Wind, N., and Sixma, T.K. (2000) The crystal structure of DNA mismatch repair protein MutS binding to a G × T mismatch. *Nature*, **407** (6805), 711–717.
101 Obmolova, G., Ban, C., Hsieh, P., and Yang, W. (2000) Crystal structures of mismatch repair protein MutS and its complex with a substrate DNA. *Nature*, **407** (6805), 703–710.
102 Gaudet, R. and Wiley, D.C. (2001) Structure of the ABC ATPase domain of human TAP1, the transporter associated with antigen processing. *The EMBO Journal*, **20** (17), 4964–4972.

103 Verdon, G., Albers, S.V., Dijkstra, B.W., Driessen, A.J., and Thunnissen, A.M. (2003) Crystal structures of the ATPase subunit of the glucose ABC transporter from *Sulfolobus solfataricus*: nucleotide-free and nucleotide-bound conformations. *Journal of Molecular Biology*, **330** (2), 343–358.

104 Schmitt, L., Benabdelhak, H., Blight, M.A., Holland, I.B., and Stubbs, M.T. (2003) Crystal structure of the nucleotide-binding domain of the ABC-transporter haemolysin B: identification of a variable region within ABC helical domains. *Journal of Molecular Biology*, **330** (2), 333–342.

105 Lewis, H.A., Buchanan, S.G., Burley, S.K., Conners, K., Dickey, M., Dorwart, M., Fowler, R., Gao, X., Guggino, W.B., Hendrickson, W.A., Hunt, J.F., Kearins, M.C., Lorimer, D., Maloney, P.C., Post, K.W., Rajashankar, K.R., Rutter, M.E., Sauder, J.M., Shriver, S., Thibodeau, P.H., Thomas, P.J., Zhang, M., Zhao, X., and Emtage, S. (2004) Structure of nucleotide-binding domain 1 of the cystic fibrosis transmembrane conductance regulator. *The EMBO Journal*, **23** (2), 282–293.

106 McKeegan, K.S., Borges-Walmsley, M.I., and Walmsley, A.R. (2003) The structure and function of drug pumps: an update. *Trends in Microbiology*, **11** (1), 21–29.

107 Davidson, A.L., Laghaeian, S.S., and Mannering, D.E. (1996) The maltose transport system of *Escherichia coli* displays positive cooperativity in ATP hydrolysis. *The Journal of Biological Chemistry*, **271** (9), 4858–4863.

108 Liu, C.E., Liu, P.Q., and Ames, G.F. (1997) Characterization of the adenosine triphosphatase activity of the periplasmic histidine permease, a traffic ATPase (ABC transporter). *The Journal of Biological Chemistry*, **272** (35), 21883–21891.

109 Stenham, D.R., Campbell, J.D., Sansom, M.S., Higgins, C.F., Kerr, I.D., and Linton, K.J. (2003) An atomic detail model for the human ATP binding cassette transporter P-glycoprotein derived from disulfide cross-linking and homology modelling. *The FASEB Journal*, **17** (15), 2287–2289.

110 Rosenberg, M.F., Callaghan, R., Modok, S., Higgins, C.F., and Ford, R.C. (2005) 3-D structure of P-glycoprotein: the transmembrane regions adopt an asymmetric configuration in the nucleotide-bound state. *The Journal of Biological Chemistry*, **280**, 2857–2862.

111 Hollenstein, K., Frei, D.C., and Locher, K.P. (2007) Structure of an ABC transporter in complex with its binding protein. *Nature*, **446**, 213–216.

112 McDevitt, C.A., Collins, R.F., Conway, M., Modok, S., Storm, J., Kerr, I.D., Ford, R.C., and Callaghan, R. (2006) Purification and 3D structural analysis of oligomeric human multidrug transporter ABCG2. *Structure*, **11**, 1623–1632.

113 Awayn, N.H., Rosenberg, M.F., Kamis, A.B., Aleksandrov, L.A., Riordan, J.R., and Ford, R.C. (2005) Crystallographic and single-particle analyses of native- and nucleotide-bound forms of the cystic fibrosis transmembrane conductance regulator (CFTR) protein. *Biochemical Society Transactions*, **33**, 996–999.

114 Rosenberg, M.F., Kamis, A.B., Callaghan, R., Higgins, C.F., and Ford, R.C. (2003) Three-dimensional structures of the mammalian multidrug resistance P-glycoprotein demonstrate major conformational changes in the transmembrane domains upon nucleotide binding. *The Journal of Biological Chemistry*, **278**, 8294–8299.

115 Chami, M., Steinfels, E., Orelle, C., Jault, J.M., Di Pietro, A., Rigaud, J.L., and Marco, S. (2002) Three-dimensional structure by cryo-electron microscopy of YvcC, an homodimeric ATP-binding cassette transporter from *Bacillus subtilis*. *Journal of Molecular Biology*, **315** (5), 1075–1085.

116 Ferreira-Pereira, A., Marco, S., Decottignies, A., Nader, J., Goffeau, A., and Rigaud, J.L. (2003) Three-dimensional reconstruction of the *Saccharomyces cerevisiae* multidrug

resistance protein Pdr5p. *The Journal of Biological Chemistry*, **278** (14), 11995–11999.

117 Kamis, A.B. (2005) Purification and electron microscopic studies of two ABC transporter proteins of clinical relevance. PhD thesis. University of Manchester, UK.

118 Mikhailov, M.V., Campbell, J.D., de Wet, H., Shimomura, K., Zadek, B., Collins, R.F., Sansom, M.S., Ford, R.C., and Ashcroft, F.M. (2005) 3-D structural and functional characterization of the purified KATP channel complex Kir6.2-SUR1. *The EMBO Journal*, **24**, 4166–4175.

119 Hofacker, M., Gompf, S., Zutz, A., Presenti, C., Haase, W., van der Does, C., Model, K., and Tampe, R. (2007) Structural and functional fingerprint of the mitochondrial ATP-binding cassette transporter Mdl1 from *Saccharomyces cerevisiae*. *The Journal of Biological Chemistry*, **282**, 3951–3961.

120 Higgins, C.F. (2007) Multiple molecular mechanisms for multidrug resistance transporters. *Nature*, **446**, 749–757.

121 Lage, H. (2003) ABC-transporters: implications on drug resistance from microorganisms to human cancers. *International Journal of Antimicrobial Agents*, **22** (3), 188–199.

122 Borst, P., Balzarini, J., Ono, N., Reid, G., de Vries, H., Wielinga, P., Wijnholds, J., and Zelcer, N. (2004) The potential impact of drug transporters on nucleoside-analog-based antiviral chemotherapy. *Antiviral Research*, **62** (1), 1–7.

123 Biedler, J.L. and Riehm, H. (1970) Cellular resistance to actinomycin D in Chinese hamster cells *in vitro*: cross-resistance, radioautographic, and cytogenetic studies. *Cancer Research*, **30** (4), 1174–1184.

124 Gottesman, M.M., Fojo, T., and Bates, S.E. (2002) Multidrug resistance in cancer: role of ATP-dependent transporters. *Nature Reviews. Cancer*, **2** (1), 48–58.

125 Van Bambeke, F., Balzi, E., and Tulkens, P.M. (2000) Antibiotic efflux pumps. *Biochemical Pharmacology*, **60** (4), 457–470.

126 Paulsen, I.T., Brown, M.H., and Skurray, R.A. (1996) Proton-dependent multidrug efflux systems. *Microbiological Reviews*, **60** (4), 575–608.

127 van Veen, H.W. and Konings, W.N. (1998) The ABC family of multidrug transporters in microorganisms. *Biochimica et Biophysica Acta*, **1365** (1–2), 31–36.

128 Lander, E.S., Linton, L.M., Birren, B., Nusbaum, C., Zody, M.C., Baldwin, J., Devon, K., Dewar, K., Doyle, M., FitzHugh, W., Funke, R., Gage, D., Harris, K., Heaford, A., Howland, J., Kann, L., Lehoczky, J., LeVine, R., McEwan, P., McKernan, K., Meldrim, J., Mesirov, J.P., Miranda, C., Morris, W., Naylor, J., Raymond, C., Rosetti, M., Santos, R., Sheridan, A., Sougnez, C., Stange-Thomann, N., Stojanovic, N., Subramanian, A., Wyman, D., Rogers, J., Sulston, J., Ainscough, R., Beck, S., Bentley, D., Burton, J., Clee, C., Carter, N., Coulson, A., Deadman, R., Deloukas, P., Dunham, A., Dunham, I., Durbin, R., French, L., Grafham, D., Gregory, S., Hubbard, T., Humphray, S., Hunt, A., Jones, M., Lloyd, C., McMurray, A., Matthews, L., Mercer, S., Milne, S., Mullikin, J.C., Mungall, A., Plumb, R., Ross, M., Shownkeen, R., Sims, S., Waterston, R.H., Wilson, R.K., Hillier, L.W., McPherson, J.D., Marra, M.A., Mardis, E.R., Fulton, L.A., Chinwalla, A.T., Pepin, K.H., Gish, W.R., Chissoe, S.L., Wendl, M.C., Delehaunty, K.D., Miner, T.L., Delehaunty, A., Kramer, J.B., Cook, L.L., Fulton, R.S., Johnson, D.L., Minx, P.J., Clifton, S.W., Hawkins, T., Branscomb, E., Predki, P., Richardson, P., Wenning, S., Slezak, T., Doggett, N., Cheng, J.F., Olsen, A., Lucas, S., Elkin, C., Uberbacher, E., Frazier, M., Gibbs, R.A., Muzny, D.M., Scherer, S.E., Bouck, J.B., Sodergren, E.J., Worley, K.C., Rives, C.M., Gorrell, J.H., Metzker, M.L., Naylor, S.L., Kucherlapati, R.S., Nelson, D.L., Weinstock, G.M., Sakaki, Y., Fujiyama, A., Hattori, M., Yada, T., Toyoda, A., Itoh, T., Kawagoe, C., Watanabe, H., Totoki, Y.,

Taylor, T., Weissenbach, J., Heilig, R., Saurin, W., Artiguenave, F., Brottier, P., Bruls, T., Pelletier, E., Robert, C., Wincker, P., Smith, D.R., Doucette-Stamm, L., Rubenfield, M., Weinstock, K., Lee, H.M., Dubois, J., Rosenthal, A., Platzer, M., Nyakatura, G., Taudien, S., Rump, A., Yang, H., Yu, J., Wang, J., Huang, G., Gu, J., Hood, L., Rowen, L., Madan, A., Qin, S., Davis, R.W., Federspiel, N.A., Abola, A.P., Proctor, M.J., Myers, R.M., Schmutz, J., Dickson, M., Grimwood, J., Cox, D.R., Olson, M.V., Kaul, R., Shimizu, N., Kawasaki, K., Minoshima, S., Evans, G.A., Athanasiou, M., Schultz, R., Roe, B.A., Chen, F., Pan, H., Ramser, J., Lehrach, H., Reinhardt, R., McCombie, W.R., de la Bastide, M., Dedhia, N., Blocker, H., Hornischer, K., Nordsiek, G., Agarwala, R., Aravind, L., Bailey, J.A., Bateman, A., Batzoglou, S., Birney, E., Bork, P., Brown, D.G., Burge, C.B., Cerutti, L., Chen, H.C., Church, D., Clamp, M., Copley, R.R., Doerks, T., Eddy, S.R., Eichler, E.E., Furey, T.S., Galagan, J., Gilbert, J.G., Harmon, C., Hayashizaki, Y., Haussler, D., Hermjakob, H., Hokamp, K., Jang, W., Johnson, L.S., Jones, T.A., Kasif, S., Kaspryzk, A., Kennedy, S., Kent, W.J., Kitts, P., Koonin, E.V., Korf, I., Kulp, D., Lancet, D., Lowe, T.M., McLysaght, A., Mikkelsen, T., Moran, J.V., Mulder, N., Pollara, V.J., Ponting, C.P., Schuler, G., Schultz, J., Slater, G., Smit, A.F., Stupka, E., Szustakowski, J., Thierry-Mieg, D., Thierry-Mieg, J., Wagner, L., Wallis, J., Wheeler, R., Williams, A., Wolf, Y.I., Wolfe, K.H., Yang, S.P., Yeh, R.F., Collins, F., Guyer, M.S., Peterson, J., Felsenfeld, A., Wetterstrand, K.A., Patrinos, A., Morgan, M.J., Szustakowki, J., de Jong, P., Catanese, J.J., Osoegawa, K., Shizuya, H., Choi, S., and Chen, Y.J., (2001) Initial sequencing and analysis of the human genome. *Nature*, **409** (6822), 860–921.

129 Venter, J.C., Adams, M.D., Myers, E.W., Li, P.W., Mural, R.J., Sutton, G.G., Smith, H.O., Yandell, M., Evans, C.A., Holt, R.A., Gocayne, J.D., Amanatides, P., Ballew, R.M., Huson, D.H., Wortman, J.R., Zhang, Q., Kodira, C.D., Zheng, X.H., Chen, L., Skupski, M., Subramanian, G., Thomas, P.D., Zhang, J., Gabor Miklos, G.L., Nelson, C., Broder, S., Clark, A.G., Nadeau, J., McKusick, V.A., Zinder, N., Levine, A.J., Roberts, R.J., Simon, M., Slayman, C., Hunkapiller, M., Bolanos, R., Delcher, A., Dew, I., Fasulo, D., Flanigan, M., Florea, L., Halpern, A., Hannenhalli, S., Kravitz, S., Levy, S., Mobarry, C., Reinert, K., Remington, K., Abu-Threideh, J., Beasley, E., Biddick, K., Bonazzi, V., Brandon, R., Cargill, M., Chandramouliswaran, I., Charlab, R., Chaturvedi, K., Deng, Z., Di Francesco, V., Dunn, P., Eilbeck, K., Evangelista, C., Gabrielian, A.E., Gan, W., Ge, W., Gong, F., Gu, Z., Guan, P., Heiman, T.J., Higgins, M.E., Ji, R.R., Ke, Z., Ketchum, K.A., Lai, Z., Lei, Y., Li, Z., Li, J., Liang, Y., Lin, X., Lu, F., Merkulov, G.V., Milshina, N., Moore, H.M., Naik, A.K., Narayan, V.A., Neelam, B., Nusskern, D., Rusch, D.B., Salzberg, S., Shao, W., Shue, B., Sun, J., Wang, Z., Wang, A., Wang, X., Wang, J., Wei, M., Wides, R., Xiao, C., Yan, C., Yao, A., Ye, J., Zhan, M., Zhang, W., Zhang, H., Zhao, Q., Zheng, L., Zhong, F., Zhong, W., Zhu, S., Zhao, S., Gilbert, D., Baumhueter, S., Spier, G., Carter, C., Cravchik, A., Woodage, T., Ali, F., An, H., Awe, A., Baldwin, D., Baden, H., Barnstead, M., Barrow, I., Beeson, K., Busam, D., Carver, A., Center, A., Cheng, M.L., Curry, L., Danaher, S., Davenport, L., Desilets, R., Dietz, S., Dodson, K., Doup, L., Ferriera, S., Garg, N., Glueksmann, A., Hart, B., Haynes, J., Haynes, C., Heiner, C., Hladun, S., Hostin, D., Houck, J., Howland, T., Ibegwam, C., Johnson, J., Kalush, F., Kline, L., Koduru, S., Love, A., Mann, F., May, D., McCawley, S., McIntosh, T., McMullen, I., Moy, M., Moy, L., Murphy, B., Nelson, K., Pfannkoch, C., Pratts, E., Puri, V., Qureshi, H.,

Reardon, M., Rodriguez, R., Rogers, Y.H., Romblad, D., Ruhfel, B., Scott, R., Sitter, C., Smallwood, M., Stewart, E., Strong, R., Suh, E., Thomas, R., Tint, N.N., Tse, S., Vech, C., Wang, G., Wetter, J., Williams, S., Williams, M., Windsor, S., Winn-Deen, E., Wolfe, K., Zaveri, J., Zaveri, K., Abril, J.F., Guigo, R., Campbell, M.J., Sjolander, K.V., Karlak, B., Kejariwal, A., Mi, H., Lazareva, B., Hatton, T., Narechania, A., Diemer, K., Muruganujan, A., Guo, N., Sato, S., Bafna, V., Istrail, S., Lippert, R., Schwartz, R., Walenz, B., Yooseph, S., Allen, D., Basu, A., Baxendale, J., Blick, L., Caminha, M., Carnes-Stine, J., Caulk, P., Chiang, Y.H., Coyne, M., Dahlke, C., Mays, A., Dombroski, M., Donnelly, M., Ely, D., Esparham, S., Fosler, C., Gire, H., Glanowski, S., Glasser, K., Glodek, A., Gorokhov, M., Graham, K., Gropman, B., Harris, M., Heil, J., Henderson, S., Hoover, J., Jennings, D., Jordan, C., Jordan, J., Kasha, J., Kagan, L., Kraft, C., Levitsky, A., Lewis, M., Liu, X., Lopez, J., Ma, D., Majoros, W., McDaniel, J., Murphy, S., Newman, M., Nguyen, T., Nguyen, N., Nodell, M., Pan, S., Peck, J., Peterson, M., Rowe, W., Sanders, R., Scott, J., Simpson, M., Smith, T., Sprague, A., Stockwell, T., Turner, R., Venter, E., Wang, M., Wen, M., Wu, D., Wu, M., Xia, A., Zandieh, A., and Zhu, X., (2001) The sequence of the human genome. *Science*, **291** (5507), 1304–1351.

130 van Tellingen, O. (2001) The importance of drug-transporting P-glycoproteins in toxicology. *Toxicology Letters*, **120** (1–3), 31–41.

131 Lin, J.H. and Yamazaki, M. (2003) Clinical relevance of P-glycoprotein in drug therapy. *Drug Metabolism Reviews*, **35** (4), 417–454.

132 Kessel, D. and Bosmann, H.B. (1970) Effects of L-asparaginase on protein and glycoprotein synthesis. *FEBS Letters*, **10** (2), 85–88.

133 Juliano, R.L. and Ling, V. (1976) A surface glycoprotein modulating drug permeability in Chinese hamster ovary cell mutants. *Biochimica et Biophysica Acta*, **455** (1), 152–162.

134 Roninson, I.B., Abelson, H.T., Housman, D.E., Howell, N., and Varshavsky, A. (1984) Amplification of specific DNA sequences correlates with multi-drug resistance in Chinese hamster cells. *Nature*, **309**, 626–628.

135 Shen, D.W., Fojo, A., Chin, J.E., Roninson, I.B., Richert, N., Pastan, I., and Gottesman, M.M. (1986) Human multidrug-resistant cell lines: increased mdr1 expression can precede gene amplification. *Science*, **232** (4750), 643–645.

136 Gros, P., Ben Neriah, Y.B., Croop, J.M., and Housman, D.E. (1986) Isolation and expression of a complementary DNA that confers multidrug resistance. *Nature*, **323** (6090), 728–731.

137 Gros, P., Croop, J., Roninson, I., Varshavsky, A., and Housman, D.E. (1986) Isolation and characterization of DNA sequences amplified in multidrug-resistant hamster cells. *Proceedings of the National Academy of Sciences of the United States of America*, **83** (2), 337–341.

138 Gros, P., Raymond, M., Bell, J., and Housman, D. (1988) Cloning and characterization of a second member of the mouse mdr gene family. *Molecular and Cellular Biology*, **8** (7), 2770–2778.

139 Ueda, K., Clark, D.P., Chen, C.J., Roninson, I.B., Gottesman, M.M., and Pastan, I. (1987) The human multidrug resistance (mdr1) gene. cDNA cloning and transcription initiation. *The Journal of Biological Chemistry*, **262** (2), 505–508.

140 Shen, D.W., Fojo, A., Roninson, I.B., Chin, J.E., Soffir, R., Pastan, I., and Gottesman, M.M. (1986) Multidrug resistance of DNA-mediated transformants is linked to transfer of the human mdr1 gene. *Molecular and Cellular Biology*, **6** (11), 4039–4045.

141 Cai, J. and Gros, P. (2003) Overexpression, purification, and functional characterization of ATP-binding cassette transporters in the yeast, *Pichia pastoris*.

Biochimica et Biophysica Acta, **1610** (1), 63–76.
142 Inaba, M. and Johnson, R.K. (1977) Decreased retention of actinomycin D as the basis for cross-resistance in anthracycline-resistant sublines of P388 leukemia. *Cancer Research*, **37** (12), 4629–4634.
143 Beck, W.T. (1987) The cell biology of multiple drug resistance. *Biochemical Pharmacology*, **36** (18), 2879–2887.
144 Beck, W.T. (1990) Multidrug resistance and its circumvention. *European Journal of Cancer*, **26** (4), 513–515.
145 Horio, M., Gottesman, M.M., and Pastan, I. (1988) ATP-dependent transport of vinblastine in vesicles from human multidrug-resistant cells. *Proceedings of the National Academy of Sciences of the United States of America*, **85** (10), 3580–3584.
146 Shapiro, A.B. and Ling, V. (1995) Reconstitution of drug transport by purified P-glycoprotein. *The Journal of Biological Chemistry*, **270** (27), 16167–16175.
147 Inaba, M., Kobayashi, H., Sakurai, Y., and Johnson, R.K. (1979) Active efflux of daunorubicin and adriamycin in sensitive and resistant sublines of P388 leukemia. *Cancer Res*, **39**, 2200–2203.
148 Schinkel, A.H. and Borst, P. (1991) Multidrug resistance mediated by P-glycoproteins. *Semin Cancer Biol*, **2**, 213–226.
149 van Zuylen, L., Verweij, J., Nooter, K., Brouwer, E., Stoter, G., and Sparreboom, A. (2000) Role of intestinal P-glycoprotein in the plasma and fecal disposition of docetaxel in humans. *Clinical Cancer Research*, **6** (7), 2598–2603.
150 Efferth, T. and Osieka, R. (1993) Clinical relevance of the MDR1 gene and its gene product, P-glycoprotein, for cancer chemotherapy: a meta-analysis. *Tumor Diagnostik und Therapie*, **14**, 238–243.
151 Jang, S.H., Wientjes, M.G., and Au, J.L. (2003) Interdependent effect of P-glycoprotein-mediated drug efflux and intracellular drug binding on intracellular paclitaxel pharmacokinetics: application of computational modelling. *The Journal of Pharmacology and Experimental Therapeutics*, **304** (2), 773–780.
152 Hendrikse, N.H., de Vries, E.G., Eriks-Fluks, L., van der Graaf, W.T., Hospers, G.A., Willemsen, A.T., Vaalburg, W., and Franssen, E.J. (1999) A new *in vivo* method to study P-glycoprotein transport in tumors and the blood–brain barrier. *Cancer Research*, **59** (10), 2411–2416.
153 Martin, C., Walker, J., Rothnie, A., and Callaghan, R. (2003) The expression of P-glycoprotein does influence the distribution of novel fluorescent compounds in solid tumour models. *British Journal of Cancer*, **89** (8), 1581–1589.
154 Cole, S.P., Bhardwaj, G., Gerlach, J.H., Mackie, J.E., Grant, C.E., Almquist, K.C., Stewart, A.J., Kurz, E.U., Duncan, A.M., and Deeley, R.G. (1992) Overexpression of a transporter gene in a multidrug-resistant human lung cancer line. *Science*, **258**, 1650–1654.
155 Jedlitschky, G., Leier, I., Buchholz, U., Center, M., and Keppler, D. (1996) Transport of glutathione glucuronate and sulfate conjugates by the MRP gene-encoded conjugate export pump. *Cancer Research*, **56**, 988–994.
156 Jedlitschky, G., Leier, I., Buchholz, U., Center, M., and Keppler, D. (1994) ATP-dependent transport of glutathione S-conjugates by multidrug resistance-associated protein. *Cancer Research*, **54**, 4833–4836.
157 Westlake, C.J., Payen, L., Gao, M., Cole, S.P., and Deeley, R.G. (2004) Identification and characterization of functionally important elements in the multidrug resistance protein 1 COOH-terminal region. *The Journal of Biological Chemistry*, **279** (51), 53571–53583.
158 Bates, S.E. (2002) Solving the problem of multidrug resistance: ABC transporters in clinical oncology, in *ABC Proteins: From Bacteria to Man* (eds K. Kuchler, S.P. Cole,

and B. Holland), Academic Press, New York, pp. 65–80.
159 Wijnholds, J., de Lange, E.C.M., Scheffer, G.L., van der Berg, D.J., Mol, C.A.A.M., van der Valk, M., Schinkel, A.H., Scheper, R.J., Breimer, D.D., and Borst, P. (2000) Multidrug resistance protein 1 protects the choroid plexus epithelium and contributes to the blood–cerebrospinal fluid barrier. *The Journal of Clinical Investigation*, **105**, 279–285.
160 Schinkel, A.H. (1997) The physiological function of drug-transporting P-glycoproteins. *Seminars in Cancer Biology*, **8** (3), 161–170.
161 Thiebaut, F., Tsuruo, T., Hamada, H., Gottesman, M.M., Pastan, I., and Willingham, M.C. (1987) Cellular localization of the multidrug-resistance gene product P-glycoprotein in normal human tissues. *Proceedings of the National Academy of Sciences of the United States of America*, **84** (21), 7735–7738.
162 Cordon-Cardo, C., O'Brien, J.P., Boccia, J., Casals, D., Bertino, J.R., and Melamed, M.R. (1990) Expression of the multidrug resistance gene product (P-glycoprotein) in human normal and tumor tissues. *The Journal of Histochemistry and Cytochemistry*, **38** (9), 1277–1287.
163 Regina, A., Koman, A., Piciotti, M., El Hafny, B., Center, M.S., Bergmann, R., Couraud, P.O., and Roux, F. (1998) Mrp1 multidrug resistance-associated protein and P-glycoprotein expression in rat brain microvessel endothelial cells. *Journal of Neurochemistry*, **71** (2), 705–715.
164 Sugawara, I., Koji, T., Ueda, K., Pastan, I., Gottesman, M.M., Nakane, P.K., and Mori, S. (1990) In situ localization of the human multidrug-resistance gene mRNA using thymine-thymine dimerized single-stranded cDNA. *Jpn J Cancer Res*, **81**, 949–955.
165 Sugawara, I., Kataoka, I., Morishita, Y., Hamada, H., Tsuruo, T., Itoyama, S., and Mori, S. (1988) Tissue distribution of P-glycoprotein encoded by a multidrug-resistant gene as revealed by a monoclonal antibody, MRK 16. *Cancer Research*, **48** (7), 1926–1929.
166 Croop, J.M., Raymond, M., Haber, D., Devault, A., Arceci, R.J., Gros, P., and Housman, D.E. (1989) The three mouse multidrug resistance (mdr) genes are expressed in a tissue-specific manner in normal mouse tissues. *Molecular and Cellular Biology*, **9** (3), 1346–1350.
167 Borst, P. and Elferink, R.O. (2002) Mammalian ABC transporters in health and disease. *Annual Review of Biochemistry*, **71**, 537–592.
168 Chaudhary, P.M. and Roninson, I.B. (1991) Expression and activity of P-glycoprotein, a multidrug efflux pump, in human hematopoietic stem cells. *Cell*, **66** (1), 85–94.
169 Wang, G., Pincheira, R., Zhang, M., and Zhang, J.T. (1997) Conformational changes of P-glycoprotein by nucleotide binding. *The Biochemical Journal*, **328** (Pt 3), 897–904.
170 Julien, M. and Gros, P. (2000) Nucleotide-induced conformational changes in P-glycoprotein and in nucleotide binding site mutants monitored by trypsin sensitivity. *Biochemistry*, **39** (15), 4559–4568.
171 Sonveaux, N., Vigano, C., Shapiro, A.B., Ling, V., and Ruysschaert, J.M. (1999) Ligand-mediated tertiary structure changes of reconstituted P-glycoprotein. A tryptophan fluorescence quenching analysis. *The Journal of Biological Chemistry*, **274** (25), 17649–17654.
172 Liu, R., Siemiarczuk, A., and Sharom, F.J. (2000) Intrinsic fluorescence of the P-glycoprotein multidrug transporter: sensitivity of tryptophan residues to binding of drugs and nucleotides. *Biochemistry*, **39** (48), 14927–14938.
173 Sonveaux, N., Shapiro, A.B., Goormaghtigh, E., Ling, V., and Ruysschaert, J.M. (1996) Secondary and tertiary structure changes of reconstituted P-glycoprotein. A Fourier transform

attenuated total reflection infrared spectroscopy analysis. *The Journal of Biological Chemistry*, **271** (40), 24617–24624.

174 Druley, T.E., Stein, W.D., and Roninson, I.B. (2001) Analysis of MDR1 P-glycoprotein conformational changes in permeabilized cells using differential immunoreactivity. *Biochemistry*, **40** (14), 4312–4322.

175 Druley, T.E., Stein, W.D., Ruth, A., and Roninson, I.B. (2001) P-glycoprotein-mediated colchicine resistance in different cell lines correlates with the effects of colchicine on P-glycoprotein conformation. *Biochemistry*, **40** (14), 4323–4331.

176 Ruth, A., Stein, W.D., Rose, E., and Roninson, I.B. (2001) Coordinate changes in drug resistance and drug-induced conformational transitions in altered-function mutants of the multidrug transporter P-glycoprotein. *Biochemistry*, **40** (14), 4332–4339.

177 Dawson, R.J. and Locher, K.P. (2007) Structure of the multidrug ABC transporter Sav1866 from *Staphylococcus aureus* in complex with AMP-PNP. *FEBS Letters*, **581**, 935–938.

178 Higgins, C.F. and Gottesman, M.M. (1992) Is the multidrug transporter a flippase? *Trends in Biochemical Sciences*, **17** (1), 18–21.

179 Homolya, L., Hollo, Z., Germann, U.A., Pastan, I., Gottesman, M.M., and Sarkadi, B. (1993) Fluorescent cellular indicators are extruded by the multidrug resistance protein. *The Journal of Biological Chemistry*, **268** (29), 21493–21496.

180 Raviv, Y., Pollard, H.B., Bruggemann, E.P., Pastan, I., and Gottesman, M.M. (1990) Photosensitized labeling of a functional multidrug transporter in living drug-resistant tumor cells. *The Journal of Biological Chemistry*, **265** (7), 3975–3980.

181 Eytan, G.D., Regev, R., Oren, G., and Assaraf, Y.G. (1996) The role of passive transbilayer drug movement in multidrug resistance and its modulation. *The Journal of Biological Chemistry*, **271** (22), 12897–12902.

182 Eytan, G.D., Regev, R., Oren, G., Hurwitz, C.D., and Assaraf, Y.G. (1997) Efficiency of P-glycoprotein-mediated exclusion of rhodamine dyes from multidrug-resistant cells is determined by their passive transmembrane movement rate. *European Journal of Biochemistry*, **248** (1), 104–112.

183 Choi, K., Chen, C.-J., Kriegler, M., and Roninson, I.B. (1988) An altered pattern of cross-resistance in multidrug resistant human cells results from spontaneous mutations in the mdr1 (P-glycoprotein) gene. *Cell*, **53**, 519–529.

184 Dhir, R., Grizzuti, K., Kajiji, S., and Gros, P. (1993) Modulatory effects on substrate specificity of independent mutations at the serine 939/941 position in predicted transmembrane domain 11 of P-glycoprotein. *Biochemistry*, **32**, 9492–9499.

185 Zhang, X., Collins, K.I., and Greenberger, L.M. (1995) Functional evidence that transmembrane 12 and the loop between transmembrane 11 and 12 form part of the drug-binding domain in P-glycoprotein encoded by MDR1. *The Journal of Biological Chemistry*, **270** (10), 5441–5448.

186 Bruggemann, E.P., Currier, S.J., Gottesman, M.M., and Pastan, I. (1992) Characterization of the azidopine and vinblastine binding site of P-glycoprotein. *The Journal of Biological Chemistry*, **267** (29), 21020–21026.

187 Demeule, M., Jodoin, J., Gingras, D., and Beliveau, R. (2000) P-glycoprotein is localized in caveolae in resistant cells and in brain capillaries. *FEBS Letters*, **466** (2–3), 219–224.

188 Demmer, A., Thole, H., Kubesch, P., Brandt, T., Raida, M., Fislage, R., and Tummler, B. (1997) Localization of the iodomycin binding site in hamster P-glycoprotein. *The Journal of Biological Chemistry*, **272** (33), 20913–20919.

189 Greenberger, L.M. (1993) Major photoaffinity labeling sites for iodoaryl azidoprazosin in P-glycoprotein are within or immediately C-terminal to transmembrane domains 6 and 12. *The Journal of Biological Chemistry*, **268**, 11417–11425.

190 Glossmann, H., Ferry, D.R., Striessnig, J., Goll, A., and Moosburger, K. (1987) Resolving the structure of the Ca^{2+} channel by photoaffinity labelling. *Trends in Pharmacological Sciences*, **8**, 95–100.

191 Tamai, I., and Safa, A.R. (1991) Azidopine noncompetitively interacts with vinblastine and cyclosporin A binding to P-glycoprotein in multidrug resistant cells. *J Biol Chem*, **266**, 16796–16800.

192 Martin, C., Berridge, G., Higgins, C.F., and Callaghan, R. (1997) The multi-drug resistance reversal agent SR33557 and modulation of vinca alkaloid binding to P-glycoprotein by an allosteric interaction. *British Journal of Pharmacology*, **122**, 765–771.

193 Ferry, D.R., Russell, M.A., and Cullen, M.H. (1992) P-glycoprotein possesses a 1,4-dihydropyridine-selective drug acceptor site which is alloserically coupled to a vinca-alkaloid-selective binding site. *Biochem Biophys Res Commun*, **188**, 440–445.

194 Malkhandi, J., Ferry, D.R., Boer, R., Gekeler, V., Ise, W., and Kerr, D.J. (1994) Dexniguldipine-HCl is a potent allosteric inhibitor of [3H]vinblastine binding to P-glycoprotein of CCRF ADR 5000 cells. *Eur J Pharmacol*, **288**, 105–114.

195 Loo, T.W., Bartlett, M.C., and Clarke, D.M. (2006) Transmembrane segment 1 of human P-glycoprotein contributes to the drug-binding pocket. *The Biochemical Journal*, **396**, 537–545.

196 Loo, T.W. and Clarke, D.M. (2000) Identification of residues within the drug-binding domain of the human multidrug resistance P-glycoprotein by cysteine-scanning mutagenesis and reaction with dibromobimane. *The Journal of Biological Chemistry*, **275**, 39272–39278.

197 Pleban, K., Kopp, S., Csaszar, E., Peer, M., Hrebicek, T., Rizzi, A., Ecker, G.F., and Chiba, P. (2005) P-glycoprotein substrate binding domains are located at the transmembrane domain/transmembrane domain interfaces: a combined photoaffinity labeling-protein homology modeling approach. *Molecular Pharmacology*, **67**, 365–374.

198 Loo, T.W., Bartlett, M.C., and Clarke, D.M. (2003) Permanent activation of the human P-glycoprotein by covalent modification of a residue in the drug-binding site. *The Journal of Biological Chemistry*, **278**, 20449–20452.

199 Loo, T.W., Bartlett, M.C., and Clarke, D.M. (2004) The drug-binding pocket of the human multidrug resistance P-glycoprotein is accessible to the aqueous medium. *Biochemistry*, **43**, 12081–12089.

200 Loo, T.W. and Clarke, D.M. (1997) Identification of residues in the drug-binding site of human P-glycoprotein using a thiol-reactive substrate. *The Journal of Biological Chemistry*, **272**, 31945–31948.

201 Rothnie, A., Storm, J., Campbell, J., Linton, K.J., Kerr, I.D., and Callaghan, R. (2004) The topography of transmembrane segment six is altered during the catalytic cycle of P-glycoprotein. *The Journal of Biological Chemistry*, **279**, 34913–34921.

202 Loo, T.W., Bartlett, M.C., and Clarke, D.M. (2006) Transmembrane segment 7 of human P-glycoprotein forms part of the drug-binding pocket. *The Biochemical Journal*, **399**, 351–359.

203 Loo, T.W. and Clarke, D.M. (1999) Identification of residues in the drug-binding domain of human P-glycoprotein. Analysis of transmembrane segment 11 by cysteine-scanning mutagenesis and inhibition by dibromobimane. *The Journal of Biological Chemistry*, **274**, 35388–35392.

204 Safa, A.R. (1993) Photoaffinity labeling of P-glycoprotein in multidrug-resistant cells. *Cancer Investigation*, **11**, 46–56.

205 Taguchi, Y., Kino, K., Morishima, M., Komano, T., Kane, S.E., and Ueda, K.

(1997) Alteration of substrate specificity by mutations at the His61 position in predicted transmembrane domain 1 of human MDR1/P-glycoprotein. *Biochemistry*, **36**, 8883–8889.

206 Taguchi, Y., Morishima, M., Komano, T., and Ueda, K. (1997) Amino acid substitutions in the first transmembrane domain (TM1) of P-glycoprotein that alter substrate specificity. *FEBS Letters*, **413**, 142–146.

207 Ma, J.F., Grant, G., and Melera, P.W. (1997) Mutations in the sixth transmembrane domain of P-glycoprotein that alter the pattern of cross-resistance also alter sensitivity to cyclosporin A reversal. *Molecular Pharmacology*, **51**, 922–930.

208 Song, J. and Melera, P.W. (2001) Transmembrane domain (TM) 9 represents a novel site in P-glycoprotein that affects drug resistance and cooperates with TM6 to mediate [^{125}I] iodoarylazidoprazosin labelling. *Molecular Pharmacology*, **60**, 254–261.

209 Hanna, M., Brault, M., Kwan, T., Kast, C., and Gros, P. (1996) Mutagenesis of transmembrane domain 11 of P-glycoprotein by alanine scanning. *Biochemistry*, **35**, 3625–3635.

210 Hafkemeyer, P., Dey, S., Ambudkar, S.V., Hrycyna, C.A., Pastan, I., and Gottesman, M.M. (1998) Contribution to substrate specificity and transport of nonconserved residues in transmembrane domain 12 of human P-glycoprotein. *Biochemistry*, **37**, 16400–16409.

211 Demeule, M., Laplante, A., Murphy, G.F., Wenger, R.M., and Beliveau, R. (1998) Identification of the cyclosporin binding site in P-glycoprotein. *Biochemistry*, **37**, 18110–18118.

212 Loo, T.W. and Clarke, D.M. (1995) Identification of residues in the drug-binding site of human P-glycoprotein. *Journal of Biological Chemistry*, **272**, 31945–31948.

213 Taylor, A.M., Storm, J., Soceneantu, L., Linton, K.J., Gabriel, M., Martin, C., Woodhouse, J., Blott, E.J., Higgins, C.F., and Callaghan, R. (2001) Detailed characterization of cysteine-less P-glycoprotein reveals subtle pharmacological differences in function from wild-type protein. *British Journal of Pharmacology*, **134**, 1609–1618.

214 Rothnie, A., Storm, J., McMahon, R., Taylor, A., Kerr, I.D., and Callaghan, R. (2005) The coupling mechanism of P-glycoprotein involves residue L339 in the sixth membrane spanning segment. *FEBS Letters*, **579**, 3984–3990.

215 Brockman, R.W. (1963) Mechanisms of resistance to anticancer agents. *Advances in Cancer Research*, **57**, 129–234.

216 Hutchison, D.J. (1963) Cross resistance and collateral sensitivity studies in cancer chemotherapy. *Advances in Cancer Research*, **57**, 235–250.

217 Didziapetris, R., Japertas, P., Avdeef, A., and Petrauskas, A. (2003) Classification analysis of P-glycoprotein substrate specificity. *Journal of Drug Targeting*, **11**, 391.

218 Agrawal, M., Abraham, J., Balis, F.M., Edgerly, M., Stein, W.D., Bates, S., Fojo, T., and Chen, C.C. (2003) Increased 99mTc-sestamibi accumulation in normal liver and drug-resistant tumors after the administration of the glycoprotein inhibitor, XR9576. *Clinical Cancer Research*, **9** (2), 650–656.

219 Ganapathi, R. and Grabowski, D. (1983) Enhancement of sensitivity to adriamycin in resistant P388 leukemia by the calmodulin inhibitor trifluoperazine. *Cancer Research*, **43** (8), 3696–3699.

220 Goldberg, H., Ling, V., Wong, P.Y., and Skorecki, K. (1988) Reduced cyclosporin accumulation in multidrug-resistant cells. *Biochemical and Biophysical Research Communications*, **152** (2), 552–558.

221 Ramu, A., Fuks, Z., Gatt, S., and Glaubiger, D. (1984) Reversal of acquired resistance to doxorubicin in P388 murine leukemia cells by perhexiline maleate. *Cancer Research*, **44** (1), 144–148.

222 Tsuruo, T., Iida, H., Kitatani, Y., Yokota, K., Tsukagoshi, S., and Sakurai, Y. (1984) Effects of quinidine and related compounds on cytotoxicity and cellular accumulation of vincristine and adriamycin in drug-resistant tumor cells. *Cancer Research*, **44** (10), 4303–4307.

223 Tsuruo, T., Iida, H., Tsukagoshi, S., and Sakurai, J. (1981) Overcoming of vincristine resistance in P388 leukemia *in vivo* and *in vitro* through enhanced cytotoxicity of vincristine and vinblastine by verapamil. *Cancer Research*, **41**, 1967–1972.

224 Benson, A.B., 3rd, Trump, D.L., Koeller, J.M., Egorin, M.I., Olman, E.A., Witte, R.S., Davis, T.E., and Tormey, D.C. (1985) Phase I study of vinblastine and verapamil given by concurrent iv infusion. *Cancer Treatment Reports*, **69** (7–8), 795–799.

225 Cairo, M.S., Siegel, S., Anas, N., and Sender, L. (1989) Clinical trial of continuous infusion verapamil, bolus vinblastine, and continuous infusion VP-16 in drug-resistant pediatric tumors. *Cancer Research*, **49** (4), 1063–1066.

226 Bartlett, N.L., Lum, B.L., Fisher, G.A., Brophy, N.A., Ehsan, M.N., Halsey, J., and Sikic, B.I. (1994) Phase I trial of doxorubicin with cyclosporine as a modulator of multidrug resistance. *Journal of Clinical Oncology*, **12** (4), 835–842.

227 Verweij, J., Herweijer, H., Oosterom, R., van der Burg, M.E., Planting, A.S., Seynaeve, C., Stoter, G., and Nooter, K. (1991) A phase II study of epidoxorubicin in colorectal cancer and the use of cyclosporin-A in an attempt to reverse multidrug resistance. *British Journal of Cancer*, **64** (2), 361–364.

228 Plumb, J.A., Milroy, R., and Kaye, S.B. (1990) The activity of verapamil as a resistance modifier *in vitro* in drug resistant human tumour cell lines is not stereospecific. *Biochemical Pharmacology*, **39** (4), 787–792.

229 Bissett, D., Kerr, D.J., Cassidy, J., Meredith, P., Traugott, U., and Kaye, S.B. (1991) Phase I and pharmacokinetic study of D-verapamil and doxorubicin. *British Journal of Cancer*, **64** (6), 1168–1171.

230 Kornek, G., Raderer, M., Schenk, T., Pidlich, J., Schulz, F., Globits, S., Tetzner, C., and Scheithauer, W. (1995) Phase I/II trial of dexverapamil, epirubicin, and granulocyte-macrophage-colony stimulating factor in patients with advanced pancreatic adenocarcinoma. *Cancer*, **76** (8), 1356–1362.

231 Twentyman, P.R. and Bleehen, N.M. (1991) Resistance modification by PSC-833, a novel non-immunosuppressive cyclosporin [corrected]. *European Journal of Cancer*, **27** (12), 1639–1642.

232 Giaccone, G., Linn, S.C., Welink, J., Catimel, G., Stieltjes, H., van der Vijgh, W.J., Eeltink, C., Vermorken, J.B., and Pinedo, H.M. (1997) A dose-finding and pharmacokinetic study of reversal of multidrug resistance with SDZ PSC 833 in combination with doxorubicin in patients with solid tumors. *Clinical Cancer Research*, **3** (11), 2005–2015.

233 Boote, D.J., Dennis, I.F., Twentyman, P.R., Osborne, R.J., Laburte, C., Hensel, S., Smyth, J.F., Brampton, M.H., and Bleehen, N.M. (1996) Phase I study of etoposide with SDZ PSC 833 as a modulator of multidrug resistance in patients with cancer. *Journal of Clinical Oncology*, **14** (2), 610–618.

234 Friedenberg, W.R., Rue, M., Blood, E.A., Dalton, W.S., Shustik, C., Larson, R.A., Sonneveld, P., and Greipp, P.R. (2006) Phase III study of PSC-833 (valspodar) in combination with vincristine, doxorubicin, and dexamethasone (valspodar/VAD) versus VAD alone in patients with recurring or refractory multiple myeloma (E1A95): a trial of the Eastern Cooperative Oncology Group. *Cancer*, **106** (4), 830–838.

235 Hyafil, F., Vergely, C., Du Vignaud, P., and Grand-Perret, T. (1993) *In vitro* and *in vivo* reversal of multidrug resistance by GF120918, an acridonecarboxamide

derivative. *Cancer Research*, **53** (19), 4595–4602.

236 Roe, M., Folkes, A., Ashworth, P., Brumwell, J., Chima, L., Hunjan, S., Pretswell, I., Dangerfield, W., Ryder, H., and Charlton, P. (1999) Reversal of P-glycoprotein mediated multidrug resistance by novel anthranilamide derivatives. *Bioorganic & Medicinal Chemistry Letters*, **9** (4), 595–600.

237 Pusztai, L., Wagner, P., Ibrahim, N., Rivera, E., Theriault, R., Booser, D., Symmans, F.W., Wong, F., Blumenschein, G., Fleming, D.R., Rouzier, R., Boniface, G., and Hortobagyi, G.N. (2005) Phase II study of tariquidar, a selective P-glycoprotein inhibitor, in patients with chemotherapy-resistant, advanced breast carcinoma. *Cancer*, **104** (4), 682–691.

238 Pearce, H.L., Safa, A.R., Bach, N.J., Winter, M.A., Cirtain, M.C., and Beck, W.T. (1989) Essential features of the P-glycoprotein pharmacophore as defined by a series of reserpine analogs that modulate multidrug resistance. *Proceedings of the National Academy of Sciences of the United States of America*, **86** (13), 5128–5132.

239 Zamora, J.M., Pearce, H.L., and Beck, W.T. (1988) Physical–chemical properties shared by compounds that modulate multidrug resistance in human leukemic cells. *Molecular Pharmacology*, **33** (4), 454–462.

240 Scala, S., Akhmed, N., Rao, U.S., Paull, K., Lan, L.B., Dickstein, B., Lee, J.S., Elgemeie, G.H., Stein, W.D., and Bates, S.E. (1997) P-glycoprotein substrates and antagonists cluster into two distinct groups. *Molecular Pharmacology*, **51** (6), 1024–1033.

241 Seelig, A. (1998) A general pattern for substrate recognition by P-glycoprotein. *European Journal of Biochemistry*, **251** (1–2), 252–261.

242 Seelig, A. and Landwojtowicz, E. (2000) Structure–activity relationship of P-glycoprotein substrates and modifiers. *European Journal of Pharmaceutical Sciences*, **12**, 31–40.

243 Raad, I., Terreux, R., Richomme, P., Matera, E.L., Dumontet, C., Raynaud, J., and Guilet, D. (2006) Structure–activity relationship of natural and synthetic coumarins inhibiting the multidrug transporter P-glycoprotein. *Bioorganic and Medicinal Chemistry*, **14** (20), 6979–6987.

244 Kaiser, D., Smiesko, M., Kopp, S., Chiba, P., and Ecker, G.F. (2005) Interaction field based and hologram based QSAR analysis of propafenone-type modulators of multidrug resistance. *Medicinal Chemistry*, **1** (5), 431–444.

245 Labrie, P., Maddaford, S.P., Fortin, S., Rakhit, S., Kotra, L.P., and Gaudreault, R.C. (2006) A comparative molecular field analysis (CoMFA) and comparative molecular similarity indices analysis (CoMSIA) of anthranilamide derivatives that are multidrug resistance modulators. *Journal of Medicinal Chemistry*, **49** (26), 7646–7660.

246 Cianchetta, G., Singleton, R.W., Zhang, M., Wildgoose, M., Giesing, D., Fravolini, A., Cruciani, G., and Vaz, R.J. (2005) A pharmacophore hypothesis for P-glycoprotein substrate recognition using GRIND-based 3D-QSAR. *Journal of Medicinal Chemistry*, **48** (8), 2927–2935.

247 Aller, S.G., Yu, J., Ward, A., Weng, Y., Chittaboina, S., Zhuo, R., Harrell, P.M., Trinh, Y.T., Zhang, Q., Urbatsch, I.L., and Chang, G. (2009) Structure of P-glycoprotein reveals a molecular basis for poly-specific drug binding. *Science*, **323**, 1718–1722.

248 Kadaba, N.S., Kaiser, J.T., Johnson, E., Lee, A., and Rees, D.C. (2008) The high-affinity E. coli methionine ABC transporter: structure and allosteric regulation. *Science*, **321**, 250–253.

249 Khare, D., Oldham, M.L., Orelle, C., Davidson, A.L., and Chen, J. (2009) Alternating access in maltose transporter mediated by rigid-body rotations. *Mol Cell*, **33**, 528–536.

2
Biochemistry, Physiology, and Pharmacology of Nucleoside and Nucleobase Transporters

Marçal Pastor-Anglada, Míriam Molina-Arcas, Pedro Cano-Soldado, and Francisco Javier Casado

2.1
Nucleoside and Nucleobase Transporters

Nucleosides are relatively hydrophilic molecules, and this feature makes the presence of specific transport proteins at the plasma membrane mandatory to grant their effective translocation. Nucleoside transporters (NTs) determine the regulation of many physiological processes by modulating nucleoside levels both inside and outside the cell. The relevance of nucleoside transport is underlined by the fact that nucleoside transporters are present in all taxonomic groups investigated so far, from bacteria to mammals. In addition to their natural role, they are also relevant in pharmacology, since they represent the entrance pathways of many cytotoxic nucleoside analogues currently used in antitumoral and antiviral chemotherapy.

Earlier, nucleoside transporters were classified on the basis of their thermodynamic and kinetic characteristics. Accordingly, seven kinetic activities have been traditionally described. Two of them correspond to facilitated diffusion, sodium-independent, equilibrative transporters, termed *es* and *ei* for their sensitivity and insensitivity, respectively, to inhibition by nanomolar concentrations of the nucleoside analogue NBTI (nitrobenzylthioinosine). The other five transport agencies kinetically described are active, concentrative transporters coupled to the transmembrane sodium gradient, termed N1 or *cif* (for *c*oncentrative, *i*nsensitive to NBTI, transporting the purine analogue *f*ormycin-B), N2 or *cit* (for *c*oncentrative, *i*nsensitive to NBTI, transporting the pyrimidine *t*hymidine), N3 or *cib* (for *c*oncentrative, *i*nsensitive to NBTI, with *b*road substrate selectivity), N4, and N5. In the mid-1990s, these transporters began to be cloned. Thereafter, a new, far more rational classification has been made (Table 2.1). There are two gene families of nucleoside transporters in humans, SLC28 and SLC29, encompassing the concentrative and equilibrative transporters, respectively. The SLC28 family consists of three members: SLC28A1 or CNT1 (corresponding to activity N2 or *cit*), SLC28A2 or CNT2 (corresponding to activity N1 or *cif*), and SLC28A3 or CNT3 (corresponding to activity N3 or *cib*). Similarly, the SLC29 family of nucleoside transporters consists

Table 2.1 Characteristics of cloned human nucleoside transporters.

			Equilibrative Nucleoside Transporters			
Activity	Protein	Gene	Human gene locus	Protein length	Homology respect ENT1	Substrates
es	hENT1	SLC29A1	6p21.1–p21.2	465 aa		Purine and pyrimidine nucleosides
ei	hENT2	SLC29A2	11q13	465 aa	46%	Purine and pyrimidine nucleosides and nucleobases
	hENT3	SLC29A3	10q22.1	475 aa	29%	Purine and pyrimidine nucleosides and some nucleobases
	hENT4	SLC29A4	7p 22.1	530 aa	18%	Adenosine

			Concentrative Nucleoside Transporters			
Activity	Protein	Gene	Human gene locus	Protein length	Homology respect CNT1	Substrates
N2 (cit)	hCNT1	SLC28A1	15q25–q26	650 aa		Pyrimidine nucleosides
N1 (cif)	hCNT2	SLC28A2	15q15	658 aa	72%	Purine nucleosides and uridine
N3 (cib)	hCNT3	SLC28A2	9q22.2	691 aa	48%	Purine and pyrimidine nucleosides

of four different members: SLC29A1 or ENT1 (corresponding to activity *es*), SLC29A2 or ENT2 (corresponding to activity *ei*), SLC29A3 or ENT3 (corresponding to a lysosomal nucleoside transporter not described previously as a kinetic entity), and SLC29A4 or ENT4/PMAT (*p*lasma *m*embrane *m*onoamine *t*ransporter) (corresponding to a monoamine transporter, again not described previously as a kinetic agency). The kinetic activities termed N4 and N5 have not been attributed to any known gene at present, although they might be the result of putative polymorphic variants of known SLC members [1].

2.1.1
Equilibrative Nucleoside Transporters

Since they catalyze a passive diffusion process, equilibrative transporters ENT1 and ENT2 are bidirectional carriers, mediating both influx and efflux of substrates. They are present in almost every cell type and show a relatively high affinity for most nucleosides, with values in the high micromolar range (100–800 µM), and a wide selectivity of substrates, accepting most purines and pyrimidines and, at least for ENT2, even some nucleobases such as hypoxanthine [2–4]. Both systems are the natural targets of vasodilation potentiators such as dipyridamole and dilazep, which act by inhibiting adenosine entry into endothelial cells and thus potentiating its vasodilation effects.

The human ENT1 was the first cloned member of the SLC29 family [5] and orthologues have been identified in mammals, yeast, nematodes, plants, and protozoa. Human equilibrative nucleoside transporter 1 (hENT1) is a glycosylated protein that contains 456 residues (50 kDa), its gene is located in chromosome 6p21.1–21.2 [6], and is expressed in a wide variety of tissues. Shortly after hENT1 cloning, hENT2 was also cloned [2, 7]. Its gene encodes a 456-amino acid protein with a 46% sequence identity with hENT1. Two new members of the SLC29 gene family have also been identified. hENT3 is a 475-amino acid protein only 29% identical to hENT1 with a very long hydrophilic N-terminus (51 residues) containing two (DE)XXXL(LI) dileucine motifs, endosomal and lysosomal targeting motifs [8]. In fact, the protein colocalizes with lysosomal markers and truncation or mutation of the dileucine motif relocates the protein to the plasma membrane [9]. In accordance with its location and function, hENT3 highly depends on pH, with an optimum at pH 5.5, while it is relatively insensitive to the classical equilibrative nucleoside transporter inhibitors NBTI, dilazep, or dipyridamole. The fourth member of the SLC29 family ENT4/PMAT codes for a protein of 530 residues, functionally quite distant from the rest of the members of the family [10, 11]. It is a monoamine transporter, hence its alternative denomination PMAT, with some capability of transporting adenosine [10–12].

ENT transporters share a topological structure composed of 11 transmembrane domains, with an intracellular amino terminus and an extracellular carboxy terminus [13] (Figure 2.1). hENT1 shows one N-glycosylation site, located in the extracellular loop between transmembrane domains 1 and 2, while hENT2 has two N-glycosylation sites in the same region. However, the glycosylation status does not seem to be essential for transporter function [14, 15]. The region between transmembrane domains 3 and 6 is responsible for the varying sensitivity to inhibitors NBTI, dipyridamole, and dilazep [16]. In this sense, substitution of Gly154 of hENT1 by the equivalent residue of hENT2 does not affect the transport capacity but abolishes NBTI sensitivity in the engineered hENT1 [8], although the reverse procedure does not render hENT2 sensitive to NBTI. Analogously, Ser160 of hENT1 seems to be involved in the sensitivity to dipyridamole [17]. Regarding substrate selectivity, transmembrane domains 1 and 6 in hENT2 seem to be responsible for the recognition of deoxynucleosides, while transmembrane domains 5 and 6 are involved in nucleobase recognition [4, 18]. In hENT1, Met89 seems to be involved in the recognition of adenosine and guanosine [17]. hENT1 is also expressed and is functional in the mitochondrial membrane, and the amino acid residues Pro71, Glu72, and Asn74 (the PEXN motif) have been described as important in mitochondrial targeting of hENT1 [19, 20].

2.1.2
Concentrative Nucleoside Transporters

The lack of high-affinity inhibitors and, for a long time, suitable antibodies has hampered the study of concentrative nucleoside transporters. They mediate the unidirectional flow of nucleosides in an active, energy-intensive process coupled to the transmembrane sodium gradient. They show a high affinity for their substrates

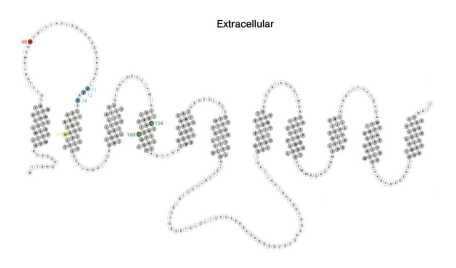

Figure 2.1 Topographical model of SLC29A1, the hENT1. *Highlighted circles* indicate amino acids that have been identified as relevant residues through chimera and mutagenesis studies. The red residue is a N-glycosylation site, the green residues are involved in NBTI and dipyridamole sensitivity, the yellow residue seems to be involved in the recognition of adenosine and guanosine, and the blue residues have been described as important mitochondrial targeting.

(most of them are in the low micromolar range, 10–100 μM), but are more selective for them than equilibrative transporters. Thus, CNT1 is the pyrimidine-preferring transporter, CNT2 is the purine-preferring transporter, and CNT3 shows a much broader selectivity.

The first concentrative transporter cloned was the rat orthologue of CNT1 [21], a 648-amino acid protein with a N2 activity. Soon, the human orthologue was cloned from kidney samples [22], and it corresponds to a 650-amino acid protein 83% identical to its rat counterpart. It has been mapped to chromosome 15q25–26. The finding that a genetic variant of human concentrative nucleoside transporter 1 (hCNT1) (Phe316His) renders this transporter sensitive to the inhibition by guanosine suggests that the kinetic agency termed N4 could be the result of the expression of this variant [1]. Also in the mid-1990s, the first gene coding for a N1 activity was cloned from rat liver [23] and the product, a protein 659-amino-acid long, was termed SPNT (from sodium-dependent, purine-preferring nucleoside transporter), later renamed as rCNT2. Again, the human orthologue was rapidly cloned [24, 25] and termed hCNT2. hCNT2 is 658 residues long and 72 and 83% identical to hCNT1 and rCNT2, respectively. The corresponding human gene has been mapped to chromosome 15 [25]. Finally, the gene coding for CNT3 was cloned from human and mouse [26]. hCNT3 is a protein of 691 amino acids, with a 79% homology to mCNT3 but only 48 and 47% identical to hCNT1 and hCNT2, respectively. The human gene has been located at chromosome 9q22.2 [27].

All members of the SLC28 gene family of concentrative nucleoside transporters share a general topology based on 13 transmembrane domains, with an extracellular carboxy terminus, at least one or two N-glycosylation site(s), and several consensus phosphorylation sites for protein kinases A and C and casein kinase II [28] (Figure 2.2). Substitution of transmembrane domains 8 and 9 in rCNT1 by the corresponding domains in rCNT2 changes a pyrimidine-preferring transporter into a purine-preferring one, while substitution of the transmembrane domain 8 alone renders a chimera with CNT3 substrate selectivity [29]; the residues responsible for these effects have been identified [30, 31]. In the human counterparts, the substitution of Ser319 in transmembrane domain 7 of hCNT1 by Gly313 of hCNT2 allows purine transport, thus converting hCNT1 into a "hCNT3-like" transporter; the additional substitution of Ser353 in transmembrane domain 8 of this chimera by the corresponding Thr247 of hCNT2 changes this "hCNT3-like" transporter into a "hCNT2-like" carrier [32].

2.2
ENT and CNT Tissue Distribution, Regulation, and Physiological Roles

Most mammalian cells express more than one type of NT, often combining members of both gene families, CNT and ENT, in a single cell type. Transcripts for CNT1,

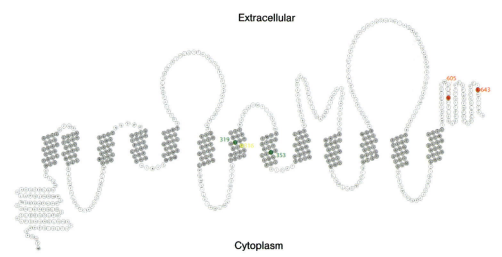

Figure 2.2 Topographical model of SLC28A1, the hCNT1. *Highlighted circles* indicate amino acids that have been identified as relevant residues through chimera and mutagenesis studies. The red residues are a N-glycosylation sites, the green residues are involved in pyrimidine and purine selectivity, and the yellow residue is a genetic variant whose effect resembles the N4-type nucleoside transporter.

CNT2, and CNT3, as well as for ENT1, ENT2, and ENT3, have been analyzed in 19 tissues of rats and mice, thus revealing broad expression for all of them, although apparently at highly variable concentrations [33]. Although this information is valuable, it does not actually reflect the complex pattern of NT expression at the protein and functional activity levels for particular cell types. More important, most NTs are localized in intracellular compartments, thus making more complicated the interpretation of tissue distribution when given merely at the mRNA level. The physiological rationale for the coexpression of several NT proteins is still a matter of controversy. From our viewpoint, tissue distribution by itself can barely explain this apparent redundancy. Actually, CNTs were initially thought to be expressed mostly in (re)absorptive epithelia, but they are now known to be broadly present in the body, including immune system cells and the central nervous system (CNS). These findings are relevant because a broader distribution anticipates a major role of these membrane proteins in the pharmacokinetics of most nucleoside-derived drugs used in anticancer and antiviral therapies, thus providing additional information about the ability of these molecules to cross selected tissue barriers, such as the placenta and the blood–brain barrier (BBB). In fact, tissue-specific regulation of NTs rather than distribution alone can probably add some valuable information about NT biology and pharmacology. Some examples of it will be reviewed subsequently (Figure 2.3). Ultimately, functional genomics will definitely provide more clues to further elucidate what particular roles a single NT protein plays in cell physiology.

2.2.1
ENT Tissue Distribution and Regulation

ENT1 and ENT2, the two equilibrative plasma membrane nucleoside transporters, show broad tissue distribution. Relatively high rat ENT1-related mRNA levels have been found in lung, heart, gonads, and blood vessels [33]. In fact, ENT1 has been reported to be a major player in the modulation of extracellular adenosine levels in vascular endothelium, a key element in the regulation of vasodilatation induced by high-glucose conditions such as diabetes [34, 35]. In human umbilical vein endothelial cells (HUVECs), hENT1 expression and activity are reduced by diabetes, an effect that seems to be associated with the reduced promoter activity of the SLC29A1 gene encoding the hENT1 protein [36]. This effect can be somehow mimicked by high glucose [35, 37]. Interestingly, hENT2 is also expressed in HUVEC and, although it is not implicated in the inhibition of adenosine transport triggered by hyperglycemia [38], it is responsible for adenosine transport recovery following insulin treatment [37], thus anticipating some sort of physiological compensation between both isoforms. Hypoxia also promotes extracellular adenosine accumulation, a phenomenon that similar to hyperglycemia associates with decreased ENT1-mediated adenosine uptake and protein expression in HUVEC and other cell types such as cardiomyocytes [39, 40]. We now know that the decrease in ENT1 expression induced by hypoxia depends on a hypoxia inducible factor-1 (HIF-1)-mediated transcriptional repression of the SLC29A1 gene promoter [41]. HIF-1α knockout (KO) mice show

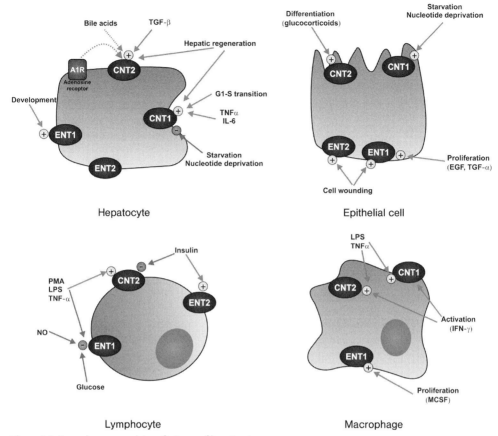

Figure 2.3 Some known regulatory features of hepatocytes, epithelial cells, B- and T-lymphocytes and macrophages.

increased epithelial ENT1 expression, thus confirming the important role this transcription factor can play *in vivo* [41]. Adenosine is also a major mediator of ethanol intoxication and ethanol itself is also known to downregulate hENT1 both acutely and chronically. In fact, ENT1 KO mice show reduced hypnotic and ataxic responses to ethanol but greater consumption than the wild-type animals [42].

In summary, ENT1 can actually be a major player in the regulation of extracellular adenosine levels in different cell types, particularly in those such as HUVEC, in which no CNT proteins appear to be coexpressed. From this viewpoint, it is a suitable pharmacological target in the treatment of cardiovascular diseases.

Nevertheless, ENT1 also seems to fulfill nucleoside salvage requirements. It is a broad selectivity, ubiquitous plasma membrane transporter that, in turn, seems to be regulated by proliferative stimuli. In murine bone marrow macrophages, ENT1 is upregulated by M-CSF, thus promoting the incorporation of extracellular nucleosides into DNA (but not into RNA), a process that, when blocked by NBTI, results in the inhibition of cell proliferation [43]. When macrophages are treated with cytokines known to promote cell growth arrest and apoptosis, such as IFN-γ and TNF-α, ENT1 activity and expression are downregulated [44, 45]. In epithelia, ENT1 is mostly located at the basolateral side, thus enabling vectorial flux of nucleosides previously concentrated by apically located CNT-type proteins [46–48]. Interestingly, ENT1 activity and expression are not regulated by agents known to promote cell differentiation in an enterocyte cell model (IEC6 cells), but it is highly sensitive to growth factors such as TGF-α and EGF, known to promote cell proliferation [49]. This effect is mimicked by cell wounding [49] and agrees with a putative role of basolateral ENT1 (facing the "blood" side of the epithelium) in facilitating nucleoside salvage for proliferation of immature enterocytes. This would also be consistent with the apparent higher expression of hENT1 in crypts than in mature human intestinal cells (unpublished observations). Moreover, as will be discussed subsequently, hENT1 is highly expressed and abundant in tumors, its occurrence being better retained than for other nucleoside transporters, such as hCNT1, whose expression is probably linked to cell differentiation and thus more easily lost during transformation [50].

In summary, hENT1 appears to some extent to be a major player in nucleoside salvage processes and probably because of this most researchers in the field anticipated hENT1 KO mice to be embryonically lethal; however, this is not the case [42]. This is the only mouse model with a knocked-down NT function available so far, although, unfortunately, its phenotype has not been comprehensively characterized. Nevertheless, the fact that it is viable highlights the possibility of complementation by other transporters or salvage processes. In this sense, hENT2 physiological roles are somehow obscure at this moment. Highly expressed in skeletal muscle, it was suggested to be a mediator of nucleobase reuptake (i.e., hypoxanthine) as a salvage process in a cell type in which the purine nucleotide cycle allows adaptation to exercise but yields a variety of nucleobases that would be otherwise irreversibly lost. ENT2 can also be upregulated in intestinal cells when incubated with proliferative agents, but the magnitude of this effect is much lower than that for hENT1 [49]. On the other hand, ENT2 is also located at the basolateral side of polarized epithelial cells [51, 52]. Considering that vectorial flux of nucleosides across epithelia is accompanied by significant intracellular metabolism, thus yielding nucleobases [46], its presence at this pole of the cell would further facilitate efficient (re)absorption of luminal nucleosides.

2.2.2
CNT Tissue Distribution and Regulation

As previously discussed, it is now well accepted that CNTs show broad tissue distribution, although it is also true that some cell types lack either any CNT or particular CNT isoform expression. Moreover, it is also common that their function-related activity gets lost easily in primary culture preparations, whereas commercially available cell lines may also lack CNT-related activities. Overall, this makes it difficult to work with, and the identification and/or generation of suitable cell models is a major bottleneck in the field.

2.2.2.1 CNTs in Absorptive Epithelia

CNT transporters, when heterologously expressed in epithelial cell models (i.e., MDCK cells), are targeted to the apical membrane [46, 48, 51] and their occurrence there confers upon the epithelium the efficient vectorial flux of nucleosides and nucleoside-derived drugs [46]. In the rat nephron, the three isoforms, CNT1, CNT2, and CNT3, show high mRNA levels in the proximal convoluted tubule (PCT) [53], thus suggesting a major role in nucleoside reabsorption. Nevertheless, CNTs (particularly CNT2 and CNT3) are also expressed at more distal segments [53], an observation that would be consistent with a putative secondary role of these transporter proteins in the uptake of secreted adenosine, known to modulate collecting duct functions. Interestingly, CNT2 and CNT3, but not CNT1, are high-affinity adenosine transporters [26, 54, 55]. In the rat intestinal epithelial cell line IEC-6, which endogenously expresses CNT1 and to a much greater extent CNT2, proliferative stimuli do not significantly modulate them although glucocorticoid treatment (i.e., dexamethasone) does increase their expressions and related functional activities [49]. This observation, along with the evidence that CNTs are barely detectable in crypts but easily found in the brush border (unpublished observations), supports the view that maturation of epithelial cells associates with CNT expression. Moreover, nutritional status also seems to modulate CNT expression (at least CNT1) because its amounts are increased in the jejunum of starved rats compared to their fed controls [56]. This effect is mimicked by feeding the animals nucleotide-free semiartificial diets, a finding that strongly suggests that it is the nucleotide content of the diet instead of the complex metabolic and endocrine changes associated with starvation that determines the most CNT expression in rat jejunum [56].

In summary, CNTs in epithelia might be major players in determining vectorial flux of nucleosides, thus contributing to absorption and reabsorption phenomena and, consequently, to whole-body nucleoside homeostasis.

2.2.2.2 CNTs in Liver Parenchymal Cells

Liver parenchymal cells are the major source of endogenous nucleotides in the body, and thus hepatocytes can probably be seen as suitable body buffers by either taking up excess nucleotides from the diet or providing them for peripheral tissues. Actually, nutritional regulation of CNT-type proteins in the liver is opposite to that found in the jejunum, CNT1 amounts being decreased in animals fed with nucleotide-free diets [56]. CNT1 was cloned from rat liver in an unexpected manner and was shown, along with CNT2, to be upregulated during liver regeneration after partial hepatectomy [57], thus explaining the nature of a previously characterized uridine transport activity that we observed to be highly sensitive to this proliferative process [58, 59]. CNT1 also seemed to be regulated in rat hepatoma FAO cells in a cell cycle-dependent manner [60]. We now know that this apparent coordinate upregulation of both transporter proteins in liver parenchymal cells is the result of different signaling processes that coordinately develop during liver regeneration after partial hepatectomy. CNT1 is indeed a target of multifunctional cytokines (TNF-α and IL-6) implicated in the priming process of hepatocytes prior to proliferation [61], whereas CNT2 is not sensitive to any of these agents but, on the contrary, is highly regulated by the proapoptotic cytokine TGF-β1, by a JNK-dependent transcriptional activation of the CNT2-encoding gene [62]. CNT2 is actually the transporter that shows the highest affinity for adenosine among all NT members. Moreover, in contrast to ENT1 transporters, CNT2 is a Na-coupled concentrative nucleoside transporter and thus a better candidate to deplete extracellular adenosine stores when required. In rat hepatoma FAO and liver parenchymal cells, CNT2 was shown to be under purinergic control via A1-type receptors, in a manner that depends on the opening of the energy-sensitive Katp channels [63]. CNT2 is abundant in intracellular structures, its insertion into the plasma membrane being upregulated by bile acids, in a microtubule-, PI3K/ERK (phosphoinositide 3-kinase/extracellular signal-related kinase)-dependent manner [64].

All these observations, taken together, argue against the conventional view that NTs exclusively promote nucleoside salvage. Moreover, although CNT2 is mostly located on the basolateral side (sinusoidal) of the membrane, CNT1 is targeted to the apical (canalicular) domain via the transcytotic pathway [65], a polarized pattern different to some extent from absorptive epithelia and not likely to be consistent with a unique salvaging role for these two transporter proteins. More conclusively, CNT1 and CNT2 proteins, although present in late fetal life, show decreased amounts in the liver compared to adult animals [66], whereas their expression can be differentially lost in rat chemically induced hepatocarcinomas or spontaneously developed in Alb-SV40 transgenic rats [67]. Although all the information available so far on liver NTs has been obtained in rodent models, we recently analyzed NT expression patterns in primary cultures of human hepatocytes [68]. Human hepatocytes express hCNT1 and hCNT2, and similar to rats, loss of the hepatic phenotype in culture is associated with a decrease in hCNT1 and hCNT2 mRNA amounts. Selected liver-enriched transcription factors (LETFs) are implicated in the regulation of SLC28 genes, again in an isoform-specific manner, HNF4α being a major determinant of CNT1 expression, whereas C/EBPα and HNF3γ modulate CNT2 [68].

Hepatocyte models are now available to further study the role of CNT proteins in liver physiology. They have also provided the first evidence of regulated intracellular transporter trafficking being used to dissect these phenomena at the molecular level.

2.2.2.3 CNTs in Immune System Cells

Expression of CNT proteins in lymphoid cells is highly variable. Although in some cases particular NT mRNA species can be amplified, functional characterization of CNT-related activities is difficult to perform, probably due to low mediated uptake. The analysis of NTs in 22 chronic lymphocytic leukemia (CLL) patients showed the expression of hENT1, hENT2, hCNT2, and hCNT3 [69]. However, only 12 patients showed Na-dependent guanosine uptake, and fludarabine accumulation was exclusively mediated by hENT1 and hENT2, thus suggesting that the only Na-dependent NT with functional activity in some CLL patients was hCNT2 [69].

In human B-cell lines, activators like phorbol esters (PMA) and bacterial lipopolysaccharide (LPS) upregulate the concentrative transporters, whereas the equilibrative transporter hENT1 is downregulated. This effect can also be produced by TNF-α, which mediates some of the functions associated with B-cell activation [70]. Moreover, the inhibition of equilibrative system triggered by the phorbol ester required sustained nitric oxide (NO) production, and NO accumulation decreased the basal uptake rates of hENT1 [71]. B- and T-lymphocytes can also be differentially regulated by insulin and glucose. Exposure of T- and B-cells to insulin results in an increase of ENT2 mRNA and a decrease of CNT2 mRNA, whereas hENT1-mediated transport was downregulated by high concentrations of glucose [72, 73].

In murine bone marrow macrophages, CNT1 and CNT2 expressions and activities are upregulated by proapoptotic agents, such as LPS, by a mechanism that partially depends on TNF-α [45]. Considering that extracellular adenosine might modulate the apoptotic response of macrophages, the increased expression of CNT2 might also be understood in the context of purinergic control of cell physiology. Activation of macrophages with interferon-gamma leads to a similar response, although the apparent transcriptional activation of both CNT1 and CNT2 in macrophages occurs in a STAT1-independent manner [44]. More recently, CNT3 has also been detected in human macrophages (unpublished data), a finding that needs further studies because, as discussed below, in contrast to CNT1 and CNT2, CNT3 is a suitable transporter for selected antiretroviral nucleosides such as AZT.

In summary, the pattern of CNT expression in immune system cells is highly heterogeneous, probably macrophages being the only cell type that shows the whole panel of NTs, measurable also at the functional activity level. Again, although little is known about their role in macrophage biology, NT coexpression cannot be explained simply by nucleoside and nucleobase salvage requirements.

2.2.2.4 CNTs in CNS

Although ENT1 and ENT2 distribution in the brain is relatively well known and its role (mostly that of hENT1) in regulating extracellular adenosine has been extensively discussed [74, 75], the occurrence of CNT-type transporters in the brain has not been so well documented until recently. CNT2 has been mapped on rat brain, mostly in neurons, by *in situ* hybridization and shown to be widespread, although it is most prevalent in the amygdala, the hippocampus, specific neocortical regions, and the cerebellum. This distribution partially overlapped that of ENT1 and was similar to that of the A1-type receptor [76]. Adenosine is a neuromodulator whose concentrations increase during sleep deprivation, playing a crucial role in the sleep/wakefulness cycle. Interestingly, sleep deprivation induces a dramatic decrease in the rat cortical amounts of CNT2 mRNA, whereas ENT1 mRNA remained unchanged. This specific decrease in CNT2 transcript suggests a new physiological role for the transporter in the modulation of extracellular adenosine levels and the sleep/wakefulness cycle. Mapping of the CNT2 protein has recently been completed (unpublished data) and correlates with that obtained by *in situ* hybridization, although in this particular case, neuron models (either in primary culture or derived cell lines) retaining consistent CNT2 functional activity are still unavailable and represent a major bottleneck in the further elucidation of the roles that CNT2 can play in CNS.

2.2.2.5 CNTs in Other Specialized Tissues

The search for nucleoside transporter expression and, particularly, for CNT-type proteins has recently focused on highly specialized tissues in which nucleoside provision can be essential or adenosine regulation particularly relevant. A CNT2 cDNA was cloned from the blood–brain barrier [77], thus anticipating that vectorial flux of nucleosides can also rely upon the heterogeneous distribution of NT proteins across the endothelium. Na^+-coupled adenosine transport in primary cultures of rat brain endothelial cells and rat choroid plexus epithelial cells is polarized apparently at the surfaces facing the interstitial and cerebrospinal fluids, respectively [78]. Adenosine is also known to be a neuromodulator implicated in both circadian clock and dark-adaptive processes in the retina [79]. The functional and molecular characterization of NTs in the blood–retinal barrier has also been recently addressed by analyzing a rat immortalized retinal capillary endothelial cell line [80]. Although mRNAs related to ENT1, ENT2, CNT2, and CNT3 were found, most of the functional activity detected in this particular cell line was equilibrative and attributable to ENT2, a finding that should be taken cautiously considering the rapid loss of CNT-type functions. Another physiological barrier with presumed high nucleoside and nucleobase needs is the blood–testis barrier. Using primary cultures of Sertoli cells, it has been reported that the whole panel of plasma membrane NTs is present in this specialized cell type (ENT1, ENT2, CNT1, CNT2, and CNT3). Na^+-coupled activity is actually present in these cultures, and this provides the first evidence for a combined role of ENTs and CNTs in providing nucleosides for spermatogenesis [81]. Also, in Sertoli cells, functional evidence for separate purine and pyrimidine nucleobase transporters has recently been provided [82]. These "kinetic agencies" are Na^+-dependent, and there is no molecular candidate responsible for these transport systems.

It is probable that CNT distribution is broader than expected due to the somehow specific role that these transporter proteins can play either in fine-tuning the extracellular adenosine concentration or in the need for a vectorial flux of nucleosides in particular tissue barriers. On the other hand, the complex pattern of NT expression, in general, does not rule out the possibility of actual redundancy, as must be the case probably in the ENT1-null mouse, as previously discussed.

2.2.3
NTs as "Transceptors"

The concept of a transporter acting as a mediator or a generator of an intracellular signal that will activate a variety of transduction pathways, thus exerting profound effects on cell physiology, is defined by the word "transceptor" (from *trans*porter and re*ceptor*). This concept was initially addressed to amino acid transporters, particularly those associated with the so-called "system A," which is actually an amino acid sensor highly responsive to amino acid deprivation [83]. As already discussed, CNT2, at least in hepatocytes and hepatoma cells, is under purinergic regulation and, to some extent, can modulate cell physiology by rapidly depleting extracellular adenosine [63]. This is not really what a transceptor is, but, interestingly,

CNT2 activation triggered by purinergic agonists is sensitive to glucose levels and depends on Katp channels with which it colocalizes in hepatocytes and hepatoma cells [63]. Thus, CNT2 function somehow "senses" the energy status. This possibility perfectly fits with recent observations suggesting that in intestinal IEC6 cells, adenosine transported via CNT2 acts as an activator of the AMP-dependent kinase (AMPK) [84]. In FAO hepatoma cells, extracellular adenosine gives a similar response [84]. AMPK is a key modulator of intracellular energy metabolism, and thus CNT2 function in this particular case exerts an effect on cell physiology that would be fully consistent with the "transceptor" concept.

In summary, NTs are drug transporters and key players in those chemotherapeutical approaches that are based on interference with nucleoside salvage and DNA/RNA synthesis (the basis for anticancer therapy), but they themselves are drug targets. They not only modulate extracellular adenosine levels but can also exert other unexpected actions, such as the control of intracellular energy metabolism via AMPK.

2.3
Nucleoside- and Nucleobase-Derived Drug Transport into Cells

Nucleosides can be structurally modified to generate pharmacologically active derivatives that, by retaining most of the metabolic properties of the parent compounds, can be transported into the cell and metabolized. They can then interfere with nucleic acid synthesis, thus promoting either antiproliferative effects or resistance to virus replication in infected cells. This is the rationale for using nucleoside derivatives in cancer and AIDS therapies.

These nucleoside-derived drugs show slight structural modifications with respect to natural nucleosides; thus, CNT and ENT are strong candidates to mediate the translocation of these compounds. However, the hypothesis that transporters belonging to the same gene family are responsible for the uptake of structurally related compounds is valid for nucleoside-derived drugs used in cancer therapy but not for all antiviral treatments.

2.3.1
Transport of Anticancer Drugs

Both pyrimidine and purine nucleoside analogues are clinically used as anticancer drugs. The purine derivatives cladribine (2-CdA, 2-chlorodeoxyadenosine) and fludarabine (F-ara-A, 9-β-D-arabinosyl-2-fluoroadenine) are extensively used in the treatment of lymphoproliferative malignancies. The most frequently used pyrimidine analogues are cytarabine (ara-C, 1-β-D-arabinosilfuranosilcytosine), gemcitabine (dFdC, 2′,2′-difluorodeoxycytidine), and capecitabine, a prodrug that yields 5-FU (5-fluorouracil) inside tumor cells, with 5′-DFUR (5′-deoxy-5-fluorouridine) being its immediate precursor. Moreover, the nucleoside analogue activity in solid tumors has triggered the synthesis of new compounds, such as troxacitabine ((−)-2′-deoxy-3′-oxacitabine) and clofarabine (2-chloro-2′-fluoro-deoxy-9-β-arabinosyladenine) [85] (Figure 2.4). All the anticancer nucleoside analogues share similar mechanisms of activation, including mediated uptake by membrane transporters, activation by kinases such as dCK, and formation of the active triphosphate metabolites. However, these compounds also possess specific properties in terms of drug–target interactions that may explain their differences in activity in various diseases [85].

Anticancer drugs

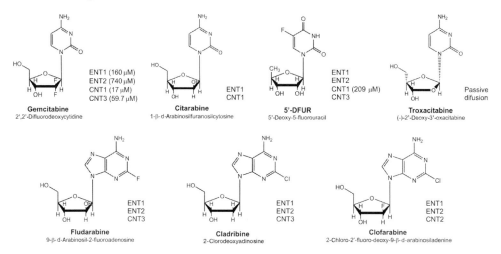

Antiviral drugs

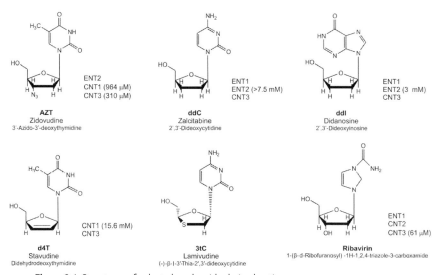

Figure 2.4 Structures of selected nucleoside-derived anticancer and antiviral drugs and their corresponding nucleoside transporter proteins. When it is known, apparent K_m value is given.

Nucleoside transporter-mediated uptake represents the major route for cellular entry of many nucleoside analogues. The pharmacological profile of NTs has been usually obtained using cross-inhibition studies. Unfortunately, transportability cannot be necessarily deduced from this type of experimental approaches. In this sense, fludarabine can inhibit hCNT2-mediated uptake but is not a suitable substrate of this transporter [86]. This pattern was also observed with natural nucleosides, since adenosine binds to hCNT1 with high affinity (K_d 14 µM) but does not translocate [55]. After the NT cloning, the ability to express a particular NT protein selectively in a null background has been particularly useful for the determination of pharmacological profiles. However, direct flux measurements need the use of labeled substrates, which often represent a bottleneck in the elucidation of the drug selectivity of a particular NT. CNT substrates can also be identified using electrophysiological approaches, as CNT function is associated with Na^+ inward currents. In particular, *Xenopus laevis* oocytes expressing human CNT isoforms can be impaled with electrodes, and using a two-electrode voltage clamp approach, selected nucleoside derivatives can be tested for their ability to generate Na^+ inward currents, which in itself is a demonstration of transportability. By combining substrate flux measurements, cross-inhibition studies, and electrophysiology, pharmacological profiles for CNTs and ENTs have been obtained, although they are still incomplete (Figure 2.4).

Substrate selectivity in the hCNT gene family is narrower than in the hENTs. For instance, gemcitabine is a high-affinity substrate for hCNT1 but is not recognized for hCNT2 and appears to be less effectively taken up by the hCNT3 isoform [26, 87]. Actually, in the SLC28 family, hCNT3 seems to be the best drug transporter as it can transport most of the pyrimidine and purine nucleoside analogues [26, 88]. Selectivity of the purine nucleoside carrier protein hCNT2 is more reduced, as it can transport clofarabine but does not take up the adenosine derivatives fludarabine and cladribine [86, 89].

As one would expect on the basis of their specificity panel for naturally occurring nucleosides, hENT1 and hENT2 show broad substrate selectivity and, when available, with apparent K_m values lower than those reported for CNTs. Moreover, gemcitabine and other nucleoside analogues used in the treatment of lymphoid malignancies also appear to be better substrates for hENT1 than for hENT2 [87].

The nucleobase 5-FU has been extensively used in the treatment of a variety of malignancies; however, very little is known about nucleobase transport in mammalian cells. Actually, hENT2 can transport nucleobases, but it is not responsible for 5-FU uptake [90]. Some authors have suggested that 5-FU is taken up by passive diffusion [91]; nevertheless, kinetic analysis of 5-FU uptake into tumor cells indicates that it is energy dependent [92, 93].

2.3.2
Transport of Antiviral Drugs

Some of the classical antiviral drugs in HIV therapy are nucleoside derivatives, and they are called nucleoside reverse transcriptase inhibitors (NRTIs). Zidovudine (azidothymidine, AZT), stavudine (dideoxythymidine, d4T), lamivudine (2′,3′-dideoxy-3′-thiacytidine, 3tC), zalcitabine (2′,3′-dideoxycytidine, ddC), and didanosine (2′,3′-dideoxyinosine, ddI) interact with nucleoside transporters (Figure 2.4).

As already discussed, some of these drugs can interact with some of these transporters, but in most cases they are not suitable substrates. The most common antiviral AZT is translocated by hCNT1, hCNT3, and hENT2 but not by hENT1, although AZT can inhibit ENT1-mediated nucleoside transport [18, 88, 94, 95]. Although CNT1 is a pyrimidine-preferring nucleoside transporter, AZT is taken up with higher affinity by CNT3. Like AZT, dideoxynucleosides such as ddC and ddI interact with the majority of nucleoside transporters. The analogues ddC and ddI are not CNT1 and CNT2 substrates, respectively, but they are able to bind the transporter with low affinity [94, 96]. However, both drugs can be recognized by the respective rat orthologue [22, 24]. The broad-specificity nucleoside transporter hCNT3 translocates ddC and ddI like equilibrative nucleoside transporters hENT1 and hENT2. hENT2 transports ddC and ddI with much higher efficiency than hENT1 [88, 97]. Nevertheless, ddC and ddI are relatively poor ENT and CNT substrates. The interactions between other NTRIs (abacavir, 3tC, and d4T) and nucleoside transporters are not well known. CNT1-mediated uridine uptake is not inhibited by 3tC, whereas d4T is recognized by CNT1 and CNT3 with relatively low affinity [88, 94].

Although these drugs share significant structural similarity with antitumoral drugs, the latter are suitable substrates of NTs as previously discussed. When comparing the structure of antitumoral and antiviral compounds, the lack of 3′-hydroxyl group of the sugar in the antiviral drugs is actually relevant for substrate recognition. This moiety plays an important role, indicating that a slight modification in nucleoside structure provokes a dramatic change in transportability, although the relevance of this shift is different for each nucleoside transporter [94, 98–101]. Thus, this finding may explain why pyrimidine derivatives such as AZT, ddC, d4T, and 3tC are not efficiently translocated by the high-affinity pyrimidine-preferring nucleoside transporter CNT1 and why they even switch recognition from the expected transporter to other carrier proteins, such as those belonging to the SLC22 gene family.

The SLC22 family comprises organic cation transporters (OCTs), zwitterion/cation transporters (OCTNs), and organic anion transporters (OATs). Transporters of the SLC22 family function in different ways: (i) as uniporters that mediate facilitated diffusion in either direction (OCTs), (ii) as anion exchangers (OATs and URAT1), and (iii) as a Na^+/L-carnitine cotransporter (OCTN2). Moreover, OATs in particular are involved in the translocation of antiviral nucleoside-derived drugs. AZT is transported by OATs with higher affinity than by the NTs [102]. The apparent K_m value is in the low micromolar range. Thus, members of this family seem to be the best candidates to mediate AZT uptake [102]. Furthermore, it has recently been proved that the other mediated routes are implicated in AZT uptake into T-lymphocytes [103]. Moreover, ddC, d4T, and 3tC are recognized as rOAT1 substrates [104].

Other well-known non-NRTI drugs such as acyclovir, gancyclovir, cidofovir, and adefovir (but not ribavirin) have been reported to be translocated by OAT1 [102, 105]. Ribavirin, a broad-spectrum antiviral agent structurally related to guanosine with activity against both DNA and RNA viruses, is a suitable substrate of ENT1, CNT2, and CNT3 [106–108]. Thus, in this particular case, NTs could contribute to the bioavailability of this drug.

2.4
Drug Transport and Responsiveness to Treatment

The clinical significance of NTs can be viewed in several ways. First, most of the anticancer nucleoside analogues need nucleoside transporters to enter the cells and reach their intracellular targets. Thus, expression of NTs in cancer cells is a prerequisite for their cytotoxicity. Second, NTs distribution in absorptive and secretory organs may influence nucleoside analogue pharmacokinetics and toxicological properties. Finally, NTs themselves could serve as drug targets. Equilibrative nucleoside transporter inhibitors such as dipyridamole and dilazep have long been used in the treatment of heart and vascular diseases. In this review, we will focus only on the first issue, that is, the role of NTs in sensitivity to nucleoside-derived drugs used in chemotherapy.

2.4.1
Analysis of the Role of NTs in Sensitivity to Nucleoside Anticancer Drugs in Cultured Cell Models

The putative role of NT function in drug uptake, bioavailability, and cytotoxicity was initially addressed in cell culture models. In 1977, analyzing the response to 5-FU of a panel of hepatoma cell lines, Greenberg *et al.* suggested that transport processes might be a limiting step in drug activation [109]. After this first evidence, a significant number of studies have demonstrated that cells lacking selected transport functions are resistant to nucleoside-derived analogues. Mutants of PK-15 cells were simultaneously resistant to tubercidin, cytosine arabinoside, and 5-fluorodeoxyuridine. Interestingly, these mutants failed to transport thymidine and uridine and had lost all high-affinity NBTI binding sites corresponding to ENT1 transporters [110]. The CCRF-CEM leukemia cell line was highly sensitive to the antiproliferative effects of troxacitabine, gemcitabine, and cytarabine, whereas a deoxycytidine kinase-deficient variant was resistant to all the three drugs. In contrast, a nucleoside transport-deficient variant was only resistant to gemcitabine and cytarabine. Actually, gemcitabine and cytarabine uptake is mediated by NTs, whereas the major route of cellular uptake of troxacitabine is passive diffusion [111]. Mackey *et al.* showed that gemcitabine required nucleoside transport to cause cytotoxicity in a study that compared gemcitabine sensitivity in cell lines with or without NT-related activity [112]. In contrast, no relationship was found between basal ENT1 levels and gemcitabine cytotoxicity in three human pancreatic cancer cell lines and one human bladder cancer cell line [113].

The pharmacological blockade of ENT-type transport activities might increase sensitivity to nucleoside derived presumably by inhibiting efflux pathways [114–116]. Actually, sensitivity is promoted when drugs reach cells prior to the inhibition of the transporter function, whereas treatment of cells after exposure to NBTI results in resistance [117]. Consistent with this, in human tumoral cell lines, sequential treatment with the ENT blocker dipyridamole 2 h after their initial exposure to cytarabine increased the cytotoxicity of this nucleoside analogue and resulted in an increase in the cellular pools of cytarabine and its metabolites [118].

The role of NTs in drug-induced cytotoxicity has also been addressed by analyzing the effect of heterologous expression of a particular NT protein on cell sensitivity to drugs. Expression of hCNT1 in Chinese hamster ovary cells induces an increase in cell sensitivity to the cytotoxic action of 5′-DFUR. This sensitization is still retained when endogenous ENTs are blocked using dipyridamole, thus suggesting that the retention of a high-affinity concentrative transporter might be a determinant of cytotoxicity by itself [119]. Moreover, slight increases in hCNT1-related function in cells derived from pancreatic adenocarcinomas also induce higher sensitivity to gemcitabine than in their parental cell lines [120]. Similarly, acquisition of hCNT2 function by gene transfer into a T-cell drug-resistant cell line results in increased sensitivity to a variety of halogenated uridine analogues [86].

The systematic analysis of NT mRNA expression in 50 cell lines did not reveal significant correlations with sensitivity to common antimetabolites such as gemcitabine, cytarabine, cladribine, and fludarabine [121]. Similarly, analysis of NT mRNA expression using oligonucleotide arrays in 60 human cancer cell lines only demonstrated a positive correlation between hENT1 and sensitivity to the nucleoside analogues azacytidine and inosine-glycodialdehyde [122]. This lack of correspondence between mRNA levels and sensitivity to nucleoside-derived drugs has subsequently been demonstrated to occur in patients with CLL (see below), highlighting the need for functional assays, suitable antibodies to NTs, or both. Therefore, in this type of studies, transporter function and protein amounts might correlate better with cytotoxicity than with mRNA levels.

Recently, different studies have used transcriptomic approaches to understand the mechanisms by which chemotherapeutical drugs cause cell death. In this sense, when analyzing the mechanism by which 5′-DFUR exerts its action on the breast cancer cell line MCF7, it was observed that short-term exposures were sufficient for transcriptional activation of selected genes, mostly implicated in apoptosis and growth arrest. Interestingly, although 5′-DFUR is taken up by both hENT1 and hENT2, inhibition of hENT1 activity using NBTI blocked most of the transcriptional changes induced by 5′-DFUR, thus evidencing a key role of hENT1, but not hENT2, in the full cytotoxic response to this drug [123]. These results highlight the relevant role of a particular transporter isoform in the nucleoside-derived triggered transcriptomic response.

2.4.2
Studies Linking NT Function to Drug Sensitivity and Clinical Outcome in Cancer Patients

Evidence obtained from *in vitro* studies (cultured cell models) clearly suggests that NTs contribute to nucleoside-derived drug cytotoxicity. Nevertheless, key issues are to determine whether these findings can be transferred to clinical settings and whether they will help to understand what roles NT proteins actually play in tumor responsiveness to nucleoside-based therapies. However, the lack, until recently, of suitable molecular tools such as isoform-specific anti-NT antibodies and the difficulty to obtain and analyze tumor samples have delayed this approach. Clinical studies suggesting a link between NT expression and drug sensitivity are summarized in Table 2.2.

Most of the clinical and *ex vivo* studies have focused on the hENT1 transporter in lymphoproliferative malignancies because hENT1 is the most abundant and widely distributed NTs in mammalian cells. Moreover, its abundance could be measured in the absence of suitable antibodies by high-affinity binding of NBTI or fluorescent nucleoside derivative SAENTA. The number of ENT1 transporters determined by NBTI binding in acute myeloid and lymphoid leukemia (AML and ALL) cells exhibited interpatient variation and correlated with intracellular accumulation of the cytarabine metabolite ara-CTP [124]. Similarly, Gati *et al.* demonstrated a correlation between the expression of hENT1 and the *in vitro* sensitivity to cytarabine and fludarabine of blasts from acute leukemia patients [125, 138]. In contrast,

Table 2.2 Clinical and *ex vivo* studies linking NT function to drug sensitivity.

Transporter	Detection method	Disease	Drug	Correlation	Reference
hENT1	NBTI binding	AML, ALL	ara-C	+	[124]
hENT1	NBTI binding	AML, ALL	ara-C, 2CdA, Fara-A	+	[125]
hENT1	NBTI binding	ALL	ara-C	+	[126]
hENT1	mRNA	AML	ara-C	+	[127, 128]
hENT1	mRNA	ALL	ara-C	+	[129]
hENT1	mRNA	AML	ara-C, 2CdA, dFdC	+	[130]
hENT2	Antibody	CLL	Fara-A	+	[131]
hENT1	mRNA and antibody	MCL	dFdC	+	[132]
hENT1	Antibody	Pancreas tumor	dFdC	+	[133]
hENT1	mRNA	Pancreas tumor	dFdC	+	[134]
hENT1	mRNA	ALL	2CdA	ns	[129]
hENT1	Antibody	HD		ns	[135]
hCNT3	Antibody	Pancreas tumor	dFdC	ns	[133]
hCNT3	Antibody	LLC	Fara-A	−	[136]
hCNT1	Antibody	Breast cancer	CMF	−	[137]

Abbreviations: HD: Hodgkin disease; ara-C: citarabine; 2-CdA: cladribine; Fara-A: fludarabine; dFdC: gemcitabine; CMF: cyclophosphamide-methotrexate-5-fluorouracil. +: positive correlation; −: negative correlation; ns: no significative correlation.

Wright et al. correlated hENT1 abundance with sensitivity to cytarabine but not to cladribine in ALL patients [126]. Moreover, a significant correlation between hENT1 mRNA levels and ara-C-induced cytotoxicity has been reported in cells from AML and ALL patients [127–130]. Together, these data strongly suggest that reduced hENT1 expression plays a significant role in clinical resistance to cytarabine in acute leukemia patients.

Cells from chronic lymphocytic leukemia patients express hENT1, hENT2, hCNT2, and hCNT3 mRNAs, whereas no hCNT1 expression has been detected [69]. No statistical correlation was found between NT mRNA levels and fludarabine transport or *ex vivo* cytotoxicity. However, a significant correlation between fludarabine uptake via ENT carriers, hENT2 protein expression, and *ex vivo* sensitivity was detected, suggesting a role of hENT2 in fludarabine responsiveness in CLL [69, 131]. Analysis of mRNA levels of hENT1, hENT2, and hCNT3 and a panel of enzymes involved in nucleotide metabolism identified two distinct populations of CLL. Surprisingly, subjects with elevated hCNT3 expression experienced a lower complete response rate to fludarabine therapy [136]. Nevertheless, in agreement with a previous report, no hCNT3-related nucleoside transport activity was detected. Indeed, all hCNT3 protein was located intracellularly [136]. Mantle cell lymphoma (MCL) cells express higher levels of hENT1 protein than CLL cells, and in contrast to CLL, a good correlation was found between protein and mRNA levels of hENT1. More important, *ex vivo* sensitivity to gemcitabine correlates with hENT1 protein and mRNA expression and drug uptake [132].

NT profiling in solid tumors is more complex than in lymphoproliferative diseases, due to the difficulty in obtaining samples and the need of suitable molecular tools, such as isoform-specific anti-NT antibodies, to analyze protein expression. Evidence for variability and selective loss of NT proteins in tumors was first provided in rat models of hepatocarcinogenesis using polyclonal antibodies raised against the rat NT orthologues rCNT1 and rCNT2 [67]. In human cancer, Mackey et al. described variability in hENT1 expression, including some hENT1-negative tumors, when analyzing a cohort of 33 breast cancer patients [139]. Some variability in hENT1 protein was also reported in Reed–Sternberg cells of Hodgkin's disease [135]. In a high-throughput analysis of gynecologic tumors, in which the abundance of selected NT proteins was assessed by immunohistochemistry, it was found that among the three studied proteins, hCNT1, hENT1, and hENT2, the most frequent loss of a particular NT protein corresponded to hCNT1. Interestingly, hCNT1 loss was associated with particular histological subtypes characterized by poor prognosis [50].

Unfortunately, studies comparing NT protein profiles with clinical parameters relevant to outcome and survival are still scarce. Spratlin et al. reported that patients with pancreatic adenocarcinoma with uniformly detectable hENT1 immunostaining have a significantly longer survival after gemcitabine chemotherapy than tumors without detectable hENT1 [133]. Similar results were obtained when analyzing hENT1-related mRNA expression in 102 pancreas cancer patients [134]. Moreover, in breast cancer patients under CMF therapy after surgery, hCNT1 alone may have prognostic value for disease-free survival and risk of relapse, with the hCNT1-positive index indicative of poor prognosis [137].

In summary, although novel therapies not based upon nucleoside derivatives are being put forward into clinics, combined treatments still rely upon nucleoside derivatives, thus making the type of studies summarized above still mandatory for a better understanding of interindividual differences in patient response to therapy. Actually, as previously discussed, nucleoside transporter expression is variable in human tumors, and evidence provided so far indicates a putative role of NTs in nucleoside-derived drug bioavailability and responsiveness. Prospective clinical studies focused on NTs as biomarkers of drug metabolism and action are required to better establish the role these membrane proteins might play in cancer chemotherapy. This would eventually lead to the analysis of NT expression patterns as suitable predictors of response to therapy and patient outcome.

2.5
Future Perspectives

From our viewpoint, major forthcoming efforts in the nucleoside and nucleobase transporter field will basically focus on selected issues of the biology and pharmacology of these membrane proteins and, maybe, of other unrelated protein families for many reasons. First, some nucleoside derivatives (i.e., those used in HAART treatment) are either weak or poor, or simply not substrates of NTs, which anticipates a growing interest on other unrelated plasma membrane transporters that can otherwise recognize some nucleoside analogues even though they might not be suitable transporters for natural nucleosides (i.e., OATs). Second, the transporter proteins implicated in the uptake of most nucleobases and nucleobase-related compounds have not been identified yet, despite the well-documented evidence that some nucleobases can actually interact with hENT2. Moreover, as discussed above, the role of NTs on the bioavailability and putative cytotoxic action of nucleoside-derived drugs in cancer treatment will require further clinical evidence but this time generated from prospective rather than from retrospective studies. Nevertheless, more important to us, nucleoside transporters appear to play very specific and somehow unexpected roles in cell physiology. This possibility, which may contribute to explain the biological rationale for the redundant expression of NTs in most mammalian cells, will also help identify selected NTs (i.e., CNT2 and/or CNT3) as novel pharmacological targets in the treatment of diseases for which NTs have never been considered suitable candidates for drug discovery and development. To fulfill these goals, efforts should be put into the study of the basic biology of these transporter proteins from a very comprehensive point of view. The concept that selected NTs can play a dual role as transporters and receptors (what we would like to call *transceptors*), based upon the growing evidence that selected NTs can also trigger intracellular responses (i.e., by modulating either intracellular or extracellular adenosine concentrations), will require further analysis. Functional genomics at the both cell and whole-animal levels will also be necessary for a better understanding of NT physiology, whereas cell biology issues related to NT trafficking, insertion, and substrate translocation regulation will be of key relevance in pharmacology.

References

1 Lai, Y., Lee, E.W., Ton, C.C., Vijay, S., Zhang, H., and Unadkat, J.D. (2005) Conserved residues F316 and G476 in the concentrative nucleoside transporter 1 (hCNT1) affect guanosine sensitivity and membrane expression, respectively. *American Journal of Physiology. Cell Physiology*, **288** (1), C39–45.

2 Crawford, C.R., Patel, D.H., Naeve, C., and Belt, J.A. (1998) Cloning of the human equilibrative, nitrobenzylmercaptopurine riboside (NBMPR)-insensitive nucleoside transporter ei by functional expression in a transport-deficient cell line. *The Journal of Biological Chemistry*, **273** (9), 5288–5293.

3 Osses, N., Pearson, J.D., Yudilevich, D.L., and Jarvis, S.M. (1996) Hypoxanthine enters human vascular endothelial cells (ECV 304) via the nitrobenzylthioinosine-insensitive equilibrative nucleoside transporter. *The Biochemical Journal*, **317** (Pt 3), 843–848.

4 Yao, S.Y., Ng, A.M., Vickers, M.F., Sundaram, M., Cass, C.E., Baldwin, S.A., and Young, J.D. (2002) Functional and molecular characterization of nucleobase transport by recombinant human and rat equilibrative nucleoside transporters 1 and 2. Chimeric constructs reveal a role for the ENT2 helix 5–6 region in nucleobase translocation. *The Journal of Biological Chemistry*, **277** (28), 24938–24948.

5 Griffiths, M., Beaumont, N., Yao, S.Y., Sundaram, M., Boumah, C.E., Davies, A., Kwong, F.Y., Coe, I., Cass, C.E., Young, J.D., and Baldwin, S.A. (1997) Cloning of a human nucleoside transporter implicated in the cellular uptake of adenosine and chemotherapeutic drugs. *Nature Medicine*, **3** (1), 89–93.

6 Coe, I.R., Griffiths, M., Young, J.D., Baldwin, S.A., and Cass, C.E. (1997) Assignment of the human equilibrative nucleoside transporter (hENT1) to 6p21.1–p21.2. *Genomics*, **45** (2), 459–460.

7 Griffiths, M., Yao, S.Y., Abidi, F., Phillips, S.E., Cass, C.E., Young, J.D., and Baldwin, S.A. (1997) Molecular cloning and characterization of a nitrobenzylthioinosine-insensitive (ei) equilibrative nucleoside transporter from human placenta. *The Biochemical Journal*, **328** (Pt 3), 739–743.

8 Hyde, R.J., Cass, C.E., Young, J.D., and Baldwin, S.A. (2001) The ENT family of eukaryote nucleoside and nucleobase transporters: recent advances in the investigation of structure/function relationships and the identification of novel isoforms. *Molecular Membrane Biology*, **18** (1), 53–63.

9 Baldwin, S.A., Yao, S.Y., Hyde, R.J., Ng, A.M., Foppolo, S., Barnes, K., Ritzel, M.W., Cass, C.E., and Young, J.D. (2005) Functional characterization of novel human and mouse equilibrative nucleoside transporters (hENT3 and mENT3) located in intracellular membranes. *The Journal of Biological Chemistry*, **280** (16), 15880–15887.

10 Baldwin, S.A., Beal, P.R., Yao, S.Y., King, A.E., Cass, C.E., and Young, J.D. (2004) The equilibrative nucleoside transporter family, SLC29. *Pflugers Archives*, **447** (5), 735–743.

11 Barnes, K., Dobrzynski, H., Foppolo, S., Beal, P.R., Ismat, F., Scullion, E.R., Sun, L., Tellez, J., Ritzel, M.W., Claycomb, W.C., Cass, C.E., Young, J.D., Billeter-Clark, R., Boyett, M.R., and Baldwin, S.A. (2006) Distribution and functional characterization of equilibrative nucleoside transporter-4, a novel cardiac adenosine transporter activated at acidic pH. *Circulation Research*, **99** (5), 510–519.

12 Engel, K., Zhou, M., and Wang, J. (2004) Identification and characterization of a novel monoamine transporter in the

human brain. *The Journal of Biological Chemistry*, **279** (48), 50042–50049.

13 Sundaram, M., Yao, S.Y., Ng, A.M., Cass, C.E., Baldwin, S.A., and Young, J.D. (2001) Equilibrative nucleoside transporters: mapping regions of interaction for the substrate analogue nitrobenzylthioinosine (NBMPR) using rat chimeric proteins. *Biochemistry*, **40** (27), 8146–8151.

14 Vickers, M.F., Mani, R.S., Sundaram, M., Hogue, D.L., Young, J.D., Baldwin, S.A., and Cass, C.E. (1999) Functional production and reconstitution of the human equilibrative nucleoside transporter (hENT1) in *Saccharomyces cerevisiae*. Interaction of inhibitors of nucleoside transport with recombinant hENT1 and a glycosylation-defective derivative (hENT1/N48Q). *The Biochemical Journal*, **339** (Pt 1), 21–32.

15 Ward, J.L., Leung, G.P., Toan, S.V., and Tse, C.M. (2003) Functional analysis of site-directed glycosylation mutants of the human equilibrative nucleoside transporter-2. *Archives of Biochemistry and Biophysics*, **411** (1), 19–26.

16 Sundaram, M., Yao, S.Y., Ng, A.M., Griffiths, M., Cass, C.E., Baldwin, S.A., and Young, J.D. (1998) Chimeric constructs between human and rat equilibrative nucleoside transporters (hENT1 and rENT1) reveal hENT1 structural domains interacting with coronary vasoactive drugs. *The Journal of Biological Chemistry*, **273** (34), 21519–21525.

17 Endres, C.J. and Unadkat, J.D. (2005) Residues Met89 and Ser160 in the human equilibrative nucleoside transporter 1 affect its affinity for adenosine, guanosine, S6-(4-nitrobenzyl)-mercaptopurine riboside, and dipyridamole. *Molecular Pharmacology*, **67** (3), 837–844.

18 Yao, S.Y., Ng, A.M., Sundaram, M., Cass, C.E., Baldwin, S.A., and Young, J.D. (2001) Transport of antiviral 3'-deoxy-nucleoside drugs by recombinant human and rat equilibrative, nitrobenzylthioinosine (NBMPR)-insensitive (ENT2) nucleoside transporter proteins produced in *Xenopus* oocytes. *Molecular Membrane Biology*, **18** (2), 161–167.

19 Lai, Y., Tse, C.M., and Unadkat, J.D. (2004) Mitochondrial expression of the human equilibrative nucleoside transporter 1 (hENT1) results in enhanced mitochondrial toxicity of antiviral drugs. *The Journal of Biological Chemistry*, **279** (6), 4490–4497.

20 Lee, E.W., Lai, Y., Zhang, H., and Unadkat, J.D. (2006) Identification of the mitochondrial targeting signal of the human equilibrative nucleoside transporter 1 (hENT1): implications for interspecies differences in mitochondrial toxicity of fialuridine. *The Journal of Biological Chemistry*, **281** (24), 16700–16706.

21 Huang, Q.Q., Yao, S.Y., Ritzel, M.W., Paterson, A.R., Cass, C.E., and Young, J.D. (1994) Cloning and functional expression of a complementary DNA encoding a mammalian nucleoside transport protein. *The Journal of Biological Chemistry*, **269** (27), 17757–17760.

22 Ritzel, M.W., Yao, S.Y., Huang, M.Y., Elliott, J.F., Cass, C.E., and Young, J.D. (1997) Molecular cloning and functional expression of cDNAs encoding a human Na^+-nucleoside cotransporter (hCNT1). *The American Journal of Physiology*, **272** (2 Pt 1), C707–C714.

23 Che, M., Ortiz, D.F., and Arias, I.M. (1995) Primary structure and functional expression of a cDNA encoding the bile canalicular, purine-specific Na^+-nucleoside cotransporter. *The Journal of Biological Chemistry*, **270** (23), 13596–13599.

24 Ritzel, M.W., Yao, S.Y., Ng, A.M., Mackey, J.R., Cass, C.E., and Young, J.D. (1998) Molecular cloning, functional expression and chromosomal localization of a cDNA encoding a human Na^+/nucleoside cotransporter (hCNT2)

selective for purine nucleosides and uridine. *Molecular Membrane Biology*, **15** (4), 203–211.

25 Wang, J., Su, S.F., Dresser, M.J., Schaner, M.E., Washington, C.B., and Giacomini, K.M. (1997) Na^+-dependent purine nucleoside transporter from human kidney: cloning and functional characterization. *The American Journal of Physiology*, **273** (6 Pt 2), F1058–F1065.

26 Ritzel, M.W., Ng, A.M., Yao, S.Y., Graham, K., Loewen, S.K., Smith, K.M., Ritzel, R.G., Mowles, D.A., Carpenter, P., Chen, X.Z., Karpinski, E., Hyde, R.J., Baldwin, S.A., Cass, C.E., and Young, J.D. (2001) Molecular identification and characterization of novel human and mouse concentrative Na^+-nucleoside cotransporter proteins (hCNT3 and mCNT3) broadly selective for purine and pyrimidine nucleosides (system cib). *The Journal of Biological Chemistry*, **276** (4), 2914–2927.

27 Ritzel, M.W., Ng, A.M., Yao, S.Y., Graham, K., Loewen, S.K., Smith, K.M., Hyde, R.J., Karpinski, E., Cass, C.E., Baldwin, S.A., and Young, J.D. (2001) Recent molecular advances in studies of the concentrative Na^+-dependent nucleoside transporter (CNT) family: identification and characterization of novel human and mouse proteins (hCNT3 and mCNT3) broadly selective for purine and pyrimidine nucleosides (system cib). *Molecular Membrane Biology*, **18** (1), 65–72.

28 Hamilton, S.R., Yao, S.Y., Ingram, J.C., Hadden, D.A., Ritzel, M.W., Gallagher, M.P., Henderson, P.J., Cass, C.E., Young, J.D., and Baldwin, S.A. (2001) Subcellular distribution and membrane topology of the mammalian concentrative Na^+-nucleoside cotransporter rCNT1. *The Journal of Biological Chemistry*, **276** (30), 27981–27988.

29 Wang, J. and Giacomini, K.M. (1997) Molecular determinants of substrate selectivity in Na^+-dependent nucleoside transporters. *The Journal of Biological Chemistry*, **272** (46), 28845–28848.

30 Wang, J. and Giacomini, K.M. (1999) Characterization of a bioengineered chimeric Na^+-nucleoside transporter. *Molecular Pharmacology*, **55** (2), 234–240.

31 Wang, J. and Giacomini, K.M. (1999) Serine 318 is essential for the pyrimidine selectivity of the N2 Na^+-nucleoside transporter. *The Journal of Biological Chemistry*, **274** (4), 2298–2302.

32 Loewen, S.K., Ng, A.M., Yao, S.Y., Cass, C.E., Baldwin, S.A., and Young, J.D. (1999) Identification of amino acid residues responsible for the pyrimidine and purine nucleoside specificities of human concentrative Na^+ nucleoside cotransporters hCNT1 and hCNT2. *The Journal of Biological Chemistry*, **274** (35), 24475–24484.

33 Lu, H., Chen, C., and Klaassen, C. (2004) Tissue distribution of concentrative and equilibrative nucleoside transporters in male and female rats and mice. *Drug Metabolism and Disposition*, **32** (12), 1455–1461.

34 Aguayo, C., Flores, C., Parodi, J., Rojas, R., Mann, G.E., Pearson, J.D., and Sobrevia, L. (2001) Modulation of adenosine transport by insulin in human umbilical artery smooth muscle cells from normal or gestational diabetic pregnancies. *The Journal of Physiology*, **534** (Pt 1), 243–254.

35 San Martin, R. and Sobrevia, L. (2006) Gestational diabetes and the adenosine/L-arginine/nitric oxide (ALANO) pathway in human umbilical vein endothelium. *Placenta*, **27** (1), 1–10.

36 Farias, M., San Martin, R., Puebla, C., Pearson, J.D., Casado, J.F., Pastor-Anglada, M., Casanello, P., and Sobrevia, L. (2006) Nitric oxide reduces adenosine transporter ENT1 gene (SLC29A1) promoter activity in human fetal endothelium from gestational diabetes. *Journal of Cellular Physiology*, **208** (2), 451–460.

37 Munoz, G., San Martin, R., Farias, M., Cea, L., Vecchiola, A., Casanello, P., and Sobrevia, L. (2006) Insulin restores

glucose inhibition of adenosine transport by increasing the expression and activity of the equilibrative nucleoside transporter 2 in human umbilical vein endothelium. *Journal of Cellular Physiology*, **209** (3), 826–835.

38 Aguayo, C., Casado, J., Gonzalez, M., Pearson, J.D., Martin, R.S., Casanello, P., Pastor-Anglada, M., and Sobrevia, L. (2005) Equilibrative nucleoside transporter 2 is expressed in human umbilical vein endothelium, but is not involved in the inhibition of adenosine transport induced by hyperglycaemia. *Placenta*, **26** (8–9), 641–653.

39 Casanello, P., Torres, A., Sanhueza, F., Gonzalez, M., Farias, M., Gallardo, V., Pastor-Anglada, M., San Martin, R., and Sobrevia, L. (2005) Equilibrative nucleoside transporter 1 expression is downregulated by hypoxia in human umbilical vein endothelium. *Circulation Research*, **97** (1), 16–24.

40 Chaudary, N., Naydenova, Z., Shuralyova, I., and Coe, I.R. (2004) Hypoxia regulates the adenosine transporter, mENT1, in the murine cardiomyocyte cell line, HL-1. *Cardiovascular Research*, **61** (4), 780–788.

41 Eltzschig, H.K., Abdulla, P., Hoffman, E., Hamilton, K.E., Daniels, D., Schonfeld, C., Loffler, M., Reyes, G., Duszenko, M., Karhausen, J., Robinson, A., Westerman, K.A., Coe, I.R., and Colgan, S.P. (2005) HIF-1-dependent repression of equilibrative nucleoside transporter (ENT) in hypoxia. *The Journal of Experimental Medicine*, **202** (11), 1493–1505.

42 Choi, D.S., Cascini, M.G., Mailliard, W., Young, H., Paredes, P., McMahon, T., Diamond, I., Bonci, A., and Messing, R.O. (2004) The type 1 equilibrative nucleoside transporter regulates ethanol intoxication and preference. *Nature Neuroscience*, **7** (8), 855–861.

43 Soler, C., Garcia-Manteiga, J., Valdes, R., Xaus, J., Comalada, M., Casado, F.J., Pastor-Anglada, M., Celada, A., and Felipe, A. (2001) Macrophages require different nucleoside transport systems for proliferation and activation. *The FASEB Journal*, **15** (11), 1979–1988.

44 Soler, C., Felipe, A., Garcia-Manteiga, J., Serra, M., Guillen-Gomez, E., Casado, F.J., MacLeod, C., Modolell, M., Pastor-Anglada, M., and Celada, A. (2003) Interferon-gamma regulates nucleoside transport systems in macrophages through signal transduction and activator of transduction factor 1 (STAT1)-dependent and -independent signalling pathways. *The Biochemical Journal*, **375** (Pt 3), 777–783.

45 Soler, C., Valdes, R., Garcia-Manteiga, J., Xaus, J., Comalada, M., Casado, F.J., Modolell, M., Nicholson, B., MacLeod, C., Felipe, A., Celada, A., and Pastor-Anglada, M. (2001) Lipopolysaccharide-induced apoptosis of macrophages determines the up-regulation of concentrative nucleoside transporters Cnt1 and Cnt2 through tumor necrosis factor-alpha-dependent and -independent mechanisms. *The Journal of Biological Chemistry*, **276** (32), 30043–30049.

46 Errasti-Murugarren, E., Pastor-Anglada, M., and Casado, F.J. (2007) Role of Cnt3 in the transepithelial flux of nucleosides and nucleoside-derived drugs. *The Journal of Physiology.*, **582** (3), 1249–1260.

47 Lai, Y., Bakken, A.H., and Unadkat, J.D. (2002) Simultaneous expression of hCNT1-CFP and hENT1-YFP in Madin-Darby canine kidney cells. Localization and vectorial transport studies. *The Journal of Biological Chemistry*, **277** (40), 37711–37717.

48 Mangravite, L.M., Lipschutz, J.H., Mostov, K.E., and Giacomini, K.M. (2001) Localization of GFP-tagged concentrative nucleoside transporters in a renal polarized epithelial cell line. *American Journal of Physiology. Renal Physiology*, **280** (5), F879–F885.

49 Aymerich, I., Pastor-Anglada, M., and Casado, F.J. (2004) Long term endocrine regulation of nucleoside transporters in

rat intestinal epithelial cells. *The Journal of General Physiology,* **124** (5), 505–512.

50 Farre, X., Guillen-Gomez, E., Sanchez, L., Hardisson, D., Plaza, Y., Lloberas, J., Casado, F.J., Palacios, J., and Pastor-Anglada, M. (2004) Expression of the nucleoside-derived drug transporters hCNT1, hENT1 and hENT2 in gynecologic tumors. *International Journal of Cancer,* **112** (6), 959–966.

51 Mangravite, L.M., Badagnani, I., and Giacomini, K.M. (2003) Nucleoside transporters in the disposition and targeting of nucleoside analogs in the kidney. *European Journal of Pharmacology,* **479** (1–3), 269–281.

52 Mangravite, L.M., Xiao, G., and Giacomini, K.M. (2003) Localization of human equilibrative nucleoside transporters, hENT1 and hENT2, in renal epithelial cells. *American Journal of Physiology. Renal Physiology,* **284** (5), F902–F910.

53 Rodriguez-Mulero, S., Errasti-Murugarren, E., Ballarin, J., Felipe, A., Doucet, A., Casado, F.J., and Pastor-Anglada, M. (2005) Expression of concentrative nucleoside transporters SLC28 (CNT1, CNT2, and CNT3) along the rat nephron: effect of diabetes. *Kidney International,* **68** (2), 665–672.

54 Gerstin, K.M., Dresser, M.J., and Giacomini, K.M. (2002) Specificity of human and rat orthologs of the concentrative nucleoside transporter, SPNT. *American Journal of Physiology. Renal Physiology,* **283** (2), F344–F349.

55 Larrayoz, I.M., Casado, F.J., Pastor-Anglada, M., and Lostao, M.P. (2004) Electrophysiological characterization of the human Na^+/nucleoside cotransporter 1 (hCNT1) and role of adenosine on hCNT1 function. *The Journal of Biological Chemistry,* **279** (10), 8999–9007.

56 Valdes, R., Ortega, M.A., Casado, F.J., Felipe, A., Gil, A., Sanchez-Pozo, A., and Pastor-Anglada, M. (2000) Nutritional regulation of nucleoside transporter expression in rat small intestine. *Gastroenterology,* **119** (6), 1623–1630.

57 Felipe, A., Valdes, R., Santo, B., Lloberas, J., Casado, J., and Pastor-Anglada, M. (1998) Na^+-dependent nucleoside transport in liver: two different isoforms from the same gene family are expressed in liver cells. *The Biochemical Journal,* **330** (Pt 2), 997–1001.

58 Ruiz-Montasell, B., Javier Casado, F., Felipe, A., and Pastor-Anglada, M. (1992) Uridine transport in basolateral plasma membrane vesicles from rat liver. *The Journal of Membrane Biology,* **128** (3), 227–233.

59 Ruiz-Montasell, B., Martinez-Mas, J.V., Enrich, C., Casado, F.J., Felipe, A., and Pastor-Anglada, M. (1993) Early induction of Na^+-dependent uridine uptake in the regenerating rat liver. *FEBS Letters,* **316** (1), 85–88.

60 Valdes, R., Casado, F.J., and Pastor-Anglada, M. (2002) Cell-cycle-dependent regulation of CNT1, a concentrative nucleoside transporter involved in the uptake of cell-cycle-dependent nucleoside-derived anticancer drugs. *Biochemical and Biophysical Research Communications,* **296** (3), 575–579.

61 Fernandez-Veledo, S., Valdes, R., Wallenius, V., Casado, F.J., and Pastor-Anglada, M. (2004) Up-regulation of the high-affinity pyrimidine-preferring nucleoside transporter concentrative nucleoside transporter 1 by tumor necrosis factor-alpha and interleukin-6 in liver parenchymal cells. *Journal of Hepatology,* **41** (4), 538–544.

62 Valdes, R., Fernandez-Veledo, S., Aymerich, I., Casado, F.J., and Pastor-Anglada, M. (2006) TGF-beta transcriptionally activates the gene encoding the high-affinity adenosine transporter CNT2 in rat liver parenchymal cells. *Cellular and Molecular Life Sciences,* **63** (21), 2527–2537.

63 Duflot, S., Riera, B., Fernandez-Veledo, S., Casado, V., Norman, R.I., Casado, F.J., Lluis, C., Franco, R., and Pastor-Anglada, M. (2004) ATP-sensitive K^+ channels regulate the concentrative adenosine transporter CNT2 following activation by A_1 adenosine receptors. *Molecular and Cellular Biology*, **24** (7), 2710–2719.

64 Fernandez-Veledo, S., Huber-Ruano, I., Aymerich, I., Duflot, S., Casado, F.J., and Pastor-Anglada, M. (2006) Bile acids alter the subcellular localization of CNT2 (concentrative nucleoside cotransporter) and increase CNT2-related transport activity in liver parenchymal cells. *The Biochemical Journal*, **395** (2), 337–344.

65 Duflot, S., Calvo, M., Casado, F.J., Enrich, C., and Pastor-Anglada, M. (2002) Concentrative nucleoside transporter (rCNT1) is targeted to the apical membrane through the hepatic transcytotic pathway. *Experimental Cell Research*, **281** (1), 77–85.

66 del Santo, B., Tarafa, G., Felipe, A., Casado, F.J., and Pastor-Anglada, M. (2001) Developmental regulation of the concentrative nucleoside transporters CNT1 and CNT2 in rat liver. *Journal of Hepatology*, **34** (6), 873–880.

67 Dragan, Y., Valdes, R., Gomez-Angelats, M., Felipe, A., Javier Casado, F., Pitot, H., and Pastor-Anglada, M. (2000) Selective loss of nucleoside carrier expression in rat hepatocarcinomas. *Hepatology*, **32** (2), 239–246.

68 Fernandez-Veledo, S., Jover, R., Casado, F.J., Gomez-Lechon, M.J., and Pastor-Anglada, M. (2007) Transcription factors involved in the expression of SLC28 genes in human liver parenchymal cells. *Biochemical and Biophysical Research Communications*, **353** (2), 381–388.

69 Molina-Arcas, M., Bellosillo, B., Casado, F.J., Montserrat, E., Gil, J., Colomer, D., and Pastor-Anglada, M. (2003) Fludarabine uptake mechanisms in B-cell chronic lymphocytic leukaemia. *Blood*, **101** (6), 2328–2334.

70 Soler, C., Felipe, A., Mata, J.F., Casado, F.J., Celada, A., and Pastor-Anglada, M. (1998) Regulation of nucleoside transport by lipopolysaccharide, phorbol esters, and tumor necrosis factor-alpha in human B-lymphocytes. *The Journal of Biological Chemistry*, **273** (41), 26939–26945.

71 Soler, C., Felipe, A., Casado, F.J., Celada, A., and Pastor-Anglada, M. (2000) Nitric oxide regulates nucleoside transport in activated B lymphocytes. *Journal of Leukocyte Biology*, **67** (3), 345–349.

72 Sakowicz, M., Szutowicz, A., and Pawelczyk, T. (2004) Insulin and glucose induced changes in expression level of nucleoside transporters and adenosine transport in rat T lymphocytes. *Biochemical Pharmacology*, **68** (7), 1309–1320.

73 Sakowicz, M., Szutowicz, A., and Pawelczyk, T. (2005) Differential effect of insulin and elevated glucose level on adenosine transport in rat B lymphocytes. *International Immunology*, **17** (2), 145–154.

74 Anderson, C.M., Baldwin, S.A., Young, J.D., Cass, C.E., and Parkinson, F.E. (1999) Distribution of mRNA encoding a nitrobenzylthioinosine-insensitive nucleoside transporter (ENT2) in rat brain. *Brain Research. Molecular Brain Research*, **70** (2), 293–297.

75 Anderson, C.M., Xiong, W., Geiger, J.D., Young, J.D., Cass, C.E., Baldwin, S.A., and Parkinson, F.E. (1999) Distribution of equilibrative, nitrobenzylthioinosine-sensitive nucleoside transporters (ENT1) in brain. *Journal of Neurochemistry*, **73** (2), 867–873.

76 Guillen-Gomez, E., Calbet, M., Casado, J., de Lecea, L., Soriano, E., Pastor-Anglada, M., and Burgaya, F. (2004) Distribution of CNT2 and ENT1 transcripts in rat brain: selective decrease of CNT2 mRNA in the

76 cerebral cortex of sleep-deprived rats. *Journal of Neurochemistry*, **90** (4), 883–893.

77 Li, J.Y., Boado, R.J., and Pardridge, W.M. (2001) Cloned blood–brain barrier adenosine transporter is identical to the rat concentrative Na$^+$ nucleoside cotransporter CNT2. *Journal of Cerebral Blood Flow and Metabolism*, **21** (8), 929–936.

78 Redzic, Z.B., Biringer, J., Barnes, K., Baldwin, S.A., Al-Sarraf, H., Nicola, P.A., Young, J.D., Cass, C.E., Barrand, M.A., and Hladky, S.B. (2005) Polarized distribution of nucleoside transporters in rat brain endothelial and choroid plexus epithelial cells. *Journal of Neurochemistry*, **94** (5), 1420–1426.

79 Ribelayga, C. and Mangel, S.C. (2005) A circadian clock and light/dark adaptation differentially regulate adenosine in the mammalian retina. *The Journal of Neuroscience*, **25** (1), 215–222.

80 Nagase, K., Tomi, M., Tachikawa, M., and Hosoya, K. (2006) Functional and molecular characterization of adenosine transport at the rat inner blood–retinal barrier. *Biochimica et Biophysica Acta*, **1758** (1), 13–19.

81 Kato, R., Maeda, T., Akaike, T., and Tamai, I. (2005) Nucleoside transport at the blood–testis barrier studied with primary-cultured sertoli cells. *The Journal of Pharmacology and Experimental Therapeutics*, **312** (2), 601–608.

82 Kato, R., Maeda, T., Akaike, T., and Tamai, I. (2006) Characterization of novel Na$^+$-dependent nucleobase transport systems at the blood–testis barrier. *American Journal of Physiology. Endocrinology and Metabolism*, **290** (5), E968–E975.

83 Hyde, R., Cwiklinski, E.L., Macaulay, K., Taylor, P.M., and Hundal, H.S. (2007) Distinct sensor pathways in the hierarchical control of SNAT2, a putative amino acid transceptor, by amino acid availability. *The Journal of Biological Chemistry*, **282** (27), 19788–19798.

84 Aymerich, I., Foufelle, F., Ferre, P., Casado, F.J., and Pastor-Anglada, M. (2006) Extracellular adenosine activates AMP-dependent protein kinase (AMPK). *Journal of Cell Science*, **119** (Pt 8), 1612–1621.

85 Galmarini, C.M., Mackey, J.R., and Dumontet, C. (2001) Nucleoside analogues: mechanisms of drug resistance and reversal strategies. *Leukemia*, **15** (6), 875–890.

86 Lang, T.T., Selner, M., Young, J.D., and Cass, C.E. (2001) Acquisition of human concentrative nucleoside transporter 2 (hcnt2) activity by gene transfer confers sensitivity to fluoropyrimidine nucleosides in drug-resistant leukemia cells. *Molecular Pharmacology*, **60** (5), 1143–1152.

87 Mackey, J.R., Yao, S.Y., Smith, K.M., Karpinski, E., Baldwin, S.A., Cass, C.E., and Young, J.D. (1999) Gemcitabine transport in *Xenopus* oocytes expressing recombinant plasma membrane mammalian nucleoside transporters. *Journal of the National Cancer Institute*, **91** (21), 1876–1881.

88 Hu, H., Endres, C.J., Chang, C., Umapathy, N.S., Lee, E.W., Fei, Y.J., Itagaki, S., Swaan, P.W., Ganapathy, V., and Unadkat, J.D. (2006) Electrophysiological characterization and modeling of the structure activity relationship of the human concentrative nucleoside transporter 3 (hCNT3). *Molecular Pharmacology*, **69** (5), 1542–1553.

89 King, K., Damaraju, V., Vickers, M., Yao, S., Lang, T., Tackaberry, T., Mowles, D., Ng, A., Young, J., and Cass, C. (2005) A comparison of the transportability, and its role in cytotoxicity, of clofarabine, cladribine and fludarabine by recombinant human nucleoside transporters produced in three model expression systems. *Molecular Pharmacology.*, **69** (1), 346–353.

90 Nagai, K., Nagasawa, K., Kihara, Y., Okuda, H., and Fujimoto, S. (2006)

Anticancer nucleobase analogues 6-mercaptopurine and 6-thioguanine are novel substrates for equilibrative nucleoside transporter 2. *International Journal of Pharmaceutics*, **333** (1–2), 56–61.

91 Nakamura, J., Horimoto, T., Hirayama, R., Mukai, T., Nakashima, M., Sasaki, H., and Nishida, K. (2003) Effect of the absorption enhancer saponin on the intrarenal distribution of 5-fluorouracil following its kidney surface application in rats. *Biological & Pharmaceutical Bulletin*, **26** (7), 1049–1051.

92 Hsu, L.S. (1982) The kinetics of uptake of 5-fluorouracil by rat liver. *British Journal of Pharmacology*, **77** (3), 413–417.

93 Ojugo, A.S., McSheehy, P.M., Stubbs, M., Alder, G., Bashford, C.L., Maxwell, R.J., Leach, M.O., Judson, I.R., and Griffiths, J.R. (1998) Influence of pH on the uptake of 5-fluorouracil into isolated tumour cells. *British Journal of Cancer*, **77** (6), 873–879.

94 Cano-Soldado, P., Lorrayoz, I.M., Molina-Arcas, M., Casado, F.J., Martinez-Picado, J., Lostao, M.P., and Pastor-Anglada, M. (2004) Interaction of nucleoside inhibitors of HIV-1 reverse transcriptase with the concentrative nucleoside transporter-1 (SLC28A1). *Antiviral Therapy*, **9** (6), 993–1002.

95 Ward, J.L., Sherali, A., Mo, Z.P., and Tse, C.M. (2000) Kinetic and pharmacological properties of cloned human equilibrative nucleoside transporters, ENT1 and ENT2, stably expressed in nucleoside transporter-deficient PK15 cells. Ent2 exhibits a low affinity for guanosine and cytidine but a high affinity for inosine. *The Journal of Biological Chemistry*, **275** (12), 8375–8381.

96 Schaner, M.E., Wang, J., Zhang, L., Su, S.F., Gerstin, K.M., and Giacomini, K.M. (1999) Functional characterization of a human purine-selective, Na^+-dependent nucleoside transporter (hSPNT1) in a mammalian expression system. *The Journal of Pharmacology and Experimental Therapeutics*, **289** (3), 1487–1491.

97 Kong, W., Engel, K., and Wang, J. (2004) Mammalian nucleoside transporters. *Current Drug Metabolism*, **5** (1), 63–84.

98 Chang, C., Swaan, P.W., Ngo, L.Y., Lum, P.Y., Patil, S.D., and Unadkat, J.D. (2004) Molecular requirements of the human nucleoside transporters hCNT1, hCNT2, and hENT1. *Molecular Pharmacology*, **65** (3), 558–570.

99 Pastor-Anglada, M., Cano-Soldado, P., Molina-Arcas, M., Lostao, M.P., Larrayoz, I., Martinez-Picado, J., and Casado, F.J. (2005) Cell entry and export of nucleoside analogues. *Virus Research*, **107** (2), 151–164.

100 Vickers, M.F., Zhang, J., Visser, F., Tackaberry, T., Robins, M.J., Nielsen, L.P., Nowak, I., Baldwin, S.A., Young, J.D., and Cass, C.E. (2004) Uridine recognition motifs of human equilibrative nucleoside transporters 1 and 2 produced in *Saccharomyces cerevisiae*. *Nucleosides, Nucleotides and Nucleic Acids*, **23** (1–2), 361–373.

101 Zhang, J., Smith, K.M., Tackaberry, T., Visser, F., Robins, M.J., Nielsen, L.P., Nowak, I., Karpinski, E., Baldwin, S.A., Young, J.D., and Cass, C.E. (2005) Uridine binding and transportability determinants of human concentrative nucleoside transporters. *Molecular Pharmacology*, **68** (3), 830–839.

102 Takeda, M., Khamdang, S., Narikawa, S., Kimura, H., Kobayashi, Y., Yamamoto, T., Cha, S.H., Sekine, T., and Endou, H. (2002) Human organic anion transporters and human organic cation transporters mediate renal antiviral transport. *The Journal of Pharmacology and Experimental Therapeutics*, **300** (3), 918–924.

103 Purcet, S., Minuesa, G., Molina-Arcas, M., Erkizia, I., Casado, F.J., Clotet, B., Martinez-Picado, J., and Pastor-Anglada, M. (2006) 3′-Azido-2′,3′-dideoxythymidine (zidovudine) uptake mechanisms in T lymphocytes. *Antiviral Therapy*, **11** (6), 803–811.

104 Wada, S., Tsuda, M., Sekine, T., Cha, S.H., Kimura, M., Kanai, Y., and Endou, H. (2000) Rat multispecific organic anion transporter 1 (rOAT1) transports zidovudine, acyclovir, and other antiviral nucleoside analogs. *The Journal of Pharmacology and Experimental Therapeutics*, **294** (3), 844–849.

105 Ho, E.S., Lin, D.C., Mendel, D.B., and Cihlar, T. (2000) Cytotoxicity of antiviral nucleotides adefovir and cidofovir is induced by the expression of human renal organic anion transporter 1. *Journal of the American Society of Nephrology*, **11** (3), 383–393.

106 Jarvis, S.M., Thorn, J.A., and Glue, P. (1998) Ribavirin uptake by human erythrocytes and the involvement of nitrobenzylthioinosine-sensitive (es)-nucleoside transporters. *British Journal of Pharmacology*, **123** (8), 1587–1592.

107 Patil, S.D., Ngo, L.Y., Glue, P., and Unadkat, J.D. (1998) Intestinal absorption of ribavirin is preferentially mediated by the Na$^+$-nucleoside purine (N1) transporter. *Pharmaceutical Research*, **15** (6), 950–952.

108 Yamamoto, T., Kuniki, K., Takekuma, Y., Hirano, T., Iseki, K., and Sugawara, M. (2007) Ribavirin uptake by cultured human choriocarcinoma (BeWo) cells and *Xenopus laevis* oocytes expressing recombinant plasma membrane human nucleoside transporters. *European Journal of Pharmacology*, **557** (1), 1–8.

109 Greenberg, N., Schumm, D.E., and Webb, T.E. (1977) Uridine kinase activities and pyrimidine nucleoside phosphorylation in fluoropyrimidine-sensitive and -resistant cell lines of the Novikoff hepatoma. *The Biochemical Journal*, **164** (2), 379–387.

110 Aran, J.M. and Plagemann, P.G. (1992) Nucleoside transport-deficient mutants of PK-15 pig kidney cell line. *Biochimica et Biophysica Acta*, **1110** (1), 51–58.

111 Gourdeau, H., Clarke, M.L., Ouellet, F., Mowles, D., Selner, M., Richard, A., Lee, N., Mackey, J.R., Young, J.D., Jolivet, J., Lafreniere, R.G., and Cass, C.E. (2001) Mechanisms of uptake and resistance to troxacitabine, a novel deoxycytidine nucleoside analogue, in human leukemic and solid tumor cell lines. *Cancer Research*, **61** (19), 7217–7224.

112 Mackey, J.R., Mani, R.S., Selner, M., Mowles, D., Young, J.D., Belt, J.A., Crawford, C.R., and Cass, C.E. (1998) Functional nucleoside transporters are required for gemcitabine influx and manifestation of toxicity in cancer cell lines. *Cancer Research*, **58** (19), 4349–4357.

113 Rauchwerger, D.R., Firby, P.S., Hedley, D.W., and Moore, M.J. (2000) Equilibrative-sensitive nucleoside transporter and its role in gemcitabine sensitivity. *Cancer Research*, **60** (21), 6075–6079.

114 Alessi-Severini, S., Gati, W.P., Belch, A.R., and Paterson, A.R. (1995) Intracellular pharmacokinetics of 2-chlorodeoxyadenosine in leukemia cells from patients with chronic lymphocytic leukaemia. *Leukemia*, **9** (10), 1674–1679.

115 Crawford, C.R., Ng, C.Y., Noel, L.D., and Belt, J.A. (1990) Nucleoside transport in L1210 murine leukemia cells. Evidence for three transporters. *The Journal of Biological Chemistry*, **265** (17), 9732–9736.

116 Wright, A.M., Gati, W.P., and Paterson, A.R. (2000) Enhancement of retention and cytotoxicity of 2-chlorodeoxyadenosine in cultured human leukemic lymphoblasts by nitrobenzylthioinosine, an inhibitor of equilibrative nucleoside transport. *Leukemia*, **14** (1), 52–60.

117 Cass, C.E., King, K.M., Montano, J.T., and Janowska-Wieczorek, A. (1992) A comparison of the abilities of nitrobenzylthioinosine, dilazep, and dipyridamole to protect human hematopoietic cells from 7-deazaadenosine (tubercidin). *Cancer Research*, **52** (21), 5879–5886.

118 Chan, T.C. (1989) Augmentation of 1-beta-D-arabinofuranosylcytosine

cytotoxicity in human tumor cells by inhibiting drug efflux. *Cancer Research*, **49** (10), 2656–2660.

119 Mata, J.F., Garcia-Manteiga, J.M., Lostao, M.P., Fernandez-Veledo, S., Guillen-Gomez, E., Larrayoz, I.M., Lloberas, J., Casado, F.J., and Pastor-Anglada, M. (2001) Role of the human concentrative nucleoside transporter (hCNT1) in the cytotoxic action of 5′-deoxy-5-fluorouridine, an active intermediate metabolite of capecitabine, a novel oral anticancer drug. *Molecular Pharmacology*, **59** (6), 1542–1548.

120 Garcia-Manteiga, J., Molina-Arcas, M., Casado, F.J., Mazo, A., and Pastor-Anglada, M. (2003) Nucleoside transporter profiles in human pancreatic cancer cells: role of hCNT1 in 2′,2′-difluorodeoxycytidine-induced cytotoxicity. *Clinical Cancer Research*, **9** (13), 5000–5008.

121 Lu, X., Gong, S., Monks, A., Zaharevitz, D., and Moscow, J.A. (2002) Correlation of nucleoside and nucleobase transporter gene expression with antimetabolite drug cytotoxicity. *Journal of Experimental Therapeutics & Oncology*, **2** (4), 200–212.

122 Huang, Y., Anderle, P., Bussey, K.J., Barbacioru, C., Shankavaram, U., Dai, Z., Reinhold, W.C., Papp, A., Weinstein, J.N., and Sadee, W. (2004) Membrane transporters and channels: role of the transportome in cancer chemosensitivity and chemoresistance. *Cancer Research*, **64** (12), 4294–4301.

123 Molina-Arcas, M., Moreno-Bueno, G., Cano-Soldado, P., Hernandez-Vargas, H., Casado, F.J., Palacios, J., and Pastor-Anglada, M. (2006) Human equilibrative nucleoside transporter-1 (hENT1) is required for the transcriptomic response of the nucleoside-derived drug 5′-DFUR in breast cancer MCF7 cells. *Biochemical Pharmacology*, **72** (12), 1646–1656.

124 Wiley, J.S., Jones, S.P., Sawyer, W.H., and Paterson, A.R. (1982) Cytosine arabinoside influx and nucleoside transport sites in acute leukaemia. *The Journal of Clinical Investigation*, **69** (2), 479–489.

125 Gati, W.P., Paterson, A.R., Larratt, L.M., Turner, A.R., and Belch, A.R. (1997) Sensitivity of acute leukemia cells to cytarabine is a correlate of cellular es nucleoside transporter site content measured by flow cytometry with SAENTA-fluorescein. *Blood*, **90** (1), 346–353.

126 Wright, A.M., Paterson, A.R., Sowa, B., Akabutu, J.J., Grundy, P.E., and Gati, W.P. (2002) Cytotoxicity of 2-chlorodeoxyadenosine and arabinosylcytosine in leukaemic lymphoblasts from paediatric patients: significance of cellular nucleoside transporter content. *British Journal of Haematology*, **116** (3), 528–537.

127 Galmarini, C.M., Thomas, X., Calvo, F., Rousselot, P., Jafaari, A.E., Cros, E., and Dumontet, C. (2002) Potential mechanisms of resistance to cytarabine in AML patients. *Leukemia Research*, **26** (7), 621–629.

128 Galmarini, C.M., Thomas, X., Calvo, F., Rousselot, P., Rabilloud, M., El Jaffari, A., Cros, E., and Dumontet, C. (2002) *AT in vivo* mechanisms of resistance to cytarabine in acute myeloid leukaemia. *British Journal of Haematology*, **117** (4), 860–868.

129 Stam, R.W., den Boer, M.L., Meijerink, J.P., Ebus, M.E., Peters, G.J., Noordhuis, P., Janka-Schaub, G.E., Armstrong, S.A., Korsmeyer, S.J., and Pieters, R. (2003) Differential mRNA expression of Ara-C-metabolizing enzymes explains Ara-C sensitivity in MLL gene-rearranged infant acute lymphoblastic leukaemia. *Blood*, **101** (4), 1270–1276.

130 Hubeek, I., Stam, R.W., Peters, G.J., Broekhuizen, R., Meijerink, J.P., van Wering, E.R., Gibson, B.E., Creutzig, U., Zwaan, C.M., Cloos, J., Kuik, D.J.,

Pieters, R., and Kaspers, G.J. (2005) The human equilibrative nucleoside transporter 1 mediates *in vitro* cytarabine sensitivity in childhood acute myeloid leukaemia. *British Journal of Cancer*, **93** (12), 1388–1394.

131 Molina-Arcas, M., Marce, S., Villamor, N., Huber-Ruano, I., Casado, F.J., Bellosillo, B., Montserrat, E., Gil, J., Colomer, D., and Pastor-Anglada, M. (2005) Equilibrative nucleoside transporter-2 (hENT2) protein expression correlates with *ex vivo* sensitivity to fludarabine in chronic lymphocytic leukemia (CLL) cells. *Leukemia*, **19** (1), 64–68.

132 Marce, S., Molina-Arcas, M., Villamor, N., Casado, F.J., Campo, E., Pastor-Anglada, M., and Colomer, D. (2006) Expression of human equilibrative nucleoside transporter 1 (hENT1) and its correlation with gemcitabine uptake and cytotoxicity in mantle cell lymphoma. *Haematologica*, **91** (7), 895–902.

133 Spratlin, J., Sangha, R., Glubrecht, D., Dabbagh, L., Young, J.D., Dumontet, C., Cass, C., Lai, R., and Mackey, J.R. (2004) The absence of human equilibrative nucleoside transporter 1 is associated with reduced survival in patients with gemcitabine-treated pancreas adenocarcinoma. *Clinical Cancer Research*, **10** (20), 6956–6961.

134 Giovannetti, E., Del Tacca, M., Mey, V., Funel, N., Nannizzi, S., Ricci, S., Orlandini, C., Boggi, U., Campani, D., Del Chiaro, M., Iannopollo, M., Bevilacqua, G., Mosca, F., and Danesi, R. (2006) Transcription analysis of human equilibrative nucleoside transporter-1 predicts survival in pancreas cancer patients treated with gemcitabine. *Cancer Research*, **66** (7), 3928–3935.

135 Reiman, T., Clarke, M.L., Dabbagh, L., Vsianska, M., Coupland, R.W., Belch, A.R., Baldwin, S.A., Young, J.D., Cass, C.E., and Mackey, J.R. (2002) Differential expression of human equilibrative nucleoside transporter 1 (hENT1) protein in the Reed–Sternberg cells of Hodgkin's disease. *Leukemia & Lymphoma*, **43** (7), 1435–1440.

136 Mackey, J.R., Galmarini, C.M., Graham, K.A., Joy, A.A., Delmer, A., Dabbagh, L., Glubrecht, D., Jewell, L.D., Lai, R., Lang, T., Hanson, J., Young, J.D., Merle-Beral, H., Binet, J.L., Cass, C.E., and Dumontet, C. (2005) Quantitative analysis of nucleoside transporter and metabolism gene expression in chronic lymphocytic leukemia (CLL): identification of fludarabine-sensitive and -insensitive populations. *Blood*, **105** (2), 767–774.

137 Gloeckner-Hofmann, K., Guillen-Gomez, E., Schmidtgen, C., Porstmann, R., Ziegler, R., Stoss, O., Casado, F.J., Ruschoff, J., and Pastor-Anglada, M. (2006) Expression of the high-affinity fluoropyrimidine-preferring nucleoside transporter hCNT1 correlates with decreased disease-free survival in breast cancer. *Oncology*, **70** (3), 238–244.

138 Gati, W.P., Paterson, A.R., Belch, A.R., Chlumecky, V., Larratt, L.M., Mant, M.J., and Turner, A.R. (1998) Es nucleoside transporter content of acute leukemia cells: role in cell sensitivity to cytarabine (araC). *Leukemia & Lymphoma*, **32** (1–2), 45–54.

139 Mackey, J.R., Jennings, L.L., Clarke, M.L., Santos, C.L., Dabbagh, L., Vsianska, M., Koski, S.L., Coupland, R.W., Baldwin, S.A., Young, J.D., and Cass, C.E. (2002) Immunohistochemical variation of human equilibrative nucleoside transporter 1 protein in primary breast cancers. *Clinical Cancer Research*, **8** (1), 110–116.

3
Organic Anion Transporting Polypeptides (Oatps/OATPs)
Mine Yarim and Meric Koksal

Abbreviations

APD-ajmalinium	N-(4,4-azo-n-pentyl)-21-deoxyajmalinium
AUC	area under curve
BSP	bromosulfophthalein
BSP-SG	glutathione-conjugated BSP
DNP-SG	dinitrophenyl-glutathione
CCK-8	cholecystokin-8
DHEAS	dehydroepiandrosterone sulfate
DPDPE	[D-penicillamine 2,5]-enkephalin
E-3-S	estrone-3-sulfate
$E_2 17\beta G$	estradiol-17β-glucuronide
Fluo-3	1-[2-amino-5-(2,7-dichloro-6-hydroxy-3-oxo-3H-xanthen-9-yl]-2-(2′-amino-5′-methylphenoxy)ethane-N,N,N,N'-tetraacetic acid tetraammonium salt
GCDCA	glycochenodeoxycholate
Gd-B 20790	gadolinium derivative
GUDCA	glycoursodeoxycholate
HMG-CoA	3-hydroxy-3-methylglutaryl-coenzyme A
LTC4	leukotriene C4
LTE4	leukotriene E4
Mrp	multidrug resistance associated protein
PGD_2	prostaglandin D_2
PGE_1	prostaglandin E_1
PGE_2	prostaglandin E_2
$PGF_{1\alpha}$	prostaglandin $F_{1\alpha}$
$PGF_{2\alpha}$	prostaglandin $F_{2\alpha}$
P-gp	P-glycoprotein
PXR	pregnane X receptor
QSAR	quantitave structure–activity relationship

Transporters as Drug Carriers: Structure, Function, Substrates. Edited by Gerhard Ecker and Peter Chiba
Copyright © 2009 WILEY-VCH Verlag GmbH & Co. KGaA, Weinheim
ISBN: 978-3-527-31661-8

Ro 48-5033	4-(2-hydroxy-1,1-dimethylethyl-N-[6-(2-hydroxyethoxy)-5-(2-methoxyphenoxy)-[2,2']bipyrimidinyl-4-yl]-benzenesulfonamide
SN-38	7-ethyl-10-hydroxycamptothecin
T_3	3,5,3'-triiodo-L-thyronine
rT_3	3,3',5'-triiodo-L-thyronine
T_4	thyroxine
TCDCA	taurochenodeoxycholate
TUDCA	tauroursodeoxycholate
TXA_2	thromboxane A_2
TXB_2	thromboxane B_2

3.1
Introduction

The organic anion transporting polypeptides (rodents: Oatps; human: OATPs) represent a family of proteins responsible for the membrane transport of a large number of endogenous and xenobiotic compounds with diverse chemical characteristics. Since the first expression of Oatp1a1 in 1994, organic anion transporting polypeptide family members have been isolated from a variety of tissues in vertebrate animal species [1].

Oatps/OATPs are multispecific sodium-independent transport proteins. They comprise at least 36 members of human, rat, mouse, and some nonmammalian species. On the basis of their phylogenetic relationships, all rodent and human Oatps/OATPs so far identified within the OATP/SLCO superfamily of solute carriers have been classified into 6 families and 13 subfamilies (Figure 3.1) [2].

3.2
Nomenclature and Classification

In the beginning, the Oatps/OATPs were given gene symbol of solute carrier family 21 (SLC) by HUGO Gene Nomenclature Committee, but the naming of each transporter was according to the group that isolated it. Since the traditional SLC21 gene classification does not permit an unequivocal and species-independent identification of genes and gene productions, all Oatps/OATPs were later classified into OATP/SLCO superfamily (Table 3.1).

Recently, the HUGO Gene Nomenclature Committee adopted a new nomenclature, the SLCO (character "O" is from the head letter of OATP). It is subdivided into families ($\geq$40% amino acid sequence identity), subfamilies ($\geq$60% amino acid sequence identity), and individual genes and gene products according to their phylogenetic relationships and chronology of identification [2].

The phylogenetic relationship was proposed as the basis of a new classification system that provides an unambiguous and species-independent nomenclature for all

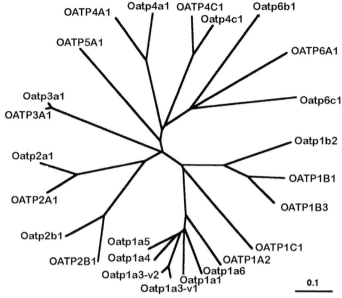

Figure 3.1 Phylogenetic tree of the Oatps/OATPs family. Multiple alignments of amino acid sequences and phylogenetic tree construction were carried out using CLUSTAL (http://www.ddbj.nig.ac.jp/Welcome-j.html).

members of the Oatps/OATP superfamily. For a comprehensive understanding of Oatps/OATPs, such a nomenclature may be in part useful in promoting this field (Table 3.1 and Figure 3.1) [3, 4]. Throughout this chapter, the novel protein names have been used in accordance with the new nomenclature system.

3.3
Tissue Distribution, Structure, and Functions

Although some important members of this transporter family are selectively expressed in rodent and human livers, where they are involved in the hepatic clearance of albumin-bound compounds from portal blood plasma [5], most Oatps/OATPs are expressed in multiple tissues, including the blood–brain barrier, choroid plexus, lung, heart, intestine, kidney, placenta, and testis [6]. Tissue distribution of Oatps/OATPs has been studied using different techniques. Consistent with their potential role in detoxification process, Oatps/OATPs are expressed in various tissues as demonstrated, for example, by RT-PCR techniques [7] in normal rat and human tissues as well as in human cancer cell lines [6]. Some transporters show a more restricted tissue expression pattern (e.g., Oatp1a1, Oatp1a5, Oatp1b2, OATP1A2, OATP1B1, OATP1B3, and OATP1C1), while others can be detected in almost every tissue that has been investigated (e.g., Oatp2b1, OATP2B1, OATP3A1,

Table 3.1 Oatp/OATP gene classification as implemented by the human and mouse gene nomenclature committees.

Novel protein name	Novel gene symbol	Former protein name	Former gene symbol
Rat Oatp			
Oatp1a1	Slco1a1	Oatp1	Slc21a1
Oatp1a3-v1, Oatp1a3-v2	Slco1a3	OAT-K1, OAT-K2	Slc21a4
Oatp1a4	Slco1a4	oatp2	Slc21a5
Oatp1a5	Slco1a5	oatp3	Slc21a7
Oatp1a6	Slco1a6	oatp5	Slc21a13
Oatp1b2	Slco1b2	oatp4/rlst-1	Slc21a10
Oatp1c1	Slco1c1	oatp14	Slc21a14
Oatp2a1	Slco2a1	rPGT	Slc21a2
Oatp2b1	Slco2b1	Oatp-9/moat1/oatp-B	Slc21a10
Oatp3a1	Slco3a1	oatp-D	Slc21a11
Oatp4a1	Slco4a1	oatp-E	Slc21a12
Oatp4c1	Slco4c1	Oatp-R	
Oatp6b1	Slco6b1	rGST-1/oatp16	Slc21a16
Oatp6c1	Slco6c1	rGST-2/oatp18	Slc21a18
Human OATP			
OATP1A2	SLCO1A2	OATP/OATP-A	SLC21A3
OATP1B1	SLCO1B1	OATP-C/LST-1/OATP2	SLC21A6
OATP1B3	SLCO1B3	OATP8/LST-2	SLC21A8
OATP1C1	SLCO1C1	OATP-F	SLC21A14
OATP2A1	SLCO2A1	hPGT	SLC21A2
OATP2B1	SLCO2B1	OATP-B/mOATP	SLC21A9
OATP3A1	SLCO3A1	OATP-D/PGT-2	SLC21A11
OATP4A1	SLCO4A1	OATP-E	SLC21A12
OATP4C1	SLCO4C1	OATP-R	SLC21A20
OATP5A1	SLCO5A1	OATP-J/OATP-RP4	SLC21A15
OATP6A1	SLCO6A1	GST/OATP-I	SLC21A19

and OATP4A1). This indicates that some Oatps/OATPs have organ-specific functions, while others might be involved in more housekeeping functions [1].

Oatps/OATPs are key membrane transporters for which crystal structures are not available. According to the hydropathy analysis, all Oatps/OATPs contain 12 transmembrane domains with both the amino and the carboxy terminal parts located intracellulary (Figure 3.2). However, the predicted 12-transmembrane domain model for any Oatp/OATP has not been proven experimentally [8].

Only a few of the Oatps/OATPs identified so far have been characterized in detail on the functional, structural, and genomic levels. Despite the fact that a larger number of endogenous compounds are known to be transported by the Oatps/OATPs, little is known about the *in vivo* physiological importance of these transporters. The exact transport mechanism(s) of the OATPs has not yet been worked out. However, studies with rat Oatps suggest that they act as organic anion exchangers [1].

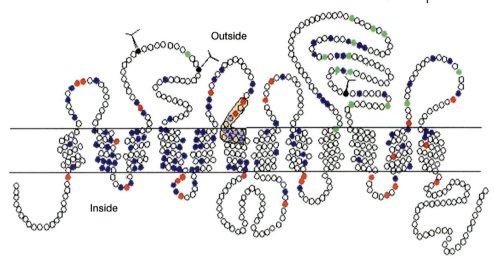

Figure 3.2 Predicted 12-transmembrane domain model of rat Oatp1a1. Conserved amino acids are indicated in blue. Conserved and charged amino acids (D, E, K, R) are given in red, and conserved cysteines (C) are marked with green. Three potential N-glycosylation sites (Y) are present on extracellular protein loops. The OATP superfamily signature is indicated at the border of the extracellular loop 3 and the transmembrane domain 6.

3.4
Substrate Spectrum

The Oatp/OATP family is expressed in various organs and its substrates comprise a broad spectrum ranging from endogenous compounds to xenobiotics. The endogenous compounds comprise bile acids and their salts, thyroid hormones (T_3, T_4), prostanoids (PGE_1, PGE_2, LT_4, and TXA_2), steroid hormones and their conjugates (DHEAS, $E_2 17\beta G$, and E-3-S), cAMP, and so on. The xenobiotics include drugs such as cardiac glycoside (digoxin), HMG-CoA reductase inhibitor (pravastatin), anticancer drug (methotrexate), angiotensin-converting enzyme inhibitors (enalapril and temocaprilat), antibiotic (benzylpenicillin), linear and cyclic oligopeptides such as endothelin receptor antagonist BQ-123, antihistamine fexofenadine, trombin inhibitor CRC220, opioid receptor agonist (DPDPE and deltrophin II), enterohepatic hormone CCK-8, some nonsteroidal anti-inflammatory drugs, and even organic cations. The hepatic Oatps/OATPs have also been shown to be responsible for the uptake of hepatotoxic exogenous cyclic peptides phalloidin, α-amanitin, and microcystin into rat and/or human livers (Table 3.2).

In general, Oatp/OATP substrates are mainly anionic amphipathic molecules with high molecular weight (>450) that under normal physiological conditions are bound to proteins (mostly albumin). More specifically, compounds with a steroid nucleus (e.g., bile salts, steroid hormones, and their conjugates) or small linear and cyclic peptides are likely candidates to be transported by certain Oatps/OATPs. These are

Table 3.2 Molecular characteristics of the members of the OATP superfamily.

Transporter	Size (amino acids)	Substrates (K_m value)	Main location	References
Rat Oatps				
Oatp1a1	670	*Bile salts*: cholate (54 µM), glycocholate (54 µM), taurocholate (19–50 µM), TCDCA (7 µM), TUDCA (13 µM), sulfotaurolithocholate (6 µM)	mRNA: liver, kidney, brain, lung, retina, skeletal muscle, proximal colon	[8, 13–26, 29, 48]
		Hormones and their conjugates: aldosterone (15 nM), cortisol (13 µM), DHEAS (5 µM), $E_217\beta G$ (3–20 µM), E-3-S (5–12 µM), T_3, rT_3, T_4	Protein: liver, kidney, choroid plexus (?)	
		Eicosanoids: LTC4 (270 nM)		
		Peptides: BQ-123 (600 µM), CRC220 (30–57 (M), deltorphin II (137 µM), DPDPE (48 µM), GSH		
		Drugs: dexamethasone, enalapril (214 µM), fexofenadine (32 µM), gadoxetate (3.3 mM), ouabain (1.7–3 mM), pravastatin (30 µM), temocaprilat (47 µM), bosentan		
		Other organic anions: monoglucuronosyl bilirubin, BSP (1–3 µM), BSP-DNP-SG (408 µM), E3040 glucuronide		
		Organic cations: APD-ajmalinium, N-methylquinidine, rocuronium		
		Toxins: ochratoxin A (17–29 µM)		
Oatp1a3-v1/v2	669/498	*Bile salts*: taurocholate (10/31 µM)	Protein: kidney	[32, 33, 35, 36]
		Hormones and their conjugates: DHEAS (8/8 µM), $E_217\beta G$ (35/45 µM), E-3-S (12/15 µM), T_3 (44/25 µM), T_4 (20/12 µM)		

		Substrates	Tissue distribution	References
Oatp1a4	661	*Eicosanoids:* PGE$_2$ *Drugs:* methotrexate (1–2 µM), zidovudine (67/76 µM) *Other organic anions:* folate *Toxins:* ochratoxin A (6/17 µM) *Bile salts:* cholate (46 µV), glycocholate (40 µM), taurocholate (35 µM), TCDCA (12 µM), TUDCA (17 µM) *Hormones and their conjugates:* DHEAS (17 µM), E$_2$17βG (3 µM), E-3-S (11 µM), T$_3$ (6 µM), T$_4$ (7 µM) *Peptides:* BQ-123 (30 µM), DPDPE (19 µM), Leu-enkephalin *Drugs:* biotin, digoxin (240 µM), fexofenadine (6 µM), ouabain (470 µM), pravastatin (38 µM), bosentan *Organic cations:* APD-ajmalinium, rocuronium	mRNA: liver, kidney, brain, retina Protein: liver, BBB, choroid plexus, retina	[4, 14, 21, 24–26, 28, 37, 38, 41, 42]
Oatp1a5	670	*Bile salts:* cholate (3 µM), glycocholate (15 µM), glycodeoxycholate (4 µM), GCDCA (6 µM), GUDCA (5 µM) taurocholate (18–30 µM) taurodeoxycholate (6 µM), TCDCA (7 µM), TUDCA (7 µM) *Hormones and their conjugates:* DHEAS (162 µM), E$_2$17βG (39 µM), E-3-S (268 µM), T$_3$ (7 µM), T$_4$ (5 µM) *Eicosanoids:* LTC4, PGE$_2$ (35 µM) *Peptides:* BQ-123 (417 µM), DPDPE (137 µM)	mRNA: retina, brain, kidney, liver, small intestine Protein: jejunum, choroid plexus (?)	[26, 41, 44, 46, 47]

(Continued)

Table 3.2 (Continued)

Transporter	Size (amino acids)	Substrates (K_m value)	Main location	References
Oatp1a6	670	Drugs: digoxin (240 μM), fexofenadine, ouabain (1.6 mM) Organic cations: rocuronium	mRNA: kidney	[48]
Oatp1b2	687	Bile salts: taurocholate (27 μM) Hormones and their conjugates: DHEAS (5 μM), $E_2 17\beta G$ (32 μM), E-3-S (37 μM), T_3, T_4 Eicosanoids: LTC4 (7 μM), PGE_2 (13 μM) Drugs: bosentan Other organic anions: BSP (1 μM) Toxins: microcystin, phalloidin (5.7 μM)	Protein: liver, eye	[1, 28, 46, 50–54]
Oatp1c1	716	Hormones and their conjugates: rT_3 (0.34 μM), T_4 (0.18 μM), $E_2 17\beta G$ (10 μM) Drugs: cerivastatin (1.3 μM), troglitazone sulfate (0.76 μM)	mRNA: brain, liver, kidney.	[55–57]
Oatp2a1	643	Eicosanoids: 6-keto-$PGF_{1\alpha}$ (6 μM), PGD_2, PGE_1 (70 nM), PGE_2 (94 nM), $PGF_{2\alpha}$ (104 nM), TXB_2 (423 nM)	Ubiquitous	[60, 61]
Oatp2b1	682	Bile salts: taurocholate (18 μM) Eicosanoids: LTC4 (3 μM), PGD_2 (36 nM), PGE_1, PGE_2, TXB_2 Drugs: iloprost	mRNA: liver, lung, heart, brain, retina, kidney	[59]

Oatp3a1	710	*Eicosanoids*: PGE$_1$, PGE$_2$, PGF$_{2\alpha}$	mRNA: brain, heart, testis	[62, 63]
Oatp4a1	722	*Bile salts*: taurocholate *Hormones*: T$_3$ *Eicosanoids*: PGE$_2$	mRNA: retina	[64]
Oatp4c1	724	*Hormones and their conjugates*: T$_3$ (1.9 μM) *Drugs*: digoxin (1.9 μM)	mRNA: kidney, lung	[65]
Oatp6b1	748	*Bile salts*: taurocholic acid (8.9 μM) *Hormones and their conjugates*: DHEAS (25.5 μM), T$_3$, T$_4$ (6.4 μM)	mRNA: testis	[51]
Oatp6c1	702	*Bile salts*: taurocholic acid (2.5 μM) *Hormones and their conjugates*: DHEAS (21 μM), T$_3$, T$_4$ (5.8 μM)	mRNA: testis	[51]
PGT2		*Hormones*: T$_3$, T$_4$	Ubiquitous	[26]
TST-1, TST-2		*Hormones*: T$_3$, T$_4$	mRNA: testis	[26]
Human OATPs				
OATP1A2	670	*Bile salts*: cholate (93 μM), glycocholate, taurocholate (60 μM), TCDCA, TUDCA (19 μM) *Hormones and their conjugates*: DHEAS (7 μM), E$_2$17βG, E-3-S (59 μM), T$_3$ (7 μM), rT$_3$, T$_4$ (8 μM) *Eicosanoids*: PGE$_2$ *Peptides*: BQ-123, CRC220, deltorphin II (330 μM), DPDPE (202 μM)	mRNA: brain, kidney, liver, lung, testis Protein: brain, liver	[16, 21, 24, 25, 52, 64, 68–70, 74–76, 78–80]

(*Continued*)

Table 3.2 (Continued)

Transporter	Size (amino acids)	Substrates (K_m value)	Main location	References
		Drugs: chlorambuciltaurocholate, fexofenadine (6 μM), Gd-B 20790, ouabain (5.5 mM), rosuvastatin		
		Other organic anions: BSP (20 μM)		
		Organic cations: APD-ajmalinium, N-methylquinine, N-methylquinidine (5 μM), rocuronium		
		Toxins: microcystin (20 ± 8 μM)		
OATP1B1	691	*Bile salts:* cholate (11 μM), glycocholate, taurocholate (10–34 μM)	Protein: liver	[1, 6, 13, 28, 52, 54, 69, 71, 79, 80, 82, 83, 85–93]
		Hormones and their conjugates: DHEAS (22 μM), $E_2 17βG$ (8–10 μM), E-3-S (13 μM), T_3 (3 μM), T_4 (3 μM)		
		Eicosanoids: LTC4, LTE4, PGE_2, TXB_2 *Peptides:* BQ-123, DPDPE		
		Drugs: benzylpenicillin, methotrexate, pravastatin (14–35 μM), fluvastatin (1–3 μM/l), simvastatin, atorvastatin, rosuvastatin, rifampicin (13 μM), troglitazone, bosentan, Ro 48-5033, SN-38		
		Other organic anions: bilirubin, monoglucuronosyl bilirubin (100 nM), bisglucoronosyl bilirubin (300 nM), BSP (100–300 nM)		
		Toxins: microcystin (7 ± 3 μM), phalloidin (17 μM)		

OATP1B3	702	*Bile salts*: glycocholate, taurocholate (6 μM)	Protein: liver	[1, 28, 52–54, 69, 78–80, 82, 85, 86, 88, 92, 93, 97–99]
		Hormones and their conjugates: DHEAS, $E_2 17\beta G$ (5 μM), E-3-S, T_3 (6 μM), rT_3, T_4		
		Eicosanoids: LTC4		
		Peptides: BQ-123, CCK-8 (11 μM), deltorphin II, DPDPE		
		Drugs: digoxin, methotrexate (25 μM), ouabain, fluvastatin, rcsuvastatin, rifampicin (2 μM), bosentan, Ro 48-5033, Fluo-3 (6.8 μM)		
		Other organic anions: monoglucuronosyl bilirubin (500 nM), BSP (0.4–3 μM)		
		Toxins: microcystin (9 ± 3 μM), phalloidin (7.5 μM), amanitin (3.7 μM)		
OATP1C1	712	*Hormones and their conjugates*: $E_2 17\beta G$, E-3-S, T_3 (128 nM), rT_3 (130 nM), T_4 (90 nM)	Protein: brain, testis	[78, 100]
		Other organic anions: BSP		
OATP2A1	643	*Eicosanoids*: PGD_2, PGE_1, PGE_2, $PGE_{2\alpha}$, 8-iso-$PGE_{2\alpha}$, TXB_2	Ubiquitous	[61, 78, 101]
OATP2B1	709	*Hormones and their conjugates*: E-3-S (1.56–6 μM), DHEAS, $E_2 17\beta G$.	mRNA: liver, placenta, spleen, lung, kidney, heart, ovary, small intestine, brain	[6, 68, 69, 78, 80, 87, 88, 92, 102, 104, 105]
		Eicosanoids: PGE_2	Protein: liver, placenta	

(Continued)

Table 3.2 (Continued)

Transporter	Size (amino acids)	Substrates (K_m value)	Main location	References
OATP3A1_v1, OATP3A1_v2	710, 666	*Drugs:* benzylpenicillin, digoxin, E-3-S, pravastatin, rosuvastatin fexofenadine, fluvastatin (1–3 μM/l), enalapril, temocaprilat, gadoxetate *Other organic anions:* BSP (0.7 μM) *Hormones and their conjugates:* T_4, E-3-S *Eicosanoids:* PGE_1 (101 ± 52 nmol/l, _v1; 219 ± 137 nmol/l, _v2) (48.5 nM), PGE_2 (218 ± 266 nmol/l, _v1; 371 ± 155 nmol/l, _v2) (55.5) nM), $PGF_{2\alpha}$ *Peptides:* BQ-123, deltorphin II, vasopressin *Drugs:* benzylpenicillin	Ubiquitous	[6, 62, 63, 78]
OATP4A1	722	*Bile salts:* Taurocholate (15 μM) *Hormones and their conjugates:* E-3-S, $E_2 17\beta G$, T_3 (1–6.5 μM), rT_3; T_4 (8.0 μM) *Eicosanoids:* PGE_2	Ubiquitous	[6, 64, 78]
OATP4C1	724	*Hormones and their conjugates:* T_3 (5.9 μM), T_4 *Drugs:* digoxin (7.8 μM), ouabain (0.38 μM), methotrexate	mRNA: kidney, fetal liver	[65]
OATP5A1				[106]
OATP6A1	719		mRNA: testis	[51]

also the attributes of compounds that are mainly excreted into bile, while products that are normally excreted into urine are represented by small and mainly hydrophilic compounds with low protein binding and known substrates of organic anion (OATs) and organic cation transporters (OCTs) [1, 9, 10].

3.5 Members of the Rodent Oatp Family

3.5.1 Oatp1a1

The first member of this carrier family, Oatp1a1, has been cloned from rat liver [8]. Oatp1a1 is an 80 kDa protein that is expressed at the basolateral membrane of hepatocytes as well as at the apical membranes of the renal proximal tubular cells and choroid plexus epithelial cells [11–14].

Oatp1a1 has broad substrate specificity, and it mediates sodium-independent uptake of conjugated and unconjugated bile salts, BSP, steroid hormones and their conjugates, thyroid hormones, certain oligopeptides, bulky organic cations, various drugs such as pravastatin and enalapril, organic cations such as APD-ajmalinium, and to a lesser degree N-methylquinine and rocuronium (Table 3.2) [15–28].

It has been shown that Oatp1a1 can mediate sinusoidal efflux of BSP, and studies on driving force of Oatp1a1 suggested the exchange of solutes taken up with intracellular anions such as HCO_3^- or glutathione [29–31].

3.5.2 Oatp1a3-v1/v2

Oatp1a3-v1 and v2 are, respectively, 72 and 65% identical to Oatp1a1 and also exhibit a kidney-specific expression [32, 33]. Oatp1a3-v2 is the short isoform of Oatp1a3-v1 and misses the first four transmembrane domains [33]. The mRNA expression of both isoforms is restricted to kidney, and the Oatp1a3-v1protein was detected in brush-border membranes [32–34]. Oatp1a3-v1 and Oatp1a3-v2 were shown to transport taurocholate, conjugated steroid hormones, thyroid hormones, ochratoxin A, methotrexate, and zidovudine (Table 3.2) [32, 35, 36].

3.5.3 Oatp1a4

Oatp1a4 was cloned initially from rat brain. It consists of 661 amino acids and is 77% identical to Oatp1a1 [37]. The protein was expressed at the basolateral membranes of hepatocytes and choroid plexus epithelial cells, at the abluminal and luminal domains of brain capillary endothelial cells, and at the apical microvilli of the retinal pigment epithelium [14, 38–40]. In addition, Oatp1a4 mRNA was detected in various regions of the brain and in retina [41].

The substrate specificity overlaps with Oatp1a1, but Oatp1a4 is a less efficient anion carrier as it does not transport BSP or leukotriene C4 (Table 3.2) [4, 21, 24–26]. The unique feature of Oatp1a4 is the high-affinity uptake of the cardiac glycoside digoxin [28, 37, 42]. Oatp1a4 can operate bidirectionally, but in contrast to Oatp1a1, a coupling of substrate uptake to glutathione efflux could not be demonstrated [43].

3.5.4
Oatp1a5

Oatp1a5 was cloned from rat retina and intestine [41, 44]. Consequently, the expression of Oatp1a5 protein could not be demonstrated unequivocally in liver or kidney. In addition, Oatp1a5 rather than Oatp1a1 is expressed at the apical plasma membrane of rat choroid plexus epithelial cells [45]. As Oatp1a1 is not expressed in the intestine, expression of Oatp1a5 could clearly be shown at the apical membranes of small intestinal epithelial cells [44]. The 670-amino acid protein shares 80% amino acid identity with Oatp1a1. Oatp1a5 can transport bile salts, steroid hormone conjugates, thyroid hormones, cardiac glycosides, and oligopeptides (Table 3.2) [26, 46, 47].

3.5.5
Oatp1a6

The tissue distribution of rat Oatp1a6 mRNA has not been investigated yet. It is indicated that Oatp1a6 is relatively kidney specific similar to the mouse orthologue. Oatp1a6 contains 670 amino acids. In contrast to mouse Oatp1a6 that has 10 transmembrane domains, rat Oatp1a6 contains 8 putative transmembrane domains. Mouse Oatp1a6 bears the highest sequence identity with rat orthologue (87.3% at the nucleotide level and 80.4% at the amino acid level) [48].

3.5.6
Oatp1b2

Oatp1b2 was isolated from rat liver in two isoforms. The first isoform was initially named rat liver-specific organic anion transporter (rlst-1) that, in comparison to full-length Oatp1b2, lacks 35 amino acids in the putative transmembrane domain IX [49, 50]. Consequently, rlst-1 encodes 652 amino acids and only 11 transmembrane spanning domains were predicted with an inverted topology of the C-terminal domain. With an amino acid identity of about 43%, the full-length Oatp1b2 and Oatp1a1 belong to different subfamilies within the OATP superfamily (Figure 3.1) [50].

The full-length Oatp1b2 is a counterpart of human OATP1B1. The overall homology with OATP1B1 was 60.2%, which is the highest among all known organic anion transporters. The Oatp1b2 mRNA is exclusively expressed in the liver. Oatp1b2 protein, like Oatp1a1 and Oatp1a4, is expressed at the basolateral domain of hepatocytes [46]. The expression of Oatp1b2 mRNA is restricted to the liver, and it was demonstrated by RNase protection assay that Oatp1b2 is the predominant transcript in rat liver [7, 49, 50]. Oatp1b2 is one of the important transporters in rat liver for the clearance of bile acid. It preferably transports taurocholate (K_m, 9.45 µmol/l) in an Na$^+$-independent manner [51]. As observed until now, rlst-1 transports only taurocholate, whereas the substrate accommodation of full-length Oatp1b2 is much broader and includes BSP, conjugated steroid hormones, PGE$_2$, LTC4, and thyroid hormones (Table 3.2) [28, 52–54]. This suggests a role of the last four transmembrane domains of full-length Oatp1b2 in nonbile salt substrate translocation [1].

3.5.7
Oatp1c1

Oatp1c1 is expressed in the border of the brain capillary endothelial cells. Northern blot analysis revealed predominant expression of Oatp1c1 in the brain and Western blot analysis revealed its expression in the brain capillary and choroid plexus. Oatp1c1 transports thyroxine as well as amphipathic organic anions such as 17β-estradiol-D-17β-glucuronide, cerivastatin, and troglitazone sulfate. Oatp1c1 can mediate a bidirectional transport of T$_4$. BSP, taurocholate, and E-3-S were potent inhibitors for Oatp1c1 (Table 3.2) [55–57].

3.5.8
Oatp2a1

The specific prostaglandin carrier Oatp2a1 was cloned from rat liver [58]. Oatp2a1 is the rat orthologue of human OATP2A1 and exhibits 37 and 42% amino acid identity to Oatp1a1 and Oatp2b1, respectively [59, 60]. Oatp2a1 mRNA was found in tissues containing epithelia, such as lung, liver, kidney, brain, stomach, ileum, jejunum, and colon. The substrate specificity of Oatp2a1 includes a series of prostanoids, but no other organic anions such as taurocholate or E$_2$17βG (Table 3.2) [60, 61].

3.5.9
Oatp2b1

Oatp2b1, originally named multispecific organic anion transporter (moat1), was cloned from rat brain. The 682-amino acid protein, which is the rat orthologue of human OATP2B1, shows an amino acid identity of 31% to Oatp1a1. Oatp2b1 mRNA was detected in brain, retina, lung, heart, liver, and kidney. Additional Northern blot and *in situ* hybridization experiments showed a wide distribution of Oatp2b1 mRNA in neuronal cells of rat brain, mainly in hippocampus and cerebellum. Oatp2b1 mediates the transport of taurocholate and various prostaglandins and is more closely related to the prostaglandin carrier Oatp2a1 than to the other Oatps (Table 3.2) [59].

3.5.10
Oatp3a1

Oatp3a1 was cloned from rat brain. It is composed of 710 amino acids and shares 97.6% identity with human OATP3A1. The expression pattern of Oatp3a1 mRNA was abundant mainly in the heart, testis, brain, and some cancer cells. Further analysis showed that it is widely expressed in vascular, renal, and reproductive systems at the protein level. Oatp3a1 plays an important role in translocating prostaglandins such as E_1, E_2, and $F_{2\alpha}$ in specialized tissues and cells [62, 63].

3.5.11
Oatp4a1

Oatp4a1 was isolated from rat retina. It transports thyroid hormone in various peripheral tissues. Oatp4a1 is composed of 722 amino acids. The overall homology between rat Oatp4a1 and human OATP4A1 was 72.6% at the amino acid level, and the transmembrane domains and their surrounding areas were highly conserved [64].

3.5.12
Oatp4c1

Oatp4c1 was isolated from rat kidney. Immunohistochemical analysis reveals that Oatp4c1 protein is localized in the basolateral membrane of the proximal tubule cell in the kidney. It consists of 724 amino acids and has 12 transmembrane domains with similar topology to human OATP4C1. The overall homology between Oatp4c1 and OATP4C1 was 80.4% at the amino acid level. They also have moderate sequence homology to other OATP family (<35%). Oatp4c1 transports digoxin and triiodothyronine. Oatp4c1 might be a first step toward the transport pathway of digoxin and various compounds into urine in the kidney [65].

3.5.13
Oatp6b1/Oatp6c1

Oatp6b1 and Oatp6c1 (gonad-specific transporters) were expressed at high level in the testis, especially in Sertoli cells, spermatogonia, and Leydig cells. Oatp6b1 and Oatp6c1 consist of 748 and 702 amino acids, respectively. Rat Oatp6b1 and Oatp6c1 show the identity of 42% at the amino acid level. Both Oatp6b1 and Oatp6c1 transport taurocholic acid, DHEAS, T_3, and T_4. Oatp6b1 and Oatp6c1 might be one of the molecular entities responsible for transporting DHEAS and thyroid hormones involved in the regulation of sex steroid transportation and spermatogenesis in the gonad [51].

3.5.14
PGT-2

The prostaglandin transporter PGT-2 was expressed in various tissues. It has 97.6% homology with OATP3A1. Thyroid hormones are general substrates for PGT-2 [26].

3.5.15
TST-1 and TST-2

Testis-specific transporters TST-1 and TST-2 have been reported to transport thyroid hormones [26]. Further studies are required to clarify the functions of TST-1 and TST-2.

3.6
Members of Nonmammalian Oatp Family

OATP/SLCO superfamily members, some of which form distinct novel families, were also identified in chicken, zebra fish, frog, fruit fly, and worm species [66]. The draft sequence of the chicken genome has finally allowed the analysis of Oatp/Slco genes in birds. DNA sequences of the chicken Oatp family were isolated using database searches. In the Oatp family, expression of the four genes (cOatp1a1, cOatp1b1, cOatp1c1, and cOatp3a2) was observed in the choroid plexus. Studies with stably expressed functional proteins in Chinese hamster ovary (CHO) cells indicate that cOatp1c1 is a high-affinity thyroid hormone transporter that could be involved in the photoperiodic response of the gonads [67].

3.7
Members of Human OATP Family

3.7.1
OATP1A2

Following the isolation of Oatp1a1, the first human OATP was cloned from human liver that showed 67% amino acid identity to Oatp1a1 [68]. The carrier was later named OATP1A2, as it appeared to exhibit tissue distribution and substrate specificity unlike any of the related rat Oatps. OATP1A2 is composed of 670 amino acids with a predicted 12-membrane domain topology [69].

Despite the fact that OATP1A2 has been reported to be expressed in various tissues such as liver, brain, testis, lung, and kidney by Northern blot analysis [68] and in liver by immunochemical analysis [70], a restricted brain distribution has been suggested by others [71]. Indeed, OATP1A2 mRNA is widespread in the brain [68] and immunodetectable protein can be found in brain capillary endothelial cells, thus indicating a role of this transporter in regulating blood–brain barrier permeability of solutes [25]. Furthermore, OATP1A2 mRNA has been detected in biliary epithelial cells [72], but conformation of protein expression and function has not been determined [69]. Whether or not OATP1A2 is also localized in the intestine requires further investigation.

OATP1A2 is capable of transporting diverse compounds, including BSP, bile acids, steroid sulfates, bulky organic cations, fexofenadine, thyroid hormones, and opioid peptides. The highest uptake rate was observed for the organic cation N-methylquinine (Table 3.2) [21, 24, 25, 52, 64, 68, 69, 73–80].

3.7.2
OATP1B1

OATP1B1 was cloned from human liver and has often been referred to as liver-specific transporter 1 (LST-1) and it was the second human OATP to be cloned. It consists of 691 amino acids, and amino acid identities between 41 and 46% were determined for OATP1B1 and OATP1A2, respectively. Together with human OATP1B3 and rodent Oatp1b2, it belongs to the OATP1B subfamily (Figure 3.1) [13, 71, 81]. Immunohistochemical analysis demonstrated OATP1B1 expression on the basolateral membrane of hepatocytes [81]. OATP1B1 supports the membrane translocation of a broad range of compounds such as bile acids, sulfate, and glucuronide conjugates, thyroid hormones, peptides, and drugs such as pravastatin and methotrexate (Table 3.2) [1, 6, 13, 28, 52, 54, 69, 71, 79–92]. More important, bilirubin and its glucuronides are known physiological substrates for OATP1B1 [93]. Overall, OATP1B1 seems to prefer negatively charged substrates [69]. Given the liver-specific tissue distribution pattern and the capacity for transporting a multiplicity of chemical structures, it is likely that OATP1B1 plays an important role in the hepatocellular elimination of drugs [84]. Furthermore, several single-nucleotide polymorphisms and naturally occurring mutations of the OATP1B1 gene were described [84, 94, 95].

OATP1B1 modulates PXR function by influencing the intracellular rifampicin concentration, which is important for the degree of induction of drug-metabolizing enzymes and transporters by this drug [96].

3.7.3
OATP1B3

OATP1B3 was cloned from human liver [82, 97]. OATP1B3 consists of 702 amino acids and it is similar to OATP1B1 at both the amino acid level (80% amino acid identity) and the liver-specific tissue distribution [97]. OATP1B3 transports BSP, steroid hormone conjugates, and thyroid hormones and exhibits high uptake rates for the anionic cyclic peptides DPDPE and BQ-123 [1, 28, 52–54, 78–80, 85, 86, 88, 92, 93, 98, 99]. Bile salts were shown to be substrates for OATP1B3 in *Xenopus laevis* oocyte system but not in transfected HEK293 cells [69, 97]. As a unique feature among human OATPs, OATP1B3 mediates the transport of the uncharged cardiac glycoside digoxin, whereas the related but more hydrophilic ouabain is taken up by both OATP1A2 and OATP1B3. Thus, in contrast to OATP1B1, OATP1B3 also transports the uncharged compounds. Until now, neither OATP1B1 nor OATP1B3 was shown to mediate the transport of organic cations (Table 3.2) [27].

Importantly, OATP1B3 is highly expressed in certain gastric, colon, and pancreatic cancers, indicating that transporter expression may alter tumor sensitivity to methotrexate treatment [82].

3.7.4
OATP1C1

OATP1C1 was cloned from human brain. The 712-amino acid protein shares 48% amino acid identity with OATP1A2, and together with its rodent orthologue Oatp1c1, it forms the OATP1C subfamily. The protein could be identified in nests of Leydig cells in testis. The mRNA could be detected in numerous brain regions with the exceptions of pons and cerebellum, but the exact subcellular localization of OATP1C1 in human brain remains to be determined. The substrate specificity of OATP1C1 seems to be quite narrow compared to other OATPs, and protein displays a noticeably high affinity for thyroxin and thus could be important for thyroid hormone disposition in brain and testis (Table 3.2) [78, 100].

3.7.5
OATP2A1

Soon after the cloning of Oatp2a1, the human orthologue OATP2A1 was cloned from human kidney. It is composed of 643 amino acids. OATP2A1 exhibits 82% amino acid identity with Oatp2a1, whereas it shares only 32% amino acid identity with OATP1A2 [1, 27]. OATP2A1 mRNA was detected in human placenta, brain, lung, liver, pancreas, kidney, spleen, prostate, ovary, small intestine, and colon. In contrast to its rat orthologue, OATP2A1 mRNA is also expressed in heart and skeletal muscle and thus not restricted to tissue containing epithelia. OATP2A1 transports prostaglandins in similar rank order as Oatp2a1 (Table 3.2) [61, 101].

3.7.6
OATP2B1

Initially cloned from human brain, OATP2B1 is now known to be expressed in a variety of other tissues, including liver, lung, kidney, placenta, brain, heart, and small intestine, based on mRNA expression [6, 69]. In liver, OATP2B1 protein is found on the basolateral membrane of hepatocytes, suggesting that this transporter functions in an uptake capacity to remove solutes from portal circulation. OATP2B1 consists of 709 amino acids and shows an amino acid identity of only 34% with OATP1A2 and therefore belongs to another OATP family [69]. OATP2B1 mediates high-affinity uptake of BSP, also transports E-3-S and DHEAS, but does not transport bile salts. Thus, OATP2B1 has narrower substrate specificity compared to the members of the OATP1 family (Table 3.2) [27, 68, 78, 80, 88, 92, 102–105].

The importance of this transporter in the hepatic elimination of drugs is uncertain. However, given its broad tissue expression, OATP2B1 may play a role in drug distribution [6, 69]. A single-nucleotide polymorphism of the OATP2B1 gene (S486F) was reported that led to a considerable decrease in transport function *in vitro* [94]. However, the possible implications of this frequent polymorphism for drug disposition have not been determined yet [27].

3.7.7
OATP3A1

OATP3A1 was cloned from human kidney [6]. It has been mainly characterized as a prostaglandin transporter with a very broad tissue expression profile [6, 62]. A unique feature of human OATP3A1 is its 97% amino acid sequence identity with its rat and mouse orthologues [2, 62]. Recently, two variants of OATP3A1, OATP3A1_v1 (710 aa) and OATP3A1_v2 (666 aa), have been identified [63]. These two variants exhibit similar transport functions but distinct cellular and subcellular expressions in testis, choroid plexus, and human brain frontal cortex. OATP3A1 transports E-3-S, PGE_1, PGE_2, thyroxin, cyclic oligopeptides BQ-123, vasopressin, and benzylpenicillin [27, 63, 78] (Table 3.2).

3.7.8
OATP4A1

OATP4A1 is ubiquitously expressed in tissues [6, 64]. It consists of 722 amino acids. Some substrates transported by OATP4A1 include E-3-S, PGE_2, and taurocholate (Table 3.2) [6, 64, 78]. The capacity for T_3 and T_4 transport and the wide tissue distribution suggest that OATP4A1 is largely responsible for the peripheral uptake of thyroid hormones [64]. Further studies are required to assess whether OATP4A1 is an important determinant of drug distribution [106].

3.7.9
OATP4C1

OATP4C1 is the first member of OATP family predominantly expressed in the kidney and consists of 724 amino acids. It transports cardiac glycosides (digoxin and ouabain), thyroid hormones (T3 and T4), cAMP, and methotrexate. It might be a first step in the transport pathway of digoxin and various compounds into urine in the kidney (Table 3.2) [4].

3.7.10
OATP5A1

OATP5A1 has been identified, but little is known about its biochemical, physiological, and pharmacological characteristics. Further studies are required to determine the impact of this transporter on the drug distribution and elimination as well as the consequences of genetic polymorphism (Table 3.2) [106].

3.7.11
OATP6A1

The human gonad-specific transporter OATP6A1 is expressed at high level in the testis. It consists of 719 amino acids. OATP6A1 is relatively close to Oatp4a1/

OATP4A1 at the amino acid level. It is the first identified organic anion transporter in human testis. The substrates for the OATP6A1 have not been elucidated yet and further investigation is needed (Table 3.2) [51].

3.8
Drug Disposition and Drug–Drug Interactions

Drug disposition highly depends on the interplay between the drug metabolism and the transport in organs such as intestine, kidney, and liver. It is now increasingly recognized that genetically determined variation in drug transporter function or expression significantly determines the intersubject variability in drug response. Oatps/OATPs are important drug transporters that, together with P-gp and the Mrps, seem to play a critical role in the overall drug absorption and drug disposition. These family members are thought to be part of the overall body detoxification system and help to remove potentially toxic endobiotics and xenobiotics from the systemic circulation [107]. For example, pravastatin, an HMG-CoA reductase inhibitor, undergoes enterohepatic circulation, which prolongs the exposure of the liver to the drug and minimizes the adverse effects on the peripheral tissues. This enterohepatic circulation is mediated by transporters in every process, from pravastatin gastrointestinal absorption to biliary transport. Pravastatin is taken up in the liver from the portal vein by OATP family proteins located on sinusoidal membrane [13, 83, 108]. After exhibiting its pharmacological action in the liver, pravastatin is then excreted into bile via MRP2 with only a minimum degree of metabolic conversion [109]. The fraction of the drug released into the duodenum is then reabsorbed by active transport [110]. Thus, efficient hepatobiliary transport by OATP and MRP2 plays an important role in the enterohepatic circulation, which is responsible for maintaining significant concentrations of this drug in the liver. Although the mechanism governing the pharmacokinetic properties of this drug was identified after their development, attempts need to be made to include this information in the design of molecules during the drug discovery process [107].

The Oatps/OATPs share some substrate overlapping specificity with other promiscuous efflux transporters such as P-gp and Mrp2, indicating a degree of coordination. Oatps/OATPs have been implicated in drug–drug interactions, as exemplified by several interactions between cerivastatin and cyclosporin A, as well as cerivastatin, gemfibrozil, and its glucuronide metabolite [107]. The hepatic drug–drug interaction via OATP1B1 has been reported. In kidney transplant recipients treated with cyclosporin, the AUC of cerivastatin was 3.8-fold higher than in healthy volunteers who were not given cyclosporin. The mild-to-moderate reduction in renal function in kidney transplant recipients compared to healthy controls is unlikely to be responsible for the observed pharmacokinetic effects because the renal clearance of cerivastatin is negligible [111]. Shitara *et al.* have examined the effect of cyclosporin on the uptake of cerivastatin into human hepatocytes to investigate the mechanism of their drug–drug interaction [112]. As a result, cyclosporin was found to inhibit transporter-mediated cerivastatin uptake in human hepatocytes with K_i values of

0.28–0.69 µM. In addition, the uptake of cerivastatin was examined in OATP1B1-expressing MDCK II cells and cerivastatin was shown to be a substrate of OATP1B1, like pravastatin [13, 83]. OATP1B1-mediated uptake of cerivastatin was also inhibited by cyclosporin A with a K_i value of 0.2 µM in transfected cells. These results suggest that the drug–drug interaction between cerivastatin and cyclosporin A can be explained by inhibition of the transporter-mediated hepatic uptake of cerivastatin and, at least in part, its OATP1B1-mediated uptake [107].

In addition, an interesting report has described an interaction between fexofenadine and grapefruit, orange, and apple juices [47]. Fexofenadine is the substrate of OATPs [21]. Grapefruit, orange, and apple juices caused a marked inhibition of OATP-mediated fexofenadine uptake in cell lines expressing OATPs. It appears that OATP-mediated fexofenadine uptake was inhibited. Since the inhibition of OATPs in the liver would reduce the biliary secretion and increase plasma fexofenadine concentrations, it appears that the fexofenadine–juice interaction is primarily the result of reduced fexofenadine absorption from the gastrointestinal tract. Fruit juices are potent inhibitors of OATPs and they can reduce oral drug bioavailability [47].

Current information regarding the molecular and cellular aspects of Oatps/OATPs has grown steadily and encouraged studies of the mechanism of drug disposition. Clarification of the role Oatps/OATPs play in drug disposition *in vivo* is vital. The information on substrate selectivity and tissue distribution of the Oatps/OATPs will aid in the prediction of the *in vivo* kinetic profile of drugs from *in vitro* data. Research on the Oatps/OATPs will lead to the development of safer and more effective drugs.

3.9
Computational Approaches

In the absence of crystal structures for many of the membrane-bound proteins involved in interactions with xenobiotics and endobiotics such as enzyme and transporters, computational approaches have been extremely useful in gaining insight into the ligand–protein interactions. However, the quality and the consistency of data sets have been a determining factor in the overall predictive value of the QSAR models to date. It has been particularly challenging to assimilate and model data acquired across species and experimental cell systems. Most QSAR studies have focused on data sets gathered from one species, cell type, and frequently one laboratory setting. The application of QSAR models for Oatps/OATPs has been limited largely due to the absence of consistent data sets. A notable exception is the study by Yarim and coworkers on rat Oatp1a5, who used comparative molecular field analysis on 18 substrates [113, 114]. An improved understanding of the structural requirements of the Oatps/OATPs may explain the mechanism underlying the reported drug–drug interactions due to transporter inhibition. Because some studies have described multiple inhibitors of uptake with EC_{50} values, it may be possible to generate similar pharmacophores for inhibitors of the respective transporters. However, the difficulty in the interpretation of whether these molecules are interacting with the same site or with sites responsible for transport is a disadvantage compared to modeling substrate K_m data [114].

Chang et al. designed a new approach called *meta-pharmacophore* modeling to study OATPs using data derived from different cell systems and laboratories. Statistically robust meta-pharmacophores for rat Oatp1a1 and human OATP1B1 were generated using measured K_m values from three different cell lines. The predictive power of each model was validated with external test sets. Both models share key pharmacophore features, that is, a large hydrophobic area flanked by two "hydrogen bond acceptor" (HBA) features. This is consistent with the degree of correlation between substrate K_m values for these two transporters. These features could infer a hydrophobic binding pocket and two "hydrogen bond donor" (HBD) features in the transporter substrate recognition site that awaits experimental verification. The application of meta-pharmacophores is no longer limited to single spaces or cell type because of the broad range of their training data set [115].

Computational modeling is an effective approach to reveal structural details of membrane transporters and direct efficient experimental designs. However, the models highly depend on empirical data. Homology model generation requires the availability of an experimentally determined template structure; substrate-based models, such as 3D-QSAR, and pharmacophores are generated based on experimentally measured activities of a set of compounds. Increasingly, the combination of wet lab experiments and computational models is playing an important role in transporter studies. While the experimental data serve as a basis for model generation, the model itself may guide the design of more efficient follow-up experiments. New results from these experiments can be used to improve the *in silico* model that, in turn, will provide better guidance for the next round of experimental design. This represents a continuous interplay between computational and experimental approaches, and this hybrid approach may eventually lead to the discovery of safer and more efficient drugs by targeting key human transporters [114].

3.10
Conclusions

The Oatps/OATPs represent important drug transporters with complementary functions in drug excretion. Their expression in tissues such as liver, kidney, blood–brain barrier, small intestine, placenta, and testis suggests the important role they play in drug distribution. The physiological role of the organic anion transporter family still remains to be unclear, except in several organs. Current information regarding the molecular and cellular aspects of Oatps/OATPs has grown steadily and has led to more studies of the mechanism of drug disposition. Clarification of the role of each transporter in drug disposition *in vivo* is of major importance. To characterize the function, it would be necessary to find new specific inhibitors of the organic anion transporting polypeptide family, which will help explain the exact physiological role of the Oatp/OATP family *in vivo*. Finally, the information on substrate selectivity and tissue distribution of this transporter system will aid in the prediction of the *in vivo* kinetic profile of drugs from *in vitro* data.

References

1 Hagenbuch, B. and Meier, P.J. (2003) The superfamily of organic anion transporting polypeptides. *Biochimica et Biophysica Acta*, **1609**, 1–18.

2 Hagenbuch, B. and Meier, P.J. (2004) Organic anion transporting polypeptides of the OATP/SLC21 family: phylogenetic classification as OATP/SLC21 superfamily, new nomenclature and molecular/functional properties. *Pflügers Archives*, **447**, 653–665.

3 Wain, H.M., Bruford, E.A., Lovering, R.C., Lush, M.J., and Wright, M.W. (2002) Guidelines for human gene nomenclature. *Genomics*, **79**, 464–470.

4 Mikkaichi, T., Suzuki, T., Tanemoto, M., Ito, S., and Abe, T. (2004) The organic anion transporter (OATP) family. *Drug Metabolism and Pharmacokinetics*, **19**, 171–179.

5 Meier, P.J. and Stieger, B. (2002) Bile salt transporters. *Annual Review of Physiology*, **64**, 635–661.

6 Tamai, I., Nezu, J., Uchino, H., Sai, Y., Oku, A., Shimane, M., and Tsuji, A. (2000) Molecular identification and characterization of novel members of the human organic anion transporter (OATP) family. *Biochemical and Biophysical Research Communications*, **273**, 251–260.

7 Li, N., Hartley, D.P., Cherrington, N.J., and Klaassen, C.D. (2002) Tissue expression, ontogeny, and inducibility of rat organic anion transporting polypeptide 4. *The Journal of Pharmacology and Experimental Therapeutics*, **301**, 551–560.

8 Jacquemin, E., Hagenbuch, B., Stieger, B., Wolkoff, A.W., and Meier, P.J. (1994) Expression cloning of a rat liver Na$^+$-independent organic anion transporter. *Proceedings of the National Academy of Sciences of the, United States of America*, **91**, 133–137.

9 Koepsell, H. (1998) Organic cation transporters in intestine, kidney, liver, and brain. *Annual Review of Physiology*, **60**, 243–266.

10 Burckhardt, G. and Wolff, N.A. (2000) Structure of renal organic anion and cation transporters. *American Journal of Physiology. Renal Physiology*, **278**, F853–F866.

11 Bergwerk, A.J., Shi, X., Ford, A.C., Kanai, N., Jacquemin, E., Burk, R.D., Bai, S., Novikoff, P.M., Stieger, B., Meier, P.J., Schuster, V.L., and Wolkoff, A.W. (1996) Immunologic distribution of an organic anion transport protein in rat liver and kidney. *American Journal of Physiology. Gastrointestinal and Liver, Physiology*, **271**, G231–G238.

12 Angeletti, R.H., Novikoff, P.M., Juvvadi, S.R., Fritschy, J.-M., Meier, P.J., and Wolkoff, A.W. (1997) The choroid plexus epithelium is the site of the organic transport protein in the brain. *Proceedings of the National Academy of Sciences of the, United States of America*, **94**, 283–286.

13 Hsiang, B., Zhu, Y., Wang, Z., Wu, Y., Sasseville, V., Yang, W.-P., and Kirchgessner, T.G. (1999) A novel human hepatic organic anion transporting polypeptide (OATP2). Identification of a liver-specific human organic anion, transporting polypeptide and identification of rat and human hydroxymethylglutaryl-CoA reductase inhibitor transporters. *The Journal of Biological Chemistry*, **274**, 37161–37168.

14 Reichel, C., Gao, B., Montfoort, J.V., Cattori, V., Rahner, C., Hagenbuch, B., Stieger, B., Kamisako, T., and Meier, P.J. (1999) Localization and function of the organic anion-transporting polypeptide Oatp2 in rat liver. *Gastroenterology*, **117**, 688–695.

15 Kullak-Ublick, G.-A., Hagenbuch, B., Stieger, B., Wolkoff, A.W., and Meier, P.J. (1994) Functional characterization of the basolateral rat liver organic anion transporting polypeptide. *Hepatology*, **20**, 411–416.

16 Bossuyt, X., Muller, M., and Meier, P.J. (1996) Multispecific amphipathic

substrate transport by an organic anion transporter of human liver. *Journal of Hepatology*, **25**, 733–738.

17 Eckhardt, U., Horz, J.A., Petzinger, E., Stuber, W., Reers, M., Dickneite, G., Daniel, H., Wagener, M., Hagenbuch, B., Stieger, B., and Meier, P.J. (1996) The peptide-based thrombin inhibitor CRC 220 is a new substrate of the basolateral rat liver organic anion transporting polypeptide. *Hepatology*, **24**, 380–384.

18 Kanai, N., Lu, R., Bao, Y., Wolkoff, A.W., Vore, M., and Schuster, V.L. (1996) Transient expression of Oatp organic anion transporter in mammalian cells: identification of candidate substrates. *American Journal of Physiology. Renal Physiology*, **270**, F326–F331.

19 Ishizuka, H., Konno, K., Naganuma, H., Nishimura, K., Kouzuki, H., Suzuki, H., Stieger, B., Meier, P.J., and Sugiyama, Y. (1998) Transport of temocaprilat into rat hepatocytes: role of organic anion transporting polypeptide. *The Journal of Pharmacology and Experimental, Therapeutics*, **287**, 37–42.

20 Pang, K.S., Wang, P.J., Chung, A., and Wolkoff, A.W. (1998) The modified dipeptide, enalapril, an angiotensin-converting enzyme inhibitor, is transported by the rat liver organic anion transport protein. *Hepatology*, **28**, 1341–1346.

21 Cvetkovic, M., Leake, B., Fromm, M.F., Wilkinson, G.R., and Kim, R.B. (1999) OATP and P-glycoprotein transporters mediate the cellular uptake and excretion of fexofenadine. *Drug Metabolism and Disposition*, **27**, 866–871.

22 Eckhardt, U., Schroeder, A., Stieger, B., Hochli, M., Landmann, L., Tynes, R., Meier, P.J., and Hagenbuch, B. (1999) Polyspecific substrate uptake by the hepatic organic anion transporter Oatp1 in stably transfected CHO cells. *American Journal of Physiology. Gastrointestinal and Liver, Physiology*, **276**, G1037–G1042.

23 Friesema, E.C., Docter, R., Moerings, E.P., Stieger, B., Hagenbuch, B., Meier, P.J., Krenning, E.P., Hennemann, G., and Visser, T.J. (1999) Identification of thyroid hormone transporters. *Biochemical and Biophysical Research Communications*, **254**, 497–501.

24 van Montfoort, J.E., Hagenbuch, B., Fattinger, K., Müller, M., Groothuis, G.M.M., Meijer, D.K.F., and Meier, P.J. (1999) Polyspecific organic anion transporting polypeptides mediate hepatic uptake of amphipathic type II organic cations. *The Journal of Pharmacology and Experimental, Therapeutics* **291**, 147–152.

25 Gao, B., Hagenbuch, B., Kullak-Ublick, G.A., Benke, D., Aguzzi, A., and Meier, P.J. (2000) Organic anion-transporting polypeptides mediate transport of opioid peptides across blood–brain barrier. *The Journal of Pharmacology and Experimental, Therapeutics*, **294**, 73–79.

26 Abe, T., Suzuki, T., Unno, M., Tokui, T., and Ito, S. (2002) Thyroid hormone transporters: recent advances. *Trends in Endocrinology and Metabolism*, **13**, 215–220.

27 van Montfoort, J.E., Hagenbuch, B., Groothuis, G.M.M., Koepsell, H., Meier, P.J., and Meijer, D.K.F. (2003) Drug uptake systems in liver and kidney. *Current Drug Metabolism*, **4**, 185–211.

28 Treiber, A., Schneiter, R., Hausler, S., and Stieger, B. (2007) Bosentan is a substrate of human OATP1B1 and OATP1B3: inhibition of hepatic uptake as the common mechanism of its interactions with cyclosporin A, rifampicin, and sildenafil. *Drug Metabolism and Disposition*, **35**, 1400–1407.

29 Shi, X.Y., Bai, S., Ford, A.C., Burk, R.D., Jacquemin, E., Hagenbuch, B., Meier, P.J., and Wolkoff, A.W. (1995) Stable inducible expression of a functional rat liver organic anion transport protein in HeLa cells. *The Journal of Biological Chemistry*, **270**, 25591–25595.

30 Satlin, L.M., Amin, V., and Wolkoff, A.W. (1997) Organic anion transporting polypeptide mediates organic anion/

HCO_3^- exchange. *The Journal of Biological Chemistry*, **272**, 26340–26345.

31 Li, L., Lee, T.K., Meier, P.J., and Ballatori, N. (1998) Identification of glutathione as a driving force and leukotriene C_4 as a substrate for Oatp1, the hepatic sinusoidal organic solute transporter. *The Journal of Biological Chemistry*, **273**, 16184–16191.

32 Saito, H., Masuda, S., and Inui, K. (1996) Cloning and functional characterization of a novel rat organic anion transporter mediating basolateral uptake of methotrexate in the kidney. *The Journal of Biological Chemistry*, **271**, 20719–20725.

33 Masuda, S., Ibaramoto, K., Takeuchi, A., Saito, H., Hashimoto, Y., and Inui, K. (1999) Cloning and functional characterization of a new multispecific organic anion transporter, OAT-K2, in rat kidney. *Molecular Pharmacology*, **55**, 743–752.

34 Masuda, S., Saito, H., Nonoguchi, H., Tomita, K., and Inui, K. (1997) mRNA distribution and membrane localization of the OAT-K1 organic anion transporter in rat renal tubules. *FEBS Letters*, **407**, 127–131.

35 Masuda, S., Takeuchi, A., Saito, H., Hashimoto, Y., and Inui, K. (1999) Functional analysis of rat renal organic anion transporter OAT-K1: bidirectional methotrexate transport in apical membrane. *FEBS Letters*, **459**, 128–132.

36 Takeuchi, A., Masuda, S., Saito, H., Abe, T., and Inui, K.-I. (2001) Multispecific substrate recognition of kidney-specific organic anion transporters OAT-K1 and OAT-K2. *The Journal of Pharmacology and Experimental, Therapeutics* **299**, 261–267.

37 Noe, B., Hagenbuch, B., Stieger, B., and Meier, P.J. (1997) Isolation of a multispecific organic anion and cardiac glycoside transporter from rat brain. *Proceedings of the National Academy of Sciences of the, United States of America*, **94**, 10346–10350.

38 Kakyo, M., Sakagami, H., Nishio, T., Nakai, D., Nakagomi, R., Tokui, T., Naitoh, T., Matsuno, S., Abe, T., and Yawo, H. (1999) Immunohistochemical distribution and functional characterization of an organic anion transporting polypeptide 2 (Oatp2). *FEBS Letters*, **445**, 343–346.

39 Gao, B., Stieger, B., Noé, B., Fritschy, J.M., and Meier, P.J. (1999) Localization of the organic anion transporting polypeptide 2 (Oatp2) in capillary endothelium and choroid plexus epithelium of rat brain. *The Journal of Histochemistry and Cytochemistry*, **47**, 1255–1264.

40 Gao, B., Wenzel, A., Grimm, C., Vavricka, S.R., Benke, D., Meier, P.J., and Remè, C.E. (2002) Localization of organic anion transport protein 2 in the apical region of rat retinal pigment epithelium. *Investigative Ophthalmology & Visual Science*, **43**, 510–514.

41 Abe, T., Kakyo, M., Sakagami, H., Tokui, T., Nishio, T., Tanemoto, M., Nomura, H., Hebert, S.C., Masuno, S., Kondo, H., and Yawo, H. (1998) Molecular characterization and tissue distribution of a new organic anion transporter subtype (Oatp3) that transports thyroid hormones and taurocholate and comparison with Oatp2. *The Journal of Biological Chemistry*, **273**, 22395–22401.

42 Tokui, T., Nakai, D., Nakagomi, R., Yawo, H., Abe, T., and Sugiyama, Y. (1999) Pravastatin, an HMG-CoA reductase inhibitor, is transported by rat organic anion transporting polypeptide, Oatp2. *Pharmaceutical Research*, **16**, 904–908.

43 Li, L., Meier, P.J., and Ballatori, N. (2000) Oatp2 mediates bidirectional organic solute transport: a role for intracellular glutathione. *Molecular Pharmacology*, **58**, 335–340.

44 Walters, H.C., Craddock, A.L., Fusegawa, H., Willingham, M.C., and Dawson, P.A. (2000) Expression, transport properties, and chromosomal location of organic anion transporter subtype 3. *The American Journal of Physiology*, **279**, 1188–1200.

45 Kusuhara, H., and Sugiyama, Y. (2004) Efflux transport systems for organic

anions and cations at the blood–CSF barrier. *Advanced Drug Delivery Reviews*, **56**, 1741–1763.

46 Cattori, V., van Montfoort, J.E., Stieger, B., Landmann, L., Meijer, D.K.F., Winterhalter, K.E., Meier, P.J., and Hagenbuch, B. (2001) Localization of organic anion transporting polypeptide 4 (Oatp4) in rat liver and comparison of its substrate specificity with Oatp1, Oatp2 and Oatp3. *Pflügers Archives*, **443**, 188–195.

47 Dresser, G.K., Bailey, D.G., Leake, B.F., Schwarz, U.I., Dawson, P.A., Freeman, D.J., and Kim, R. B. (2002) Fruit juices inhibit organic anion transporting polypeptide-mediated drug uptake to decrease the oral availability of fexofenadine. *Clinical Pharmacology and Therapeutics*, **71**, 11–20.

48 Choudhuri, S., Ogura, K., and Klaassen, C.D. (2001) Cloning, expression, and ontogeny of mouse organic anion transporting polypeptide-5, a kidney-specific organic anion transporter. *Biochemical and Biophysical Research Communications*, **280**, 92–98.

49 Kakyo, M., Uno, M., Tokui, T., Nakagomi, R., Nihio, T., Iwasashi, H., Nakai, D., Seki, M., Suzuki, M., Naitoh, T., Matsuno, S., Yawa, H., and Abe, T. (1999) Molecular characterization and functional regulation of a novel rat liver-specific organic anion transporter rlst-1. *Gastroenterology*, **117**, 770–775.

50 Cattori, V., Hagenbuch, B., Hagenbuch, N., Stieger, B., Ha, R., Winterhalter, K.E., and Meier, P.J. (2000) Identification of organic anion transporting polypeptide 4 (Oatp4) as a major full-length isoform of the liver-specific transporter-1 (rlst-1) in rat liver. *FEBS Letters*, **474**, 242–245.

51 Suzuki, T., Onogawa, T., Asano, N., Mizutamari, H., Mikkaichi, T., Tanemoto, M., Abe, M., Satoh, F., Unno, M., Nunoki, K., Suzuki, M., Hishinuma, T., Goto, J., Shimosegawa, T., Matsuno, S., Ito, S., and Abe, T. (2003) Identification and characterization of novel rat and human gonad-specific organic anion transporters. *Molecular Endocrinology*, **17**, 1203–1215.

52 Fischer, W.J., van Montfoort, J.E., Cattori, V., Meier, P.J., Dietrich, D.R., and Hagenbuch, B. (2001) Organic anion transporting polypeptides (OATPs) mediate uptake of microcystin into brain and liver. *Journal of Toxicology. Clinical Toxicology*, **39**, A565.

53 Ismair, M.G., Stieger, B., Cattori, V., Hagenbuch, B., Fried, M., Meier, P.J., and Kullak-Ublick, G.A. (2001) Hepatic uptake of cholecystokinin octapeptide (CCK-8) by organic anion transporting polypeptides Oatp4 (Slc21a6) and OATP8 (SLC21A8) of rat and human liver. *Gastroenterology*, **121**, 1185–1190.

54 Meier-Abt, F., Faulstich, H., and Hagenbuch, B. (2004) Identification of phalloidin uptake system of rat and human liver. *Biochimica et Biophysica Acta*, **1664**, 64–69.

55 Li, J.Y., Boado, R.J., and Pardridge, W.M. (2001) Blood–brain barrier genomics. *Journal of Cerebral Blood Flow and Metabolism*, **21**, 61–68.

56 Sugiyama, D., Kusuhara, H., Taniguchi, H., Ishikawa, S., Nozaki, Y., Aburatanis, H., and Sugiyama, Y. (2003) Functional characterization of rat brain-specific organic anion transporter (Oatp14) at the blood–brain barrier. *The Journal of Biological Chemistry*, **278**, 43489–43495.

57 Tohyama, K., Kusuhara, H., and Sugiyama, Y. (2004) Involvement of multispecific organic anion transporter, Oatp14 (*Slc21a14*), in the transport of thyroxine across the blood–brain barrier. *Endocrinology*, **145**, 4384–4391.

58 Hakes, D.J. and Berezney, R. (1991) Molecular cloning of matrin F/G: a DNA binding protein of the nuclear matrix that contains putative zinc finger motifs. *Proceedings of the National Academy of Sciences of the, United States of America*, **88**, 6186–6190.

59 Nishio, T., Adachi, H., Nakagomi, R., Tokui, T., Sato, E., Tanemoto, M.,

Fujiwara, K., Okabe, M., Onogawa, T., Suzuki, T., Nakai, D., Shiiba, K., Suzuki, M., Ohtani, H., Kondo, Y., Unno, M., Ito, S., Iinuma, K., Nunoki, K., Matsuno, S., and Abe, T. (2000) Molecular identification of a rat novel organic anion transporter moat1, which transports prostaglandin D(2), leukotriene C(4), and taurocholate. *Biochemical and Biophysical Research Communications*, **275**, 831–838.

60 Kanai, N., Lu, R., Satriano, J.A., Bao, Y., Wolkoff, A.W., and Schuster, V.L. (1995) Identification and characterization of a prostaglandin transporter. *Science*, **268**, 866–869.

61 Lu, R., Kanai, N., Bao, Y., and Schuster, V.L. (1996) Cloning, *in vitro* expression, and tissue distribution of a human prostaglandin transporter cDNA (hPGT). *The Journal of Clinical Investigation*, **98**, 1142–1149.

62 Adachi, H., Suzuki, T., Abe, M., Asano, N., Mizutamari, H., Tanemoto, M., Nishio, T., Onogawa, T., Toyohara, T., Kasai, S., Satoh, F., Suzuki, M., Tokui, T., Unno, M., Shimosegawa, T., Matsuno, S., Ito, S., and Abe, T. (2003) Molecular characterization of human and rat organic anion transporter OATP-D. *American Journal of Physiology. Renal Physiology*, **285**, F1188–F1197.

63 Huber, R.D., Gao, B., Pfändler, M.S., Zhang-Fu, W., Leuthold, S., Hagenbuch, B., Folkers, G., Meier, P.J., and Stieger, B. (2007) Characterization of two splice variants of human organic anion transporting polypeptide 3A1 isolated from human brain. *American Journal of Physiology. Cell Physiology*, **292**, C795–C806.

64 Fujiwara, K., Adachi, H., Nishio, T., Unno, M., Tokui, T., Okabe, M., Onogawa, T., Suzuki, T., Asano, N., Tanemoto, M., Seki, M., Shiiba, K., Suzuki, M., Kondo, Y., Nunoki, K., Shimosegawa, T., Iinuma, K., Ito, S., Matsuno, S., and Abe, T. (2001) Identification of thyroid hormone transporters in humans: different molecules are involved in a tissue-specific manner. *Endocrinology*, **142**, 2005–2012.

65 Mikkaichi, T., Suzuki, T., Onogawa, T., Tanemoto, M., Mizutamari, H., Okada, M., Chaki, T., Masuda, S., Tokui, T., Eto, N., Abe, M., Satoh, F., Unno, M., Hishinuma, T., Inui, K.-I., Ito, S., Goto, J., and Abe, T. (2004) Isolation and characterization of a digoxin transporter and its rat homologue expressed in the kidney. *Proceedings of the National Academy of Sciences of the, United States of America*, **101**, 3569–3574.

66 Meier-Abt, F., Mokrab, Y., and Mizuguchi, K. (2006) Organic anion transporting polypeptides of the OATP/*SLCO* superfamily: identification of new members in nonmammalian species, comparative modeling and a potential transport mode. *Journal of Membrane Biology*, **208**, 213–227.

67 Nakao, N., Takagi, T., Iigo, M., Tsukamoto, T., Yasuo, S., Masuda, T., Yanagisawa, T., Ebihara, S., and Yoshimura, T. (2006) Possible involvement of organic anion transporting polypeptide 1c1 in the photoperiodic response of gonads in birds. *Endocrinology*, **147**, 1067–1073.

68 Kullak-Ublick, G.A., Hagenbuch, B., Stieger, B., Schteingart, C.D., Hofmann, A.F., Wolkoff, A.W., and Meier, P.J. (1995) Molecular and functional characterization of an organic anion transporting polypeptide cloned from human liver. *Gastroenterology*, **109**, 1274–1282.

69 Kullak-Ublick, G.A., Ismair, M.G., Stieger, B., Landmann, L., Huber, R., Pizzagalli, F., Fattinger, K., Meier, P.J., and Hagenbuch, B. (2001) Organic anion-transporting polypeptide B (OATP-B) and its functional comparison with three other OATPs of human liver. *Gastroenterology*, **120**, 525–533.

70 Kullak-Ublick, G.A., Glasa, J., Boker, C., Oswald, M., Grutzner, U., Hagenbuch, B., Stieger, B., Meier, P.J., Beuers, U., Kramer, W., Wess, G., and Paumgartner, G. (1997) Chlorambucil-taurocholate is transported by bile acid carriers expressed in human hepatocellular carcinomas. *Gastroenterology*, **113**, 1295–1305.

71 Abe, T., Kakyo, M., Tokui, T., Nakagomi, R., Nishio, T., Nakai, D., Nomura, H., Unno, M., Suzuki, M., Naitoh, T., Matsuno, S., and Yawo, H. (1999) Identification of a novel gene family encoding human liver-specific organic anion transporter LST-1. *The Journal of Biological Chemistry*, **274**, 17159–17163.

72 Chignard, N., Mergey, M., Veissiere, D., Parc, R., Capeau, J., Poupon, R., Paul, A., and Housset, C. (2001) Bile acid transport and regulating functions in the human biliary epithelium. *Hepatology*, **33**, 496–503.

73 Bossuyt, X., Muller, M., Hagenbuch, B., and Meier, P.J. (1996) Polyspecific drug and steroid clearance by an organic anion transporter of mammalian liver. *The Journal of Pharmacology and Experimental, Therapeutics* **276**, 891–896.

74 Meier, P.J., Eckhardt, U., Schroeder, A., Hagenbuch, B., and Stieger, B. (1997) Substrate specificity of sinusoidal bile acid and organic anion uptake systems in rat and human liver. *Hepatology*, **26**, 1667–1677.

75 Kullak-Ublick, G.A., Fisch, T., Oswald, M., Hagenbuch, B., Meier, P.J., Beuers, U., and Paumgartner, G. (1998) Dehydroepiandrosterone sulfate (DHEAS): identification of a carrier protein in human liver and brain. *FEBS Letters*, **424**, 173–176.

76 Pascolo, L., Cupelli, F., Anelli, P.L., Lorusso, V., Visigalli, M., Uggeri, F., and Tiribelli, C. (1999) Molecular mechanisms for the hepatic uptake of magnetic resonance imaging contrast agents. *Biochemical and Biophysical Research Communications*, **257**, 746–752.

77 van Montfoort, J.E., Müller, M., Groothuis, G.M.M., Meijer, D.K.F., Koepsell, H., and Meier, P.J. (2001) Comparison of "type I" and "type II" organic cation transport by organic cation transporters and organic anion-transporting polypeptides. *The Journal of Pharmacology and Experimental, Therapeutics* **298**, 110–115.

78 Sai, Y. and Tsuji, A. (2004) Transporter-mediated drug delivery: recent progress and experimental approaches. *Drug Discovery Today*, **9**, 712–720.

79 Fischer, W.J., Altheimer, S., Cattori, V., Meier, P.J., Dietrich, D.R., and Hagenbuch, B. (2005) Organic anion transporting polypeptides expressed in liver and brain mediate uptake of microcystin. *Toxicology and Applied Pharmacology*, **203**, 257–263.

80 Ho, R.H., Tirona, R.G., Leake, B.F., Glaeser, H., Lee, W., Lemke, C.J., Wang, Y., and Kim, R.B. (2006) Drug and bile acid transporters in rosuvastatin hepatic uptake: function, expression, and pharmacogenetics. *Gastroenterology*, **130**, 1793–1806.

81 König, J., Cui, Y., Nies, A.T., and Keppler, D. (2000) A novel human organic anion transporting polypeptide localized to the basolateral hepatocyte membrane. *American Journal of Physiology. Gastrointestinal and Liver, Physiology*, **278**, 156–164.

82 Abe, T., Unno, M., Onogawa, T., Tokui, T., Kondo, T.N., Nakagomi, R., Adachi, H., Fujiwara, K., Okabe, M., Suzuki, T., Nunoki, K., Sato, E., Kakyo, M., Nishio, T., Sugita, J., Asano, N., Tanemoto, M., Seki, M., Date, F., Ono, K., Kondo, Y., Shiiba, K., Suzuki, M., Ohtani, H., Shimosegawa, T., Iinuma, K., Nagura, H., Ito, S., and Matsuno, S. (2001) LST-2, a human liver-specific organic anion transporter, determines methotrexate sensitivity in gastrointestinal cancers. *Gastroenterology*, **120**, 1689–1699.

83 Nakai, D., Nakagomi, R., Furuta, Y., Tokui, T., Abe, T., Ikeda, T., and Nishimura, K. (2001) Human liver-specific organic anion transporter, LST-1, mediates uptake of pravastatin by human hepatocytes. *The Journal of Pharmacology and Experimental, Therapeutics* **297**, 861–867.

84 Tirona, R.G., Leake, B.F., Merino, G., and Kim, R.B. (2001) Polymorphisms in OATP-C. Identification of multiple allelic

variants associated with, altered transport activity among European- and African-Americans. *The Journal of Biological Chemistry*, **276**, 35669–35675.

85 Vavricka, S.R., van Montfoort, J., Ha, H.R., Meier, P.J., and Fattinger, K. (2002) Interactions of rifamycin SV and rifampicin with organic anion uptake systems of human liver. *Hepatology*, **36**, 164–172.

86 Briz, O., Serrano, M.A., Macias, R.I.R., Gonzalez-Gallego, J., and Marin, J.J.G. (2003) Role of organic anion-transporting polypeptides, OATP-A, OATP-C and OATP-8, in human placenta-maternal liver tandem excretory pathway for foetal bilirubin. *The Biochemical Journal*, **371**, 897–905.

87 Nozawa, T., Sugiura, S., Nakajima, M., Goto, A., Yokoi, T., Nezu, J., Tsuji, A., and Tamai, I. (2004) Involvement of organic anion transporting polypeptides in the transport of troglitazone sulfate: implications for understanding troglitazone hepatotoxicity. *Drug Metabolism and Disposition*, **32**, 291–294.

88 Ho, R.H. and Kim, R.B. (2005) Transporters and drug therapy: implications for drug disposition and disease. *Clinical Pharmacology and Therapeutics*, **78**, 260–277.

89 Nozawa, T., Minami, H., Sugiura, S., Tsuji, A., and Tamai, I. (2005) Role of organic anion transporter OATP1B1 (OATP-C) in hepatic uptake of irinotecan and its active metabolite, 7-ethyl-10-hydroxycamptothecin: in vitro evidence and, effect of single nucleotide polymorphisms. *Drug Metabolism and Disposition*, **33**, 434–439.

90 Endes, C.J., Hsiao, P., Chung, F.S., and Unadkat, J.D. (2006) The role of transporters in drug interactions. *European Journal of Pharmaceutical Sciences*, **27**, 501–517.

91 Grube, M., Köck, K., Oswald, S., Draber, K., Meissner, K., Eckel, L., Böhm, M., Felix, S.B., Vogelgesang, S., Jedlitschky, G., Siegmund, W., Warzok, R., and Kroemer, H.K. (2006) Organic anion transporting polypeptide 2B1 is a high-affinity transporter for atorvastatin and is expressed in the human heart. *Clinical Pharmacology and Therapeutics*, **80**, 607–620.

92 Noe, J., Portmann, R., Brun, M., and Funk, C. (2007) Substrate dependent drug–drug interactions between gemfibrozil, fluvastatin and other OATP substrates on OATP1B1, OATP2B1 and OATP1B3. *Drug Metabolism and Disposition*, **35**, 1308–1314.

93 Cui, Y., König, J., Leier, I., Buchholz, U., and Keppler, D. (2001) Hepatic uptake of bilirubin and its conjugates by the human organic anion transporter SLC21A6. *The Journal of Biological Chemistry*, **276**, 9626–9630.

94 Nozawa, T., Nakajima, M., Tamai, I., Noda, K., Nezu, J., Sai, Y., Tsuji, A., and Yokoi, T. (2002) Genetic polymorphisms of human organic anion transporters OATP-C (SLC21A6) and OATP-B (SLC21A9): allele frequencies in the Japanese population and functional analysis. *The Journal of Pharmacology and Experimental, Therapeutics* **302**, 804–813.

95 Michalski, C., Cui, Y., Nies, A.T., Nuessler, A.K., Neuhaus, P., Zanger, U.M., Klein, K., Eichelbaum, M., Keppler, D., and König, J. (2002) A naturally occurring mutation in the SLC21A6 gene causing impaired membrane localization of the hepatocyte uptake transporter. *The Journal of Biological Chemistry*, **277**, 43058–43063.

96 Tirona, R.G., Leake, B.F., Wolkoff, A.W., and Kim, R.B. (2003) Human organic anion transporting polypeptide-C (SLC21A6) is a major determinant of rifampin-mediated pregnane X receptor activation. *The Journal of Pharmacology and Experimental, Therapeutics* **304**, 223–228.

97 König, J., Cui, Y., Nies, A.T., and Keppler, D. (2000) Localization and genomic organization of a new hepatocellular

organic anion transporting polypeptide. *The Journal of Biological Chemistry*, **275**, 23161–23168.

98 Baldes, C., Koenig, P., Neumann, D., Lenhof, H.P., Kohlbacher, O., and Lehr, M.C. (2006) Development of a fluorescence-based assay for screening of modulators of human organic anion transporter 1B3 (OATP1B3). *European Journal of Pharmaceutics and Biopharmaceutics*, **62**, 39–63.

99 Letschert, K., Faulstich, H., Keller, D., and Keppler, D. (2006) Molecular characterization and inhibition of amanitin uptake into human hepatocytes. *Toxicological Sciences*, **91**, 140–149.

100 Pizzagalli, F., Hagenbuch, B., Stieger, B., Klenk, U., Folkers, G., and Meier, P.J. (2002) Identification of a novel human organic anion transporting polypeptide as a high affinity thyroxine transporter. *Molecular Endocrinology*, **16**, 2283–2296.

101 Schuster, V.L. (1998) Molecular mechanisms of prostaglandin transport. *Annual Review of Physiology*, **60**, 221–242.

102 Kobayashi, D., Nozawa, T., Imai, K., Nezu, J.-I., Tsuji, A., and Tamai, I. (2003) Involvement of human organic anion transporting polypeptide OATP-B (SLC21A9) in pH-dependent transport across intestinal apical membrane. *The Journal of Pharmacology and Experimental, Therapeutics* **306**, 703–708.

103 Nozawa, T., Imai, K., Nezu, J.-I., Tsuji, A., and Tamai, I. (2004) Functional characterization of pH-sensitive organic anion transporting polypeptide OATP-B in human. *The Journal of Pharmacology and Experimental, Therapeutics* **308**, 438–445.

104 Steffansen, B., Nielsen, C.U., Brodin, B., Eriksson, A.H., Andersen, R., and Frokjaer, S. (2004) Intestinal solute carriers: an overview of trends and strategies for improving oral drug absorption. *European Journal of Pharmaceutical Sciences*, **21**, 3–16.

105 Sai, Y., Kaneko, Y., Ito, S., Mitsuoka, K., Kato, Y., Tamai, I., Artursson, P., and Tsuji, A. (2006) Predominant contribution of organic anion transporting polypeptide OATP-B (OATP2B1) to apical uptake of estrone-3-sulfate by human intestinal Caco-2 cells. *Drug Metabolism and Disposition*, **34**, 1423–1431.

106 Tirona, R.G. and Kim, R.B. (2002) Pharmacogenomics of organic anion-transporting polypeptides (OATP). *Advanced Drug Delivery Reviews*, **54**, 1343–1352.

107 Mizuno, N., Niwa, T., Yotsumoto, Y., and Sugiyama, Y. (2003) Impact of drug transporter studies on drug discovery and development. *Pharmacological Reviews*, **55**, 425–461.

108 Sasaki, M., Suzuki, H., Ito, K., Abe, T., and Sugiyama, Y. (2002) Transcellular transport of organic anions across a double-transfected Madin-Darby canine kidney II cell monolayer expressing both human organic anion-transporting polypeptide (OATP2/SLC21A6) and multidrug resistance-associated protein 2 (MRP2/ABCC2). *The Journal of Biological Chemistry*, **277**, 6497–6503.

109 Yamazaki, M., Suzuki, H., and Sugiyama, Y. (1996) Recent advances in carrier-mediated hepatic uptake and biliary excretion of xenobiotics. *Pharmaceutical Research*, **13**, 497–513.

110 Tamai, I., Takanaga, H., Meada, H., Ogihara, T., Yoneda, M., and Tsuji, A. (1995) Proton-cotransport of pravastatin across intestinal brush-border membrane. *Pharmaceutical Research*, **12**, 1727–1732.

111 Muck, W., Ritter, W., Ochmann, K., Unger, S., Ahr, G., Wingerder, W., and Kuhlmann, J. (1997) Absolute and relative bioavailability of HMG-CoA reductase inhibitor cerivastatin. *International Journal of Clinical Pharmacology and, Therapeutics* **35**, 255–260.

112 Shitara, Y., Itoh, T., Sato, H., Li, A.P., and Sugiyama, Y. (2003) Inhibition of transporter-mediated hepatic uptake as a mechanism for drug–drug interaction between cerivastatin and cyclosporin A. *The Journal of Pharmacology and Experimental, Therapeutics* **304**, 610–616.

113 Yarim, M., Moro, S., Huber, R., Meier, P.J., Kaseda, C., Kashima, T., Hagenbuch, B., and Folkers, G. (2005) Application of QSAR analysis to organic anion transporting polypeptide 1a5 (Oatp1a5) substrates. *Bioorganic and Medicinal Chemistry*, **13**, 463–471.

114 Chang, C., and Swaan, P.W. (2006) Computational approaches to modelling drug transporters. *European Journal of Pharmaceutical Sciences*, **27**, 411–424.

115 Chang, C., Pang, K.S., Swaan, P.W., and Ekins, S. (2005) Comparative pharmacophore modeling of organic anion transporting polypeptides: a meta-analysis of rat Oatp1a1 and human OATP1B1. *The Journal of Pharmacology and Experimental Therapeutics* **314**, 533–541.

4
CNS - Transporters as Drug Targets

Klaus Gundertofte

4.1
Introduction

Nerve cells are widely distributed in the mammalians, particularly in the central nervous system (CNS). Nerve cells synthesize, store, and eventually release their neurotransmitters into the synaptic cleft between the presynaptic and the postsynaptic neuron. Nature is conservative, hence evolution has provided specialized membrane proteins to allow the organism to reuse the neurotransmitters. These membrane transport proteins strictly control the concentration of neurotransmitters in the synaptic cleft.

Two major classes of neurotransmitter transporters are present in the cell membrane of neurons: the solute carrier 1 (SLC1) transporter family [1] and the SLC6 transporter family [2]. SLC1 includes the Na^+-dependent transporters that reuptake glutamate. The SLC6 family consists of transporters for the Na^+-dependent uptake of dopamine, serotonin (5-HT), noradrenaline (norepinephrine), glycine, and GABA. A number of other solute carrier families also exist; one example discussed below is the SLC2 family, transporting glucose.

The neurotransmitter transporters are responsible for terminating or modulating the action of neurotransmitters released from the presynaptic neuron. The transport of neurotransmitters takes place against a concentration gradient; thus, transport is coupled to the downhill transport of Na^+. Furthermore, the SLC1 family requires K^+ transport, while SLC6 requires a cotransport of Cl^-.

4.2
Structure of Transporters

The structures of thousands of proteins have been resolved through X-ray crystallography. Membrane proteins represent a special class in this respect, as they are not easily crystallizable. Thus, ligand-based approaches for drug design (e.g.,

Transporters as Drug Carriers: Structure, Function, Substrates. Edited by Gerhard Ecker and Peter Chiba
Copyright © 2009 WILEY-VCH Verlag GmbH & Co. KGaA, Weinheim
ISBN: 978-3-527-31661-8

pharmacophore modeling) have been the preferred tool until recently. As the first member of the SLC1 family, the structure of a glutamate transporter homologue from the bacteria *Pyrococcus hirikoshii* with a sequence identity to the glutamate transporter GLT-1 of 37% was crystallized and resolved to a resolution of 3.5 Å in 2004 [3]. The leucine transporter LeuT$_{Aa}$ from the bacteria *Aquifex aeolicus* was resolved a year later at an impressive resolution of 1.65 Å [4]. The LeuT$_{Aa}$ is a member of the SLC6 family with 20–25% sequence identity to mammalian transporters. The two structures are quite different. The former shows a trimeric complex forming a bowl with the putative binding site for glutamate in the bottom of the bowl, while the LeuT$_{Aa}$ structure reveals a dimeric structure consisting of protomers with 12 transmembrane helices. The binding site for leucine is buried inside the TM region. The crystallization of the members of the SLC1 and SLC6 families has already initiated homology modeling [5], opening the way to further understanding of the structure–activity relationships and the mechanistic aspects of the transporters.

Being essential regulators of the neural transmission, the neurotransmitter transport proteins represent obvious targets of a broad range of neurological and psychiatric diseases. This is supported by the association of certain diseases with mutations in genes coding for various transporters. Examples include a mutation in the noradrenaline transporter (NAT) leading to orthostatic hypotension and a mutation in GLT-1 leading to amyotrophic lateral sclerosis [6].

4.3
Monoamine Transporters

The plasma membrane transporters that clear extracellular serotonin and noradrenaline, that is, serotonin transporters (SERTs) and NATs, have received considerable attention since the 1950s because of their role in amine neurotransmitter inactivation. They are the major targets for most antidepressants including the tricyclic antidepressants and the selective serotonin uptake inhibitors. The classical antidepressant drugs interact mainly with noradrenaline and serotonin transporters and are still on the market, even though they were developed more than 50 years ago. They are, however, associated with severe side effects such as dry mouth caused by their anticholinergic component. In the late 1960s, the importance of serotonin as a mood regulator was recognized. This initiated research in many laboratories, aiming at selective serotonin reuptake inhibitors (SSRIs), and a large number of compounds were investigated and patented in the early 1970s. Fluoxetine (Prozac) (**1**) and zimelidine (**2**) were the first to be discovered in this new class of drugs, showing a clearly improved side effect pattern. This improvement triggered the inclusion of a much larger patient population and served as the fundament for the massive success of this drug class, which also includes citalopram/escitalopram (**3**) (the active enantiomer of citalopram), paroxetine, sertraline, and fluvoxamine. Persisting problems with this treatment are, however, the slow onset of action and nonresponse by up to 30% of the patients. Various add-on treatments and augmentation strategies have been pursued. Apart from the combined serotonin–noradrenaline reuptake inhibitors – for example, velnafaxine and duloxetine – none of these attempts has until now resulted in drugs as efficacious as the SSRIs [7].

An allosteric binding site on the serotonin transporter has been identified. It is a low-affinity binding site distinct from the high-affinity site. Only escitalopram, and to a lesser extent paroxetine, interacts with this site. It is concluded that the allosteric binding site is independent of the high-affinity binding site. It may therefore represent a new drug target and the term ASRI (allosteric serotonin reuptake inhibitors) has been ascribed to those new compounds demonstrating dual activity.

The low-affinity allosteric site influences the dissociation of uptake inhibitors, such as paroxetine, and citalopram from the primary site, when it is occupied by serotonin. Escitalopram has been shown to stabilize its own binding to the primary site, an effect counteracted by R-citalopram, which is therefore by no means an inactive compound. Clinical data have demonstrated a faster onset of escitalopram action compared to SSRIs. The superiority of escitalopram may be ascribed to this unique interaction with the allosteric site [8].

In addition to the noradrenaline and serotonin transporters, the dopamine transporter (DAT) is a noticeable target. It may be a target in neurodegenerative and psychiatric disorders such as Parkinson's disease and attention deficit hyperactivity disorder (ADHD). It plays a key role in mediating the actions of psychostimulants such as cocaine and the amphetamines. The main caveat of this target is therefore the abuse liability. No drugs with the primary interaction with DAT have made it to the market. A few compounds, for example, GBR12909, did undergo clinical studies but were withdrawn due to their toxicological effects [9].

In the central nervous system, vesicular monoamine transporter 2 (VMAT2) moves cytoplasmic dopamine into synaptic vesicles for storage and subsequent exocytotic release. Agents enhancing dopamine sequestration by VMAT2, thereby preventing the oxidation of dopamine in the cytoplasm, form a potential strategy for the treatment of diseases such as Parkinson's disease. Furthermore, VMAT2 may be a potential target for developing drugs for the treatment of drug abuse.

4.4
Transporters for Amino Acids

Extensive medicinal chemistry research on inhibitory amino acid transporters has been performed. The only clinically used compound so far is the antiepileptic drug tiagabine (4) acting as a GABA uptake inhibitor on GABA transporter-1 (GAT1). Inhibition of neuronal GABA uptake of GABA is highly correlated with anticonvulsant activity [10]. Thus, tiagabine has confirmed the important role that GABA transporters play in the control of CNS excitability. Tiagabine is selective for GAT1, but lacks cell type selectivity being an equipotent inhibitor of neuronal and glial GAT1. Three other GABA transporters (GAT2–4) and a vesicular GABA transporter (VGAT) have been cloned. Heterogeneity of GAT plays a role in the control of CNS function. The lack of selective inhibitors for these subtypes is, however, hampering medicinal chemistry research due to the lack of knowledge on the role of these subtypes. The clinical success of tiagabine may stimulate further research on these interesting new drug targets. New clinical studies on tiagabine are underway aiming at uncovering the potential use of the drug in anxiety, sleep disorders, and pain.

The observation that NMDA receptor antagonists induce symptoms like the ones observed in schizophrenic patients in the clinic gave rise to the NMDA receptor hypofunction hypothesis in schizophrenia [11]. Glycine is an obligatory coagonist at the NMDA receptors, and the inhibition of glycine transporter-1 (GlyT1), yet another member of the SLC6 family, augments NMDA signaling and theoretically relieves the schizophrenic symptoms. Based on the number of publications from academia and industry, a substantial interest in developing selective GlyT1 inhibitors is evident. It remains to be seen, however, if the hypothesis is valid, since no clinical studies have determined the efficacy of these inhibitors in schizophrenia.

The second major class of transporters includes the excitatory amino acid transporters (EAATs) [12]. These are essential for the termination of signal transmission mediated by glutamine. Furthermore, these transporters serve to prevent neurotoxicity mediated by glutamine, since inadequate clearance from the synaptic cleft and from the extrasynaptic space causes glutamate to act as a potent neurotoxin. It may be related to several neurodegenerative pathologies including epilepsy, ischemia, amyotrophic lateral sclerosis, and Alzheimer's disease. None of the currently marketed drugs exerts its effect through glutamine transporters. Rather, a limited medicinal chemistry work has been done on this target, which may be due to a limited therapeutic potential, at least for the time being.

4.5
Nonneurotransmitter Transporters

Insulin regulates the concentration of glucose in muscle and fat cells by facilitating glucose transport through a rapid gain in surface-bound glucose transporters (GLUT) belonging to SLC2. Among the GLUT family isoforms, GLUT4 is particularly important for maintaining the glucose metabolism homeostasis [13].

Thus, this target together with the sodium–glucose cotransporter (SGLT) may be important in diabetes. Inhibitors of the latter have been shown to lower the blood sugar concentration and are regarded as a novel and promising agent for the treatment of diabetes mellitus.

4.6
Concluding Remarks

As can be seen from the above discussion, numerous targets within the superfamily of transporters are still to be investigated. Drugs have been marketed exerting their effect on surprisingly few targets in these classes. The recently published structures of homologues of the SLC1 and SLC6 families have opened up the area and made a new beginning in research efforts toward the development of specific drugs interacting with new important targets. This in turn may lead to substantial improvements in the treatment of patients suffering from psychiatric or neurological diseases.

References

1 Kanai, Y. and Hediger, M.A. (2004) The glutamate/neutral amino acid transporter family SLC1: molecular, physiological and pharmacological aspects. *Pflugers Archives*, **447**, 469–479.

2 Chen, N.H. *et al.* (2004) Synaptic uptake and beyond: the sodium- and chloride-dependent neurotransmitter transporter family SLC6. *Pflugers Archives*, **447**, 519–531.

3 Yernool, D., Boudker, O., Jin, Y., and Gouaux, E. (2004) Structure of a glutamate transporter homologue from *Pyrococcus horikoshii*. *Nature*, **431**, 811–818.

4 Yamashita, A., Singh, S.K., Kawate, T., Jin, Y., and Gouaux, E. (2005) Crystal structure of a bacterial homologue of Na^+/Cl^--dependent neurotransmitter transporters. *Nature*, **437**, 215–223.

5 Jorgensen, A.M., Tagmose, L., Jorgensen, A.M.M., Topiol, S., Sabio, M., Gundertofte, K., Bogeso, K.P., and Peters, G.H. (2007) Homology modeling of the serotonin transporter: insights into the primary escitalopram-binding site. *ChemMedChem*, **2**, 815–826.

6 Trotti, D., Aoki, M., Pasinelli, P., Berger, U.V., Danbolt, N.C., Brown, R.H., and Hediger, M.A. (2001) Amyotrophic lateral sclerosis-linked glutamate transporter mutant has impaired glutamate clearance capacity. *Journal of Biological Chemistry*, **276**, 576–582.

7 Moltzen, E.K. and Bang-Andersen, B. (2006) Serotonin reuptake inhibitors: the corner stone in treatment of depression for half a century – a medicinal chemistry survey. *Current Topics in Medicinal Chemistry*, **6**, 1801–1823.

8 Wiborg, O. and Sanchez, C. (2003) *R*-Citalopram decreases the association of [^{3}H]-*S*-citalopram with the human serotonin transporter by an allosteric mechanism. *European Neuropsychopharmacology*, **13**, S217–S217.

9 Herman, B.H., Elkashef, A., and Vocci, F. (2005) Therapeutic strategies. *Drug Discovery Today*, **2**, 87–92.

10 Borden, L.A., Dhar, T.G.M., Smith, K.E., Weinshank, R.I., Branchek, T.A., and Gluchowski, C. (1994) Tiagabine, SKF 89976-A, CI-966, and NNC-711 are selective for the cloned GABA transporter GAT-1. *European Journal of Pharmacology. Molecular Pharmacology*, **269**, 219–224.

11 Lindsley, C.W., Wolkenberg, S.E., and Kinney, G.G. (2006) Progress in the preparation and testing of glycine transporter type-1 (GlyT1) inhibitors. *Current Topics in Medicinal Chemistry*, **6**, 1883–1896.

12 Hinoi, E., Takarada, T., Tsuchihashi, Y., and Yoneda, Y. (2005) Glutamate transporters as drug targets. *Current Drug Targets: CNS & Neurological Disorders*, **4**, 211–220.

13 Asano, T., Ogihara, T., Katagiri, H., Sakoda, H., Ono, H., Fujishiro, M., Anai, M., Kurihara, H., and Uchijima, Y. (2004) Glucose transporter and Na^+/glucose cotransporter as molecular targets of anti-diabetic drugs. *Current Medicinal Chemistry*, **11**, 2717–2724.

Part Two:
Drug Transport in Microorganisms and Fungi

5
Bacterial Multidrug Transporters: Molecular and Clinical Aspects
Olga Lomovskaya, Helen I. Zgurskaya, and Keith Bostian

5.1
Introduction

5.1.1
The Multiple Antibiotic Resistance Problem

Antimicrobial drug resistance is the leading challenge in the management of infectious diseases [1, 2]. The realization that bacteria might become resistant to antibiotics is almost as old as the discovery of antibiotics itself. But for some time, antibiotic resistance was not particularly alarming since more and more diverse drugs were being successfully introduced into clinical practice. The 1940s uncovered the magic of aminoglycosides and β-lactams, followed by the even more fruitful 1950s, which brought about chloramphenicol, tetracycline, macrolides, glycopeptides, streptogramins, and lincosamides. However, this "golden era of discovery" of new antibiotic classes ended some time in 1960s, with the advent of rifamycin and the man-made nalidixic acid. Only two completely new classes of therapeutically useful antibiotics, the oxazolidinones (linezolid) and the lipopeptides (daptomycin), have been discovered since then, in 1980s. It took some 20 years to introduce them into clinical practice. During the decades of the 1970s and 1980s, a "golden age of antibiotic medicinal chemistry," many useful modifications to known antibiotics were devised in order to improve their efficacy and safety and to overcome growing resistance. However, during the past 25 years the pace of introduction of new antibiotics into the clinic has significantly slowed down, not surprisingly, concomitant with a dramatic increase in the prevalence of untreatable pan-resistant pathogens [3].

5.1.2
The Superbugs

There are two types of these so-called superbugs. Some of them, such as the notoriously known methicillin-resistant *Staphylococcus aureus* (MRSA), classified in the same genera or even species as normal human commensal flora, have become more dangerous [4, 5]. They have become more virulent and have acquired plasmids and transposons frequently containing multiple resistance genes, each responsible for a specific biochemical resistance mechanism. These mechanisms are numerous and efficient. As a result, antibiotics might be degraded, inactivated, or sequestered, their targets might be protected or modified so that binding affinity is reduced, or moreover, sensitive targets might be replaced altogether with insensitive ones [6, 7].

Superbugs of a different type are opportunistic pathogens that typically infect sick or immunocompromised patients and intrinsically more difficult to treat. One of them is *Pseudomonas aeruginosa*, a frequent and deadly nosocomial pathogen [8]. The natural ecological niches of this and similar bacteria are soil and water, a diverse environment crammed with multiple chemical challenges to overcome. For these superbugs, multidrug resistance appears to be their natural state [3]. It is achieved by the cooperation of two attributes essential for their survival in natural habitats: a low-permeability cell envelope that provides a physical barrier to the entry of both lipophilic and hydrophilic molecules and efflux systems consisting of membrane transporters with unusually broad substrate specificity [9].

5.1.3
The Multidrug Resistance Transporters

Multidrug resistance or polyspecific transporters are present in all living systems, be it bacteria or man. Anticancer drugs that are aimed at fighting tumors, drugs that are meant to cure various brain or liver disorders, drugs that are much more preferably taken orally than through a needle more often than not are intercepted by these versatile proteins. They provide a frontline nonspecific defense, independent of an attacked target, which then might be further enhanced by much more toxin-specific mechanisms, such as toxin metabolism or desensitization of the toxin's target [10].

Bacteria, which have been around for a couple of billion years longer than eukaryotic organisms, appear to be particularly endowed with polyspecific transporters. According to predictions from the bioinformatic analysis of more than 200 available bacterial genomes, putative MDRs comprise 2–7% of the total bacterial protein complement [11]. Such putative MDRs are identified based on sequence similarity with experimentally confirmed transporters able to handle multiple "drug-like" substrates. Most of these substrates are hydrophobic or amphipathic molecules frequently containing weakly basic moieties. Other substrates are organic cations whose permanent charge is distributed over a large hydrophobic surface [12, 13]. These general substrate characteristics parallel physiochemical, "drug-like," features required for crossing biological membranes and must be present in both natural toxins and man-made drugs [14].

Obviously, not all predicted transporters are being confirmed to be polyspecific, but quite a few of them are [15], underscoring the impressive scale of the bacterial "resistome." We should consider ourselves lucky that not all of these polyspecific transporters, frequently dubbed MDRs, are in fact clinically relevant. While they effectively expel dyes, quaternary ammonium compounds, and other organic chemicals, efflux of therapeutically useful antibiotics either cannot be detected or is not often significant enough to confer resistance associated with treatment failures. However, enough of them do their job so well that they frequently leave us completely armless against some bacterial pathogens.

In this by no means comprehensive overview, we examine the families of bacterial polyspecific transporters and features of their structure and transport mechanisms that make them particularly efficient in preventing antibiotics access to their targets. We discuss their impact on antibiotic effectiveness in clinical settings and on anti-infective drug discovery. Finally, we describe some humble efforts to outsmart mother nature.

An enormous amount of information on bacterial efflux transporters has been summarized in several recent excellent review articles, addressing various aspects of the efflux problem, such as the origin and evolution of efflux pump [16], mechanisms of multidrug recognition [10, 17], regulation of expression of MDR pumps [18–20], natural functions of MDR pumps [21], and clinical aspects of MDR [22–24].

5.2
Diversity of Bacterial MDR Efflux Systems

Functional studies and subsequent phylogenetic analysis demonstrated that bacterial MDR transporters could be organized into five evolutionary distinct protein families that significantly differ in bioenergetics, structure, and transport mechanism [11]. The important implication is that during evolution polyspecific transporters appear to have arisen independently several times. This structural and functional diversity apparently strengthens antitoxin barriers. By the same token, however, it also significantly complicates the task of penetrating them on demand [16].

Most of MDRs are found in three large and diverse superfamilies: ABC (ATP binding cassette) [25], MFS (major facilitator superfamily) [26], and RND (resistance-nodulation-cell division) [27]. In addition, some MDRs form a core of smaller superfamilies: SMR (small multidrug resistance) family (now part of the DMT (drug/metabolite transporter) superfamily) [28, 29] and MATE (multidrug and toxic extrusion) family (recently joined the MOP (multidrug/oligosaccharidyl-lipid/polysaccharide) superfamily) [30]. A relational database (TransportDB) that classifies and predicts drug efflux systems from organisms whose genomes have been sequenced is available at http://membranetransport.org/.

Genomic studies indicate that the total number of MDR systems is approximately proportional to the total number of all of the transport systems identified in a given organism [31]. The complement of MDR transporters in an organism appears to correlate with both evolutionary history and overall physiology and lifestyles of the organism. Accordingly, intracellular pathogens seem to contain a somewhat limited repertoire of MDRs, presumably due to their stable environment. On the contrary, soil- and plant-associated bacteria generally appear to encode a larger variety and number of MDR transporters, perhaps reflecting much more variable conditions of their own ecological niches [31].

ABC MDRs (as all other members of this superfamily) are primary active transporters that couple substrate translocation to binding and hydrolysis of ATP. MDRs in all the other superfamilies are secondary transporters that use electrochemical gradients of ions (most frequently protons but sometime sodium) to transport their diverse substrates. Both primary and secondary transporters are ubiquitous in bacteria; however, their relative presence seems to correlate with energy generation: fermentative bacteria tend to rely more on the primary transporters while aerobic bacteria contain somewhat more secondary transporters in their genomes [32, 33].

While being polyspecific, bacterial MDRs from different families are still vastly different in the level of promiscuity toward various substrates. Nonetheless, some substrates such as hydrophobic cations, for example, ethidium bromide, can be handled by a majority of them. Bacteria have apparently evolved with special care to ensure protection against such compounds. Such compounds have the potential to be particularly harmful. If not intercepted, they can accumulate intracellularly, driven by the inside negative membrane potential. Perhaps not by coincidence but exactly because of this extra precaution taken by bacteria long ago, hydrophobic cations are conspicuously absent from the list of clinically useful antibiotics [34].

A large body of biochemical and genetic studies coupled with recent advances in structural biology has provided a solid basic understanding of the architectural principles for an MDR binding pocket that enables recognition of structurally diverse drugs. Still much is left to be learned about specific coupling mechanisms and the nature of corresponding conformational changes.

5.2.1
ABC Transporters

Putative drug transporters belonging to the ABC family appear to be one of the most abundant drug efflux families in prokaryotic genomes [35]. Both antibiotic-specific and polyspecific ABC transporters have been identified. Until recently, there were no examples of ABC transporters mediating clinically relevant antibiotic resistance in bacterial pathogens. Emerging data, however, might significantly change this perception, at least for Gram-positive bacteria (see below).

Most of the drug-specific ABCs have been found in antibiotic-producing soil bacteria, such as *Streptomyces*, where they participate in antibiotic export. They are also an essential component for protection of bacteria from their own antibiotic [20, 36]. An intriguing possibility is that they might also be used by the antibiotic-producing organisms as a mechanism for programmed cell death. By the same token, the transporters might serve as potential targets for assisted suicide, performed by environmental cohabitants, which are armed with appropriate transporter inhibitors.

All ABC transporters contain four essential modules, two nucleotide binding domains (NBDs) and two transmembrane domains (TMDs). These four modules can be encoded by four separate genes or fused pairwise in all possible combinations [35, 37]. Bacterial MDRs are usually homo- or heterodimers in which one NBD is fused to one TMD. This is the case for the most well-studied bacterial ABC MDRs, LmrA [38] and LmrCD [39], both from *Lactococcus lactis*.

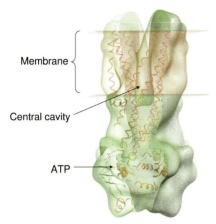

Figure 5.1 Structure of Sav1866 (PDB code 2onj). The simplified shape of the protein surface was generated in ICM (Molsoft) using an FFT-based smoothing algorithm.

Vast amount of biochemical, genetic, and structural information on purified NBDs has been amassed during past two decades [40, 41]. These studies clearly establish the existence of two ATP binding sites at the interface between the two interacting NBDs, so that each bound ATP molecule interacts with amino acid residues from both. Drug binding pocket(s) is localized to TMDs based on a variety of biochemical, mutagenesis, and cross-linking experimental data [42, 43]. The first high-resolution crystal structures, derived from the putative MDR ABC transporter, Sav1866 from *S. aureus*, were recently determined, containing either bound ADP or ATP [44, 45] (Figure 5.1). Both structures in fact are very similar, reflective of the nucleotide phosphate-bound state, with the NBDs in close contact. The two transmembrane α-helices (TMHs) form a central chamber that is shielded from the inner leaflet of the lipid bilayer and from the cytoplasm, but exposed to the outer leaflet and the external space. Though the crystals lack substrate, it is conceivable that this chamber constitutes a drug binding pocket for the MDR ABC transporters. This architecture also indicates that the residues for all transmembrane helices might potentially contribute to substrate binding, providing the structural basis for a single, large, and versatile pocket that can accommodate many structurally diverse substrates.

The structure of Sav1866 is consistent with an "alternating access and release" mechanism of the ATP-driven ABC-mediated efflux. This mechanism invokes two alternative states: an inward-facing conformation with the high-affinity binding site accessible from the inner membrane leaflet and an outward-facing low-affinity extrusion pocket [37]. According to the "the ATP switch model" proposed by Linton and Higgins [46], it is ATP binding and hydrolysis that converts one state into the other. In the transport process, high-affinity substrate binding induces high-affinity ATP binding, consequent conformational changes in the TMHs, resulting in reorientation of the binding site, and reduction of drug binding affinity so that drug can be released into the extracellular milieu, which in turn induces ATP hydrolysis and subsequent release of ADP and P_i to reset the system. Recent structural studies of substrate-specific ABC transporters provided further support for the "ATP switch" model and highlighted conformational changes in transporters associated with this reaction cycle [44, 47–50]. However, the nature of conformational changes is not known.

5.2.2
MFS Transporters

MFS is the largest superfamily of transporters involved in symport, uniport, or antiport of various small solutes [26, 51]. Examples include sugars, neurotransmitters, amino acids, Krebs cycle metabolites, and importantly for this chapter, drugs. Both drug-specific and MDR proteins are among members of this large and diverse superfamily [52], and both play significant roles in drug resistance in clinical settings. Selective examples include tetracycline and macrolide-specific efflux pumps and efflux pumps conferring resistance to fluoroquinolones in Gram-positive bacteria [20, 22, 53] (see below). Most are 400–600-amino acid residues in size and possess either 12 or 14 putative transmembrane domains. Based on sequence similarity between the N-terminal and C-terminal halves, it appears that MFS transporters arose from an internal gene duplication event [26].

In general, hydrophobic cationic compounds are the preferred but not only substrates of MFS MDRs. MdfA from *Escherichia coli* [54] can extrude neutral compounds such as chloramphenicol in addition to various cationic compounds. MdfA was also shown to pump out an artificial substrate of β-galactosidase, isopropyl-β-D-galactoside [55].

Based on phylogenetic analyses, it appears that specific and MDR transporters in the MFS family are scattered randomly on the evolutionary tree, indicating that the broadening and narrowing of specificity toward particular drugs occurred repeatedly during evolution. Indeed, it has been shown that a single amino acid change in the *S. aureus* QacB transporter enables it to recognize not only monovalent but also divalent cationic compounds [52]. A similar result was obtained for LmrP from *L. lactis* where two acidic residues were required to recognize a bivalent cation, Hoechst 33342, while a single negatively charged residue was sufficient for binding the monovalent cation ethidium [56]. In the case of MdfA, where a membrane-embedded negatively charged glutamic acid residue was replaced with positively charged lysine, the ability to transport cationic substrates was lost, while the ability to efflux chloramphenicol was retained [57]. The modulation of substrate specificity may very well occur in response to strong selective pressure applied by antibiotics in the clinic.

The crystal structure for the first MFS MDR transporter, EmrD from *E. coli*, was recently determined at 3.5 Å resolution and it shows a cavity formed by transmembrane helices within the membrane [58] (Figure 5.2). The overall architecture is similar to that found in earlier studies for two substrate-specific transporters, LacY [59] and GlpT [60] from *E. coli*, somewhat surprising given the fact that LacY is a symporter of protons and galactosides and GlpT is an organic phosphate/inorganic phosphate antiporter. The external helices for all of these proteins adopt similar configurations. However, the cavity formed by EmrD is larger in size and aligned mainly by hydrophobic and aromatic amino acid residues, while LacY and

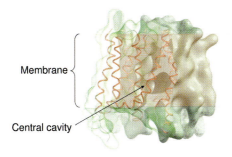

Figure 5.2 Structure of EmrD (PDB code 2gfp).

GlpT have much more hydrophilic interiors. EmrD has not been characterized biochemically, and as such its structure was solved without a substrate. Nevertheless, comparison with much better characterized MFS MDRs (LmrP and MdfA) indicates that the EmrD cavity indeed might be involved in binding of multiple drugs [58].

Based on structural similarities with the extensively studied LacY transporter, it is proposed that EmrD and other MFS transporters may be acting by a similar "alternate access and release" mechanism [61, 62]. Accordingly, the substrate binding site has alternating access to either side of the membrane as a result of the opening and closing of the binding cavity to either membrane side. The proton and substrate, in the case of LacY, lactose, seem to use distinct pathways. Binding of a proton followed by substrate binding initiates a series of conformational changes resulting in reorientation of the binding pocket, followed by release of substrate and proton at the other side of the membrane, with a conformational change corresponding to a switch back to the basal configuration. Overall, this mechanism is very similar to the above-discussed mechanism for ABC transporters; the main difference is that it is protonation/deprotonation of acidic amino acid residues rather than ATP binding and hydrolysis that control the conversion of a transporter from one state to the other.

In MFS transporters, negatively charged amino acid residues located in TMHs appear to play a critical role in the protonation/deprotonation step during transport. They are essential for transporter activity. Unexpectedly, analyses of the two MFS MDRs, LmrP [63] and MdfA [64], revealed that no single membrane-embedded acidic residue is critical for drug transport. This is in contrast not only to other MFS proteins but also to many other secondary transporters including the MDRs belonging to the RND and SMR families, where negatively charged membrane-embedded residues have been shown to play irreplaceable roles in proton-coupled transport reactions. Thus, a seemingly well-studied field remains wide open for further perusal.

5.2.3
MATE MDR Transporters

Transporters from the MATE family are secondary transporters driven by Na^+ or H^+ ion gradients [65]. Most members of this family consist of 400–550-residue polypeptides with 12 putative TMHs and sharing about 40% sequence similarity. Currently, no high-resolution structures are available for this family of transporters. Secondary structure predictions showed the symmetric repetition of conserved regions in the N- and C-terminal halves of the proteins suggesting a gene duplication event. Little is known about the molecular mechanism underlying their organic cation transport. Similarities with the MFS secondary transporters can be anticipated on the basis of common structural features such as 12 TMHs without large loops. In addition, a recent site-directed mutagenesis study has revealed the importance of negatively charged residues Asp32, Glu251, and Asp367 located in the TMH1 of the NorM transporter from *Vibrio parahaemolyticus* responsible for Na^+-driven organic cation export [66].

Although not abundant, multiple copies of MATE orthologues are usually present in bacterial genomes [51]. While their preferred substrates are hydrophobic cations, recent findings implicate the MATE transporter MepA from *S. aureus* in a two- to four fold level of resistance to the antibiotic tigecycline [67], the most recent antibiotic approved by the FDA for the treatment of some multidrug-resistant Gram-negative and Gram-positive bacteria, including MSRA [68].

5.2.4
SMR MDR Transporters

The SMR family includes small multidrug transporters widespread among eubacteria [69]. These proteins are about 100-amino acid residues long, with four TMHs. The family archetype is EmrE, originally identified as the genetic locus in *E. coli* that encoded a protein conferring bacterial resistance to ethidium bromide and methyl viologen [70]. Extensive biochemical studies have established that EmrE functions as an oligomer, probably a dimer. It is presumed that all other members of the SMR family function similarly as homo- or heterodimers, and it is hypothesized that heterodimers were selected as a means to expand substrate specificity. Labeling and cross-linking studies, and more recently the use of genetically fused EmrE monomers, unequivocally show that the monomers have the same orientation in the membrane [71, 72]. However, genome-wide topology predictions and structural data suggest that some SMR heterodimers can assume opposite orientations in membrane [73, 74].

The small size and complete functionality of EmrE in detergents made it an attractive model for studying the structural/functional aspects of transport reactions in ion-coupled processes. SMR proteins appear to employ the simplest coupling mechanism. Glu-14 is the only charged residue in the putative membrane domain of EmrE and is highly conserved in other SMR proteins. It plays an indispensable role in multidrug transport by SMRs. In detergent solution, Glu-14 appears to have an unusually elevated pK_a, at around 8.3–8.5 instead of 4.25 in an aqueous environment. In the absence of substrate, this elevated pH appears to be stabilized by at least three aromatic residues, all tryptophans. It was demonstrated that deprotonation is essential for substrate binding. Substrate binding in turn induces proton release. Therefore, there is an overlap of substrate and the proton binding site, and transport appears to occur by "alternative occupancy" or a "time-share" mechanism. The mutual exclusivity of substrate and proton binding provides the basis of the coupling mechanism. Importantly, similar to other transporters, the binding site is rich in hydrophobic residues (at least six of them, three from each monomer) [70, 72].

5.2.5
RND MDR Transporters

The RND transporters are responsible for the high intrinsic antibiotic resistance seen in Gram-negative bacteria, one component of the notorious "natural superbug" phenotype (see below). They are also found in Gram-positive bacteria where their functions are largely unknown [27]. Numerous RND transporters from *Mycobacterium tuberculosis* are all related to metabolism of lipids and cell wall [75], though a recent report implicates one of these transporters in efflux of the first-line antimycobacterial drug, isonazide [76].

RND transporters possess an astonishing breadth of substrate specificity and in this respect surpass even the ABC transporters, which are a major hurdle in effective anticancer therapeutics [77]. RND pumps can recognize and extrude positive, negative, or neutral charged molecules, substances as hydrophobic as organic solvents and lipids, and compounds as hydrophilic as aminoglycoside antibiotics. They are a ubiquitous family whose members are distributed across various kingdoms [27]. Several representatives of the RND-permease superfamily are encoded in the human genome, though the similarity to bacterial RNDs is negligible (16% identity). Examples of human RNDs include the NPC1 protein, localized in lysosomal membranes and apparently involved in intracellular cholesterol transport [78, 79], and the homologue of *Drosophila* morphogen receptor, Patched, thought to be crucial in the suppression of basal cell carcinoma [80].

An extraordinary amount of structural information on bacterial RND transporters has been published in the last few years (Figure 5.3). Several high-resolution structures of the AcrB efflux pump from *E. coli*, with and without cocrystallized substrates, as well of several mutant AcrBs, have emerged from several laboratories across the world [81–86]. As discussed below, RND transporters from Gram-negative bacteria function with two other proteins membrane fusion protein (MFP) and TolC in a large complex. X-ray structures of the other proteins have also become available [87–91], and only the structure of the tripartite complex remains to be determined (Figure 5.4 depicts the fitted model of the *E. coli* AcrAB–TolC tripartite efflux complex). The output from analysis of this detailed structural information should dramatically facilitate discovery of inhibitors of RND transporters, an industry goal for improving the efficacy of antibiotics against problematic Gram-negative bacteria that are the cause of many life-threatening infections.

The high-resolution structure of *E. coli* AcrB highlights the structural features of RND transporters responsible for their ability to bind multiple drugs and associate with the accessory proteins. RND transporters function as homo- or heterotrimers consisting of protomers of about 1100-amino acid residues in length. Each protomer consists of the 12 TMHs (TMH1–TMH12) and a large periplasmic domain. The

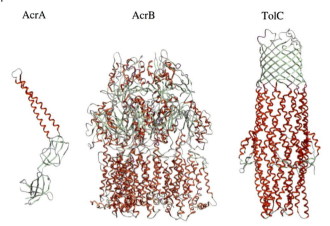

Figure 5.3 Structure of AcrA (PBD code 2f1m), AcrB (PBD code 1ek9), and TolC (PBD code 2dhh) proteins from *E. coli*. α-Helices and β-sheets are colored in red and green, respectively. The figure shows a monomer of AcrA and trimers of AcrB and TolC. All proteins are shown to scale.

periplasmic domains assemble through multiple contacts between protomers into a "mushroom"-like structure with a central channel, three cavities and three vestibules, which is also called the pore domain. In contrast, inside the membrane, the three protomers have very limited contact with each other.

Mutagenesis and structural studies have established that the periplasmic domains of RND transporters play a dual role [92–95]. On the one hand, these domains are

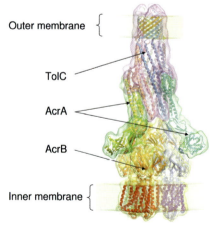

Figure 5.4 Fitted model of the *E. coli* AcrAB–TolC tripartite efflux complex. The model is based on 1 : 1 : 1 stoichiometry of AcrA: AcrB:TolC suggested in Ref. [113]. In this model, three AcrAs do not form a complete ring that would seal TolC/AcrB connection as was suggested in earlier models.

implicated in substrate recognition and binding. Several substrate binding pockets have been mapped in the pore domain of the transporter by structural and genetic studies. On the other hand, the periplasmic domains provide RND transporters with structural extensions that allow them to traverse halfway through the periplasm to gain contact with the outer membrane channel. The pore domain contains deep cavities, sites of possible interaction with a periplasmic protein, MFP. The contact between these proteins is believed to be extensive, involving the entire periplasmic length of the RND protein. In contrast, contact with the OM channel is very limited, with the two proteins barely touching each other.

AcrB-drug cocrystals have revealed drugs interacting with the protein in several, almost too numerous locations. In the first report, several structurally unrelated drugs were seen in the central cavity on the membrane–periplasm interface, prompting a model that drug substrates first intercalate into the phospholipid bilayer and then laterally diffuse into the central cavity of AcrB [82]. This central cavity is very similar to the binding chamber described earlier for P-gp transporters, a voluminous space formed at the interface of interacting subunits containing multiple aromatic residues capable of hydrophobic and stacking interactions. However, in the central RND cavity, residues involved in substrate binding are highly conserved, and binding of substrates in this site does not explain the documented differences in substrate specificities for different RNDs. Mutational analyses of AcrB and the related EmhB from *Pseudomonas fluorescens* nevertheless show that amino acid residues in the central cavity do have an impact on transporter-mediated antibiotic resistance [83, 96].

In the second cocrystal structure of the same protein, an additional drug binding pocket was detected in the prominent cleft on the surface of the periplasmic domain. This site might be fully exposed to the periplasm with easy access for drug binding. A problem with this site is that it is not a clear place for the drug to go. Mutations altering several amino acids within this site have also been reported to impact RND-related antibiotic resistance [83, 93, 97]. One intriguing possibility is that this site might play a role in the regulation of the activity of the transporter by its substrates, rather than mediating the actual transport process directly.

Finally, a third cocrystal structure of AcrB has revealed yet another, nonoverlapping, multidrug binding pocket, located deep inside the periplasmic domain [85]. This voluminous pocket is extremely rich in aromatic amino acid residues capable of hydrophobic and stacking interactions. There are also a few polar residues that can form hydrogen bonds. Interestingly, the cocrystallized substrates doxorubicin and minocycline were found to be interacting with different sets of amino acid residues (Figure 5.5a and b). This binding pocket is accessible through an uptake channel, a channel branching from a vestibule that runs parallel to the membrane plane and reaches the central cavity. There is also a clear path from this binding pocket to the top of the funnel and therefore to TolC. Mutagenesis studies hold several amino acid residues responsible for substrate specificity in this site [93, 95]. While it seems that this particular pocket combines all proper features for a multisubstrate binding site, more studies are required to clarify the roles of all three observed sites in binding and/or transport of multiple drugs.

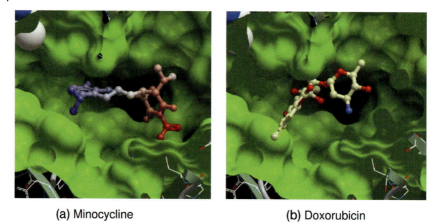

(a) Minocycline (b) Doxorubicin

Figure 5.5 Periplasmic substrate binding pocket of AcrB. The cocrystallized substrates minocycline (a) and doxorubicin (b) were found to be interacting with different sets of amino acid residues.

No substrate competition was observed in recent accumulation assays using whole cells of *E. coli* and cocktails of up to seven diverse substrates or competing substrates of similar structural classes. In addition, very high concentrations of substrates significantly exceeding concentrations required for cell growth inhibition failed to saturate transporter in the same experimental system [98]. Interpretation of these experiments is complicated by the presence of the outer membrane that retards the uptake of various compounds to varying degrees. However, despite the fact that these experiments might not provide rigorous biochemical evidence of, for example, simultaneous binding of different substrates at the same binding pocket, such experiments provide valuable clues in support of current models for RND transporter function. The very fact that they were performed with whole cells and were guided by specific antibiotic susceptibility data makes this experimental setup particularly relevant to the "real-bug" situation and helps to explain some results concerning the effect of RND transporters on antibiotic resistance in clinical settings (see below).

Besides transporter architecture and location of potential substrate binding pockets, crystal structures provided further exciting and unexpected insight into the possible transport mechanisms by RND transporters [85, 86]. The first X-ray structure of the RND pump AcrB was presented as a perfectly symmetric trimer [81]. It gave rise to the so-called "elevator mechanism" of transport wherein it was proposed that substrates accumulating in the central cavity are actively transported into the upper portal space via a channel that opens along the central axis of the structure. However, for this model to work, a very significant conformational change associated with channel opening would have to be coupled with proton transport via the transmembrane domain in order to accommodate the passage of substrates. In contrast, two new structures of AcrB trimers, while symmetric overall, show each protomer in a distinct conformation [85, 86]. Although only one of these "new generation" structures contains cocrystallized substrates [85], the conformations of protomers in both are strikingly similar. This shows that substrate is not needed to induce asymmetry.

In the periplasmic portion, the main differences between the old and new structures are in the substrate binding pocket. First, the substrate is present in only one of the protomers, dubbed the "binding" protomer (B). The spacious drug binding pocket described earlier is open to the periplasm and expands to almost reach the TolC docking funnel. The exit from the pocket into the funnel is blocked by an inclined α-helix of the central pore from the adjacent protomer. The binding pocket of this second protomer, called the "extrusion" protomer (E), is closed to the periplasm, significantly reduced in size, and opened toward the funnel. The binding cavity of the third "access" protomer (A) is largely inaccessible from either the periplasm or the exit funnel.

On the basis of this asymmetric structure, a new mechanism of drug transport has been proposed. This "alternate occupancy" model implies that each protomer cycles through three consecutive conformations, named after the nomenclature for F_1F_0-ATPase, as loose (L), tight (T), and open (O), corresponding to three phases of efflux [86]. This cycling is sequential, rather than synchronous, such that at any given time each protomer exists in a different phase. Affinity of the substrate to the periplasm-accessible conformation of the pump subunit cavity is expected to be higher than the affinity to the funnel-opened conformation. The transition of the ligand-bound protomer from a high-affinity to low-affinity state should require energy input, which is evidently provided by the coupled proton transport in the transmembrane domains. While the details of the mechanism remain to be clarified, existing studies provide some initial clues.

Mutagenesis data indicate that AcrB has four electrostatically interactive residues that constitute the putative proton relay pathway, located in TMH4, TMH10, and TMH11 [99, 100]. In the binding and access protomers, Lys940 of TM10 and Asp407 and Asp408 of TM4 are coordinated by salt bridges. However, in the extrusion protomer, Lys940 is turned toward Thr978 and the salt bridges are absent. This in turn causes twisting of TMH4 and TMH10. More data are required to determine whether these subtle changes in TMHs are sufficient to produce the large conformational changes in the porter domain, resulting in ultimate efflux. However, what is absolutely clear is that the residues involved in proton and substrate translocation are, as expected, far apart.

5.2.6
Diversity on a Theme

Based on recent results from very different polyspecific transporters, it is becoming apparent that all share common principles of multidrug recognition. The very first insight into the multidrug binding mechanism became possible after solving the high-resolution structures for several soluble multidrug binding regulatory proteins bound to various lipophilic ligands (BmrR [101, 102], QacR [103, 104], and PXR [105]). In all cases, ligands were found in large drug binding pockets formed mainly by hydrophobic and aromatic residues that participate in a number of van der Waals interactions with the ligand, augmented by electrostatic attraction between positively charged ligand moieties and negatively charged amino acid residues in the pocket.

It was insightfully proposed by the late Alex Neyfakh [17] that relatively strong binding based on such interactions is possible without precise spatial match between protein and its ligand. It was argued that the classical "lock-and-key" principle developed for enzymes interacting with hydrophilic substrates does not apply. With "lock-and-key" binding, a precise match makes it possible to form a collection of hydrogen and electrostatic bonds between enzymes and substrates. Only then is it possible to outcompete multiple water molecules that are present in the reaction environment. Moreover, the precise special match will automatically result in high substrate specificity. In contrast, lipophilic substrates that are driven into hydrophobic pockets simply by a hydrophobic effect do not need to overcome such obstacles. The lack of a "lock-and-key interaction" provides the basis for binding promiscuity and a minimum requirement for a substrate to just fit into the pocket.

This concept for binding of lipophilic substrates in hydrophobic pockets is the basis for a rapidly emerging paradigm for versatile multispecific recognition pockets in polyspecific transporters. The paradigm is consistent with both the limited structural data and the vast amount of biochemical and genetic data on MDR transporters demonstrating that (i) different substrates can use different residues to bind in the same pocket; (ii) the same substrates can assume multiple positions in the pocket; (iii) two substrates can be bound in a single pocket simultaneously; and (iv) this binding can give rise to negative or positive cooperativity. Elegant studies on the human P-gp transporter, though not at the structural level, further illustrate such versatility through the concept of the "induced best fit," by which a substrate can provoke rearrangements in the pocket during binding [106, 107]. In general, provided that appropriate selective pressure is applied experimentally, it is relatively easy to modulate the substrate specificity in mutants of a polyspecific transporter, forcing them to either a broader or more limited substrate range. This further stresses their flexibility. However, it also underscores the potential for bacterial transporters to become major impediments to infectious disease control.

5.3
Accessory Proteins from Gram-Negative Bacteria

In eukaryotic and Gram-positive prokaryotic cells, a single transporter located in the cytoplasmic membrane is sufficient to facilitate efflux. In contrast, the majority of multidrug transporters from Gram-negative bacteria, with its outer membrane, require two additional proteins, the periplasmic protein belonging to the MFP family and the outer membrane channel, a member of the outer membrane factor (OMF) family. In fact, the majority of the characterized RND, MFS, and ABC transporters involved in drug efflux in Gram-negative bacteria depend on accessory proteins for their activities. There are some exceptions: the MFS transporter MdfA and characterized members of SMR and MATE families appear to function alone.

Efflux is most effective when working in cooperation with other resistance mechanisms. Reduced uptake across the outer membrane of Gram-negative bacteria, which is a significant permeability barrier for both hydrophilic and hydrophobic compounds, constitutes such a mechanism [108]. To take full advantage of the reduced uptake, the MDR pumps in Gram-negative bacteria engage the periplasmic and OM proteins in drug efflux. This allows to efficiently bypass the periplasm so that drugs are extruded directly into the external medium.

The structural features of the accessory proteins are consistent with their role in extending drug efflux across the outer membrane. The 3D structures of three OMFs, TolC, OprM, and VceC from *E. coli*, *P. aeruginosa*, and *Vibrio cholerae*, respectively, have been recently solved [89, 91, 109]. Despite very little sequence similarity, they are structurally conserved. Like AcrB, they form stable trimers organized into two-barrel structures. A 12-stranded 40 Å long β-barrel inserts into the outer membrane to form an open pore 30 Å in diameter. The unusuall α-helical barrel 100 Å in length protrudes deep into the periplasm, where it reaches the TolC docking domain of AcrB. The lower half of this barrel is bound by an equatorial domain of mixed α/β-structure. The tip of the periplasmic end of the channel is closed in an iris-like manner by interacting loops of α-helices.

Biochemical and genetic data demonstrate that MFPs interact with both the RND pump and the OM channel [110, 111]. It is therefore proposed that the MFP stabilizes weak RND–OMF interactions and promotes and maintains the tripartite complex. The recently determined structures of MexA and AcrA, from *P. aeruginosa* and *E. coli*, respectively [87, 88, 90], are consistent with such a function. These MFPs appear to have a modular structure, with a β-barrel domain connected to a lipoyl domain that in turn is attached to a long periplasmic α-helical hairpin. Mutagenesis and cross-linking studies of site-specific TolC and AcrA cysteine variants identified the interface between two proteins [112, 113]. This interface is formed by residues on the lower α-helical barrel domain of TolC and within the N-terminal α-helix of the AcrA coiled coil. In addition, the MFP appears to possess significant conformational flexibility [114], which might be important to ensure the most advantageous interaction with TolC. In modeling an "open state" of the TolC entrance, there appears to be a perfect fit with the funnel-like opening of the TolC docking domain of AcrB [115]. The possibility exists that the MFP plays an active role in the opening of the TolC channel during drug transport. It is also possible that the "open" state is the result of the AcrB–TolC interaction and that the role of the MFP is to keep both proteins in this fixed state.

The oligomeric state of MFPs still remains controversial. Soluble forms of AcrA and MexA have been found to be monomeric *in vitro*, but cross-linking of AcrA *in vivo* suggests that the MFP works as a trimer with its two other partners [13, 116].

There are no obvious structural differences between transporters from the same superfamily that function either alone or in complex with accessory proteins. No specific structural determinants were identified that enable a transporter to engage the accessory proteins. In contrast, periplasmic MFPs are adapted to the structural diversity of transporters. The major differences in MFPs are in the N-terminal domains and the length of the periplasmic coiled-coil helical regions [117]. MFPs that function with transporters lacking large periplasmic extensions, such as ABC and MFS transporters, contain a single N-terminal transmembrane α-helix, which is essential for protein–protein interactions within the membrane. In contrast, MFPs engaged by the RND transporters, which contain large periplasmic domains, undergo processing followed by lipid modification of their N-terminal cysteine residues. Therefore, interacting surfaces of MFPs and RND transporters are located in the periplasm. The central hairpin domains follow the same pattern. The central domains of MFPs engaged by ABC and MFS transporters are notably larger than those engaged by RND transporters.

It is presently unclear whether all or only some of the MFPs interact with substrates during transport. The periplasmic portions of the MFPs associated with ABC and MFS transporters are fully exposed to the periplasm and are expected to come into a direct contact with substrates during translocation across the periplasm. In the case of RND pumps, however, the entire MFP seems to be engaged into protein–protein interactions without any freedom to bind substrates on its own [115, 118, 119].

The direct interaction with substrates was demonstrated biochemically or genetically for MFPs associated with all three types of transporters. The periplasmic domain of the EmrA protein associated with the multidrug MFS transporter EmrB was reported to bind multiple drugs [114]. Genetic data suggested that both the cytoplasmic and the periplasmic domains of HlyD, a component of the ABC-dependent type I hemolysin secretion, are important in substrate recognition [120]. Specific amino acid residues in both N- and C-terminal halves of the periplasmic domain of HlyD are important not only for the secretion of hemolysin but also for the proper folding of this toxin after the translocation across the outer membrane. CusB, a component of the RND-type copper/silver efflux complex, was reported to bind silver [121]. Furthermore, substrate binding causes conformational changes in CusB. Thus, substrate binding could be a universal feature of MFPs.

Binding of substrates suggests that MFPs might be directly involved in transporting substrates across the periplasm and the outer membrane. The *in vitro* reconstitution studies support this idea. AcrA stimulates the transport activities of the reconstituted RND transporters AcrB and AcrD [95, 116]. Another MFP MacA is absolutely required for the ATPase activity of MacB, the ABC-type macrolide efflux transporter [122].

While many details remain to be clarified, the emerging architecture of the trimeric MDR transporter complex provides a structural basis for understanding transenvelope efflux (Figure 5.4). A substrate enters the tripartite transporter through the appropriate interenvelope "substrate gate" and exits into the extracellular space through the "exhaust pipe" of TolC.

5.4
Efflux and Antibiotic Resistance

Efflux plays a significant role as a mechanism for antibiotic resistance in many important human pathogens. However, the prevalence of efflux-mediated antibiotic resistance varies significantly depending upon specific antibiotic and bacterium. Both antibiotic-specific and MDR pumps are known to confer clinically relevant resistance [22, 53]. The former are carried on plasmids and transposons. While they can be rapidly spread between various strains, they are at hand only in a proportion of a total population of a given species. Tet and Mef pumps, which confer resistance to tetracycline and macrolides, respectively, in both Gram-positive and Gram-negative bacteria, are clinically important and highly prevalent antibiotic-specific transporters. In fact, it is the wide distribution of Tet transporters that led to the demise of tetracycline, one of the only few truly broad-spectrum antibiotics. Over the years, significant efforts have been made to modify both tetracycline and macrolide scaffolds to identify derivatives no longer recognized by these transporters. These multiyear efforts have been a success, with the introduction of tigecycline and telithromycin, which avoid efflux by multiple Tet and Mef transporters, respectively [123], and are the latest addition to approved antibiotics with activity against tetracycline and azithromycin/erythromycin (macrolides) resistant pathogens. However, the activity and spectrum of these antibiotics are still significantly reduced by MDR transporters present in some important Gram-negative pathogens [53].

MDR transporters are usually encoded by housekeeping genes as normal constituents of bacterial chromosome and are present in the whole population of a given bacterial species. The basal level of expression of nonspecific multidrug efflux pumps in "wild-type" cells determines the basal level of antibiotic susceptibility. This innate "resistance" may still be low enough such that bacteria are susceptible to therapy with a given antibiotic.

5.4.1
Gram-Positive Efflux Resistance

Efflux-mediated clinical intrinsic resistance is generally not seen with Gram-positive bacteria. There are only a few examples when antibiotic nonsusceptibility is caused by efflux. One such example emerged relatively recently in the case of *Enterococcus faecalis*. This bacterial pathogen paved its way to the recent list of "superbugs" developed by the Infectious Disease Society of America [2]. Enterococci are part of the diverse bacterial community inhabiting the gastrointestinal tract. They are normally benign commensals [100]. However, they can cause serious hospital infections such as bacterial endocarditis and meningitis, in particular in immunocompromised patients. For a Gram-positive bacterium, *E. faecalis* has an unusually high level of intrinsic antibiotic resistance. It is not susceptible to lincosamides (lincomycin and clindamycin) and significantly less susceptible to aminoglycosides, macrolides, and fluoroquinolones. *E. faecalis* is also intrinsically resistant to the recently developed antibiotic Synercid (the synergistic mixture of the two semisynthetic derivatives of the streptogramin-class pristinamycin, quinupristin and dalfopristin) [124]. Systematic genetic disruption of 23 genes encoding putative ABC transporters showed that two were involved in conferring high-level intrinsic antibiotic resistance in this organism. Deletion of abc16 resulted in a 10-fold increase in susceptibility to macrolides, while deletion of abc23 made *E. faecalis* 50–200-fold more susceptible to lincosamides and Synercid [125]. No significant reduction in the resistance to fluoroquinolones was observed. It is quite possible that other, non-ABC-type MDRs contribute to fluoroquinolone resistance in this organism.

Fluoroquinoles are one of the most widely used classes of antibiotics. They are the rare example of a purely synthetic antibiotic (not derived from naturally occurring microbial metabolites) and represent a great success in the application of medicinal chemistry to antibacterial drug development. The antibacterial activity of fluoroquinolones is based on the inhibition of type II topoisomerases, DNA gyrase (encoded by the *gyrA* and *gyrB* genes) and DNA topoisomerase IV (encoded by the *parC* and *parE* genes). Owing to their synthetic origin, the only relevant mechanisms of resistance are target modification and efflux by MDR transporters [126, 127]. They are one of the very few classes of broad-spectrum antibiotics and are frontline agents indicated for the treatment of many bacterial pathogens. Among them are *S. aureus* and *Streptococcus pneumoniae*. Both of these Gram-positive pathogens are the leading causative agents for many community and hospital infections. Both contain constitutively expressed MDR efflux pumps defining the basal level of resistance to some so-called "second-generation" fluoroquinolones, such as ciprofloxacin [128]. The MFS transporter NorA is a constitutive efflux pump in *S. aureus* [129, 130], whereas the recently discovered heterodimeric ABC transporter SP2073/SP2075 protects *S. pneumoniae*. Inactivation of either of these transporters results in fourfold increased susceptibility [131, 132].

Efflux pumps might become overexpressed as a result of induction or mutations in regulatory genes. The resulting reduced susceptibility might still not be high enough to be viewed as clinical resistance if an approved antibiotic regimen is expected to treat the corresponding strains. However, clinical resistance can subsequently be rapidly developed due to acquisition of additional resistance mechanisms, for example, target mutations. This process is well documented in case of resistance to fluoroquinolones in *S. aureus*, *S. pneumoniae*, *E. coli*, *P. aeruginosa*, and many other pathogens [20, 22]. Target mutations make the target of the antibiotic less sensitive to antibiotic action, and as a result higher internal and consequently external concentrations are needed to inhibit bacterial growth. Importantly, the contribution of efflux to resistance appears to be added on top of this, and the same regardless of the presence of even multiple target mutations, indicating that transporters can still maintain an antibiotic concentration gradient even when external concentrations are significantly elevated. An experimental demonstration of this high capacity of efflux pumps was described earlier in this chapter, exemplified by the AcrAB–TolC RND transporter from *E. coli*, which is capable of preventing drug accumulation at external drug concentrations at least 100-fold higher than that required to inhibit bacterial growth [98].

It is also worth noting that, in general, the frequency of efflux-mediated resistance is higher than the frequency of resistance based on target alterations [133]. This is easily understood considering the fact that many efflux genes are under the control of negative regulators and their overexpression occurs as a result of loss-of-function mutations in corresponding repressor genes. The frequency of loss-of-function mutations is usually much higher than the frequency of change-of-function mutations. The loss of function will occur due to many nonsense and frame-shift mutations, while only rare mutations in specific regions of an essential target gene product will maintain protein function with a decreased affinity to an inhibitor compound.

For the above reasons, a dramatic reduction in the emergence of fluoroquinolone (ciprofloxacin) resistance is seen when bacterial strains lacking efflux pumps are used for resistance selection or when selection experiments are performed in the presence of an efflux pump inhibitor (EPI) [133–136], the opposite of what is seen clinically.

Efflux resistance in clinical isolates of *S. pneumoniae* results from overproduction of its constitutive ABC transporter and an additional MDR, PmrA, from the MFS family [137–139]. In *S. aureus*, efflux resistance in clinical isolates results from the overexpression of at least four different MFS MDR pumps: NorA, NorB, NorC, and MdeA [140]. Importantly, the newer generation fluoroquinolones, including levofloxacin, moxifloxacin, and gemifloxacin, are affected by the multidrug resistance pumps in *S. aureus* and *S. pneumoniae* to a much less degree than ciprofloxacin [137, 139, 141]. Among all of the mentioned transporters, it is only NorB from *S. aureus* that is capable of somewhat increasing resistance to moxifloxacin (twofold) [142]. It is believed that the newer fluoroquinolone derivatives are sufficiently hydrophobic so that their rapid passive uptake overwhelms active efflux from the cell. Another possibility is that structural modifications resulting in increased hydrophobicity might directly contribute to an altered affinity for the transporter itself [143]. The altered affinity could be either reduced or enhanced. In the former case, the pump protein might not recognize the substrate, and in the latter case, the substrate might not efficiently dissociate from its binding site. The result, however, would be the same, a decreased effect of efflux on antibiotic susceptibility. Indeed, one specific fluoroquinolone, sparfloxacin, has been shown to noncompetitively inhibit the NorA transporter, which may be the reason why susceptibility to sparfloxacin is unaffected by the NorA pump [144]. Of note is that all these new fluoroquinolones are less effective against Gram-negative bacteria compared to ciprofloxacin. This decreased activity can be generally traced to the constitutively expressed RND transporters (see below).

Thus, while efflux-mediated resistance to some fluoroquinolones is highly prevalent in Gram-positive pathogens, other derivatives appear to overcome this efflux, at least for now. Superbugs such as MRSA are spreading widely in hospitals and now in the community [5, 100]. However, MRSA remains *S. aureus*, a relatively susceptible Gram-positive bacterium that professed acquisition of multiple resistance mechanisms. But due to the natural susceptibility of *S. aureus*, there are many reasons to expect that yet new class of antibiotics will be brought about to fight MRSA infections. Indeed, most of the recently approved antibiotics, such as linezolid, daptomycin, tigecycline, and several novel β-lactams now undergoing clinical development, have excellent activity against MRSA [68].

5.4.2
Gram-Negative Efflux Resistance

The situation is quite different for Gram-negative bacteria, some more than others, and in particular *P. aeruginosa*, the Gram-negative superbug. *P. aeruginosa* is a common pathogen associated with hospital-acquired infections, the causative agent of many life-threatening conditions and the major reason for the shortened life span of people with cystic fibrosis. Pseudomonas infections can be successfully treated by only a few specific representatives of the fluoroquinolones, β-lactams, or aminoglycosides [145]. Other classes of currently available antibiotics (macrolides/ketolides, oxazolidinones, lipopeptides glycopeptides, rifamycins, streptogramins, etc.), in general, are considered to be "Gram-positive only," since they lack any utility against *P. aeruginosa* and other Gram-negative bacteria. RND transporters are the major reason for this intrinsic nonsusceptibility [146].

That intrinsic efflux is the cause of nonsusceptibility to antibiotics in these bacteria is evident from the fact that deletion of the major constitutively expressed MDR efflux pumps in *P. aeruginosa* and many other Gram-negative bacteria renders them as sensitive to antibiotics as the much more susceptible Gram-positive pathogens [53, 147]. The converse, enhanced efflux due to overexpression of MDR pumps, as might be expected and as well documented [22], plays a prominent role in acquired resistance. For example, overexpression of at least four tripartite RND MDR systems, MexAB–OprM, MexXY–OprM, MexCD–OprJ, and MexEF–OprN renders *P. aeruginosa* clinically resistant to fluoroquinlones. MexAB–OprM and MexXY–OprM are also implicated in resistance to β-lactams and aminoglycosides, respectively [20]. Importantly, this phenomenon is seen not only for antibiotics in clinical use but also for drugs in clinical development and for compounds at various stages of preclinical research [23]. This general trend in the lack of Gram-negative activity is rather remarkable considering the plethora of drug targets covered, both old and new [148]. The pervasiveness of this efflux problem in Gram-negative pathogens is particularly unfortunate given that most antibacterial compounds have the potential to be truly broad spectrum, since they inhibit the activity of targets that are well conserved in both Gram-positive and Gram-negative organisms.

Those involved in the earliest stage of antibiotic drug discovery, primary antibacterial screening using large libraries of synthetic compounds or natural products, will assert the fact that the hit rate in such screenings is 1000-fold lower against Gram-negative than Gram-positive bacteria. The Gram-negative hit rate is significantly increased when screening is performed with mutants lacking efflux pumps [149]. Thus, it is clear that MDRs from Gram-negative bacteria can successfully handle thousands of different substrates. The result of this intrinsic efflux is an inadequate number of primary hits, which are the starting point for the discovery of new agents.

In summary, efflux significantly reduces the number of therapeutic options available among antibiotics in current clinical practice and has prevented the efficient discovery and development of new agents. Furthermore, due to the nature of the envelope of Gram-negative bacteria, and considering the promiscuity of RND transporters, there is not much hope for avoiding the impact of efflux with rapidly diffusing antibiotic derivatives.

These considerations provide a strong argument and rationale for discovery and development of inhibitors of RND transporters from Gram-negative bacteria that could be used in conjunction with antibiotics. Such inhibitors would increase antibacterial potency, expand the spectrum of antibacterial activity, reverse resistance and dramatically reduce the rate of resistance development, and could be applied to new or existing therapeutic agents. Based on the emerging significance of efflux pumps in bacterial pathogenesis [21], one can imagine an additional benefit of some EPIs (efflux pump inhibitors) as antivirulence agents.

5.4.3
Inhibiting Gram-Negative RND Pumps

While RND pumps present an attractive target for the development of antibiotic potentiators, it is important to realize that efflux is often only one of the several mechanisms of resistance in place for a given antibiotic. Thus, its inhibition will have a significant therapeutic effect only for those antibiotics and in those organisms where efflux is a major contributor to resistance. For example, resistance to fluoroquinolones can arise from both target site mutations and efflux via RND-type pumps. In organisms harboring both, the loss of efflux alone dramatically reduces resistance [123, 133]. Thus, inhibition of efflux should improve the effectiveness of fluoroquinolones against a majority of clinical isolates. Resistance to β-lactams in *P. aeruginosa* is mediated by the RND family MexAB–OprM efflux system and by β-lactamases, capable of degrading these antibiotics. It has been shown that the loss of efflux in strains that express both mechanisms fails to overcome resistance [150]. Efflux inhibition would not, therefore, be an effective means of countering β-lactam resistance in *P. aeruginosa*.

Another consideration is that a single bacterium may contain multiple RND systems able to confer resistance to a given antibiotic, as is the case for fluoroquinolones in *P. aeruginosa* and various Enterobacteriaceae. As a result, in such a setting, only a broad-spectrum inhibitor able to interfere with multiple transporters would be effective. In the case of aminoglycoside (e.g., tobramycin) use in maintenance therapy to prevent acute exacerbation of *P. aeruginosa* lung infections in cystic fibrosis patients, however, the major determinant of resistance is a single RND-type efflux system, MexXY–OprM [53], and disrupting this single pump would be expected to be clinically effective.

Based on the elucidation of the structure and the efflux mechanism of RND transporters, there may be several potential ways to interfere with their function.

Inhibitors could target the substrate binding pocket. This voluminous pocket can accommodate an extraordinary variety of structures; hence, many distinct pharmacophore classes can be foreseen. Importantly, the binding site of the RND transporter is located in the periplasm, so that an inhibitor need only cross one membrane.

As was evident in the most recent structural study [85], various substrates can occupy different locations in the large substrate binding pocket. It is therefore expected that some inhibitors will interact with and potentiate only a subset of substrate antibiotics. This is the case with MC-207110, the very first efflux pump inhibitor to be identified, which is active against multiple RND pumps in a variety of Gram-negative pathogens [135]. This compound (Figure 5.6a) was originally identified in a screen for potentiators of the fluoroquinolone levofloxacin in an RND pump-expressing organism *P. aeruginosa* and is now routinely used as a research tool to

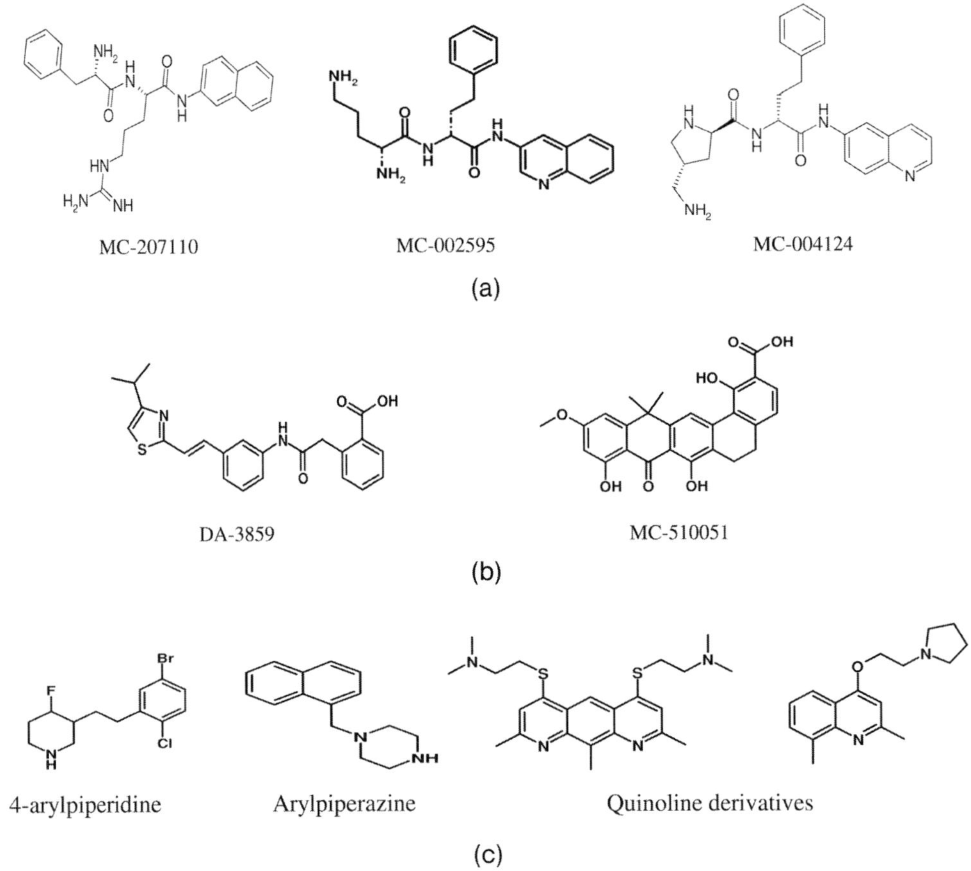

Figure 5.6 Various inhibitors of RND transporters. (a) Broad-spectrum EPIs with activity against multiple RND pumps, including those in *P. aeruginosa*. These inhibitors are themselves substrates of efflux pumps and most probably interact with the transporter via a substrate binding pocket. (b) Narrow-spectrum EPIs with selective activity against the MexAB–OprM complex from *P. aeruginosa*. These inhibitors might interact with the transporter at the allosteric "modulator" site. (c) EPIs with activity against transporters from various species of Enterobacteriaceae. Their mode of action is mostly uncharacterized.

evaluate the contribution of efflux in various Gram-negative bacteria [151]. Consistent with targeting the substrate binding region of the pumps, MC-207110 potentiates the activity of only some antibiotic pump substrates (e.g., fluoroquinolones, macrolides, tetracyclines, etc., but not β-lactams) and was itself shown to be exported by RND systems [152]. This may also explain why attempts to isolate target-based mutations conferring resistance on MC-207110 (making efflux pump nonsusceptible to inhibition) were unsuccessful (Lomovskaya, unpublished results). Most likely, such mutations would render the pump incapable of interacting with other substrates and hence be observed as inactive. This being the case, specific targeting of the pump substrate binding site may be a viable future strategy to design alternative or improved efflux pump inhibitors. It also became clear during the course of the program that in any empirical search for efflux pump inhibitors, it is very important to identify and use specific partner antibiotics.

More advanced analogues of this compound were also identified in a collaboration between Microcide Pharmaceuticals and Daiichi Pharmaceuticals (e.g., MC-002595 [153] and MC-004124 [154] (Figure 5.6a), and Mpex Pharmaceuticals (www.mpexpharma.com) is now pursuing an EPI program based on inhibitors from this class of compounds [23]. Recently, a series of arylpiperazines (Figure 5.6c) were identified as inhibitors of at least two RND systems from *E. coli* (AcrAB–TolC and AcrEF–TolC [155]) as well as other RND pumps in enterobacteria [156]. As with MC-207110, they appear to potentiate the activity of some but not all of the antimicrobial substrates of these transporters. Interestingly, they differ from MC-207110 in the spectrum of antibiotics affected [156].

Efflux inhibitors might also act at sites distinct from those involved in substrate binding, but whose disruption impacts overall pump activity. Such allosteric inhibitors would be expected to inhibit efflux of all substrates and therefore proportionally potentiate the activity of multiple antibiotics. A series of structurally diverse inhibitors with high selectivity toward the MexAB–OprM efflux pump in *P. aeruginosa* have been identified [157–162] (Figure 5.6b), which negatively impact the export of all MexAB–OprM antimicrobial substrates equally. It has been hypothesized that these EPIs bind not to substrate binding sites on the pump, but rather to site(s) that modulates pump activity (i.e., modulation sites). This hypothesis is based on findings that (i) mutations that make MexB nonsusceptible to inhibition had been identified, (ii) these mutations did not affect substrate specificity of MexB, and (iii) mutations were not cross-resistant with MC-207110 (Lomovskaya, unpublished results). Several alkoxy- and alkylaminoquinoline EPIs (Figure 5.6c) showing activity against clinical strains of *Enterobacter aerogenes* have also been reported [151, 163–166] that equally potentiate the activities of all antimicrobial substrates tested, consistent with action at a modulation site of an RND-type efflux system. At present, it is unclear whether the RND transporters do, in fact, have a "dedicated" modulation site, but the empirical observation of a link between the ability to potentiate multiple substrates ("modulator mode") and high selectivity toward specific RND transporters is suggestive of such a feature.

Other possibilities for interfering with MDR transporters include targeting the assembly of the pump components and blocking the TolC-like tunnel. At present, these are purely hypothetical. There are no reports of molecules with such activity.

5.5
The Search for Efflux Inhibitors

The structural elucidation of efflux pumps will undoubtedly facilitate the future search and optimization of EPIs. It will also help clarify the mode of action of EPIs identified in the past using more traditional approaches. However, higher resolution structures of multiple conformations will be required before structural information can be applied to the rational design of inhibitors.

Up to now, most inhibitors of efflux pumps have been discovered through traditional random screening of synthetic compounds or natural products libraries. The assays that are used are very simple and easily adapted to high-throughput formats [135]. For example, MC-207110 was identified in a levofloxacin potentiation assay using a strain of *P. aeruginosa* overexpressing MexAB–OprM, while (3-phenethyl)piperidines [167] were found by scientists at Pharmacia (now Pfizer) in a novobiocin potentiation assay using a strain of *E. coli* overexpressing AcrAB–TolC. Numerous inhibitors of NorA were found using random screening as well [168, 169].

An alternative approach is to screen libraries of known drugs. The identification of a novel mode of action in an approved drug could significantly shorten the development pathway and mitigate the risks inherent with an NCE (new chemical entity). However, for such a proposition to be practical, the EPI activity needs to be much more potent than the original pharmacological activity of the drug. One exception may be anti-infective drugs other than antibacterials that would have low pharmacological activity on human targets. Pentamidine, a drug used for prophylaxis and treatment of *Pneumocystis carinii* pneumonia, was identified as an RND EPI by scientists at Mpex. This compound entered phase I clinical trials in cystic fibrosis patients, but the program is currently on hold due to concerns around drug tolerability in this patient population (Lomovskaya and Bostian, unpublished results).

Recently, several efficient pharmacoinformatic methods have been reported for the identification of new inhibitors of the P-gp [170] transporter. Leads were discovered through ligand-based virtual screening of large, commercially available libraries of compounds. These approaches are based on machine learning algorithms and require a relatively large number of known inhibitors to be used as a training set. In the case of P-gp, many structurally unrelated compounds are available for this purpose. It is expected that as a greater number of more diverse bacterial EPIs are identified through "wet" screening, the more feasible alternative screening methods, including various virtual screening protocols, will become possible.

Finally, based on the characteristics of the binding sites of RND pumps, another approach might be the synthesis of flexible molecules carrying multiple aromatic moieties that could bind with high affinity into the recognition cavities by inducing the best fit in the binding pocket. A complementary approach might be the synthesis of inhibitor dimers [171] and substrate–inhibitor hybrids [172].

5.6
Challenges and Perspectives

Many challenges are encountered on the path to conversion of a drug lead with an attractive new mode of action into a clinically useful therapeutic agent. As well appreciated by drug hunters, more risk is associated with the development of an NCE than an improved representative from an already widely used and proven class of antibiotics (with the combination of clinical benefits and appropriate toxicological profile validated by thousands of patients). Hence, major efforts continue to develop yet more improved β-lactams, fluoroquinolones, macrolides, and glycopeptides. An EPI unfortunately is likely to be an NCE and subject to the same higher risks. Moreover, the development of combination therapy brings additional complexity due to the necessity of precision tailoring of the pharmacokinetics of both agents to achieve the desired pharmacodynamic effect.

It is also essential that the antibiotic and the EPI should not engage in drug–drug interactions. In this respect, some lessons may be learned from the clinical experience with inhibitors of P-gp and other ABC transporters as reversing agents for combination with anticancer drugs. The search for such compounds started in the mid-1970s, almost concomitant with the discovery of P-gp [173]. Several P-gp inhibitors have since failed in clinical trials. Perhaps, the main reason for this is that P-gp and other ABC transporters have distinct physiological functions in the human body, protecting various cells from endogenous toxic metabolites and xenobiotics, as well as the cytotoxic anticancer drugs themselves. In addition, they participate in drug disposition. As a result, in the presence of P-pg inhibitors, exposure to the coadministered cytotoxic drug in normal cells is increased, resulting in toxicity. Several more potent and selective agents are undergoing clinical development that may overcome this potential problem [174].

It is expected that the introduction of inhibitors of bacterial RND transporters as anti-infective agents might be more expeditious than for MDR-reversing agents for cancer therapy, since no close human homologues exist and therefore no target-based toxicity is expected.

Additional challenges are associated with the fact that antimicrobial therapy with EPIs is a combination therapy by its very nature. In order to provide the maximum pharmacodynamic benefit, the pharmacokinetics of the EPI should be appropriately tailored to the pharmacokinetics of the antibiotic component of the combination. This pharmacokinetic tailoring to establish the optimal ratio and dosing regimens to create the best possible efficacy while maintaining appropriate toxicological profiles represents a nonarguable challenge. On the positive side is the fact that animal models with engineered strains lacking efflux pumps can be used to very precisely define the pharmacokinetic/pharmacodynamic (PK/PD) targets associated with the best impact for the EPIs on the efficacy of the potential partner antibiotics [175].

Additional challenges lie in the design of clinical trials and are of a regulatory character. For example, the major beneficial consequence of combining an EPI with a fluoroquinolone is reduction in the rate of resistance development, which is best demonstrated by bacteriologic end points and by using PK/PD *in vitro* and animal models of infection. However, it is clinical end points that are relevant in the FDA approval process. Some changes in regulatory decision making regarding resistance would be very helpful in order to facilitate approval based on prevention of resistance development. Another important benefit provided by EPIs is actual reversal of resistance. To demonstrate this benefit, large clinical trials would need to be performed in order to enroll a sufficient number of patients carrying resistant strains. This problem might be mitigated if the results of preclinical *in vitro* and *in vivo* PK/PD studies could be used to demonstrate that a drug has similar activity against strains that are susceptible or resistant to an antibiotic component of the combination. Based on availability of these results, clinical data against susceptible strains may be supportive of efficacy against resistant strains (although some clinical data against resistant strains still will be necessary). It appears that many significant innovations are being considered by the FDA to facilitate and stimulate the development of new antibiotics (www.fda.gov/cder/drug/antimicrobial/FDA_IDSA_ISAP_Presentations.htm).

Continuing efforts in both academic and applied research should help introduce these important agents into clinical practice. Demonstration of their multifactorial benefits in clinical settings will provide the ultimate validation of the EPI-based combination approach.

References

1 Levy, S.B. and Marshall, B. (2004) Antibacterial resistance worldwide: causes, challenges and responses. *Nature Medicine*, **10** (12 Suppl), S122–S129.

2 Talbot, G.H., Bradley, J., Edwards, J.E., Jr., Gilbert, D., Scheld, M., and Bartlett, J.G. (2006) Bad bugs need drugs: an update on the development pipeline from the Antimicrobial Availability Task Force of the Infectious Diseases Society of America. *Clinical Infectious Diseases*, **42** (5), 657–668.

3 Wright, G.D. (2007) The antibiotic resistome: the nexus of chemical and genetic diversity. *Nature Reviews. Microbiology*, **5** (3), 175–186.

4 Nordmann, P., Naas, T., Fortineau, N., and Poirel, L. (2007) Superbugs in the coming new decade: multidrug resistance and prospects for treatment of *Staphylococcus aureus*, *Enterococcus* spp. and *Pseudomonas aeruginosa* in 2010. *Current Opinion in Microbiology*, **10** (5), 436–440.

5 de Lencastre, H., Oliveira, D., and Tomasz, A. (2007) Antibiotic resistant *Staphylococcus aureus*: a paradigm of adaptive power. *Current Opinion in Microbiology*, **10** (5), 428–435.

6 Tenover, F.C. (2006) Mechanisms of antimicrobial resistance in bacteria. *American Journal of Infection Control*, **34** (5 Suppl 1), S3–S10; discussion S64–S73.

7 Woodford, N. (2005) Biological counterstrike: antibiotic resistance mechanisms of Gram-positive cocci. *Clinical Microbiology and Infection*, **11** (Suppl 3), 2–21.

8 Rice, L.B. (2007) Emerging issues in the management of infections caused by multidrug-resistant Gram-negative bacteria. *Cleveland Clinic Journal of Medicine*, **74** (Suppl 4), S12–S20.

9 Nikaido, H. (1994) Prevention of drug access to bacterial targets: permeability barriers and active efflux. *Science*, **264** (5157), 382–388.

10 Higgins, C.F. (2007) Multiple molecular mechanisms for multidrug resistance transporters. *Nature*, **446** (7137), 749–757.

11 Saier, M.H., Jr., and Paulsen, I.T. (2001) Phylogeny of multidrug transporters. *Seminars in Cell & Developmental Biology*, **12** (3), 205–213.

12 Nikaido, H. (1998) Multiple antibiotic resistance and efflux. *Current Opinion in Microbiology*, **1** (5), 516–523.

13 Zgurskaya, H.I. and Nikaido, H. (2000) Multidrug resistance mechanisms: drug efflux across two membranes. *Molecular Microbiology*, **37** (2), 219–225.

14 Higgins, C.F. and Gottesman, M.M. (1992) Is the multidrug transporter a

flippase? *Trends in Biochemical Sciences*, **17** (1), 18–21.

15 Nishino, K. and Yamaguchi, A. (2001) Analysis of a complete library of putative drug transporter genes in *Escherichia coli*. *Journal of Bacteriology*, **183** (20), 5803–5812.

16 Saier, M.H., Jr., Paulsen, I.T., Sliwinski, M.K., Pao, S.S., Skurray, R.A., and Nikaido, H. (1998) Evolutionary origins of multidrug and drug-specific efflux pumps in bacteria. *The FASEB Journal*, **12** (3), 265–274.

17 Neyfakh, A.A. (2002) Mystery of multidrug transporters: the answer can be simple. *Molecular Microbiology*, **44** (5), 1123–1130.

18 Grkovic, S., Brown, M.H., and Skurray, R.A. (2001) Transcriptional regulation of multidrug efflux pumps in bacteria. *Seminars in Cell & Developmental Biology*, **12** (3), 225–237.

19 Grkovic, S., Brown, M.H., and Skurray, R.A. (2002) Regulation of bacterial drug export systems. *Microbiology and Molecular Biology Reviews*, **66** (4), 671–701.

20 Poole, K. (2005) Efflux-mediated antimicrobial resistance. *The Journal of Antimicrobial Chemotherapy*, **56** (1), 20–51.

21 Piddock, L.J. (2006) Multidrug-resistance efflux pumps – not just for resistance. *Nature Reviews. Microbiology*, **4** (8), 629–636.

22 Piddock, L.J. (2006) Clinically relevant chromosomally encoded multidrug resistance efflux pumps in bacteria. *Clinical Microbiology Reviews*, **19** (2), 382–402.

23 Lomovskaya, O. and Bostian, K.A. (2006) Practical applications and feasibility of efflux pump inhibitors in the clinic – a vision for applied use. *Biochemical Pharmacology*, **71** (7), 910–918.

24 Lomovskaya, O. and Watkins, W.J. (2001) Efflux pumps: their role in antibacterial drug discovery. *Current Medicinal Chemistry*, **8** (14), 1699–1711.

25 Higgins, C.F. (2001) ABC transporters: physiology, structure and mechanism – an overview. *Research in Microbiology*, **152** (3–4), 205–210.

26 Saier, M.H., Jr., Beatty, J.T., Goffeau, A., Harley, K.T., Heijne, W.H., Huang, S.C., Jack, D.L., Jahn, P.S., Lew, K., Liu, J., Pao, S.S., Paulsen, I.T., Tseng, T.T., and Virk, P.S. (1999) The major facilitator superfamily. *Journal of Molecular Microbiology and Biotechnology*, **1** (2), 257–279.

27 Tseng, T.T., Gratwick, K.S., Kollman, J., Park, D., Nies, D.H., Goffeau, A., and Saier, M.H., Jr. (1999) The RND permease superfamily: an ancient, ubiquitous and diverse family that includes human disease and development proteins. *Journal of Molecular Microbiology and Biotechnology*, **1** (1), 107–125.

28 Chung, Y.J. and Saier, M.H., Jr. (2001) SMR-type multidrug resistance pumps. *Current Opinion in Drug Discovery & Development*, **4** (2), 237–245.

29 Jack, D.L., Yang, N.M., and Saier, M.H., Jr. (2001) The drug/metabolite transporter superfamily. *European Journal of Biochemistry*, **268** (13), 3620–3639.

30 Hvorup, R.N., Winnen, B., Chang, A.B., Jiang, Y., Zhou, X.F., and Saier, M.H., Jr. (2003) The multidrug/oligosaccharidyl-lipid/polysaccharide (MOP) exporter superfamily. *European Journal of Biochemistry*, **270** (5), 799–813.

31 Ren, Q. and Paulsen, I.T. (2007) Large-scale comparative genomic analyses of cytoplasmic membrane transport systems in prokaryotes. *Journal of Molecular Microbiology and Biotechnology*, **12** (3–4), 165–179.

32 Paulsen, I.T., Nguyen, L., Sliwinski, M.K., Rabus, R., and Saier, M.H., Jr. (2000) Microbial genome analyses: comparative transport capabilities in eighteen prokaryotes. *Journal of Molecular Biology*, **301** (1), 75–100.

33 Paulsen, I.T., Sliwinski, M.K., and Saier, M.H., Jr. (1998) Microbial genome analyses: global comparisons of transport capabilities based on phylogenies, bioenergetics and substrate specificities.

Journal of Molecular Biology, **277** (3), 573–592.

34 Lewis, K. (2001) In search of natural substrates and inhibitors of MDR pumps. *Journal of Molecular Microbiology and Biotechnology*, **3** (2), 247–254.

35 Lubelski, J., Konings, W.N., and Driessen, A.J. (2007) Distribution and physiology of ABC-type transporters contributing to multidrug resistance in bacteria. *Microbiology and Molecular Biology Reviews*, **71** (3), 463–476.

36 Martin, J.F., Casqueiro, J., and Liras, P. (2005) Secretion systems for secondary metabolites: how producer cells send out messages of intercellular communication. *Current Opinion in Microbiology*, **8** (3), 282–293.

37 Hollenstein, K., Dawson, R.J., and Locher, K.P. (2007) Structure and mechanism of ABC transporter proteins. *Current Opinion in Structural Biology*, **17** (4), 412–418.

38 van Veen, H.W., Margolles, A., Muller, M., Higgins, C.F., and Konings, W.N. (2000) The homodimeric ATP-binding cassette transporter LmrA mediates multidrug transport by an alternating two-site (two-cylinder engine) mechanism. *The EMBO Journal*, **19** (11), 2503–2514.

39 Lubelski, J., de Jong, A., van Merkerk, R., Agustiandari, H., Kuipers, O.P., Kok, J., and Driessen, A.J. (2006) LmrCD is a major multidrug resistance transporter in *Lactococcus lactis*. *Molecular Microbiology*, **61** (3), 771–781.

40 Moody, J.E. and Thomas, P.J. (2005) Nucleotide binding domain interactions during the mechanochemical reaction cycle of ATP-binding cassette transporters. *Journal of Bioenergetics and Biomembranes*, **37** (6), 475–479.

41 Pedersen, P.L. (2005) Transport ATPases: structure, motors, mechanism and medicine: a brief overview. *Journal of Bioenergetics and Biomembranes*, **37** (6), 349–357.

42 Loo, T.W. and Clarke, D.M. (2005) Recent progress in understanding the mechanism of P-glycoprotein-mediated drug efflux. *The Journal of Membrane Biology*, **206** (3), 173–185.

43 Pleban, K., Kopp, S., Csaszar, E., Peer, M., Hrebicek, T., Rizzi, A., Ecker, G.F., and Chiba, P. (2005) P-glycoprotein substrate binding domains are located at the transmembrane domain/transmembrane domain interfaces: a combined photoaffinity labeling–protein homology modeling approach. *Molecular Pharmacology*, **67** (2), 365–374.

44 Dawson, R.J. and Locher, K.P. (2007) Structure of the multidrug ABC transporter Sav1866 from *Staphylococcus aureus* in complex with AMP-PNP. *FEBS Letters*, **581** (5), 935–938.

45 Dawson, R.J. and Locher, K.P. (2006) Structure of a bacterial multidrug ABC transporter. *Nature*, **443** (7108), 180–185.

46 Linton, K.J. and Higgins, C.F. (2007) Structure and function of ABC transporters: the ATP switch provides flexible control. *Pflugers Archives*, **453** (5), 555–567.

47 Hollenstein, K., Frei, D.C., and Locher, K.P. (2007) Structure of an ABC transporter in complex with its binding protein. *Nature*, **446** (7132), 213–216.

48 Locher, K.P., Lee, A.T., and Rees, D.C. (2002) The *E. coli* BtuCD structure: a framework for ABC transporter architecture and mechanism. *Science*, **296** (5570), 1091–1098.

49 Oldham, M.L., Khare, D., Quiocho, F.A., Davidson, A.L., and Chen, J. (2007) Crystal structure of a catalytic intermediate of the maltose transporter. *Nature*, **450** (7169), 515–521.

50 Pinkett, H.W., Lee, A.T., Lum, P., Locher, K.P., and Rees, D.C. (2007) An inward-facing conformation of a putative metal-chelate-type ABC transporter. *Science*, **315** (5810), 373–377.

51 Saidijam, M., Benedetti, G., Ren, Q., Xu, Z., Hoyle, C.J., Palmer, S.L., Ward, A., Bettaney, K.E., Szakonyi, G., Meuller, J., Morrison, S., Pos, M.K., Butaye, P., Walravens, K., Langton, K., Herbert, R.B.,

Skurray, R.A., Paulsen, I.T., O'Reilly, J., Rutherford, N.G., Brown, M.H., Bill, R.M., and Henderson, P.J. (2006) Microbial drug efflux proteins of the major facilitator superfamily. *Current Drug Targets*, **7** (7), 793–811.

52 Paulsen, I.T., Brown, M.H., and Skurray, R.A. (1996) Proton-dependent multidrug efflux systems. *Microbiological Reviews*, **60** (4), 575–608.

53 Poole, K. (2007) Efflux pumps as antimicrobial resistance mechanisms. *Annals of Medicine*, **39** (3), 162–176.

54 Bibi, E., Adler, J., Lewinson, O., and Edgar, R. (2001) MdfA, an interesting model protein for studying multidrug transport. *Journal of Molecular Microbiology and Biotechnology*, **3** (2), 171–177.

55 Bohn, C. and Bouloc, P. (1998) The *Escherichia coli* cmlA gene encodes the multidrug efflux pump Cmr/MdfA and is responsible for isopropyl-beta-D-thiogalactopyranoside exclusion and spectinomycin sensitivity. *Journal of Bacteriology*, **180** (22), 6072–6075.

56 Mazurkiewicz, P., Konings, W.N., and Poelarends, G.J. (2002) Acidic residues in the lactococcal multidrug efflux pump LmrP play critical roles in transport of lipophilic cationic compounds. *The Journal of Biological Chemistry*, **277** (29), 26081–26088.

57 Edgar, R. and Bibi, E. (1999) A single membrane-embedded negative charge is critical for recognizing positively charged drugs by the *Escherichia coli* multidrug resistance protein MdfA. *The EMBO Journal*, **18** (4), 822–832.

58 Yin, Y., He, X., Szewczyk, P., Nguyen, T., and Chang, G. (2006) Structure of the multidrug transporter EmrD from *Escherichia coli*. *Science*, **312** (5774), 741–744.

59 Abramson, J., Smirnova, I., Kasho, V., Verner, G., Iwata, S., and Kaback, H.R. (2003) The lactose permease of *Escherichia coli*: overall structure, the sugar-binding site and the alternating access model for transport. *FEBS Letters*, **555** (1), 96–101.

60 Lemieux, M.J., Huang, Y., and Wang, D.N. (2004) The structural basis of substrate translocation by the *Escherichia coli* glycerol-3-phosphate transporter: a member of the major facilitator superfamily. *Current Opinion in Structural Biology*, **14** (4), 405–412.

61 Abramson, J., Smirnova, I., Kasho, V., Verner, G., Kaback, H.R., and Iwata, S. (2003) Structure and mechanism of the lactose permease of *Escherichia coli*. *Science*, **301** (5633), 610–615.

62 Abramson, J., Iwata, S., and Kaback, H.R. (2004) Lactose permease as a paradigm for membrane transport proteins (review). *Molecular Membrane Biology*, **21** (4), 227–236.

63 Mazurkiewicz, P., Poelarends, G.J., Driessen, A.J., and Konings, W.N. (2004) Facilitated drug influx by an energy-uncoupled secondary multidrug transporter. *The Journal of Biological Chemistry*, **279** (1), 103–108.

64 Sigal, N., Molshanski-Mor, S., and Bibi, E. (2006) No single irreplaceable acidic residues in the *Escherichia coli* secondary multidrug transporter MdfA. *Journal of Bacteriology*, **188** (15), 5635–5639.

65 Omote, H., Hiasa, M., Matsumoto, T., Otsuka, M., and Moriyama, Y. (2006) The MATE proteins as fundamental transporters of metabolic and xenobiotic organic cations. *Trends in Pharmacological Sciences*, **27** (11), 587–593.

66 Otsuka, M., Yasuda, M., Morita, Y., Otsuka, C., Tsuchiya, T., Omote, H., and Moriyama, Y. (2005) Identification of essential amino acid residues of the NorM Na^+/multidrug antiporter in *Vibrio parahaemolyticus*. *Journal of Bacteriology*, **187** (5), 1552–1558.

67 McAleese, F., Petersen, P., Ruzin, A., Dunman, P.M., Murphy, E., Projan, S.J., and Bradford, P.A. (2005) A novel MATE family efflux pump contributes to the reduced susceptibility of laboratory-derived *Staphylococcus aureus* mutants to

tigecycline. *Antimicrobial Agents and Chemotherapy*, **49** (5), 1865–1871.

68 Lentino, J.R., Narita, M., and Yu, V.L. (2008) New antimicrobial agents as therapy for resistant Gram-positive cocci. *European Journal of Clinical Microbiology & Infectious Diseases*, **27** (1), 3–15.

69 Bay, D.C., Rommens, K.L., and Turner, R.J. (2008) Small multidrug resistance proteins: a multidrug transporter family that continues to grow. *Biochimica et Biophysica Acta*, **1778** (9), 1814–1838.

70 Schuldiner, S., Granot, D., Steiner, S., Ninio, S., Rotem, D., Soskin, M., and Yerushalmi, H. (2001) Precious things come in little packages. *Journal of Molecular Microbiology and Biotechnology*, **3** (2), 155–162.

71 Steiner-Mordoch, S., Soskine, M., Solomon, D., Rotem, D., Gold, A., Yechieli, M., Adam, Y., and Schuldiner, S. (2008) Parallel topology of genetically fused EmrE homodimers. *The EMBO Journal*, **27** (1), 17–26.

72 Schuldiner, S. (2007) When biochemistry meets structural biology: the cautionary tale of EmrE. *Trends in Biochemical Sciences*, **32** (6), 252–258.

73 Schuldiner, S. (2007) Controversy over EmrE structure. *Science*, **317** (5839), 748–751; author reply 748–751.

74 Rapp, M., Seppala, S., Granseth, E., and von Heijne, G. (2007) Emulating membrane protein evolution by rational design. *Science*, **315** (5816), 1282–1284.

75 Domenech, P., Reed, M.B., and Barry, C.E. 3rd (2005) Contribution of the *Mycobacterium tuberculosis* MmpL protein family to virulence and drug resistance. *Infection and Immunity*, **73** (6), 3492–3501.

76 Pasca, M.R., Guglierame, P., De Rossi, E., Zara, F., and Riccardi, G. (2005) mmpL7 gene of *Mycobacterium tuberculosis* is responsible for isoniazid efflux in *Mycobacterium smegmatis*. *Antimicrobial Agents and Chemotherapy*, **49** (11), 4775–4777.

77 Zgurskaya, H.I. and Nikaido, H. (2002) Mechanistic parallels in bacterial and human multidrug efflux transporters. *Current Protein & Peptide Science*, **3** (5), 531–540.

78 Scott, C. and Ioannou, Y.A. (2004) The NPC1 protein: structure implies function. *Biochimica et Biophysica Acta*, **1685** (1–3), 8–13.

79 Ko, D.C., Gordon, M.D., Jin, J.Y., and Scott, M.P. (2001) Dynamic movements of organelles containing Niemann-Pick C1 protein: NPC1 involvement in late endocytic events. *Molecular Biology of the Cell*, **12** (3), 601–614.

80 Bale, A.E. and Yu, K.P. (2001) The hedgehog pathway and basal cell carcinomas. *Human Molecular Genetics*, **10** (7), 757–762.

81 Murakami, S., Nakashima, R., Yamashita, E., and Yamaguchi, A. (2002) Crystal structure of bacterial multidrug efflux transporter AcrB. *Nature*, **419** (6907), 587–593.

82 Yu, E.W., McDermott, G., Zgurskaya, H.I., Nikaido, H., and Koshland, D.E., Jr. (2003) Structural basis of multiple drug-binding capacity of the AcrB multidrug efflux pump. *Science*, **300** (5621), 976–980.

83 Yu, E.W., Aires, J.R., McDermott, G., and Nikaido, H. (2005) A periplasmic drug-binding site of the AcrB multidrug efflux pump: a crystallographic and site-directed mutagenesis study. *Journal of Bacteriology*, **187** (19), 6804–6815.

84 Su, C.C., Li, M., Gu, R., Takatsuka, Y., McDermott, G., Nikaido, H., and Yu, E.W. (2006) Conformation of the AcrB multidrug efflux pump in mutants of the putative proton relay pathway. *Journal of Bacteriology*, **188** (20), 7290–7296.

85 Murakami, S., Nakashima, R., Yamashita, E., Matsumoto, T., and Yamaguchi, A. (2006) Crystal structures of a multidrug transporter reveal a functionally rotating mechanism. *Nature*, **443** (7108), 173–179.

86 Seeger, M.A., Schiefner, A., Eicher, T., Verrey, F., Diederichs, K., and Pos, K.M. (2006) Structural asymmetry of AcrB trimer suggests a peristaltic pump

87 Akama, H., Matsuura, T., Kashiwagi, S., Yoneyama, H., Narita, S., Tsukihara, T., Nakagawa, A., and Nakae, T. (2004) Crystal structure of the membrane fusion protein, MexA, of the multidrug transporter in *Pseudomonas aeruginosa*. *The Journal of Biological Chemistry*, **279** (25), 25939–25942.

88 Higgins, M.K., Bokma, E., Koronakis, E., Hughes, C., and Koronakis, V. (2004) Structure of the periplasmic component of a bacterial drug efflux pump. *Proceedings of the National Academy of Sciences of the United States of America*, **101** (27), 9994–9999.

89 Koronakis, V., Sharff, A., Koronakis, E., Luisi, B., and Hughes, C. (2000) Crystal structure of the bacterial membrane protein TolC central to multidrug efflux and protein export. *Nature*, **405** (6789), 914–919.

90 Mikolosko, J., Bobyk, K., Zgurskaya, H.I., and Ghosh, P. (2006) Conformational flexibility in the multidrug efflux system protein AcrA. *Structure*, **14** (3), 577–587.

91 Akama, H., Kanemaki, M., Yoshimura, M., Tsukihara, T., Kashiwagi, T., Yoneyama, H., Narita, S., Nakagawa, A., and Nakae, T. (2004) Crystal structure of the drug discharge outer membrane protein, OprM, of *Pseudomonas aeruginosa*: dual modes of membrane anchoring and occluded cavity end. *The Journal of Biological Chemistry*, **279** (51), 52816–52819.

92 Elkins, C.A. and Nikaido, H. (2002) Substrate specificity of the RND-type multidrug efflux pumps AcrB and AcrD of *Escherichia coli* is determined predominantly by two large periplasmic loops. *Journal of Bacteriology*, **184** (23), 6490–6498.

93 Mao, W., Warren, M.S., Black, D.S., Satou, T., Murata, T., Nishino, T., Gotoh, N., and Lomovskaya, O. (2002) On the mechanism of substrate specificity by resistance nodulation division (RND)-type multidrug resistance pumps: the large periplasmic loops of MexD from *Pseudomonas aeruginosa* are involved in substrate recognition. *Molecular Microbiology*, **46** (3), 889–901.

94 Tikhonova, E.B., Wang, Q., and Zgurskaya, H.I. (2002) Chimeric analysis of the multicomponent multidrug efflux transporters from Gram-negative bacteria. *Journal of Bacteriology*, **184** (23), 6499–6507.

95 Aires, J.R. and Nikaido, H. (2005) Aminoglycosides are captured from both periplasm and cytoplasm by the AcrD multidrug efflux transporter of *Escherichia coli*. *Journal of Bacteriology*, **187** (6), 1923–1929.

96 Hearn, E.M., Gray, M.R., and Foght, J.M. (2006) Mutations in the central cavity and periplasmic domain affect efflux activity of the resistance-nodulation-division pump EmhB from *Pseudomonas fluorescens* cLP6a. *Journal of Bacteriology*, **188** (1), 115–123.

97 Middlemiss, J.K. and Poole, K. (2004) Differential impact of MexB mutations on substrate selectivity of the MexAB-OprM multidrug efflux pump of *Pseudomonas aeruginosa*. *Journal of Bacteriology*, **186** (5), 1258–1269.

98 Elkins, C.A. and Mullis, L.B. (2007) Substrate competition studies using whole-cell accumulation assays with the major tripartite multidrug efflux pumps of *Escherichia coli*. *Antimicrobial Agents and Chemotherapy*, **51** (3), 923–929.

99 Goldberg, M., Pribyl, T., Juhnke, S., and Nies, D.H. (1999) Energetics and topology of CzcA, a cation/proton antiporter of the resistance-nodulation-cell division protein family. *The Journal of Biological Chemistry*, **274** (37), 26065–26070.

100 Alekshun, M.N. and Levy, S.B. (2006) Commensals upon us. *Biochemical Pharmacology*, **71** (7), 893–900.

101 Heldwein, E.E. and Brennan, R.G. (2001) Crystal structure of the transcription activator BmrR bound to DNA and a drug. *Nature*, **409** (6818), 378–382.

102 Zheleznova, E.E., Markham, P.N., Neyfakh, A.A., and Brennan, R.G. (1999) Structural basis of multidrug recognition by BmrR, a transcription activator of a multidrug transporter. *Cell*, **96** (3), 353–362.

103 Grkovic, S., Hardie, K.M., Brown, M.H., and Skurray, R.A. (2003) Interactions of the QacR multidrug-binding protein with structurally diverse ligands: implications for the evolution of the binding pocket. *Biochemistry*, **42** (51), 15226–15236.

104 Schumacher, M.A. and Brennan, R.G. (2002) Structural mechanisms of multidrug recognition and regulation by bacterial multidrug transcription factors. *Molecular Microbiology*, **45** (4), 885–893.

105 Watkins, R.E., Wisely, G.B., Moore, L.B., Collins, J.L., Lambert, M.H., Williams, S.P., Willson, T.M., Kliewer, S.A., and Redinbo, M.R. (2001) The human nuclear xenobiotic receptor PXR: structural determinants of directed promiscuity. *Science*, **292** (5525), 2329–2333.

106 Loo, T.W., Bartlett, M.C., and Clarke, D.M. (2003) Simultaneous binding of two different drugs in the binding pocket of the human multidrug resistance P-glycoprotein. *The Journal of Biological Chemistry*, **278** (41), 39706–39710.

107 Loo, T.W., Bartlett, M.C., and Clarke, D.M. (2003) Substrate-induced conformational changes in the transmembrane segments of human P-glycoprotein. Direct evidence for the substrate-induced fit mechanism for drug binding. *The Journal of Biological Chemistry*, **278** (16), 13603–13606.

108 Nikaido, H. (2003) Molecular basis of bacterial outer membrane permeability revisited. *Microbiology and Molecular Biology Reviews*, **67** (4), 593–656.

109 Federici, L., Du, D., Walas, F., Matsumura, H., Fernandez-Recio, J., McKeegan, K.S., Borges-Walmsley, M.I., Luisi, B.F., and Walmsley, A.R. (2005) The crystal structure of the outer membrane protein VceC from the bacterial pathogen *Vibrio cholerae* at 1.8 Å resolution. *The Journal of Biological Chemistry*, **280** (15), 15307–15314.

110 Tikhonova, E.B. and Zgurskaya, H.I. (2004) AcrA, AcrB, and TolC of *Escherichia coli* form a stable intermembrane multidrug efflux complex. *The Journal of Biological Chemistry*, **279** (31), 32116–32124.

111 Touze, T., Eswaran, J., Bokma, E., Koronakis, E., Hughes, C., and Koronakis, V. (2004) Interactions underlying assembly of the *Escherichia coli* AcrAB-TolC multidrug efflux system. *Molecular Microbiology*, **53** (2), 697–706.

112 Bokma, E., Koronakis, E., Lobedanz, S., Hughes, C., and Koronakis, V. (2006) Directed evolution of a bacterial efflux pump: adaptation of the *E. coli* TolC exit duct to the *Pseudomonas* MexAB translocase. *FEBS Letters*, **580** (22), 5339–5343.

113 Lobedanz, S., Bokma, E., Symmons, M.F., Koronakis, E., Hughes, C., and Koronakis, V. (2007) A periplasmic coiled-coil interface underlying TolC recruitment and the assembly of bacterial drug efflux pumps. *Proceedings of the National Academy of Sciences of the United States of America*, **104** (11), 4612–4617.

114 Borges-Walmsley, M.I., Beauchamp, J., Kelly, S.M., Jumel, K., Candlish, D., Harding, S.E., Price, N.C., and Walmsley, A.R. (2003) Identification of oligomerization and drug-binding domains of the membrane fusion protein EmrA. *The Journal of Biological Chemistry*, **278** (15), 12903–12912.

115 Fernandez-Recio, J., Walas, F., Federici, L., Venkatesh Pratap, J., Bavro, V.N., Miguel, R.N., Mizuguchi, K., and Luisi, B. (2004) A model of a transmembrane drug-efflux pump from Gram-negative bacteria. *FEBS Letters*, **578** (1–2), 5–9.

116 Zgurskaya, H.I. and Nikaido, H. (1999) AcrA is a highly asymmetric protein capable of spanning the periplasm. *Journal of Molecular Biology*, **285**, 409–420.

117 Johnson, J.M. and Church, G.M. (1999) Alignment and structure prediction of

divergent protein families: periplasmic and outer membrane proteins of bacterial efflux pumps. *Journal of Molecular Biology*, **287** (3), 695–715.

118 Elkins, C.A. and Nikaido, H. (2003) Chimeric analysis of AcrA function reveals the importance of its C-terminal domain in its interaction with the AcrB multidrug efflux pump. *Journal of Bacteriology*, **185** (18), 5349–5356.

119 Krishnamoorthy, G., Tikhonova, E.B., and Zgurskaya, H.I. (2008) Fitting periplasmic membrane fusion proteins to inner membrane transporters: mutations that enable *Escherichia coli* AcrA to function with *Pseudomonas aeruginosa* MexB. *Journal of Bacteriology*, **190** (2), 691–698.

120 Pimenta, A.L., Racher, K., Jamieson, L., Blight, M.A., and Holland, I.B. (2005) Mutations in HlyD, part of the type 1 translocator for hemolysin secretion, affect the folding of the secreted toxin. *Journal of Bacteriology*, **187** (21), 7471–7480.

121 Bagai, I., Liu, W., Rensing, C., Blackburn, N.J., and McEvoy, M.M. (2007) Substrate-linked conformational change in the periplasmic component of a Cu(I)/Ag(I) efflux system. *The Journal of Biological Chemistry*, **282** (49), 35695–35702.

122 Tikhonova, E.B., Devroy, V.K., Lau, S.Y., and Zgurskaya, H.I. (2007) Reconstitution of the *Escherichia coli* macrolide transporter: the periplasmic membrane fusion protein MacA stimulates the ATPase activity of MacB. *Molecular Microbiology*, **63** (3), 895–910.

123 Lomovskaya, O. and Watkins, W. (2001) Inhibition of efflux pumps as a novel approach to combat drug resistance in bacteria. *Journal of Molecular Microbiology and Biotechnology*, **3** (2), 225–236.

124 Tendolkar, P.M., Baghdayan, A.S., and Shankar, N. (2003) Pathogenic enterococci: new developments in the 21st century. *Cellular and Molecular Life Sciences*, **60** (12), 2622–2636.

125 Davis, D.R., McAlpine, J.B., Pazoles, C.J., Talbot, M.K., Alder, E.A., White, C., Jonas, B.M., Murray, B.E., Weinstock, G.M., and Rogers, B.L. (2001) *Enterococcus faecalis* multi-drug resistance transporters: application for antibiotic discovery. *Journal of Molecular Microbiology and Biotechnology*, **3** (2), 179–184.

126 Hooper, D.C. (2001) Mechanisms of action of antimicrobials: focus on fluoroquinolones. *Clinical Infectious Diseases*, **32** (Suppl 1), S9–S15.

127 Hooper, D.C. (2000) Mechanisms of action and resistance of older and newer fluoroquinolones. *Clinical Infectious Diseases*, **31** (Suppl 2), S24–S28.

128 Hassan, K.A., Skurray, R.A., and Brown, M.H. (2007) Active export proteins mediating drug resistance in staphylococci. *Journal of Molecular Microbiology and Biotechnology*, **12** (3–4), 180–196.

129 Aeschlimann, J.R., Kaatz, G.W., and Rybak, M.J. (1999) The effects of NorA inhibition on the activities of levofloxacin, ciprofloxacin and norfloxacin against two genetically related strains of *Staphylococcus aureus* in an in-vitro infection model. *The Journal of Antimicrobial Chemotherapy*, **44** (3), 343–349.

130 Kaatz, G.W., Seo, S.M., and Ruble, C.A. (1993) Efflux-mediated fluoroquinolone resistance in *Staphylococcus aureus*. *Antimicrobial Agents and Chemotherapy*, **37** (5), 1086–1094.

131 Marrer, E., Schad, K., Satoh, A.T., Page, M.G., Johnson, M.M., and Piddock, L.J. (2006) Involvement of the putative ATP-dependent efflux proteins PatA and PatB in fluoroquinolone resistance of a multidrug-resistant mutant of *Streptococcus pneumoniae*. *Antimicrobial Agents and Chemotherapy*, **50** (2), 685–693.

132 Robertson, G.T., Doyle, T.B., and Lynch, A.S. (2005) Use of an efflux-deficient *Streptococcus pneumoniae* strain panel to identify ABC-class multidrug

transporters involved in intrinsic resistance to antimicrobial agents. *Antimicrobial Agents and Chemotherapy*, **49** (11), 4781–4783.

133 Lomovskaya, O., Lee, A., Hoshino, K., Ishida, H., Mistry, A., Warren, M.S., Boyer, E., Chamberland, S., and Lee, V.J. (1999) Use of a genetic approach to evaluate the consequences of inhibition of efflux pumps in *Pseudomonas aeruginosa*. *Antimicrobial Agents and Chemotherapy*, **43** (6), 1340–1346.

134 Markham, P.N. (1999) Inhibition of the emergence of ciprofloxacin resistance in *Streptococcus pneumoniae* by the multidrug efflux inhibitor reserpine. *Antimicrobial Agents and Chemotherapy*, **43** (4), 988–989.

135 Lomovskaya, O., Warren, M.S., Lee, A., Galazzo, J., Fronko, R., Lee, M., Blais, J., Cho, D., Chamberland, S., Renau, T., Leger, R., Hecker, S., Watkins, W., Hoshino, K., Ishida, H., and Lee, V.J. (2001) Identification and characterization of inhibitors of multidrug resistance efflux pumps in *Pseudomonas aeruginosa*: novel agents for combination therapy. *Antimicrobial Agents and Chemotherapy*, **45** (1), 105–116.

136 Ricci, V., Tzakas, P., Buckley, A., and Piddock, L.J. (2006) Ciprofloxacin-resistant *Salmonella enterica* serovar Typhimurium strains are difficult to select in the absence of AcrB and TolC. *Antimicrobial Agents and Chemotherapy*, **50** (1), 38–42.

137 Gill, M.J., Brenwald, N.P., and Wise, R. (1999) Identification of an efflux pump gene, pmrA, associated with fluoroquinolone resistance in *Streptococcus pneumoniae*. *Antimicrobial Agents and Chemotherapy*, **43** (1), 187–189.

138 Piddock, L.J., Johnson, M.M., Simjee, S., and Pumbwe, L. (2002) Expression of efflux pump gene pmrA in fluoroquinolone-resistant and -susceptible clinical isolates of *Streptococcus pneumoniae*. *Antimicrobial Agents and Chemotherapy*, **46** (3), 808–812.

139 Avrain, L., Garvey, M., Mesaros, N., Glupczynski, Y., Mingeot-Leclercq, M.P., Piddock, L.J., Tulkens, P.M., Vanhoof, R., and Van Bambeke, F. (2007) Selection of quinolone resistance in *Streptococcus pneumoniae* exposed *in vitro* to subinhibitory drug concentrations. *The Journal of Antimicrobial Chemotherapy*, **60** (5), 965–972.

140 DeMarco, C.E., Cushing, L.A., Frempong-Manso, E., Seo, S.M., Jaravaza, T.A., and Kaatz, G.W. (2007) Efflux-related resistance to norfloxacin, dyes, and biocides in bloodstream isolates of *Staphylococcus aureus*. *Antimicrobial Agents and Chemotherapy*, **51** (9), 3235–3239.

141 Truong-Bolduc, Q.C., Strahilevitz, J., and Hooper, D.C. (2006) NorC, a new efflux pump regulated by MgrA of *Staphylococcus aureus*. *Antimicrobial Agents and Chemotherapy*, **50** (3), 1104–1107.

142 Truong-Bolduc, Q.C., Dunman, P.M., Strahilevitz, J., Projan, S.J., and Hooper, D.C. (2005) MgrA is a multiple regulator of two new efflux pumps in *Staphylococcus aureus*. *Journal of Bacteriology*, **187** (7), 2395–2405.

143 Takenouchi, T., Tabata, F., Iwata, Y., Hanzawa, H., Sugawara, M., and Ohya, S. (1996) Hydrophilicity of quinolones is not an exclusive factor for decreased activity in efflux-mediated resistant mutants of *Staphylococcus aureus*. *Antimicrobial Agents and Chemotherapy*, **40** (8), 1835–1842.

144 Yu, J.L., Grinius, L., and Hooper, D.C. (2002) NorA functions as a multidrug efflux protein in both cytoplasmic membrane vesicles and reconstituted proteoliposomes. *Journal of Bacteriology*, **184** (5), 1370–1377.

145 Mesaros, N., Nordmann, P., Plesiat, P., Roussel-Delvallez, M., Van Eldere, J., Glupczynski, Y., Van Laethem, Y., Jacobs, F., Lebecque, P., Malfroot, A., Tulkens, P.M., and Van Bambeke, F. (2007) *Pseudomonas aeruginosa*: resistance and

therapeutic options at the turn of the new millennium. *Clinical Microbiology and Infection*, **13** (6), 560–578.

146 Lomovskaya, O., Zgurskaya, H.I., Totrov, M., and Watkins, W.J. (2007) Waltzing transporters and 'the dance macabre' between humans and bacteria. *Nature Reviews. Drug Discovery*, **6** (1), 56–65.

147 De Kievit, T.R., Parkins, M.D., Gillis, R.J., Srikumar, R., Ceri, H., Poole, K., Iglewski, B.H., and Storey, D.G. (2001) Multidrug efflux pumps: expression patterns and contribution to antibiotic resistance in *Pseudomonas aeruginosa* biofilms. *Antimicrobial Agents and Chemotherapy*, **45** (6), 1761–1770.

148 Payne, D.J., Gwynn, M.N., Holmes, D.J., and Pompliano, D.L. (2007) Drugs for bad bugs: confronting the challenges of antibacterial discovery. *Nature Reviews. Drug Discovery*, **6** (1), 29–40.

149 Hsieh, P.C., Siegel, S.A., Rogers, B., Davis, D., and Lewis, K. (1998) Bacteria lacking a multidrug pump: a sensitive tool for drug discovery. *Proceedings of the National Academy of Sciences of the United States of America*, **95** (12), 6602–6606.

150 Nakae, T., Nakajima, A., Ono, T., Saito, K., and Yoneyama, H. (1999) Resistance to beta-lactam antibiotics in *Pseudomonas aeruginosa* due to interplay between the MexAB-OprM efflux pump and beta-lactamase. *Antimicrobial Agents and Chemotherapy*, **43** (5), 1301–1303.

151 Pages, J.M., Masi, M., and Barbe, J. (2005) Inhibitors of efflux pumps in Gram-negative bacteria. *Trends in Molecular Medicine*, **11** (8), 382–389.

152 Mao, W., Warren, M.S., Lee, A., Mistry, A., and Lomovskaya, O. (2001) MexXY-OprM efflux pump is required for antagonism of aminoglycosides by divalent cations in *Pseudomonas aeruginosa*. *Antimicrobial Agents and Chemotherapy*, **45** (7), 2001–2007.

153 Renau, T.E., Leger, R., Filonova, L., Flamme, E.M., Wang, M., Yen, R., Madsen, D., Griffith, D., Chamberland, S., Dudley, M.N., Lee, V.J., Lomovskaya, O., Watkins, W.J., Ohta, T., Nakayama, K., and Ishida, Y. (2003) Conformationally-restricted analogues of efflux pump inhibitors that potentiate the activity of levofloxacin in *Pseudomonas aeruginosa*. *Bioorganic & Medicinal Chemistry Letters*, **13** (16), 2755–2758.

154 Watkins, W.J., Landaverry, Y., Leger, R., Litman, R., Renau, T.E., Williams, N., Yen, R., Zhang, J.Z., Chamberland, S., Madsen, D., Griffith, D., Tembe, V., Huie, K., and Dudley, M.N. (2003) The relationship between physicochemical properties, *in vitro* activity and pharmacokinetic profiles of analogues of diamine-containing efflux pump inhibitors. *Bioorganic & Medicinal Chemistry Letters*, **13** (23), 4241–4244.

155 Bohnert, J.A. and Kern, W.V. (2005) Selected arylpiperazines are capable of reversing multidrug resistance in *Escherichia coli* overexpressing RND efflux pumps. *Antimicrobial Agents and Chemotherapy*, **49** (2), 849–852.

156 Schumacher, A., Steinke, P., Bohnert, J.A., Akova, M., Jonas, D., and Kern, W.V. (2006) Effect of 1-(1-naphthylmethyl)-piperazine, a novel putative efflux pump inhibitor, on antimicrobial drug susceptibility in clinical isolates of Enterobacteriaceae other than *Escherichia coli*. *The Journal of Antimicrobial Chemotherapy*, **57** (2), 344–348.

157 Nakayama, K., Ishida, Y., Ohtsuka, M., Kawato, H., Yoshida, K., Yokomizo, Y., Hosono, S., Ohta, T., Hoshino, K., Ishida, H., Renau, T.E., Leger, R., Zhang, J.Z., Lee, V.J., and Watkins, W.J. (2003) MexAB-OprM-specific efflux pump inhibitors in *Pseudomonas aeruginosa*. Part 1. Discovery and early strategies for lead optimization. *Bioorganic & Medicinal Chemistry Letters*, **13** (23), 4201–4204.

158 Nakayama, K., Ishida, Y., Ohtsuka, M., Kawato, H., Yoshida, K., Yokomizo, Y., Ohta, T., Hoshino, K., Otani, T., Kurosaka, Y., Ishida, H., Lee, V.J., Renau, T.E., and Watkins, W.J. (2003) MexAB-OprM specific efflux pump inhibitors in

Pseudomonas aeruginosa. Part 2. Achieving activity *in vivo* through the use of alternative scaffolds. *Bioorganic & Medicinal Chemistry Letters*, **13** (23), 4205–4208.

159 Nakayama, K., Kuru, N., Ohtsuka, M., Yokomizo, Y., Sakamoto, A., Kawato, H., Yoshida, K., Ohta, T., Hoshino, K., Akimoto, K., Itoh, J., Ishida, H., Cho, A., Palme, M.H., Zhang, J.Z., Lee, V.J., and Watkins, W.J. (2004) MexAB-OprM specific efflux pump inhibitors in *Pseudomonas aeruginosa*. Part 4. Addressing the problem of poor stability due to photoisomerization of an acrylic acid moiety. *Bioorganic & Medicinal Chemistry Letters*, **14** (10), 2493–2497.

160 Yoshida, K.I., Nakayama, K., Yokomizo, Y., Ohtsuka, M., Takemura, M., Hoshino, K., Kanda, H., Namba, K., Nitanai, H., Zhang, J.Z., Lee, V.J., and Watkins, W.J. (2006) MexAB-OprM specific efflux pump inhibitors in *Pseudomonas aeruginosa*. Part 6. Exploration of aromatic substituents. *Bioorganic and Medicinal Chemistry*, **14** (24), 8506–8518.

161 Yoshida, K., Nakayama, K., Kuru, N., Kobayashi, S., Ohtsuka, M., Takemura, M., Hoshino, K., Kanda, H., Zhang, J.Z., Lee, V.J., and Watkins, W.J. (2006) MexAB-OprM specific efflux pump inhibitors in *Pseudomonas aeruginosa*. Part 5. Carbon-substituted analogues at the C-2 position. *Bioorganic and Medicinal Chemistry*, **14** (6), 1993–2004.

162 Nakayama, K., Kawato, H., Watanabe, J., Ohtsuka, M., Yoshida, K., Yokomizo, Y., Sakamoto, A., Kuru, N., Ohta, T., Hoshino, K., Yoshida, K., Ishida, H., Cho, A., Palme, M.H., Zhang, J.Z., Lee, V.J., and Watkins, W.J. (2004) MexAB-OprM specific efflux pump inhibitors in *Pseudomonas aeruginosa*. Part 3. Optimization of potency in the pyridopyrimidine series through the application of a pharmacophore model. *Bioorganic & Medicinal Chemistry Letters*, **14** (2), 475–479.

163 Mallea, M., Mahamoud, A., Chevalier, J., Alibert-Franco, S., Brouant, P., Barbe, J., and Pages, J.M. (2003) Alkylaminoquinolines inhibit the bacterial antibiotic efflux pump in multidrug-resistant clinical isolates. *The Biochemical Journal*, **376** (Pt 3), 801–805.

164 Chevalier, J., Bredin, J., Mahamoud, A., Mallea, M., Barbe, J., and Pages, J.M. (2004) Inhibitors of antibiotic efflux in resistant *Enterobacter aerogenes* and *Klebsiella pneumoniae* strains. *Antimicrobial Agents and Chemotherapy*, **48** (3), 1043–1046.

165 Mallea, M., Chevalier, J., Eyraud, A., and Pages, J.M. (2002) Inhibitors of antibiotic efflux pump in resistant *Enterobacter aerogenes* strains. *Biochemical and Biophysical Research Communications*, **293** (5), 1370–1373.

166 Chevalier, J., Atifi, S., Eyraud, A., Mahamoud, A., Barbe, J., and Pages, J.M. (2001) New pyridoquinoline derivatives as potential inhibitors of the fluoroquinolone efflux pump in resistant *Enterobacter aerogenes* strains. *Journal of Medicinal Chemistry*, **44** (23), 4023–4026.

167 Thorarensen, A., Presley-Bodnar, A.L., Marotti, K.R., Boyle, T.P., Heckaman, C.L., Bohanon, M.J., Tomich, P.K., Zurenko, G.E., Sweeney, M.T., and Yagi, B.H. (2001) 3-Arylpiperidines as potentiators of existing antibacterial agents. *Bioorganic & Medicinal Chemistry Letters*, **11** (14), 1903–1906.

168 Stavri, M., Piddock, L.J., and Gibbons, S. (2007) Bacterial efflux pump inhibitors from natural sources. *The Journal of Antimicrobial Chemotherapy*, **59** (6), 1247–1260.

169 Kaatz, G.W. (2005) Bacterial efflux pump inhibition. *Current Opinion in Investigational Drugs*, **6** (2), 191–198.

170 Pleban, K., Kaiser, D., Kopp, S., Peer, M., Chiba, P., and Ecker, G.F. (2005) Targeting drug-efflux pumps – a pharmacoinformatic approach. *Acta Biochimica Polonica*, **52** (3), 737–740.

171 Sauna, Z.E., Andrus, M.B., Turner, T.M., and Ambudkar, S.V. (2004) Biochemical basis of polyvalency as a strategy for enhancing the efficacy of P-glycoprotein (ABCB1) modulators: stipiamide homodimers separated with defined-length spacers reverse drug efflux with greater efficacy. *Biochemistry*, **43** (8), 2262–2271.

172 Ball, A.R., Casadei, G., Samosorn, S., Bremner, J.B., Ausubel, F.M., Moy, T.I., and Lewis, K. (2006) Conjugating berberine to a multidrug efflux pump inhibitor creates an effective antimicrobial. *ACS Chemical Biology*, **1** (9), 594–600.

173 Szakacs, G., Paterson, J.K., Ludwig, J.A., Booth-Genthe, C., and Gottesman, M.M. (2006) Targeting multidrug resistance in cancer. *Nature Reviews. Drug Discovery*, **5** (3), 219–234.

174 Liang, X.J. and Aszalos, A. (2006) Multidrug transporters as drug targets. *Current Drug Targets*, **7** (8), 911–921.

175 Drusano, G.L. (2004) Antimicrobial pharmacodynamics: critical interactions of 'bug and drug'. *Nature Reviews. Microbiology*, **2** (4), 289–300.

6
Membrane Transporters in Pleiotropic Drug Resistance and Stress Response in Yeast and Fungal Pathogens

Tobias Schwarzmüller, Cornelia Klein, Martin Valachovic, Walter Glaser, and Karl Kuchler

Abbreviations

ABC	ATP binding cassette
ALDP	adrenoleukodystrophy protein
CFTR	cystic fibrosis transmembrane conductance regulator
DRE	drug response element
ER	endoplasmic reticulum
ERG	ergosterol
MDR	multidrug resistance
MFS	major facilitator superfamily
MRP	multidrug resistance-related protein
NBD	nucleotide binding domain
NTE	N-terminal extension
PDR	pleiotropic drug resistance
PDRE	pleiotropic drug resistance element
SRE	sterol regulatory element
STRE	stress response element
TF	transcription factor
TMD	transmembrane domain
TMS	transmembrane-spanning segment
WARE	weak acid response element
YRE	Yap1 response element
YRRE	Yrr1 response element

6.1
Introduction

During the past decades, the prevalence of opportunistic fungal infections has gained importance, particularly because clinical therapy is sometimes hampered

Transporters as Drug Carriers: Structure, Function, Substrates. Edited by Gerhard Ecker and Peter Chiba
Copyright © 2009 WILEY-VCH Verlag GmbH & Co. KGaA, Weinheim
ISBN: 978-3-527-31661-8

by antifungal drug resistance [1–4]. Notably, the molecular understanding of mechanisms causing antifungal resistance demonstrated that similar molecular mechanisms are the basis of multidrug resistance (MDR) in cancer cells [5, 6] as well as microbial pathogens. Indeed, multidrug resistance occurs in all organisms, including mammals, fungal pathogens, parasites, and, of course, bacteria [7–9]. This phenomenon is mainly mediated by ATP binding cassette (ABC) proteins [10], which constitute a large family of proteins sharing similar structural features [7]. Notably, they are involved not only in transmembrane transport but also in several other fundamental cellular processes, and they are connected to prominent genetic diseases [7, 11, 12].

The phenomenon of pleiotropic drug resistance (PDR) has also been described in the baker's yeast *Saccharomyces cerevisiae*, which serves as a valuable model system to study MDR/PDR and to gain further insight into the complex mechanisms and functional specificities of ABC proteins. Certain yeast ABC genes represent orthologues of mammalian disease genes such as the cystic fibrosis transmembrane conductance regulator (CFTR) [13]. In addition, several yeast ABC genes are orthologous to those in clinically relevant pathogens, including *Candida albicans*, *C. glabrata*, and *Aspergillus fumigatus* [14–16]. Access to the full genome sequences and protein databases of *S. cerevisiae* (http://www.yeastgenome.org/), *C. albicans* (http://www.candidagenome.org/), *C. glabrata* (http://cbi.labri.fr/Genolevures/index.php), and other fungal species has facilitated new approaches and tools to study drug carriers of the ABC transporter family and their transcriptional regulators in molecular detail.

Like most ABC proteins, fungal ABC transporters share a similar domain organization (Figure 6.1), consisting of two nucleotide binding domains (NBD) and two transmembrane domains (TMD). The highly conserved 200–220 residues containing NBDs are characterized by different motifs of which the Walker A and Walker B are common to all ATP binding proteins. A consensus sequence unique to ABC transporters is the C-loop or signature motif (LSGGQ) [17, 18]. Several other motifs are located in proximity to the Walker A and Walker B motifs. They are referred to as the center motifs, A-loop and D-loop [17]. These motifs bind ATP and coordinate Mg^{2+} and H_2O at the binding site [19]. Each NBD binds a single ATP molecule orientated in a sandwich-like, head-to-tail dimer arrangement. Through ATP hydrolysis, the NBDs drive translocation of substrates. In contrast, TMDs share less homology within different ABC proteins. This might be explained by their roles as substrate binding sites. A TMD consists of a bundle of α-helices usually building six predicted transmembrane-spanning segments (Figure 6.1). TMDs are important for the overall architecture, although the overall structure and perhaps even topology is certainly influenced by conformational states of the NBDs. In Figure 6.1, two possible topologies of fungal ABC proteins, the forward arrangement $(TMS_6\text{-}NBD)_2$ and the reverse topology $(NBD\text{-}TMS_6)_2$ are depicted. The latter is present in most full-size transporters of the PDR family. Conversely, members of the MRP/CFTR subfamily are arranged in a forward orientation, including an additional TMD at the N-terminus, known as N-terminal extension (NTE). The half-size transporters are found in the MDR and ALDP family. Two half-size transporters build a functional

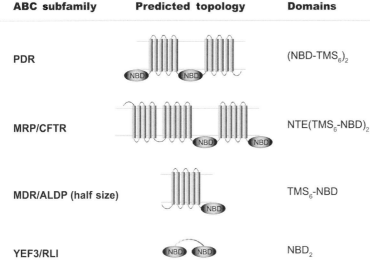

Figure 6.1 Predicted topology and domain organization of fungal ABC protein subfamilies. The figure depicts the predicted membrane topology and domain organization of all subfamilies encoding yeast ABC proteins (see text for details). NBD, nucleotidebinding domain; NTE, N-terminal extension; TMS, transmembrane segment.

protein through homo- or heterodimerization. Interestingly, some ABC proteins consist of only two NBD domains but no TMDs (Figure 6.1).

Most ABC transporters, also known as ABC efflux pumps, are found in membranes of distinct cellular compartments (Figure 6.2). They localize in the plasma membrane, the vacuolar membrane, peroxisomes, and the inner mitochondrial membrane, although none seems to reside in the nuclear envelope or the ER. Notably, they mediate membrane transport of numerous molecules, including ions, heavy metals, sugars and amino acids, drugs, xenobiotics, bile acids, steroids and glucocorticoids, GS-conjugates, lipids, fluorescent dyes, and even whole proteins [20]. However, the broad substrate specificity and molecular transport mechanisms still remain a mystery [21]. Likewise, it is still an enigma why some ABC transporters show narrow substrate specificity, whereas others are quite unspecific, implicated in many different cellular processes, including pheromone transport, peroxisome biogenesis, maturation of cytosolic Fe/S proteins, mitochondrial functions, lipid bilayer homeostasis, and stress response [20, 22]. Remarkably, phenotypes linked to PDR are usually well documented in the literature, whereas physiological functions of many ABC transporters remain undisclosed.

The yeast PDR network (Figure 6.3) consists of several ABC transporters distinctly regulated by dedicated transcription factors. Indeed, the expression of PDR genes has been tightly linked to drug resistance. PDR, as well as MDR, is best described as the ability to develop resistance to a single toxic compound, followed by the appearance of cross-resistance to a great variety of structurally and functionally unrelated

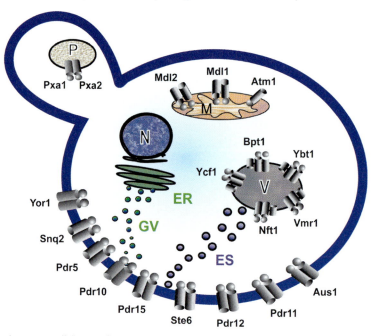

Figure 6.2 Cellular membranes in yeast harboring ABC proteins. The figure depicts the subcellular localization of prominent membrane ABC transporters at the cell surface, in the vacuole, in mitochondria, and in peroxisomes. Only ABC transporters whose functions have been studied beyond sequencing are depicted (see text for details). N, nucleus; V, vacuole; ER, endoplasmic reticulum; GV, Golgi vesicles; ES, endosomes; M, mitochondrion; P, peroxisome.

substances. PDR is the consequence of several independent yet overlapping mechanisms, all of which contribute to a compound PDR phenotype. First, mutations in target genes, their transcriptional activators, or specific inhibitors may alter the response to the drug. Thus, expression levels of transporter genes or their transcriptional regulators may be affected. Consequently, the intracellular drug concentration is reduced due to the increased efflux. Overexpression of ABC pumps is indeed a major cause for acquired drug resistance in fungal pathogens. The second cause is the removal of toxic compounds through vacuolar sequestration. Furthermore, PDR can be achieved by intracellular drug inactivation. The reduced drug uptake and changes in the composition of the permeability of the plasma membrane can also contribute to the development of PDR. Finally, signaling, stress response, and target alteration through mutation are additional components contributing to PDR [23–25].

6.2
ABC Protein Genes in *S. cerevisiae*

The genome of *S. cerevisiae* contains some 30 distinct genes encoding ABC proteins (see Table 6.1). Based on evolutionary relationships, these genes have been grouped

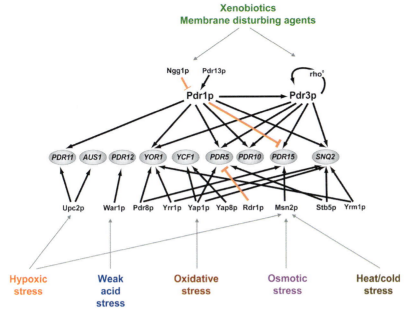

Figure 6.3 The PDR regulatory network in yeast. Genes in the centerline represent target genes of transcriptional regulators depicted above and below. Note that the figure lists only genes of the ABC family. The yeast PDR network also contains non-ABC genes whose function has not been established in many cases (see text for further details). Red lines indicate a negative regulatory impact, while black lines ending with an arrow indicate positive regulation.

into five subfamilies, which are referred to as the PDR, MRP/CFTR, MDR, ALDP, and YEF3/Rli families [26, 27]. As shown in Table 6.1, there are ABC proteins residing in various cellular organelles such as mitochondria and peroxisomes. However, none of these ABC proteins has been implicated in obvious drug resistance phenomenon. Therefore, we shall not discuss them here for the sake of brevity and refer to several recent reviews addressing their functions in detail [23, 26, 28–32].

6.2.1
The PDR Family

This family contains two extensively studied yeast ABC transporters, namely, Pdr5 and Snq2. Other members include Pdr10, Pdr11, Pdr12, Pdr15, Adp1, Aus1, and a putative ABC transporter encoded by the gene YNR070W. Adp1 is the only member in this group that is not a full-size transporter. They all share the so-called reverse topology, where the NBD is followed by a predicted TM region. Snq2 was the first ABC transporter implicated in drug resistance in yeast. It confers resistance to mutagens such as 4-nitroquinoline-N-oxide (4-NQO), triaziquone, and the chemicals sulfometuron methyl and phenanthroline [33, 34]. Snq2 is also linked to cation resistance because *snq2* null mutants exhibit increased sensitivities to NaCl, LiCl, and $MnCl_2$ [35].

Table 6.1 ABC proteins in the yeast *S. cerevisiae*.

ABC protein	Size	Topology	Local	Substrates/function	References
PDR family					
Pdr5	1511	(NBD-TMS$_6$)$_2$	PM	PDR, drugs, lipids?	[37–39]
Snq2	1501	(NBD-TMS$_6$)$_2$	PM	PDR	[34]
Pdr10	1564	(NBD-TMS$_6$)$_2$	PM	?	[49]
Pdr11	1411	(NBD-TMS$_6$)$_2$	PM	Sterol uptake	[56]
Pdr12	1511	(NBD-TMS$_6$)$_2$	PM	Weak organic acids	[52]
Pdr15	1529	(NBD-TMS$_6$)$_2$	PM	General stress response	[49]
Aus1	1394	(NBD-TMS$_6$)$_2$	PM	Sterol uptake	[56]
Adp1	1049	TMS$_2$-NBD-TMS$_7$	?	?	[22]
YNR070w	1333	(NBD-TMS$_6$)$_2$	?	?	[22]
YOL075c	1294	(NBD-TMS$_6$)$_2$	?	?	[22]
MRP/CFTR family					
Yor1	1477	NTE-(TMS$_6$-NBD)$_2$	PM	PDR, lipid transport?	[214]
Ycf1	1515	NTE-(TMS$_6$-R-NBD)$_2$	V	Detoxification	[65]
Ybt1	1661	NTE-(TMS$_6$-NBD)$_2$	V	Bile acid transport	[75]
Bpt1	1559	NTE-(TMS$_6$-NBD)$_2$	V	Detoxification	[72]
Nft1	1218	(TMS$_6$-NBD)$_2$	?	Detoxification?	[76]
Vmr1/YHL035c	1592	NTE-(TMS$_6$-NBD)$_2$	V?	?	Wawrzycka, D., Bartosz, G., Ulaszewski, S., and Goffeau, A., (2003) A novel vacuolar multidrug resistance (*VMR1*) ABC transporter in *Saccharomyces cerevisiae*, personal communication to SGD

MDR family					
Ste6	1290	(TMS$_6$-NBD)$_2$	PM	a-factor	[215, 216]
Atm1	690	TMS$_6$-NBD	M	Fe/S cluster precursor	[217]
Mdl1	695	TMS$_6$-NBD	M	Peptide transport	[188, 218]
Mdl2	773	TMS$_6$-NBD	M	?	[218]
ALDP family					
Pxa1	870	TMS$_6$-NBD	P	LCFA	[219]
Pxa2	853	TMS$_6$-NBD	P	LCFA	[220]
YEF3/RLI1 family					
Yef3	1044	NBD$_2$	C	Translational EF	[221]
Hef3	1044	NBD$_2$	C	Translational EF	[222, 223]
Gcn20	752	NBD$_2$	C	Positive regulator of Gcn2	[224]
New1	1196	TMS$_3$-NBD$_2$	C	?	[225]
Arb1	610	NBD$_2$	C	Ribosome biogenesis	[226]
Rli1	608	NBD$_2$	C	Translation initiation	[227]
Caf16	289	NBD	C?	?	[228]
YDR061w	539	NBD	M?	?	

Abbreviations: NBD, nucleotide binding domain; TMS, transmembrane-spanning segment; NTE, N-terminal extension; PDR, pleiotropic drug resistance; PM, plasma membrane; V, vacuole; M, mitochondrion; P, peroxisome; C, cytoplasm; LCFA, long chain fatty acid; EF, elongation factor.

One of the best studied ABC transporters is Pdr5, a functional homologue of the mammalian P-glycoprotein (*MDR1*) [36]. The gene *PDR5* was cloned and characterized by several laboratories, independently identifying Pdr5 as a transporter mediating resistance to mycotoxins [37], cycloheximide [38], cerulenin [39], and glucocorticoids [40]. The number of potential substrates increased continuously. At present, a broad spectrum of more than 100 compounds is known for Pdr5. Different mutagens, mycotoxins, ions, and heavy metals appear effluxed by Pdr5 as well as anticancer drugs, azoles, herbicides, antibiotics, detergents, bile acids, steroids, and many others [41–44]. Mutagenesis studies have revealed the importance of TM10 in modulating substrate specificity as well as susceptibility to inhibitors of Pdr5 transport activity. Strikingly, mutation of the residue S1360 to different amino acids could lead to either nonresponsiveness or hypersensitivity [45–47]. Very recent studies have also suggested influence of the NBDs on Pdr5 substrate specificity [48].

PDR10 and *PDR15* encode full-size ABC proteins sharing a very high identity with *PDR5* [49]. In contrast to Pdr5, limited data are available on these two ABC pumps. Pdr15 is assumed to be involved in both general stress and membrane stress responses. It contributes to resistance to different membrane-perturbing agents such as polyoxyethylene-9-lauryl ether as well as to chloramphenicol [43, 50]. However, the expression of Pdr15 is strongly induced by various stresses such as low pH, heat shock, high osmolarity, starvation, and weak acids [43]. In addition, it was found that *PDR15* expression increases rapidly when cells exit the exponential growth phase and it stays highly induced during the stationary phase. In contrast, *PDR5* levels are high during exponential growth but disappear after reaching the stationary phase [43, 51]. The role of Pdr10 still needs to be elucidated, but it may be involved in membrane lipid homeostasis [23].

Sharing about 60% homology, Pdr12 is most closely related to Snq2. Induction of *PDR12* is mediated by carboxylate anions such as benzoic, sorbic, and propionic acids rather than by hydrophobic compounds [52]. In addition to these weak organic acids, fluorescein is a substrate of Pdr12 [52, 53]. Weak acids are of special interest for the food industry because they are used as food preservatives inhibiting the growth of spoilage microorganisms [54]. Indeed, the Pdr12 involvement in the metabolic network of *S. cerevisiae* is suggested by the fact that Pdr12 exports aromatic and branched-chain organic acids such as the fusel acids, phenylacetate, indoleacetate, and 2-methylbutanoate, all of which are unwanted by-products of amino acid metabolism [55].

No PDR phenotype has been disclosed for Aus1 and Pdr11. Interestingly, Aus1 and Pdr11 are implicated in the import of sterol into the cell. Deletion of any of the two genes encoding Aus1 and Pdr11 significantly decreased sterol uptake while simultaneous deletion of *AUS1* and *PDR11* completely diminished sterol import, suggesting partially overlapping functions [56]. Recently, *AUS1* was shown to be upregulated in response to azole treatment [57], suggesting that Aus1 might counteract the depletion of ergosterol in the membrane, which is caused by the azole treatment.

ADP1 and YNR070w and the closely related YOL075c gene product are PDR gene family members of unknown function. Adp1 is the only PDR member that differs in the molecular architecture. Although all other members are full-size transporters of the reverse $(NBD-TMS_6)_2$ topology, Adp1, whose function remains elusive, consists of only one NBD, flanked by two predicted transmembrane segments, upstream TMS_2 and downstream TMS_7 [22].

6.2.2
The MRP/CFTR Family

Besides the major drug efflux pumps Pdr5 and Snq2, the plasma membrane Yor1 oligomycin transporter is the third major driver of PDR in S. cerevisiae [58]. Based on their common topology $NTE-(TMS_6-NBD)_2$, Yor1 belongs to the MRP/CFTR family, which also comprises the vacuolar ABC transporters Ycf1, Ybt1, and Bpt1. A *yor1* deletion mutant displays hypersensitivity to a variety of xenobiotics, including oligomycin, reveromycin A, Cd^{2+}, the antibiotics tetracycline and erythromycin, the anticancer drugs daunorubicin and doxorubicin, different azoles, and acetic, benzoic, and propionic acids [59, 60]. As for Pdr5 [61] and other yeast ABC transporters [62], certain amino acid residues are also essential for the correct trafficking and localization of Yor1 to the plasma membrane, a prerequisite for a function of detoxification. Deletion of the phenylalanine residue at position 670 in Yor1 results in the ER retention of the mutant protein, leading to a loss of oligomycin resistance [58]. Interestingly, truncations in the N-terminal part of Yor1 result in similar defects in plasma membrane localization. Deletion of the residues 2–38 results in a mutant in which Yor1 is still able to confer oligomycin resistance, whereas deletion of the residues 60–77 traps the nonfunctional protein in the ER [63, 64]. These studies clarified that N- and C-terminal regions are essential for proper trafficking, and thus play a crucial role for a functional ABC transporter.

The transporters Ycf1, Ybt1, and Bpt1 contribute to PDR by the so-called vacuolar sequestration of conjugated and unconjugated molecules or heavy metals. This mechanism enables cells to tolerate high drug concentrations by sequestering toxic compounds into the vacuole. Ycf1 was identified by its ability to mediate resistance to the heavy metal cadmium. Interestingly, Ycf1 is related to the mammalian MRP1 and MRP2 transporters, and moreover shows 45% similarity to the human CFTR transporter [22, 65]. At present, several Ycf1 substrates are known, including glutathione, glutathione-S conjugates, and the red pigment accumulating in *ade2* mutant cells. Furthermore, Ycf1 confers resistance to arsenite, mercury, and diazaborine [66–70]. Notably, resistance to diazaborine can also be mediated by the Snq2 and Pdr5 transporters.

Bpt1 is a homologue of Ycf1 and overlaps in substrate specificity with Ycf1 [71–73]. Surprisingly, a double deletion strain lacking both *YCF1* and *BPT1* leads to acetaminophen resistance. The reasons for this observation may be the increased toxicity of accumulating GSH conjugates or a feedback inhibition of the enzymes responsible for the production of the toxic intermediates [74]. In contrast, the overexpression of the ABC pump Snq2 and the MFS permease Flr1 renders cells more tolerant to acetaminophen [74], indicating that different ABC pumps act together to develop resistance to a compound, yet each contributing to a different extent.

YBT1, also known as BAT1, was isolated using degenerated primers matching conserved regions of ABC transporters. The deletion of YBT1 results in the defective transport of bile acids into the vacuole [75]. Two additional transporters of the MRP/CFTR family are represented by the NFT1 and VMR1 genes. Nft1 is a full-size transporter displaying a typical MRP/CFTR topology; it might be involved in resistance to cadmium or arsenite [76]. Nothing is known about the function of Vmr1.

6.3
Orchestrating Pleiotropic Drug Resistance: The PDR Network

As noted previously, multidrug or pleiotropic drug resistance is a highly conserved phenomenon in all organisms ranging from bacteria to mammals [7]. One common cause of hyperresistance is increased drug efflux due to the overexpression of drug transporters. PDR is tightly regulated by dedicated TFs falling into two major groups: the zinc cluster and the bZIP family of transcriptional regulators (Table 6.2 and Figure 6.3). The most prominent TFs of the PDR network are Pdr1 and Pdr3, which were the first genes to be identified by genetic screens for drug-resistant strains [77, 78]. These strains harbored hyperactive alleles such as pdr1–3 [77], pdr1–8 [79], pdr1–12, or pdr3–33 [70]. Constitutive hyperactivity was caused by single amino acid substitutions clustering in defined regions of the proteins such as the central regulatory region or the C-terminal activation domain [79–81]. Both Pdr1 and Pdr3 are binuclear $Zn(II)_2$-Cys_6 zinc cluster proteins [82–84] containing the conserved DNA binding motif Cys-X_2-Cys-X_6-Cys-X_{5-16}-Cys-X_2-Cys-X_{6-8}-Cys [85], a C-terminal activation domain, and the so-called middle homology domain [82, 86]. The best studied prototype family member is the TF Gal4 [87, 88]. Other regulator genes controlling PDR listed in Table 6.2 are YRR1 [89, 90], RDR1 [91], STB5 [84], PDR8 [92], and YRM1 [93].

The homologous regulators Pdr1 and Pdr3 control expression of several ABC transporters, including Pdr5, Snq2, Yor1, Pdr10, and Pdr15 [94–98], as well as the major facilitator superfamily (MFS) proteins Hxt9 and Hxt11 [99]. Pdr1 and Pdr3 act on cis-acting motifs, the so-called PDREs (pleiotropic drug resistance elements). This motif consists of the consensus sequence 5′-TCCGCGGA-3′ [95, 100], including the everted CGG repeat recognized by Gal4-like proteins [91]. PDREs are found in the promoter regions of the ABC genes PDR5, PDR10, PDR15, SNQ2, and YOR1 [101], as well as in a number of other potential PDR target genes [96]. In contrast to PDR1, PDR3 is subject to autoregulation [90, 100, 102]. Although a single PDRE is necessary and sufficient for regulation of a target gene, several PDREs may be present in the promoter. In addition, Pdr1 and Pdr3 can both positively and negatively regulate the expression of target genes [103], implying that additional factors are required for fine-tuning the regulation of this complex network (Figure 6.3). Indeed, Pdr1 and Pdr3 are able to form homo- and heterodimers with each other, as well as with other members of the zinc cluster protein family [104]. A recent study also identified several loss-of-function pdr3 alleles whose expression sensitized cells to known Pdr5 substrates [105]. Interestingly, Pdr3 also plays a role in retrograde activation of PDR5

Table 6.2 Modulators and regulators of the yeast PDR network.

Transcription factor	Topology	Function	Target genes	References
S. cerevisiae				
Pdr1	Zn(II)$_2$Cys$_6$	Regulation of PDR	*PDR5, SNQ2, YOR1, HXT9, HXT11*	[77]
Pdr3	Zn(II)$_2$Cys$_6$	Regulation of PDR	*PDR5, SNQ2, YOR1, HXT9, HXT11, PDR3*	[78]
Yrr1	Zn(II)$_2$Cys$_6$	Regulation of PDR	*SNQ2, YOR1, AZR1, FLR1, SNG1*	[90]
Stb5	Zn(II)$_2$Cys$_6$	Regulation of PDR	Pentose phosphate genes	[84]
Yrm1	Zn(II)$_2$Cys$_6$	Regulation of PDR	*YRR1*	[93]
Rdr1	Zn(II)$_2$Cys$_6$	Transcriptional repressor	*PDR5*	[91]
Pdr8	Zn(II)$_2$Cys$_6$	Regulation of PDR	*PDR15, YOR1, AZR1, SNG1*	[92]
War1	Zn(II)$_2$Cys$_6$	Weak acid stress response	*PDR12, FUN34, TFB2*	[109]
Upc2	Zn(II)$_2$Cys$_6$	Regulation of PDR and ergosterol biosynthesis	*AUS1, PDR11, ERG* genes	[112]
Ecm22	Zn(II)$_2$Cys$_6$	Regulation of PDR and ergosterol biosynthesis	*AUS1, PDR11, ERG* genes	[112]
Yap1	bZip	Oxidative stress response	*SNQ2, YCF1, PDR5, ATR1, FLR1*	[125]
Yap2	bZip TF	Cadmium resistance	?	[130]
Yap8	bZip TF	Arsenite resistance	?	[123]
Ngg1	TF	Inhibition of Pdr1 activity	*PDR1*	[135]
Msn2/Msn4	Cys$_2$His$_2$	General stress response	*PDR15*	[116, 117]
C. albicans				
Cap1	bZip TF	Oxidative stress response	*CaMDR1*	[160]
Tac1	Zn(II)$_2$Cys$_6$	Transcriptional activator CDR1 and CDR2	*CDR1, CDR2*	[143]
Fcr1	Zn(II)$_2$Cys$_6$	Negative regulator of drug resistance	?	[171]
Upc2	Zn(II)$_2$Cys$_6$	Regulation of sterol uptake and ergosterol biosynthesis	*ERG* genes?	[148]

Abbreviations: TF, transcription factor; PDR, pleiotropic drug resistance; bZip, basic leucine zipper.

transcription and activation of the PDR network. In rho^0 cells lacking mitochondrial DNA, a signal from defective mitochondria leads to posttranslational modification of Pdr3 and the subsequent induction of *PDR5* expression in a Pdr1-independent manner [106] through the PDREs in the *PDR5* promoter.

In contrast to Pdr1/Pdr3 heterodimers, Yrr1 functions as a homodimer [107]. Furthermore, Pdr1, Pdr3, and Stb5 share only a few common target genes. Stb5 binds to genes of the pentose phosphate pathway and those involved in NADPH production. A *stb5* deletion strain is sensitive to diamide and H_2O_2, suggesting that Stb5 is required for protection against oxidative stress or reactive oxygen species. Stb5 acts as an activator and inhibitor, confirming a putative role of Stb5 in oxidative stress tolerance [108]. Yrr1 shares overlapping functions with Pdr1 and Pdr3 by controlling *SNQ2* and *YOR1* [89, 90]. The promoters of these genes bear an YRRE (Yrr1-response element) regulatory sequence sharing similarities with the PDRE consensus motif [89]. Yrr1 regulates resistance to reveromycin A, oligomycin, and 4-nitroquinoline-1-oxide [102]. Like *PDR3*, *YRR1* appears to be autoregulated [93]. In addition, *YRR1* expression is modulated by Yrm1 [93]. *YRM1* (yeast reveromycin resistance modulator) encodes another $Zn(II)_2Cys_6$ transcriptional factor acting as a specific inhibitor of Yrr1. In an *yrr1* deletion strain, Yrm1 activates transcription of genes that are otherwise direct targets of Yrr1 [93].

Rdr1, another binuclear regulator, acts as a repressor of *PDR5* in a PDRE-dependent manner by forming heterodimers with Pdr1/Pdr3. Another explanation may be that Rdr1 competes with Pdr1 and Pdr3 for binding to PDREs [91]. Further, microarray profiling identified the putative binuclear zinc cluster transcription factor Pdr8 as yet another regulator of PDR [92]. Taken together, the great variety of different transcription factors reflects the high level of complexity of the regulatory processes controlling PDR in yeast.

Notably, not all yeast ABC transporters are regulated by the major players of the PDR network. For instance, War1 mediates weak acid resistance via induction of the Pdr12 ABC transporter [109]. The War1 zinc cluster protein is extensively phosphorylated in response to weak organic acid stress. *In vivo* footprinting revealed that War1 decorates the *cis*-acting weak acid response element (WARE) present in the *PDR12* promoter [109]. A nonphosphorylated *war1* allele fails to induce weak acid stress response, arguing for a tight coupling of posttranslational modification with War1 activity [110]. Noteworthy, microarray analysis identified a rather small War1-regulon with Pdr12 as the major target [111].

Upc2 is a transcription factor regulating expression of the ABC genes *AUS1* and *PDR11* encoding ABC transporters involved in sterol influx [56]. In addition, Upc2 controls transcription of sterol biosynthetic genes such as *ERG2* and *ERG3*, as well as several genes encoding anaerobic cell-wall manoproteins of the DAN/TIR family [112, 113]. Regulation requires the so-called sterol regulatory element (SRE), a conserved 11-bp (5'-CTCGTATAAGC-3') motif present in the promoters of many ergosterol pathway genes [112, 114]. A homologue of Upc2, Ecm22, shares 45% identical residues and overlapping functions [115]. Notably, disruption of *UPC2* leads to a loss of ketoconazole resistance, whereas an *ecm22* knockout strain is sensitive to cycloheximide, thus connecting both genes to drug resistance [84].

Two members of the C$_2$H$_2$ zinc finger family, Msn2 and Msn4, are major TFs of the general stress response also involved in the PDR network. The so-called stress response elements (STREs) with the consensus sequence 5'-CCCCT-3' are recognized and decorated by these stress regulators [116, 117]. Msn2 is indispensable for the induction of *PDR15*, which is upregulated upon various adverse conditions, showing that the general stress response and pleiotropic drug resistance share common regulators and effectors [43]. Msn2 functions as the downstream effector of the high-osmolarity glycerol (HOG) pathway [118]. Interestingly, induction of *PDR15* through Msn2 appears independent of the HOG pathway, implying a new upstream branch of the HOG pathway or an as yet unknown Msn2 activator [43]. Both TFs are also essential for the recovery of cells from damage due to freezing conditions [119].

The bZIP protein family is the third prominent group of TFs playing a vital role in PDR. Of the several YAP genes in yeast, the basic leucine zipper transcription factors Yap1, Yap2, and Yap8 are linked to oxidative stress response [98, 120], vacuolar detoxification, and heavy metal tolerance.

The best characterized protein is Yap1, the major regulator of oxidative stress response. It mediates response to oxidative and osmotic stresses, as well as to heat shock [121]. Overexpression of *YAP1* renders cells resistant to 4-nitroquinoline-*N*-oxide, triaziquone, and cycloheximide [33], and a mutant lacking *YAP1* causes hypersensitivity to the heavy metal cadmium [122, 123]. Yap1 modulates expression of the *SNQ2*, *YCF1* [122, 124, 125], and *PDR5* genes [35]. A Yap1 response element (YRE, consensus 5'-TTAC/GTAA-3') is present in several target genes [126, 127]. Upon stress, cytoplasmic Yap1 accumulates in the nucleus due to the disrupted interaction with the nuclear export factor Crm1 [128, 129]. Two additional genes of the Yap protein family seem to affect drug resistance. Yap2 and Yap8 confer cadmium and arsenate resistance phenotypes [123, 130, 131]. Yap2 seems to be clearly involved in resistance toward many toxic compounds. A recent study showed that the regulation through Yap8 depends on the ubiquitin–proteasome pathway, as arsenite stabilizes Yap8, which leads to expression of target genes, whereas Yap8 is degraded in unstressed cells [132].

Finally, two proteins known to influence expression of PDR genes or interacting directly with members of the network are Pdr13 and Ngg1. *PDR13/SSZ1* encodes an Hsp70 protein modulating *PDR5* expression and drug resistance in a Pdr1-dependent, but Pdr3-independent, manner [133]. Moreover, the Hsp70 protein Ssa1 negatively regulates Pdr3 but has no influence on Pdr1-responsive genes [134]. Ngg1 interacts with the C-terminal region of Pdr1 to decrease the transcriptional activity [135, 136].

6.4
ABC Drug Transporters of Human Fungal Pathogens

The specialization for a distinct niche, various morphological forms, and the ability to rapidly acquire resistance to antifungal drugs may explain the virulence of fungal pathogens. Fungal infections can often cause life-threatening infections in immunocompromised patients. Hence, prolonged hospitalization, general immunosuppression, AIDS, organ transplantation, chemotherapy, or antimicrobial therapy dramatically increases the risk of fungal diseases [137, 138].

ABC transporters represent an essential part of the fungal defense mechanisms leading to the development of resistance against antifungals (Table 6.3). Overexpression of ABC pumps such as Cdr1 drives resistance in several azole-resistant clinical isolates of C. albicans [139, 140]. *CDR1*, the gene encoding for the first ABC transporter identified in C. albicans, was cloned by homology to *PDR5* [15, 16]. Although five *CDR* genes exist, only two close homologues *CDR1* and *CDR2* are involved in drug resistance, showing broad substrate specificity. Overexpression of both proteins confers resistance to different azoles, such as fluconazole, ketoconazole, or itraconazole, and other substances, such as cycloheximide, rhodamine 6G, and cerulenin [15, 16, 141]. Interestingly, a strain lacking *CDR2* does not render cells hypersusceptible when compared to a *cdr1* mutant. However, a double deletion strain is more sensitive to azoles than a single *cdr1* mutant [142]. Furthermore, *CDR2* expression is strongly upregulated in clinically resistant isolates, implying that *CDR2* is involved in drug resistance [142]. Levels of Cdr1 and Cdr2 are controlled by the C. albicans Tac1 regulator, which plays the same role as Pdr1 and Pdr3 in yeast. Tac1 mediates resistance to antifungal drugs by driving a drug response element (DRE) in the promoters of *CDR1* and *CDR2* [143]. This DRE involves the conserved 5′-CGG-3′ cognate motif for Zn(II)$_2$Cys$_6$ cluster regulators. A similar motif appears in the S. cerevisiae promoter regions decorated by Pdr1, Pdr3 [95, 100], and War1 [109]. Moreover, a steroid-responsive element is present in the upstream region of *CDR1*, which may be relevant for the host situation *in vivo* [144–146]. Notably, progesterone can regulate *CDR1* expression, thereby increasing resistance to fluconazole, miconazole, and 5-fluorouracil [147]. Interestingly, an orthologue of the S. cerevisiae sterol biosynthesis regulator Upc2 was discovered in C. albicans. Strains lacking Upc2 are more susceptible to ketoconazole and fluconazole and other cytotoxic compounds. Interestingly, in this deletion strain, drug-induced expression of *ERG* genes is abolished [148, 149]. Hence, as in S. cerevisiae, a connection exists between ergosterol biosynthesis and drug resistance, perhaps involving changes in membrane lipid homeostasis and permeability. Thus, a plausible physiological function of ABC transporters might be the translocation of phospholipids, as it has been shown for the human MDR2 transporter, the S. cerevisiae Pdr5 protein, and the C. albicans pumps Cdr1–Cdr3 (reviewed in Ref. [150]). This hypothesis is further supported by the recent demonstration that purified and reconstituted C. albicans Cdr1 transports fluorescent-labeled phospholipids in an ATP-dependent manner [151].

The mutational analysis of Cdr1 and Cdr2 (reviewed in Ref. [152]), including a replacement of conserved residues, permitted insights into ATP hydrolysis and substrate binding [145, 153, 154]. For example, the Cys193 and Lys901 in Cdr1 were mutated showing their requirement for ATP hydrolysis. Both residues are localized in the Walker A domain of NBD1 and NBD2. Although the Lys901Cys mutation rendered cells hypersensitive to drugs, the Cys193Lys mutant behaved like wild type

Table 6.3 Antifungal resistance-associated transporters in pathogenic fungi.

Species	Type	Size	Topology	Location	Function	References
C. albicans						
Cdr1	ABC	1501	$(NBD-TMS_6)_2$	PM	Drug efflux, lipid translocation	[15, 16]
Cdr2	ABC	1499	$(NBD-TMS_6)_2$	PM	Drug efflux, lipid translocation	[16]
Cdr3	ABC	1502	$(NBD-TMS_6)_2$	PM	Lipid translocation, opaque-phase specific	[157]
Cdr4	ABC	1491	$(NBD-TMS_6)_2$	PM	?	[158]
Cdr11/Cdr5	ABC	1530	?	?	?	www.candidagenome.org/
CaMdr1	MFS	564	TMS	PM	Drug efflux	[165]
Flu1	MFS	610	TMS	PM?	Drug efflux	[164]
C. glabrata						
CgCdr1	ABC	1499	$(NBD-TMS_6)_2$	?	Drug efflux	[174]
Pdh1	ABC	1542	$(NBD-TMS_6)_2$	?	Drug efflux	[173]
C. dubliniensis						
CdCdr1	ABC	1501	$(NBD-TMS_6)_2$	?	Drug efflux	[186]
CdCdr2	ABC	1500	$(NBD-TMS_6)_2$	?	Drug efflux	[186]
CdMdr1	MFS	557	12TMS	?		[186]
C. krusei						
Abc1	ABC	?	?	?	Drug efflux	[187]
Abc2	ABC	?	?	?	Drug efflux	[187]
A. fumigatus						
AfuMDR1	ABC	1349	$(TMS_6-NBD)_2$	PM?	Drug efflux	[14]
AfuMDR2	ABC	791	TMS_6-NBD	PM?	Protein transport	[14]
AfuMDR3	MFS	515	14TMS	PM?	Drug efflux	[190]
AfuMDR4	ABC	1344	?	PM?	Drug efflux	[190]
AtrF	ABC	1547	$(TMS_6-NBD)_2$	PM?	Drug efflux	[189]
A. flavus						
AflMDR1	ABC	1307	$(TMS_6-NBD)_2$	PM?	?	[14]
C. neoformans						
CnAfr1	ABC	1543	$(TMS_6-NBD)_2$	?	Drug efflux	[196]
T. rubrum						
TruMDR1	ABC	1612	?	PM?	Drug efflux	[199]
TruMDR2	ABC	1350	$(TMS_6-NBD)_2$	PM?	Drug efflux	[200]

Abbreviations: ABC, ATP- binding cassette MFS, major facilitator superfamily; PM, plasma membrane; NBD, nucleotide binding domain; TMS, transmembrane-spanning segment; CGD, *Candida* genome database.

expressing a native form of Cdr1 [155]. Mutation of the Walker B residue Trp326 to an alanine residue showed that it is important for nucleotide binding [154]. Deletions in the TMS6, TMS11, and TMS12 demonstrated that these regions are necessary for correct folding and drug transport [145, 156].

Cdr3 and Cdr4 share a high degree of homology not only with each other but also with Cdr1 and Cdr2. Interestingly enough, both the proteins do not appear to be involved in drug resistance. Neither overexpression nor deletion of any of them affected the drug susceptibility [157, 158]. The *Candida* Genome Database lists a fifth CDR gene, designated *CDR11* or *CDR5*, whose function has not been addressed yet. Further homologues of yeast ABC genes in *C. albicans* have been described, including *SNQ2*, *YCF1*, and *YOR1* [159, 160].

In addition to the ABC transporters, the MFS family transporters play an important role in clinical drug resistance [161]. The most prominent ones are Mdr1 [162, 163] and Flu1 [164]. Ca*MDR1*, formerly known as BEN^r, confers resistance to a variety of different compounds. Cells lacking Ca*MDR1* are susceptible to 4-nitro-quinoline-*N*-oxide, methotrexate, and cycloheximide [165, 166]. The closest yeast homologue of Ca*MDR1* is *FLR1*. As mentioned above, Flr1 is known to be involved in drug transport and it mediates resistance to the same spectrum of drugs as Ca*MDR1*.

The *C. albicans* transcription factor Cap1, a homologue of yeast Yap1, is linked to oxidative stress response and drug resistance [160, 167, 168]. Disruption of *CAP1* causes increased Ca*MDR1* expression [160], suggesting that Cap1 is its negative regulator. However, conflicting data on Ca*MDR1* regulation by Cap1 [169] exist in the literature. Ca*MDR1* expression depends on a sequence motif similar to YRE in *S. cerevisiae*, which positively responds to the oxidizing agent *tert*-butyl hydrogen peroxide [170]. Cap1 might also influence the regulation of Ca*YCF1* [160]. Fcr1, a third putative transcriptional regulator, might also be involved in pleiotropic drug resistance. The deletion of *FCR1* rendered *C. albicans* cells hyperresistant to fluconazole, indicating that it acts as a negative regulator [171, 172].

The MFS permease Flu1 mediates resistance to fluconazole and cycloheximide. Although disruption does not change susceptibility to azoles, it increases mycophenolic acid sensitivity. However, loss of *FLU1* in a strain lacking *CDR1*, *CDR2*, and Ca*MDR1* led to enhanced susceptibility to azoles, suggesting at least some synergy between these transporters. This contrasts data showing that azole-susceptible and azole-resistant clinical isolates do not display significant changes in the Flu1 expression levels. Thus, Flu1 may only indirectly modulate azole resistance [164].

C. glabrata is the second most frequent cause of *Candida* infections. An important and interesting aspect is its high resistance to antifungal drugs, which may be at least in part a consequence of ABC transporters. The genome organization of *C. glabrata* is related to the one in *S. cerevisiae* and harbors several ABC transporters linked to PDR. Cg*CDR1* and *PDH1*, also designated Cg*CDR2*, encode typical ABC transporters with a (NBD-TMS$_6$)$_2$ topology [173, 174] with 73% identity, conferring resistance to several azoles and rhodamine [174]. Similar to the situation in *C. albicans*, disruption of *PDH1* alone does not affect azole susceptibility. However, a *cdr1 pdh1* double deletion results in increased drug susceptibility [175]. Heterologous expression of *PDH1* in a *S. cerevisiae pdr5* mutant complemented the deletion and restored resistance to chloramphenicol [176]. The regulation of transporters in *C. glabrata* is likely to involve a dedicated transcription factor, CgPdr1, which is 40% similar to ScPdr1 [177]. Elevated expression of Cg*CDR1* and *PDH1* results from a Cg*PDR1* gain-of-function mutation. Microarray studies comparing the Cg*PDR1* gain-of-function mutant with another azole-resistant strain indicate that several *C. glabrata* genes are regulated by CgPdr1, including *CDR1*, *PDH1*, *YOR1*, *YBT1*, *QDR2*, and *PDR1* itself [178]. Notably, downregulated genes are homologous to *PDR12* (weak acid response), and the MFS permeases *ZRT1* (zinc transporter) and *FLR1* of *S. cerevisiae*. Moreover, a *C. glabrata* PDRE motif 5'-TCC(AG)(TC)G(GC)(AG)-3' may share conserved functions with the yeast counterparts concerning autoregulation and control of a similar PDR network in *C. glabrata* [177]. Heterologous expression of Cg*PDR1* in a *pdr1* deletion strain of *S. cerevisiae* complemented the loss when cells were exposed to fluconazole or rhodamine. Reintroduction of Cg*PDR1* into the *C. glabrata pdr1* disruptant restored resistance to the azoles fluconazole, voriconazole, and itraconazole [177].

An interesting drug resistance mechanism in *C. glabrata* is the upregulation of the ABC efflux pumps in mutants showing a petite phenotype with deficiencies in respiration. In *S. cerevisiae*, a connection exists between upregulation of *PDR5* and overexpression of Pdr3 in petite mutants [102]. Mutations in mitochondrial DNA causing respiratory deficiency appear associated with azole resistance, providing an explanation for the petite phenotype [179, 180]. However, this connection is seen only after exposure to azoles *in vitro*. However, petite mutants are less virulent than the corresponding wild-type strain in a murine model of systemic infection [181], but it is not clear if this is of clinical relevance.

The regulation of the drug efflux in *C. glabrata* also depends on the phosphorylation of the ABC transporters CgCdr1 and Pdh1 [182, 183]. Phosphorylation of Pdh1 affects drug efflux in a PKA-dependent manner and ATPase activity is glucose dependent [182]. Notably, a single mutation of a residue located in the NBD1 influenced ATPase and pumping activity, suggesting that NBD1 plays a role as a sensor [183].

Azole resistance in *C. glabrata* may also occur through a different mechanism. In *S. cerevisiae*, the ABC transporters Aus1 and Pdr11 are required for sterol uptake from media under anaerobic conditions. Interestingly, the *C. glabrata AUS1* homologue seems to function as a sterol transporter. However, in contrast to *S. cerevisiae*, *C. glabrata* seems to take up exogenous sterols under aerobic conditions in the presence of serum. This mechanism may represent a new way of *in vivo* protection against azole drugs [184].

A recent study identified Cg*AP1*, a *C. glabrata* homologue of *YAP1*, and two ORFs encoding *FLR1* homologues. The *C. glabrata* disruption strain is hypersensitive to H_2O_2, cadmium, and 4-NQO [185], and Cg*AP1* complements a loss of *YAP1* in *S. cerevisiae*. CgAp1 acts as a transcriptional activator of *FLR1* upon benomyl stress confirming that both proteins play a role in resistance [185].

ABC pumps and membrane permeases involved in drug efflux have also been found in other *Candida* strains of clinical relevance. Cd*CDR1* and Cd*CDR2* encode

ABC transporters in *C. dubliniensis*. Cd*MDR1* encodes an MFS transporter [186]. *ABC1* and *ABC2* encode ABC transporters in *C. krusei* [187].

Human fungal pathogens with emerging clinical relevance include different *Aspergillus* species, *Cryptococcus neoformans*, and *Trichophyton rubrum* [2–4]. In the filamentous fungus *A. fumigatus*, several genes are linked to multidrug resistance, since four ABC transporters and an MFS permease might mediate PDR. The ABC transporter genes Afu*MDR1* and Afu*MDR2* were identified by a PCR-based homology cloning strategy [14]. Notably, heterologous expression of Afu*MDR1* in *S. cerevisiae* increases resistance to cilofungin [14]. Afu*MDR2* is a protein similar to the ABC transporters Mdl1 and Mdl2 of *S. cerevisiae*, which are involved in protein translocation across the inner mitochondrial membrane [188]. AtrF is an ABC pump correlated with itraconazole resistance [189]. Afu*MDR4* encodes a new *A. fumigatus* ABC pump sharing 33% identity with Afu*MDR1* and 20% with AtrF. Afu*mdr4* mutants are highly sensitive to itraconazole, and their overexpression causes resistance [190]. A second upregulated gene in *A. fumigatus* encodes AfuMdr3, a MFS-type protein with 14 predicted TMS, sharing 33% similarity with Atr1, the *S. cerevisiae* MFS permease conferring resistance to aminotriazole [190]. Several ABC transporters also operate in *A. nidulans*. The genes *atrA* and *atrB* [191], *atrC* [192–195], *atrD*, and *atrE* and *atrF* [193] were identified based on homology searches [192]. The *A. flavus* gene Afl*MDR1* encodes a protein containing two homologous halves, each consisting of six predicted TMS and an NBD. It is a very close homologue of *A. fumigatus* AfuMDR1, showing even an conserved intron organization [14].

C. neoformans is an encapsulated yeast-like fungus, causing pulmonary infections and meningitis with a high mortality. So far, a single ABC transporter, CnAFR1, has been shown to confer resistance to fluconazole. Reintroduction of the intact gene in cells lacking CnAfr1 restored the wild-type drug sensitivity [196]. Interestingly, Cn*AFR1* seems to be implicated in azole resistance, as it enhances virulence of *C. neoformans* [197].

Finally, dermatophytes cause skin diseases and are quite commonly known. A prominent member is the filamentous fungus *T. rubrum*, causing the majority of dermatomycoses [198]. Although only little data are available about molecular resistance mechanisms in *T. rubrum*, two genes, Tru*MDR1* and Tru*MDR2*, may be involved in drug resistance [199, 200]. The TruMdr2 protein is closely related to the *A. fumigatus* AfuMdr1 ABC transporter. A disruption mutant is viable but displays increased sensitivity to terbinafine, 4-nitroquinoline-*N*-oxide, and ethidium bromide [200].

6.5
Physiological Roles of Drug Transporting ABC Proteins – Search for Substrates

Despite many studies on fungal ABC proteins, the physiological function and natural substrates have only been identified for a small number of transporters. *S. cerevisiae* and pathogenic fungi, such as *C. albicans* or *A. fumigatus*, have not always been exposed to toxic compounds, yet their ABC transporters can efficiently handle hundreds of substrates, raising the question about the physiological processes of ABC pumps involved.

Naturally, a major task appears to be cellular detoxification under normal and adverse growth conditions. For example, Pdr12 functions in the disposal of toxic weak acid catabolites that accumulate when cells are approaching the stationary growth phase [52, 55]. Pdr12 contributes to the export of catabolic products such as phenylacetate or other carboxylic acids derived from amino acid catabolism. In this context, Pdr12 may be considered as an integral part of the so-called Ehrlich pathway, in which carboxylic acids are decarboxylated to the corresponding aldehyde [55, 201]. Pdr12 is highly induced in response to weak organic acids such as sorbic, benzoic, and propionic acids.

Several yeast and mammalian [202] ABC transporters do have hydrophobic and lipophilic substrates. For instance, Aus1 and Pdr11 may function in sterol uptake. Translocation of membrane phospholipids was also demonstrated for Pdr5 and Yor1, the *C. albicans* drug transporters Cdr1–3, as well as several mammalian ABC transporters [202, 203]. Mutant strains expressing alleles *PDR1–11* and *PDR3–11* display altered phospholipid accumulation and membrane asymmetry [49, 204]. This is consistent with the idea of ABC transporters acting as membrane lipid flippases [205] or being implicated in membrane lipid homeostasis [50, 206]. This notion is also supported by the finding that *IPT1*, encoding an enzyme involved in sphingolipid biosynthesis, appears to be regulated by Pdr1/Pdr3 [133]. Pdr1 and Pdr3 also control the ABC transporters Pdr5 [204] and Pdr15 [43], which may be involved in the maintenance of membrane bilayer function by removing toxic lipid-like compounds [50]. Hence, yeast ABC transporters may be involved in lipid transport and may even contribute to the assembly and maintenance of the asymmetric lipid bilayer distribution. Alternatively, ABC proteins may detoxify cellular membranes from unwanted and toxic breakdown products, many of which are structurally similar to xenobiotic compounds.

6.6
Conclusions and Perspectives

Many years of ABC transporter research literally uncovered hundreds of xenobiotic compounds and other molecules transported by ABC pumps. Despite the large number of substrates and numerous genetic and biochemical studies on the function of ABC transporters, many mysteries remain open. How do ABC transporters recognize and translocate such a great variety of unrelated compounds although their general domain structure is highly conserved during evolution? What are the true physiological substrates? In the past years, promising progress has been made in studying the function of ABC pumps in more detail. For example, it is possible to efficiently overexpress, purify, and reconstitute some of the membrane-located ABC proteins. This will allow the dissection of their molecular transport mechanisms. The next step will be to obtain crystal structures of eukaryotic ABC transporters, as this will be helpful in understanding the molecular organization, catalytic cycle, and mode of action. The availability of crystal structures for several bacterial ABC proteins [207–211] and some structures for mammalian NBDs [212, 213] will also aid similar approaches for eukaryotic and, especially, mammalian transporters. Structural approaches of drug resistance transporters may lead to a better understanding of how fungal pathogens develop resistance through these pumps. The process of substrate recognition and translocation, the catalytic cycle, and the identification of proteins interacting with predicted intracellular loops that connect NBD domains of an ABC transporter with their membrane units will further advance the field.

Acknowledgments

We thank our laboratory members for critical reading of the manuscript. We are thankful to Helmut Jungwirth and Christoph Schüller for helping with graphics and artwork. Research in our laboratory is supported by grants from the Christian Doppler Research Society, the EraNet Pathogenomics project FunPath (FWF-I-0125), the Austrian Science Foundation (FWF-SFB35-04), the FP6 Marie Curie Training Networks *CanTrain* (MRTN-CT-2004-512481) and *Flippases* (MRTN-CT-2004-5330), the FP6 EURESFUN project (STREP-2004-CT-PL518199), the Austrian GenAU research program (SysMo-MOSES), the Herzfelder Family Foundation, and the Vienna Biocenter PhD program.

References

1 Anaissie, E. (1992) Opportunistic mycoses in the immunocompromised host: experience at a cancer center and review. *Clinical Infectious Diseases*, **14** (Suppl 1), S43–S53.

2 Ponton, J., Ruchel, R., Clemons, K.V., Coleman, D.C., Grillot, R., Guarro, J., Aldebert, D., Ambroise-Thomas, P., Cano, J., Carrillo-Munoz, A.J., Gene, J., Pinel, C., Stevens, D.A., and Sullivan, D.J. (2000) Emerging pathogens. *Medical Mycology*, **38** (Suppl 1), 225–236.

3 Fleming, R.V., Walsh, T.J., and Anaissie, E.J. (2002) Emerging and less common fungal pathogens. *Infectious Disease Clinics of North America*, **16**, 915–933; vi–vii.

4 Baddley, J.W. and Moser, S.A. (2004) Emerging fungal resistance. *Clinical Laboratory Medicine*, **24**, 721–735; vii.

5 Litman, T., Druley, T.E., Stein, W.D., and Bates, S.E. (2001) From MDR to MXR: new understanding of multidrug resistance systems, their properties and clinical significance. *Cellular and Molecular Life Sciences*, **58**, 931–959.

6 Dean, M. and Annilo, T. (2005) Evolution of the ATP-binding cassette (ABC) transporter superfamily in vertebrates. *Annual Review of Genomics and Human Genetics*, **6**, 123–142.

7 Holland, I.B., Cole, S.P., Kuchler, K., and Higgins, C.F. (2003) *ABC Proteins: From Bacteria to Man*, Academic Press/Elsevier Science, Amsterdam.

8 Li, X.Z. and Nikaido, H. (2004) Efflux-mediated drug resistance in bacteria. *Drugs*, **64**, 159–204.

9 Ouellette, M., Legare, D., and Papadopoulou, B. (2001) Multidrug

resistance and ABC transporters in parasitic protozoa. *Journal of Molecular Microbiology and Biotechnology*, **3**, 201–206.

10 Wolfger, H., Mamnun, Y.M., and Kuchler, K. (2001) Fungal ABC proteins: pleiotropic drug resistance, stress response and cellular detoxification. *Research in Microbiology*, **152**, 375–389.

11 Dean, M. (2005) The genetics of ATP-binding cassette transporters. *Methods in Enzymology*, **400**, 409–429.

12 Nikles, D. and Tampe, R. (2007) Targeted degradation of ABC transporters in health and disease. *Journal of Bioenergetics and Biomembranes*, **39**, 489–497.

13 Riordan, J.R. (1993) The cystic fibrosis transmembrane conductance regulator. *Annual Review of Physiology*, **55**, 609–630.

14 Tobin, M.B., Peery, R.B., and Skatrud, P.L. (1997) Genes encoding multiple drug resistance-like proteins in *Aspergillus fumigatus* and *Aspergillus flavus*. *Gene*, **200**, 11–23.

15 Prasad, R., De Wergifosse, P., Goffeau, A., and Balzi, E. (1995) Molecular cloning and characterization of a novel gene of *Candida albicans*, *CDR1*, conferring multiple resistance to drugs and antifungals. *Current Genetics*, **27**, 320–329.

16 Sanglard, D., Kuchler, K., Ischer, F., Pagani, J.L., Monod, M., and Bille, J. (1995) Mechanisms of resistance to azole antifungal agents in *Candida albicans* isolates from AIDS patients involve specific multidrug transporters. *Antimicrobial Agents and Chemotherapy*, **39**, 2378–2386.

17 Walker, J.E., Saraste, M., Runswick, M.J., and Gay, N.J. (1982) Distantly related sequences in the alpha- and beta-subunits of ATP synthase, myosin, kinases and other ATP-requiring enzymes and a common nucleotide binding fold. *The EMBO Journal*, **1**, 945–951.

18 Higgins, C.F. (1992) ABC transporters: from microorganisms to man. *Annual Review of Cell Biology*, **8**, 67–113.

19 Urbatsch, I.L., Gimi, K., Wilke-Mounts, S., and Senior, A.E. (2000) Investigation of the role of glutamine-471 and glutamine-1114 in the two catalytic sites of P-glycoprotein. *Biochemistry*, **39**, 11921–11927.

20 Bauer, B.E., Schüller, C., and Kuchler, K. (2003) Inventory and evolution of fungal ABC protein genes, in *ABC Proteins: From Bacteria to Man* (eds I.B. Holland, K. Kuchler, and C.F. Higgins), Academic Press/Elsevier Science, Amsterdam.

21 Golin, J., Ambudkar, S.V., and May, L. (2007) The yeast Pdr5p multidrug transporter: how does it recognize so many substrates? *Biochemical and Biophysical Research Communications*, **356**, 1–5.

22 Schüller, C., Bauer, B.E., and Kuchler, K. (2003) Inventory and evolution of fungal ABC protein genes, in *ABC Proteins: From Bacteria to Man* (eds I.B. Holland, K. Kuchler, and C.F. Higgins), Academic Press/Elsevier Science, Amsterdam, pp. 279–293.

23 Sipos, G. and Kuchler, K. (2006) Fungal ATP-binding cassette (ABC) transporters in drug resistance & detoxification. *Current Drug Targets*, **7**, 471–481.

24 Cowen, L.E. and Steinbach, W.J. (2008) Stress, drugs, and evolution: the role of cellular signaling in fungal drug resistance. *Eukaryotic Cell*, **7**, 747–764.

25 Cowen, L.E. (2008) The evolution of fungal drug resistance: modulating the trajectory from genotype to phenotype. *Nature Reviews Microbiology*, **6**, 187–198.

26 Jungwirth, H. and Kuchler, K. (2006) Yeast ABC transporters – a tale of sex, stress, drugs and aging. *FEBS Letters*, **580**, 1131–1138.

27 Decottignies, A. and Goffeau, A. (1997) Complete inventory of the yeast ABC proteins. *Nature Genetics*, **15**, 137–145.

28 Lill, R. and Muhlenhoff, U. (2008) Maturation of iron–sulfur proteins in eukaryotes: mechanisms, connected processes, and diseases. *Annual Review of Biochemistry*, **77**, 669–700.

29 Burke, M.A. and Ardehali, H. (2007) Mitochondrial ATP-binding cassette proteins. *Translational Research*, **150**, 73–80.

30 Nakatsukasa, K., Huyer, G., Michaelis, S., and Brodsky, J.L. (2008) Dissecting the ER-associated degradation of a misfolded polytopic membrane protein. *Cell*, **132**, 101–112.

31 Moye-Rowley, W.S. (2005) Retrograde regulation of multidrug resistance in *Saccharomyces cerevisiae*. *Gene*, **354**, 15–21.

32 Wanders, R.J. and Waterham, H.R. (2006) Biochemistry of mammalian peroxisomes revisited. *Annual Review of Biochemistry*, **75**, 295–332.

33 Haase, E., Servos, J., and Brendel, M. (1992) Isolation and characterization of additional genes influencing resistance to various mutagens in the yeast *Saccharomyces cerevisiae*. *Current Genetics*, **21**, 319–324.

34 Servos, J., Haase, E., and Brendel, M. (1993) Gene *SNQ2* of *Saccharomyces cerevisiae*, which confers resistance to 4-nitroquinoline-*N*-oxide and other chemicals, encodes a 169 kDa protein homologous to ATP-dependent permeases. *Molecular & General Genetics*, **236**, 214–218.

35 Miyahara, K., Mizunuma, M., Hirata, D., Tsuchiya, E., and Miyakawa, T. (1996) The involvement of the *Saccharomyces cerevisiae* multidrug resistance transporters Pdr5p and Snq2p in cation resistance. *FEBS Letters*, **399**, 317–320.

36 Callaghan, R., Crowley, E., Potter, S., and Kerr, I.D. (2008) P-glycoprotein: so many ways to turn it on. *Journal of Clinical Pharmacology*, **48**, 365–378.

37 Bissinger, P.H. and Kuchler, K. (1994) Molecular cloning and expression of the *Saccharomyces cerevisiae STS1* gene product. A yeast ABC transporter conferring mycotoxin resistance. *The Journal of Biological Chemistry*, **269**, 4180–4186.

38 Balzi, E., Wang, M., Leterme, S., Van Dyck, L., and Goffeau, A. (1994) PDR5, a novel yeast multidrug resistance conferring transporter controlled by the transcription regulator *PDR1*. *The Journal of Biological Chemistry*, **269**, 2206–2214.

39 Hirata, D., Yano, K., Miyahara, K., and Miyakawa, T. (1994) *Saccharomyces cerevisiae* YDR1, which encodes a member of the ATP-binding cassette (ABC) superfamily, is required for multidrug resistance. *Current Genetics*, **26**, 285–294.

40 Kralli, A., Bohen, S.P., and Yamamoto, K.R. (1995) Lem1p, an ATP-binding-cassette transporter, selectively modulates the biological potency of steroid hormones. *Proceedings of the National Academy of Sciences of the United States of America*, **92**, 4701–4705.

41 Mahe, Y., Lemoine, Y., and Kuchler, K. (1996) The ATP binding cassette transporters Pdr5p and Snq2p of *Saccharomyces cerevisiae* can mediate transport of steroids *in vivo*. *The Journal of Biological Chemistry*, **271**, 25167–25172.

42 Kolaczkowski, M., Kolaczowska, A., Luczynski, J., Witek, S., and Goffeau, A. (1998) *In vivo* characterization of the drug resistance profile of the major ABC transporters and other components of the yeast pleiotropic drug resistance network. *Microbial Drug Resistance*, **4**, 143–158.

43 Wolfger, H., Mamnun, Y.M., and Kuchler, K. (2004) The yeast Pdr15p ATP-binding cassette (ABC) protein is a general stress response factor implicated in cellular detoxification. *The Journal of Biological Chemistry*, **279**, 11593–11599.

44 Alenquer, M., Tenreiro, S., and Sa-Correia, I. (2006) Adaptive response to the antimalarial drug artesunate in yeast involves Pdr1p/Pdr3p-mediated transcriptional activation of the resistance determinants *TPO1* and *PDR5*. *FEMS Yeast Research*, **6**, 1130–1139.

45 Egner, R., Bauer, B.E., and Kuchler, K. (2000) The transmembrane domain 10 of the yeast Pdr5p ABC antifungal efflux pump determines both substrate specificity and inhibitor susceptibility. *Molecular Microbiology*, **35**, 1255–1263.

46 Egner, R., Rosenthal, F.E., Kralli, A., Sanglard, D., and Kuchler, K. (1998) Genetic separation of FK506 susceptibility and drug transport in the yeast Pdr5 ATP-binding cassette multidrug resistance transporter. *Molecular Biology of the Cell*, **9**, 523–543.

47 Tutulan-Cunita, A.C., Mikoshi, M., Mizunuma, M., Hirata, D., and Miyakawa, T. (2005) Mutational analysis of the yeast multidrug resistance ABC transporter Pdr5p with altered drug specificity. *Genes to Cells*, **10**, 409–420.

48 Ernst, R., Kueppers, P., Klein, C.M., Schwarzmueller, T., Kuchler, K., and Schmitt, L. (2008) A mutation of the H-loop selectively affects rhodamine transport by the yeast multidrug ABC transporter Pdr5. *Proceedings of the National Academy of Sciences of the United States of America*, **105**, 5069–5074.

49 Decottignies, A., Grant, A.M., Nichols, J.W., de Wet, H., McIntosh, D.B., and Goffeau, A. (1998) ATPase and multidrug transport activities of the overexpressed yeast ABC protein Yor1p. *The Journal of Biological Chemistry*, **273**, 12612–12622.

50 Schüller, C., Mamnun, Y.M., Wolfger, H., Rockwell, N., Thorner, J., and Kuchler, K. (2007) Membrane-active compounds activate the transcription factors Pdr1 and Pdr3 connecting pleiotropic drug resistance and membrane lipid homeostasis in *Saccharomyces cerevisiae*. *Molecular Biology of the Cell*, **18**, 4932–4944.

51 Mamnun, Y.M., Schuller, C., and Kuchler, K. (2004) Expression regulation of the yeast *PDR5* ATP-binding cassette (ABC) transporter suggests a role in cellular detoxification during the exponential growth phase. *FEBS Letters*, **559**, 111–117.

52 Piper, P., Mahe, Y., Thompson, S., Pandjaitan, R., Holyoak, C., Egner, R., Muhlbauer, M., Coote, P., and Kuchler, K. (1998) The Pdr12p ABC transporter is required for the development of weak organic acid resistance in yeast. *The EMBO Journal*, **17**, 4257–4265.

53 Holyoak, C.D., Bracey, D., Piper, P.W., Kuchler, K., and Coote, P.J. (1999) The *Saccharomyces cerevisiae* weak-acid-inducible ABC transporter Pdr12p transports fluorescein and preservative anions from the cytosol by an energy-dependent mechanism. *Journal of Bacteriology*, **181**, 4644–4652.

54 Papadimitriou, M.N., Resende, C., Kuchler, K., and Brul, S. (2007) High Pdr12 levels in spoilage yeast (*Saccharomyces cerevisiae*) correlate directly with sorbic acid levels in the culture medium but are not sufficient to provide cells with acquired resistance to the food preservative. *International Journal of Food Microbiology*, **113**, 173–179.

55 Hazelwood, L.A., Tai, S.L., Boer, V.M., de Winde, J.H., Pronk, J.T., and Daran, J.M. (2006) A new physiological role for Pdr12p in *Saccharomyces cerevisiae*: export of aromatic and branched-chain organic acids produced in amino acid catabolism. *FEMS Yeast Research*, **6**, 937–945.

56 Wilcox, L.J., Balderes, D.A., Wharton, B., Tinkelenberg, A.H., Rao, G., and Sturley, S.L. (2002) Transcriptional profiling identifies two members of the ATP-binding cassette transporter superfamily required for sterol uptake in yeast. *The Journal of Biological Chemistry*, **277**, 32466–32472.

57 Barker, K.S., Pearson, M.M., and Rogers, P.D. (2003) Identification of genes differentially expressed in association with reduced azole susceptibility in *Saccharomyces cerevisiae*. *The Journal of Antimicrobial Chemotherapy*, **51**, 1131–1140.

58 Katzmann, D.J., Epping, E.A., and Moye-Rowley, W.S. (1999) Mutational disruption of plasma membrane trafficking of *Saccharomyces cerevisiae* Yor1p, a homologue of mammalian multidrug resistance protein.

59 Nagy, Z., Montigny, C., Leverrier, P., Yeh, S., Goffeau, A., Garrigos, M., and Falson, P. (2006) Role of the yeast ABC transporter Yor1p in cadmium detoxification. *Biochimie*, **88** (11), 1665–1671.

60 Cui, Z., Hirata, D., Tsuchiya, E., Osada, H., and Miyakawa, T. (1996) The multidrug resistance-associated protein (MRP) subfamily (Yrs1p/Yor1p) of *Saccharomyces cerevisiae* is important for the tolerance to a broad range of organic anions. *The Journal of Biological Chemistry*, **271**, 14712–14716.

61 de Thozee, C.P., Cronin, S., Goj, A., Golin, J., and Ghislain, M. (2007) Subcellular trafficking of the yeast plasma membrane ABC transporter, Pdr5, is impaired by a mutation in the N-terminal nucleotide-binding fold. *Molecular Microbiology*, **63**, 811–825.

62 Falcon-Perez, J.M., Mazon, M.J., Molano, J., and Eraso, P. (1999) Functional domain analysis of the yeast ABC transporter Ycf1p by site-directed mutagenesis. *The Journal of Biological Chemistry*, **274**, 23584–23590.

63 Epping, E.A. and Moye-Rowley, W.S. (2002) Identification of interdependent signals required for anterograde traffic of the ATP-binding cassette transporter protein Yor1p. *The Journal of Biological Chemistry*, **277**, 34860–34869.

64 Pagant, S., Kung, L., Dorrington, M., Lee, M.C., and Miller, E.A. (2007) Inhibiting endoplasmic reticulum (ER)-associated degradation of misfolded Yor1p does not permit ER export despite the presence of a diacidic sorting signal. *Molecular Biology of the Cell*, **18**, 3398–3413.

65 Szczypka, M.S., Wemmie, J.A., Moye-Rowley, W.S., and Thiele, D.J. (1994) A yeast metal resistance protein similar to human cystic fibrosis transmembrane conductance regulator (CFTR) and multidrug resistance-associated protein. *The Journal of Biological Chemistry*, **269**, 22853–22857.

66 Gueldry, O., Lazard, M., Delort, F., Dauplais, M., Grigoras, I., Blanquet, S., and Plateau, P. (2003) Ycf1p-dependent Hg(II) detoxification in *Saccharomyces cerevisiae*. *European Journal of Biochemistry*, **270**, 2486–2496.

67 Ghosh, M., Shen, J., and Rosen, B.P. (1999) Pathways of As(III) detoxification in *Saccharomyces cerevisiae*. *Proceedings of the National Academy of Sciences of the United States of America*, **96**, 5001–5006.

68 Sharma, K.G., Kaur, R., and Bachhawat, A.K. (2003) The glutathione-mediated detoxification pathway in yeast: an analysis using the red pigment that accumulates in certain adenine biosynthetic mutants of yeasts reveals the involvement of novel genes. *Archives of Microbiology*, **180**, 108–117.

69 Li, Z.S., Szczypka, M., Lu, Y.P., Thiele, D.J., and Rea, P.A. (1996) The yeast cadmium factor protein (Ycf1p) is a vacuolar glutathione S-conjugate pump. *The Journal of Biological Chemistry*, **271**, 6509–6517.

70 Wehrschutz-Sigl, E., Jungwirth, H., Bergler, H., and Hogenauer, G. (2004) The transporters Pdr5p and Snq2p mediate diazaborine resistance and are under the control of the gain-of-function allele *PDR1–12*. *European Journal of Biochemistry*, **271**, 1145–1152.

71 Klein, M., Mamnun, Y.M., Eggmann, T., Schuller, C., Wolfger, H., Martinoia, E., and Kuchler, K. (2002) The ATP-binding cassette (ABC) transporter Bpt1p mediates vacuolar sequestration of glutathione conjugates in yeast. *FEBS Letters*, **520**, 63–67.

72 Petrovic, S., Pascolo, L., Gallo, R., Cupelli, F., Ostrow, J.D., Goffeau, A., Tiribelli, C., and Bruschi, C.V. (2000) The products of *YCF1* and YLL015w (*BPT1*) cooperate for the ATP-dependent vacuolar transport of unconjugated bilirubin in *Saccharomyces cerevisiae*. *Yeast*, **16**, 561–571.

73 Sharma, K.G., Mason, D.L., Liu, G., Rea, P.A., Bachhawat, A.K., and Michaelis, S. (2002) Localization, regulation, and substrate transport properties of Bpt1p, a *Saccharomyces cerevisiae* MRP-type ABC transporter. *Eukaryotic Cell*, **1**, 391–400.

74 Srikanth, C.V., Chakraborti, A.K., and Bachhawat, A.K. (2005) Acetaminophen toxicity and resistance in the yeast *Saccharomyces cerevisiae*. *Microbiology*, **151**, 99–111.

75 Ortiz, D.F., St Pierre, M.V., Abdulmessih, A., and Arias, I.M. (1997) A yeast ATP-binding cassette-type protein mediating ATP-dependent bile acid transport. *The Journal of Biological Chemistry*, **272**, 15358–15365.

76 Mason, D.L., Mallampalli, M.P., Huyer, G., and Michaelis, S. (2003) A region within a lumenal loop of *Saccharomyces cerevisiae* Ycf1p directs proteolytic processing and substrate specificity. *Eukaryotic Cell*, **2**, 588–598.

77 Balzi, E., Chen, W., Ulaszewski, S., Capieaux, E., and Goffeau, A. (1987) The multidrug resistance gene *PDR1* from *Saccharomyces cerevisiae*. *The Journal of Biological Chemistry*, **262**, 16871–16879.

78 Delaveau, T., Delahodde, A., Carvajal, E., Subik, J., and Jacq, C. (1994) AT *PDR3*, a new yeast regulatory gene, is homologous to *PDR1* and controls the multidrug resistance phenomenon. *Molecular & General Genetics*, **244**, 501–511.

79 Carvajal, E., van den Hazel, H.B., Cybularz-Kolaczkowska, A., Balzi, E., and Goffeau, A. (1997) Molecular and phenotypic characterization of yeast *PDR1* mutants that show hyperactive transcription of various ABC multidrug transporter genes. *Molecular & General Genetics*, **256**, 406–415.

80 Nourani, A., Papajova, D., Delahodde, A., Jacq, C., and Subik, J. (1997) Clustered amino acid substitutions in the yeast transcription regulator Pdr3p increase pleiotropic drug resistance and identify a new central regulatory domain. *Molecular & General Genetics*, **256**, 397–405.

81 Simonics, T., Kozovska, Z., Michalkova-Papajova, D., Delahodde, A., Jacq, C., and Subik, J. (2000) Isolation and molecular characterization of the carboxy-terminal *pdr3* mutants in *Saccharomyces cerevisiae*. *Current Genetics*, **38**, 248–255.

82 Schjerling, P. and Holmberg, S. (1996) Comparative amino acid sequence analysis of the C6 zinc cluster family of transcriptional regulators. *Nucleic Acids Research*, **24**, 4599–4607.

83 Akache, B., Wu, K., and Turcotte, B. (2001) Phenotypic analysis of genes encoding yeast zinc cluster proteins. *Nucleic Acids Research*, **29**, 2181–2190.

84 Akache, B. and Turcotte, B. (2002) New regulators of drug sensitivity in the family of yeast zinc cluster proteins. *The Journal of Biological Chemistry*, **277**, 21254–21260.

85 Vallee, B.L., Coleman, J.E., and Auld, D.S. (1991) Zinc fingers, zinc clusters, and zinc twists in DNA-binding protein domains. *Proceedings of the National Academy of Sciences of the United States of America*, **88**, 999–1003.

86 Leuther, K.K., Salmeron, J.M., and Johnston, S.A. (1993) Genetic evidence that an activation domain of GAL4 does not require acidity and may form a beta sheet. *Cell*, **72**, 575–585.

87 Johnston, S.A. and Hopper, J.E. (1982) Isolation of the yeast regulatory gene *GAL4* and analysis of its dosage effects on the galactose/melibiose regulon. *Proceedings of the National Academy of Sciences of the United States of America*, **79**, 6971–6975.

88 Laughon, A. and Gesteland, R.F. (1982) Isolation and preliminary characterization of the *GAL4* gene, a positive regulator of transcription in yeast. *Proceedings of the National Academy of Sciences of the United States of America*, **79**, 6827–6831.

89 Le Crom, S., Devaux, F., Marc, P., Zhang, X., Moye-Rowley, W.S., and Jacq, C. (2002) New insights into the

pleiotropic drug resistance network from genome-wide characterization of the *YRR1* transcription factor regulation system. *Molecular and Cellular Biology,* **22**, 2642–2649.

90 Cui, Z., Shiraki, T., Hirata, D., and Miyakawa, T. (1998) Yeast gene *YRR1,* which is required for resistance to 4-nitroquinoline N-oxide, mediates transcriptional activation of the multidrug resistance transporter gene *SNQ2. Molecular Microbiology,* **29**, 1307–1315.

91 Hellauer, K., Akache, B., MacPherson, S., Sirard, E., and Turcotte, B. (2002) Zinc cluster protein Rdr1p is a transcriptional repressor of the *PDR5* gene encoding a multidrug transporter. *The Journal of Biological Chemistry,* **277**, 17671–17676.

92 Hikkel, I., Lucau-Danila, A., Delaveau, T., Marc, P., Devaux, F., and Jacq, C. (2003) A general strategy to uncover transcription factor properties identifies a new regulator of drug resistance in yeast. *The Journal of Biological Chemistry,* **278**, 11427–11432.

93 Lucau-Danila, A., Delaveau, T., Lelandais, G., Devaux, F., and Jacq, C. (2003) Competitive promoter occupancy by two yeast paralogous transcription factors controlling the multidrug resistance phenomenon. *The Journal of Biological Chemistry,* **278**, 52641–52650.

94 Mahe, Y., Parle-McDermott, A., Nourani, A., Delahodde, A., Lamprecht, A., and Kuchler, K. (1996) The ATP-binding cassette multidrug transporter Snq2p of *Saccharomyces cerevisiae*: a novel target for the transcription factors Pdr1p and Pdr3p. *Molecular Microbiology,* **20**, 109–117.

95 Katzmann, D.J., Burnett, P.E., Golin, J., Mahe, Y., and Moye-Rowley, W.S. (1994) Transcriptional control of the yeast *PDR5* gene by the *PDR3* gene product. *Molecular and Cellular Biology,* **14**, 4653–4661.

96 DeRisi, J., van den Hazel, B., Marc, P., Balzi, E., Brown, P., Jacq, C., and Goffeau, A. (2000) Genome microarray analysis of transcriptional activation in multidrug resistance yeast mutants. *FEBS Letters,* **470**, 156–160.

97 Decottignies, A., Lambert, L., Catty, P., Degand, H., Epping, E.A., Moye-Rowley, W.S., Balzi, E., and Goffeau, A. (1995) Identification and characterization of *SNQ2*, a new multidrug ATP binding cassette transporter of the yeast plasma membrane. *The Journal of Biological Chemistry,* **270**, 18150–18157.

98 Moye-Rowley, W.S. (2003) Transcriptional control of multidrug resistance in the yeast *Saccharomyces. Progress in Nucleic Acid Research and Molecular Biology,* **73**, 251–279.

99 Nourani, A., Wesolowski-Louvel, M., Delaveau, T., Jacq, C., and Delahodde, A. (1997) Multiple-drug-resistance phenomenon in the yeast *Saccharomyces cerevisiae*: involvement of two hexose transporters. *Molecular and Cellular Biology,* **17**, 5453–5460.

100 Delahodde, A., Delaveau, T., and Jacq, C. (1995) Positive autoregulation of the yeast transcription factor Pdr3p, which is involved in control of drug resistance. *Molecular and Cellular Biology,* **15**, 4043–4051.

101 Bauer, B.E., Wolfger, H., and Kuchler, K. (1999) Inventory and function of yeast ABC proteins: about sex, stress, pleiotropic drug and heavy metal resistance. *Biochimica et Biophysica Acta,* **1461**, 217–236.

102 Zhang, X. and Moye-Rowley, W.S. (2001) *Saccharomyces cerevisiae* multidrug resistance gene expression inversely correlates with the status of the F(0) component of the mitochondrial ATPase. *The Journal of Biological Chemistry,* **276**, 47844–47852.

103 Wolfger, H., Mahe, Y., Parle-McDermott, A., Delahodde, A., and Kuchler, K. (1997) The yeast ATP binding cassette (ABC) protein genes *PDR10* and *PDR15* are novel targets for the Pdr1p and Pdr3p transcriptional regulators. *FEBS Letters,* **418**, 269–274.

104 Mamnun, Y.M., Pandjaitan, R., Mahe, Y., Delahodde, A., and Kuchler, K. (2002) The yeast zinc finger regulators Pdr1p and Pdr3p control pleiotropic drug resistance (PDR) as homo- and heterodimers *in vivo*. *Molecular Microbiology*, **46**, 1429–1440.

105 Sidorova, M., Drobna, E., Dzugasova, V., Hikkel, I., and Subik, J. (2007) Loss-of-function *pdr3* mutations convert the Pdr3p transcription activator to a protein suppressing multidrug resistance in *Saccharomyces cerevisiae*. *FEMS Yeast Research*, **7**, 254–264.

106 Hallström, T.C. and Moye-Rowley, W.S. (2000) Multiple signals from dysfunctional mitochondria activate the pleiotropic drug resistance pathway in *Saccharomyces cerevisiae*. *The Journal of Biological Chemistry*, **275**, 37347–37356.

107 Akache, B., MacPherson, S., Sylvain, M.A., and Turcotte, B. (2004) Complex interplay among regulators of drug resistance genes in *Saccharomyces cerevisiae*. *The Journal of Biological Chemistry*, **279**, 27855–27860.

108 Larochelle, M., Drouin, S., Robert, F., and Turcotte, B. (2006) Oxidative stress-activated zinc cluster protein Stb5p has dual activator/repressor functions required for pentose phosphate pathway regulation and NADPH production. *Molecular and Cellular Biology*, **26**, 6690–6701.

109 Kren, A., Mamnun, Y.M., Bauer, B.E., Schuller, C., Wolfger, H., Hatzixanthis, K., Mollapour, M., Gregori, C., Piper, P., and Kuchler, K. (2003) War1p, a novel transcription factor controlling weak acid stress response in yeast. *Molecular and Cellular Biology*, **23**, 1775–1785.

110 Gregori, C., Schuller, C., Frohner, I.E., Ammerer, G., and Kuchler, K. (2008) Weak organic acids trigger conformational changes of the yeast transcription factor War1 *in vivo* to elicit stress adaptation. *The Journal of Biological Chemistry*, **283** (37), 25752–25764.

111 Schüller, C., Mamnun, Y.M., Mollapour, M., Krapf, G., Schuster, M., Bauer, B.E., Piper, P.W., and Kuchler, K. (2004) Global phenotypic analysis and transcriptional profiling defines the weak acid stress response regulon in *Saccharomyces cerevisiae*. *Molecular Biology of the Cell*, **15**, 706–720.

112 Vik, A. and Rine, J. (2001) Upc2p and Ecm22p, dual regulators of sterol biosynthesis in *Saccharomyces cerevisiae*. *Molecular and Cellular Biology*, **21**, 6395–6405.

113 Abramova, N.E., Cohen, B.D., Sertil, O., Kapoor, R., Davies, K.J., and Lowry, C.V. (2001) Regulatory mechanisms controlling expression of the *DAN/TIR* mannoprotein genes during anaerobic remodeling of the cell wall in *Saccharomyces cerevisiae*. *Genetics*, **157**, 1169–1177.

114 Cliften, P., Sudarsanam, P., Desikan, A., Fulton, L., Fulton, B., Majors, J., Waterston, R., Cohen, B.A., and Johnston, M. (2003) Finding functional features in *Saccharomyces* genomes by phylogenetic footprinting. *Science*, **301**, 71–76.

115 Shianna, K.V., Dotson, W.D., Tove, S., and Parks, L.W. (2001) Identification of a *UPC2* homolog in *Saccharomyces cerevisiae* and its involvement in aerobic sterol uptake. *Journal of Bacteriology*, **183**, 830–834.

116 Martinez-Pastor, M.T., Marchler, G., Schuller, C., Marchler-Bauer, A., Ruis, H., and Estruch, F. (1996) The *Saccharomyces cerevisiae* zinc finger proteins Msn2p and Msn4p are required for transcriptional induction through the stress response element (STRE). *The EMBO Journal*, **15**, 2227–2235.

117 Ruis, H. and Schuller, C. (1995) Stress signaling in yeast. *Bioessays*, **17**, 959–965.

118 Gorner, W., Durchschlag, E., Martinez-Pastor, M.T., Estruch, F., Ammerer, G., Hamilton, B., Ruis, H., and Schuller, C. (1998) Nuclear localization of the C2H2 zinc finger protein Msn2p is regulated by stress and protein kinase A activity. *Genes and Development*, **12**, 586–597.

119 Izawa, S., Ikeda, K., Ohdate, T., and Inoue, Y. (2007) Msn2p/Msn4p-activation is essential for the recovery from freezing stress in yeast. *Biochemical and Biophysical Research Communications*, **352**, 750–755.

120 Moye-Rowley, W.S. (2003) Regulation of the transcriptional response to oxidative stress in fungi: similarities and differences. *Eukaryotic Cell*, **2**, 381–389.

121 Rodrigues-Pousada, C.A., Nevitt, T., Menezes, R., Azevedo, D., Pereira, J., and Amaral, C. (2004) Yeast activator proteins and stress response: an overview. *FEBS Letters*, **567**, 80–85.

122 Wemmie, J.A., Wu, A.L., Harshman, K.D., Parker, C.S., and Moye-Rowley, W.S. (1994) Transcriptional activation mediated by the yeast *AP-1* protein is required for normal cadmium tolerance. *The Journal of Biological Chemistry*, **269**, 14690–14697.

123 Wu, A., Wemmie, J.A., Edgington, N.P., Goebl, M., Guevara, J.L., and Moye-Rowley, W.S. (1993) Yeast bZip proteins mediate pleiotropic drug and metal resistance. *The Journal of Biological Chemistry*, **268**, 18850–18858.

124 Wemmie, J.A., Szczypka, M.S., Thiele, D.J., and Moye-Rowley, W.S. (1994) Cadmium tolerance mediated by the yeast AP-1 protein requires the presence of an ATP-binding cassette transporter-encoding gene, YCF1. *The Journal of Biological Chemistry*, **269**, 32592–32597.

125 Moye-Rowley, W.S., Harshman, K.D., and Parker, C.S. (1989) Yeast *YAP1* encodes a novel form of the jun family of transcriptional activator proteins. *Genes and Development*, **3**, 283–292.

126 Kuge, S. and Jones, N. (1994) *YAP1* dependent activation of *TRX2* is essential for the response of *Saccharomyces cerevisiae* to oxidative stress by hydroperoxides. *The EMBO Journal*, **13**, 655–664.

127 Fernandes, L., Rodrigues-Pousada, C., and Struhl, K. (1997) Yap, a novel family of eight bZIP proteins in *Saccharomyces cerevisiae* with distinct biological functions. *Molecular and Cellular Biology*, **17**, 6982–6993.

128 Yan, C., Lee, L.H., and Davis, L.I. (1998) Crm1p mediates regulated nuclear export of a yeast AP-1-like transcription factor. *The EMBO Journal*, **17**, 7416–7429.

129 Kuge, S., Toda, T., Iizuka, N., and Nomoto, A. (1998) Crm1 (XpoI) dependent nuclear export of the budding yeast transcription factor yAP-1 is sensitive to oxidative stress. *Genes to Cells*, **3**, 521–532.

130 Bobrowicz, P., Wysocki, R., Owsianik, G., Goffeau, A., and Ulaszewski, S. (1997) Isolation of three contiguous genes, *ACR1*, *ACR2* and *ACR3*, involved in resistance to arsenic compounds in the yeast *Saccharomyces cerevisiae*. *Yeast*, **13**, 819–828.

131 Wysocki, R., Fortier, P.K., Maciaszczyk, E., Thorsen, M., Leduc, A., Odhagen, A., Owsianik, G., Ulaszewski, S., Ramotar, D., and Tamas, M.J. (2004) Transcriptional activation of metalloid tolerance genes in *Saccharomyces cerevisiae* requires the AP-1-like proteins Yap1p and Yap8p. *Molecular Biology of the Cell*, **15**, 2049–2060.

132 Menezes, R.A., Amaral, C., Batista-Nascimento, L., Santos, C., Ferreira, R.B., Devaux, F., Eleutherio, E.C., and Rodrigues-Pousada, C. (2008) Contribution of Yap1 towards *S. cerevisiae* adaptation to arsenic mediated oxidative stress. *The Biochemical Journal*, **414** (2), 301–311.

133 Hallström, T.C., Lambert, L., Schorling, S., Balzi, E., Goffeau, A., and Moye-Rowley, W.S. (2001) Coordinate control of sphingolipid biosynthesis and multidrug resistance in *Saccharomyces cerevisiae*. *The Journal of Biological Chemistry*, **276**, 23674–23680.

134 Shahi, P., Gulshan, K., and Moye-Rowley, W.S. (2007) Negative transcriptional regulation of multidrug

135. Saleh, A., Lang, V., Cook, R., and Brandl, C.J. (1997) Identification of native complexes containing the yeast coactivator/repressor proteins Ngg1p/Ada3p and Ada2p. *The Journal of Biological Chemistry*, **272**, 5571–5578.
136. Martens, J.A., Genereaux, J., Saleh, A., and Brandl, C.J. (1996) Transcriptional activation by yeast Pdr1p is inhibited by its association with Ngg1p/Ada3p. *The Journal of Biological Chemistry*, **271**, 15884–15890.
137. Pfaller, M.A. and Diekema, D.J. (2007) Epidemiology of invasive candidiasis: a persistent public health problem. *Clinical Microbiology Reviews*, **20**, 133–163.
138. Ruhnke, M. (2006) Epidemiology of *Candida albicans* infections and role of non-*Candida-albicans* yeasts. *Current Drug Targets*, **7**, 495–504.
139. Cannon, R.D., Lamping, E., Holmes, A.R., Niimi, K., Tanabe, K., Niimi, M., and Monk, B.C. (2007) *Candida albicans* drug resistance another way to cope with stress. *Microbiology*, **153**, 3211–3217.
140. Prasad, R., Gaur, N.A., Gaur, M., and Komath, S.S. (2006) Efflux pumps in drug resistance of *Candida*. *Infectious Disorders Drug Targets*, **6**, 69–83.
141. Sanglard, D., Ischer, F., Monod, M., and Bille, J. (1996) Susceptibilities of *Candida albicans* multidrug transporter mutants to various antifungal agents and other metabolic inhibitors. *Antimicrobial Agents and Chemotherapy*, **40**, 2300–2305.
142. Sanglard, D., Ischer, F., Monod, M., and Bille, J. (1997) Cloning of *Candida albicans* genes conferring resistance to azole antifungal agents: characterization of *CDR2*, a new multidrug ABC transporter gene. *Microbiology*, **143** (Pt 2), 405–416.
143. Coste, A.T., Karababa, M., Ischer, F., Bille, J., and Sanglard, D. (2004) AT *TAC1*, transcriptional activator of CDR genes, is a new transcription factor involved in the regulation of *Candida albicans* ABC transporters *CDR1* and *CDR2*. *Eukaryotic Cell*, **3**, 1639–1652.
144. Karnani, N., Gaur, N.A., Jha, S., Puri, N., Krishnamurthy, S., Goswami, S.K., Mukhopadhyay, G., and Prasad, R. (2004) SRE1 and SRE2 are two specific steroid-responsive modules of *Candida* drug resistance gene 1 (*CDR1*) promoter. *Yeast*, **21**, 219–239.
145. Krishnamurthy, S., Chatterjee, U., Gupta, V., Prasad, R., Das, P., Snehlata, P., and Hasnain, S.E. (1998) Deletion of transmembrane domain 12 of Cdr1p, a multidrug transporter from *Candida albicans*, leads to altered drug specificity: expression of a yeast multidrug transporter in baculovirus expression system. *Yeast*, **14**, 535–550.
146. Dogra, S., Krishnamurthy, S., Gupta, V., Dixit, B.L., Gupta, C.M., Sanglard, D., and Prasad, R. (1999) Asymmetric distribution of phosphatidylethanolamine in *C. albicans*: possible mediation by *CDR1*, a multidrug transporter belonging to ATP binding cassette (ABC) superfamily. *Yeast*, **15**, 111–121.
147. Larsen, B., Anderson, S., Brockman, A., Essmann, M., and Schmidt, M. (2006) Key physiological differences in *Candida albicans CDR1* induction by steroid hormones and antifungal drugs. *Yeast*, **23**, 795–802.
148. Silver, P.M., Oliver, B.G., and White, T.C. (2004) Role of *Candida albicans* transcription factor Upc2p in drug resistance and sterol metabolism. *Eukaryotic Cell*, **3**, 1391–1397.
149. MacPherson, S., Akache, B., Weber, S., De Deken, X., Raymond, M., and Turcotte, B. (2005) *Candida albicans* zinc cluster protein Upc2p confers resistance to antifungal drugs and is an activator of ergosterol biosynthetic genes. *Antimicrobial Agents and Chemotherapy*, **49**, 1745–1752.
150. Pohl, A., Devaux, P.F., and Herrmann, A. (2005) Function of prokaryotic and

eukaryotic ABC proteins in lipid transport. *Biochimica et Biophysica Acta*, **1733**, 29–52.

151 Shukla, S., Rai, V., Saini, P., Banerjee, D., Menon, A.K., and Prasad, R. (2007) Candida drug resistance protein 1, a major multidrug ATP binding cassette transporter of *Candida albicans*, translocates fluorescent phospholipids in a reconstituted system. *Biochemistry*, **46**, 12081–12090.

152 Niimi, K., Maki, K., Ikeda, F., Holmes, A.R., Lamping, E., Niimi, M., Monk, B.C., and Cannon, R.D. (2006) Overexpression of *Candida albicans CDR1, CDR2*, or *MDR1* does not produce significant changes in echinocandin susceptibility. *Antimicrobial Agents and Chemotherapy*, **50**, 1148–1155.

153 Jha, S., Karnani, N., Dhar, S.K., Mukhopadhayay, K., Shukla, S., Saini, P., Mukhopadhayay, G., and Prasad, R. (2003) Purification and characterization of the N-terminal nucleotide binding domain of an ABC drug transporter of *Candida albicans*: uncommon cysteine 193 of Walker A is critical for ATP hydrolysis. *Biochemistry*, **42**, 10822–10832.

154 Rai, V., Shukla, S., Jha, S., Komath, S.S., and Prasad, R. (2005) Functional characterization of N-terminal nucleotide binding domain (NBD-1) of a major ABC drug transporter Cdr1p of *Candida albicans*: uncommon but conserved Trp326 of Walker B is important for ATP binding. *Biochemistry*, **44**, 6650–6661.

155 Jha, S., Dabas, N., Karnani, N., Saini, P., and Prasad, R. (2004) ABC multidrug transporter Cdr1p of *Candida albicans* has divergent nucleotide-binding domains which display functional asymmetry. *FEMS Yeast Research*, **5**, 63–72.

156 Shukla, S., Saini, P., Jha, S.S, Ambudkar, S.V., and Prasad, R. (2003) Functional characterization of *Candida albicans* ABC transporter Cdr1p. *Eukaryotic Cell*, **2**, 1361–1375.

157 Balan, I., Alarco, A.M., and Raymond, M. (1997) The *Candida albicans CDR3* gene codes for an opaque-phase ABC transporter. *Journal of Bacteriology*, **179**, 7210–7218.

158 Franz, R., Michel, S., and Morschhauser, J. (1998) A fourth gene from the *Candida albicans CDR* family of ABC transporters. *Gene*, **220**, 91–98.

159 Ogawa, A., Hashida-Okado, T., Endo, M., Yoshioka, H., Tsuruo, T., Takesako, K., and Kato, I. (1998) Role of ABC transporters in aureobasidin A resistance. *Antimicrobial Agents and Chemotherapy*, **42**, 755–761.

160 Alarco, A.M. and Raymond, M. (1999) The bZip transcription factor Cap1p is involved in multidrug resistance and oxidative stress response in *Candida albicans*. *Journal of Bacteriology*, **181**, 700–708.

161 White, T.C., Holleman, S., Dy, F., Mirels, L.F., and Stevens, D.A. (2002) Resistance mechanisms in clinical isolates of *Candida albicans*. *Antimicrobial Agents and Chemotherapy*, **46**, 1704–1713.

162 Fling, M.E., Kopf, J., Tamarkin, A., Gorman, J.A., Smith, H.A., and Koltin, Y. (1991) Analysis of a *Candida albicans* gene that encodes a novel mechanism for resistance to benomyl and methotrexate. *Molecular & General Genetics*, **227**, 318–329.

163 Ben-Yaacov, R., Knoller, S., Caldwell, G.A., Becker, J.M., and Koltin, Y. (1994) *Candida albicans* gene encoding resistance to benomyl and methotrexate is a multidrug resistance gene. *Antimicrobial Agents and Chemotherapy*, **38**, 648–652.

164 Calabrese, D., Bille, J., and Sanglard, D. (2000) A novel multidrug efflux transporter gene of the major facilitator superfamily from *Candida albicans* (*FLU1*) conferring resistance to fluconazole. *Microbiology*, **146** (Pt 11), 2743–2754.

165 Gupta, V., Kohli, A., Krishnamurthy, S., Puri, N., Aalamgeer, S.A., Panwar, S., and Prasad, R. (1998) Identification of polymorphic mutant alleles of Ca*MDR1*, a major facilitator of *Candida albicans*

which confers multidrug resistance, and its *in vitro* transcriptional activation. *Current Genetics*, **34**, 192–199.
166 Goldway, M., Teff, D., Schmidt, R., Oppenheim, A.B., and Koltin, Y. (1995) Multidrug resistance in *Candida albicans*: disruption of the *BENr* gene. *Antimicrobial Agents and Chemotherapy*, **39**, 422–426.
167 Zhang, X., De Micheli, M., Coleman, S.T., Sanglard, D., and Moye-Rowley, W.S. (2000) Analysis of the oxidative stress regulation of the *Candida albicans* transcription factor, Cap1p. *Molecular Microbiology*, **36**, 618–629.
168 Wang, Y., Cao, Y.Y., Jia, X.M., Cao, Y.B., Gao, P.H., Fu, X.P., Ying, K., Chen, W.S., and Jiang, Y.Y. (2006) Cap1p is involved in multiple pathways of oxidative stress response in *Candida albicans*. *Free Radical Biology & Medicine*, **40**, 1201–1209.
169 DeMicheli, M., Bille, J., and Sanglard, D. (1999) Regulation of ATP-binding cassette (ABC) transporter genes in *Candida albicans*. Presented at the ASM Conference on Candida and Candidiasis, Charleston, SC.
170 Harry, J.B., Oliver, B.G., Song, J.L., Silver, P.M., Little, J.T., Choiniere, J., and White, T.C. (2005) Drug-induced regulation of the *MDR1* promoter in *Candida albicans*. *Antimicrobial Agents and Chemotherapy*, **49**, 2785–2792.
171 Talibi, D. and Raymond, M. (1999) Isolation of a putative *Candida albicans* transcriptional regulator involved in pleiotropic drug resistance by functional complementation of a *pdr1 pdr3* mutation in *Saccharomyces cerevisiae*. *Journal of Bacteriology*, **181**, 231–240.
172 Shen, H., An, M.M., Wang de, J., Xu, Z., Zhang, J.D., Gao, P.H., Cao, Y.Y., Cao, Y.B., and Jiang, Y.Y. (2007) Fcr1p inhibits development of fluconazole resistance in *Candida albicans* by abolishing *CDR1* induction. *Biological & Pharmaceutical Bulletin*, **30**, 68–73.
173 Miyazaki, H., Miyazaki, Y., Geber, A., Parkinson, T., Hitchcock, C., Falconer, D.J., Ward, D.J., Marsden, K., and Bennett, J.E. (1998) Fluconazole resistance associated with drug efflux and increased transcription of a drug transporter gene, *PDH1*, in *Candida glabrata*. *Antimicrobial Agents and Chemotherapy*, **42**, 1695–1701.
174 Sanglard, D., Ischer, F., Calabrese, D., Majcherczyk, P.A., and Bille, J. (1999) The ATP binding cassette transporter gene *CgCDR1* from *Candida glabrata* is involved in the resistance of clinical isolates to azole antifungal agents. *Antimicrobial Agents and Chemotherapy*, **43**, 2753–2765.
175 Sanglard, D., Ischer, F., and Bille, J. (2001) Role of ATP-binding-cassette transporter genes in high-frequency acquisition of resistance to azole antifungals in *Candida glabrata*. *Antimicrobial Agents and Chemotherapy*, **45**, 1174–1183.
176 Izumikawa, K., Kakeya, H., Tsai, H.F., Grimberg, B., and Bennett, J.E. (2003) Function of *Candida glabrata* ABC transporter gene, PDH1. *Yeast*, **20**, 249–261.
177 Tsai, H.F., Krol, A.A., Sarti, K.E., and Bennett, J.F. (2006) *Candida glabrata PDR1*, a transcriptional regulator of a pleiotropic drug resistance network, mediates azole resistance in clinical isolates and petite mutants. *Antimicrobial Agents and Chemotherapy*, **50**, 1384–1392.
178 Vermitsky, J.P., Earhart, K.D., Smith, W.L., Homayouni, R., Edlind, T.D., and Rogers, P.D. (2006) Pdr1p regulates multidrug resistance in *Candida glabrata*: gene disruption and genome-wide expression studies. *Molecular Microbiology*, **61**, 704–722.
179 Brun, S., Aubry, C., Lima, O., Filmon, R., Berges, T., Chabasse, D., and Bouchara, J.P. (2003) Relationships between respiration and susceptibility to azole antifungals in *Candida glabrata*. *Antimicrobial Agents and Chemotherapy*, **47**, 847–853.
180 Brun, S., Berges, T., Poupard, P., Vauzelle-Moreau, C., Renier, G.,

Chabasse, D., and Bouchara, J.P. (2004) Mechanisms of azole resistance in petite mutants of *Candida glabrata*. *Antimicrobial Agents and Chemotherapy*, **48**, 1788–1796.

181 Brun, S., Dalle, F., Saulnier, P., Renier, G., Bonnin, A., Chabasse, D., and Bouchara, J.P. (2005) Biological consequences of petite mutations in *Candida glabrata*. *The Journal of Antimicrobial Chemotherapy*, **56**, 307–314.

182 Wada, S., Niimi, M., Niimi, K., Holmes, A.R., Monk, B.C., Cannon, R.D., and Uehara, Y. (2002) *Candida glabrata* ATP-binding cassette transporters Cdr1p and Pdh1p expressed in a *Saccharomyces cerevisiae* strain deficient in membrane transporters show phosphorylation-dependent pumping properties. *The Journal of Biological Chemistry*, **277**, 46809–46821.

183 Wada, S., Tanabe, K., Yamazaki, A., Niimi, M., Uehara, Y., Niimi, K., Lamping, E., Cannon, R.D., and Monk, B.C. (2005) Phosphorylation of *Candida glabrata* ATP-binding cassette transporter Cdr1p regulates drug efflux activity and ATPase stability. *The Journal of Biological Chemistry*, **280**, 94–103.

184 Nakayama, H., Tanabe, K., Bard, M., Hodgson, W., Wu, S., Takemori, D., Aoyama, T., Kumaraswami, N.S., Metzler, L., Takano, Y., Chibana, H., and Niimi, M. (2007) The *Candida glabrata* putative sterol transporter gene *CgAUS1* protects cells against azoles in the presence of serum. *The Journal of Antimicrobial Chemotherapy*, **60**, 1264–1272.

185 Cheng, S., Clancy, C.J., Nguyen, K.T., Clapp, W., and Nguyen, M.H. (2007) A *Candida albicans* petite mutant strain with uncoupled oxidative phosphorylation overexpresses *MDR1* and has diminished susceptibility to fluconazole and voriconazole. *Antimicrobial Agents and Chemotherapy*, **51**, 1855–1858.

186 Moran, G.P., Sanglard, D., Donnelly, S.M., Shanley, D.B., Sullivan, D.J., and Coleman, D.C. (1998) Identification and expression of multidrug transporters responsible for fluconazole resistance in *Candida dubliniensis*. *Antimicrobial Agents and Chemotherapy*, **42**, 1819–1830.

187 Katiyar, S.K. and Edlind, T.D. (2001) Identification and expression of multidrug resistance-related ABC transporter genes in *Candida krusei*. *Medical Mycology*, **39**, 109–116.

188 Young, L., Leonhard, K., Tatsuta, T., Trowsdale, J., and Langer, T. (2001) Role of the ABC transporter Mdl1p in peptide export from mitochondria. *Science*, **291**, 2135–2138.

189 Slaven, J.W., Anderson, M.J., Sanglard, D., Dixon, G.K., Bille, J., Roberts, I.S., and Denning, D.W. (2002) Increased expression of a novel *Aspergillus fumigatus* ABC transporter gene, *atrF*, in the presence of itraconazole in an itraconazole resistant clinical isolate. *Fungal Genetics and Biology*, **36**, 199–206.

190 Nascimento, A.M., Goldman, G.H., Park, S., Marras, S.A., Delmas, G., Oza, U., Lolans, K., Dudley, M.N., Mann, P.A., and Perlin, D.S. (2003) Multiple resistance mechanisms among *Aspergillus fumigatus* mutants with high-level resistance to itraconazole. *Antimicrobial Agents and Chemotherapy*, **47**, 1719–1726.

191 Del Sorbo, G., Andrade, A.C., Van Nistelrooy, J.G., Van Kan, J.A., Balzi, E., and De Waard, M.A. (1997) Multidrug resistance in *Aspergillus nidulans* involves novel ATP-binding cassette transporters. *Molecular & General Genetics*, **254**, 417–426.

192 Angermayr, K., Parson, W., Stoffler, G., and Haas, H. (1999) Expression of *atrC* – encoding a novel member of the ATP binding cassette transporter family in *Aspergillus nidulans* – is sensitive to cycloheximide. *Biochimica et Biophysica Acta*, **1453**, 304–310.

193 do Nascimento, A.M., Goldman, M.H., and Goldman, G.H. (2002) Molecular

characterization of ABC transporter-encoding genes in *Aspergillus nidulans*. *Genetics and Molecular Research*, **1**, 337–349.

194 de Waard, M.A., Andrade, A.C., Hayashi, K., Schoonbeek, H.J., Stergiopoulos, I., and Zwiers, L.H. (2006) Impact of fungal drug transporters on fungicide sensitivity, multidrug resistance and virulence. *Pest Management Science*, **62**, 195–207.

195 Andrade, A.C., Del Sorbo, G., Van Nistelrooy, J.G., and Waard, M.A. (2000) The ABC transporter *AtrB* from *Aspergillus nidulans* mediates resistance to all major classes of fungicides and some natural toxic compounds. *Microbiology*, **146** (Pt 8), 1987–1997.

196 Posteraro, B., Sanguinetti, M., Sanglard, D., La Sorda, M., Boccia, S., Romano, L., Morace, G., and Fadda, G. (2003) Identification and characterization of a *Cryptococcus neoformans* ATP binding cassette (ABC) transporter-encoding gene, *CnAFR1*, involved in the resistance to fluconazole. *Molecular Microbiology*, **47**, 357–371.

197 Sanguinetti, M., Posteraro, B., La Sorda, M., Torelli, R., Fiori, B., Santangelo, R., Delogu, G., and Fadda, G. (2006) Role of *AFR1*, an ABC transporter-encoding gene, in the *in vivo* response to fluconazole and virulence of *Cryptococcus neoformans*. *Infection and Immunity*, **74**, 1352–1359.

198 Fernandez-Torres, B., Vazquez-Veiga, H., Llovo, X., Pereiro, M., Jr., and Guarro, J. (2000) *In vitro* susceptibility to itraconazole, clotrimazole, ketoconazole and terbinafine of 100 isolates of *Trichophyton rubrum*. *Chemotherapy*, **46**, 390–394.

199 Cervelatti, E.P., Fachin, A.L., Ferreira-Nozawa, M.S., and Martinez-Rossi, N.M. (2006) Molecular cloning and characterization of a novel ABC transporter gene in the human pathogen *Trichophyton rubrum*. *Medical Mycology*, **44**, 141–147.

200 Fachin, A.L., Ferreira-Nozawa, M.S., Maccheroni, W., Jr., and Martinez-Rossi, N.M. (2006) Role of the ABC transporter *TruMDR2* in terbinafine, 4-nitroquinoline N-oxide and ethidium bromide susceptibility in *Trichophyton rubrum*. *Journal of Medical Microbiology*, **55**, 1093–1099.

201 Ehrlich, F. (1907) Über die Bedingungen der Fuselölbildung und über ihren Zusammenhang mit dem Eiweißaufbau der Hefe. *Berichte der Deutschen Chemischen Gesellschaft*, **40**, 1027–1047.

202 Deeley, R.G., Westlake, C., and Cole, S.P. (2006) Transmembrane transport of endo- and xenobiotics by mammalian ATP-binding cassette multidrug resistance proteins. *Physiological Reviews*, **86**, 849–899.

203 van Helvoort, A., Smith, A.J., Sprong, H., Fritzsche, I., Schinkel, A.H., Borst, P., and van Meer, G. (1996) *MDR1* P-glycoprotein is a lipid translocase of broad specificity, while *MDR3* P-glycoprotein specifically translocates phosphatidylcholine. *Cell*, **87**, 507–517.

204 Kean, L.S., Grant, A.M., Angeletti, C., Mahe, Y., Kuchler, K., Fuller, R.S., and Nichols, J.W. (1997) Plasma membrane translocation of fluorescent-labeled phosphatidylethanolamine is controlled by transcription regulators, *PDR1* and *PDR3*. *The Journal of Cell Biology*, **138**, 255–270.

205 Higgins, C.F. and Gottesman, M.M. (1992) Is the multidrug transporter a flippase? *Trends in Biochemical Sciences*, **17**, 18–21.

206 Kihara, A. and Igarashi, Y. (2004) Cross talk between sphingolipids and glycerophospholipids in the establishment of plasma membrane asymmetry. *Molecular Biology of the Cell*, **15**, 4949–4959.

207 Kadaba, N.S., Kaiser, J.T., Johnson, E., Lee, A., and Rees, D.C. (2008) The high-affinity *E. coli* methionine ABC transporter: structure and allosteric regulation. *Science*, **321**, 250–253.

208 Hollenstein, K., Frei, D.C., and Locher, K.P. (2007) Structure of an ABC transporter in complex with its binding protein. *Nature*, **446**, 213–216.

209 Oldham, M.L., Khare, D., Quiocho, F.A., Davidson, A.L., and Chen, J. (2007) Crystal structure of a catalytic intermediate of the maltose transporter. *Nature*, **450**, 515–521.

210 Pinkett, H.W., Lee, A.T., Lum, P., Locher, K.P., and Rees, D.C. (2007) An inward-facing conformation of a putative metal-chelate-type ABC transporter. *Science*, **315**, 373–377.

211 Dawson, R.J. and Locher, K.P. (2006) Structure of a bacterial multidrug ABC transporter. *Nature*, **443**, 180–185.

212 Gaudet, R. and Wiley, D.C. (2001) Structure of the ABC ATPase domain of human TAP1, the transporter associated with antigen processing. *The EMBO Journal*, **20**, 4964–4972.

213 Lewis, H.A., Buchanan, S.G., Burley, S.K., Conners, K., Dickey, M., Dorwart, M., Fowler, R., Gao, X., Guggino, W.B., Hendrickson, W.A., Hunt, J.F., Kearins, M.C., Lorimer, D., Maloney, P.C., Post, K.W., Rajashankar, K.R., Rutter, M.E., Sauder, J.M., Shriver, S., Thibodeau, P.H., Thomas, P.J., Zhang, M., Zhao, X., and Emtage, S. (2004) Structure of nucleotide-binding domain 1 of the cystic fibrosis transmembrane conductance regulator. *The EMBO Journal*, **23**, 282–293.

214 Katzmann, D.J., Hallstrom, T.C., Voet, M., Wysock, W., Golin, J., Volckaert, G., and Moye-Rowley, W.S. (1995) Expression of an ATP-binding cassette transporter-encoding gene (*YOR1*) is required for oligomycin resistance in *Saccharomyces cerevisiae*. *Molecular and Cellular Biology*, **15**, 6875–6883.

215 Kuchler, K., Sterne, R.E., and Thorner, J. (1989) *Saccharomyces cerevisiae STE6* gene product: a novel pathway for protein export in eukaryotic cells. *The EMBO Journal*, **8**, 3973–3984.

216 McGrath, J.P. and Varshavsky, A. (1989) The yeast *STE6* gene encodes a homologue of the mammalian multidrug resistance P-glycoprotein. *Nature*, **340**, 400–404.

217 Leighton, J. (1995) ATP-binding cassette transporter in *Saccharomyces cerevisiae* mitochondria. *Methods in Enzymology*, **260**, 389–396.

218 Dean, M., Allikmets, R., Gerrard, B., Stewart, C., Kistler, A., Shafer, B., Michaelis, S., and Strathern, J. (1994) Mapping and sequencing of two yeast genes belonging to the ATP-binding cassette superfamily. *Yeast*, **10**, 377–383.

219 Shani, N., Watkins, P.A., and Valle, D. (1995) AT *PXA1*, a possible *Saccharomyces cerevisiae* ortholog of the human adrenoleukodystrophy gene. *Proceedings of the National Academy of Sciences of the United States of America*, **92**, 6012–6016.

220 Shani, N. and Valle, D. (1996) A *Saccharomyces cerevisiae* homolog of the human adrenoleukodystrophy transporter is a heterodimer of two half ATP-binding cassette transporters. *Proceedings of the National Academy of Sciences of the United States of America*, **93**, 11901–11906.

221 Kamath, A. and Chakraburtty, K. (1989) Role of yeast elongation factor 3 in the elongation cycle. *The Journal of Biological Chemistry*, **264**, 15423–15428.

222 Maurice, T.C., Mazzucco, C.E., Ramanathan, C.S., Ryan, B.M., Warr, G.A., and Puziss, J.W. (1998) A highly conserved intraspecies homolog of the *Saccharomyces cerevisiae* elongation factor-3 encoded by the *HEF3* gene. *Yeast*, **14**, 1105–1113.

223 Sarthy, A.V., McGonigal, T., Capobianco, J.O., Schmidt, M., Green, S.R., Moehle, C.M., and Goldman, R.C. (1998) Identification and kinetic analysis of a functional homolog of elongation factor 3, *YEF3* in *Saccharomyces cerevisiae*. *Yeast*, **14**, 239–253.

224 Garcia-Barrio, M., Dong, J., Ufano, S., and Hinnebusch, A.G. (2000) Association of *GCN1–GCN20* regulatory complex with the N-terminus of eIF2alpha kinase

Gcn2p is required for Gcn2p activation. *The EMBO Journal*, **19**, 1887–1899.

225 Osherovich, L.Z. and Weissman, J.S. (2001) Multiple Gln/Asn-rich prion domains confer susceptibility to induction of the yeast [PSI$^+$] prion. *Cell*, **106**, 183–194.

226 Dong, J., Lai, R., Jennings, J.L., Link, A.J., and Hinnebusch, A.G. (2005) The novel ATP-binding cassette protein Arb1p is a shuttling factor that stimulates 40S and 60S ribosome biogenesis. *Molecular and Cellular Biology*, **25**, 9859–9873.

227 Valasek, L., Phan, L., Schoenfeld, L.W., Valaskova, V., and Hinnebusch, A.G. (2001) Related eIF3 subunits Tif32p and Hcr1p interact with an RNA recognition motif in Prt1p required for eIF3 integrity and ribosome binding. *The EMBO Journal*, **20**, 891–904.

228 Liu, H.Y., Chiang, Y.C., Pan, J., Chen, J., Salvadore, C., Audino, D.C., Badarinarayana, V., Palaniswamy, V., Anderson, B., and Denis, C.L. (2001) Characterization of *CAF4* and *CAF16* reveals a functional connection between the CCR4-NOT complex and a subset of SRB proteins of the RNA polymerase II holoenzyme. *The Journal of Biological Chemistry*, **276**, 7541–7548.

Part Three:
Structure Activity Relationship Studies on ABC Transporter

7
QSAR Studies on ABC Transporter – How to Deal with Polyspecificity
Gerhard F. Ecker

7.1
The Problem of Polyspecificity/Promiscuity

With a deeper understanding of the processes involved in ADMET, the concept of avoiding interaction with drug transporters has gained increasing awareness. Several key ABC proteins identified so far in the ADMET cascade show a broad and structurally unrelated substrate and inhibitor pattern. We will term this as polyspecificity throughout this chapter. We prefer this term to promiscuity, as these transporters still show specificity toward distinct structural scaffolds, and predictive QSAR models can also be obtained. At the molecular level, polyspecificity might have several fundamental causes. These include binding sites (or "binding zones") accommodating more than one ligand, multiple separate (maybe in part overlapping) binding sites, and high protein flexibility. The current methods used in the computational drug design field are only in part suited to deal with this type of complex phenomena. Traditional QSAR methods assume distinct ligand binding conformations and are generally suited only for homologous series of compounds. However, there have been considerable modeling efforts to target polyspecific proteins using basically the full armory of available methods, including pharmacophore modeling and machine learning approaches. Although most of them show good to excellent performance, generally applicable models are still missing.

7.2
QSAR Approaches to Design Inhibitors of P-glycoprotein (ABCB1)

It was in 1976, when the group of Victor Ling identified P-glycoprotein (ABCB1) as being responsible for the reduced drug accumulation in multidrug-resistant Chinese hamster ovary cells [1]. ABCB1 functions as a membrane-bound, ATP-dependent efflux pump extruding a wide variety of functionally and structurally diverse natural toxins out of mammalian cells [2]. Overexpression of the protein in tumor cells thus leads to multiresistance to cytotoxic agents. Five years later, Tsuruo *et al.* identified

---

Transporters as Drug Carriers: Structure, Function, Substrates. Edited by Gerhard Ecker and Peter Chiba
Copyright © 2009 WILEY-VCH Verlag GmbH & Co. KGaA, Weinheim
ISBN: 978-3-527-31661-8

verapamil as first inhibitor of P-gp-mediated transport, with hundreds of compounds to follow in the subsequent years [3–5]. Blocking P-gp restores sensitivity of multidrug-resistant cells to chemotherapeutic agents and thus represents a versatile approach for overcoming drug resistance.

In lead optimization programs, numerous QSAR studies on structurally homologous series of compounds have been performed. Especially, verapamil analogues, triazines, acridonecarboxamides, phenothiazines, thioxanthenes, flavones, dihydropyridines, propafenones, and cyclosporin derivatives have been extensively studied, and the results are summarized in several excellent reviews [6, 7]. These studies pinpoint the importance of H-bond acceptors and their strength, of the distance between aromatic moieties and H-bond acceptors, and the influence of global physicochemical parameters, such as lipophilicity and molar refractivity. Systematic quantitative structure–activity relationship studies have been performed mainly on phenothiazines and propafenones [8]. The latter studies have been carried out using Hansch and Free-Wilson analyses [9], hologram QSAR, CoMFA, and CoMSIA [10] as well as nonlinear methods [11] and similarity-based approaches [12]. Hansch-type correlation analyses normally lead to excellent correlations between lipophilicity and pIC_{50} values within structurally homologous series of compounds (Figures 7.1 and 7.2). However, this is not surprising as the interaction of ligands with P-gp is supposed to take place in the membrane bilayer. Thus, lipophilicity of the compounds triggers their concentration at the binding site, rather than being a parameter important for ligand–protein interaction. Different intercepts of correlation lines and outliers point to altered pharmacophoric patterns. This is especially exemplified in Figure 7.1 showing the log P/pIC_{50} correlation for series of propafenones and

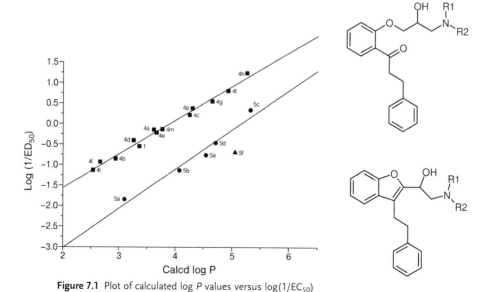

Figure 7.1 Plot of calculated log P values versus log($1/EC_{50}$) values for a series of propafenones (■) and analogous benzofurans (●).

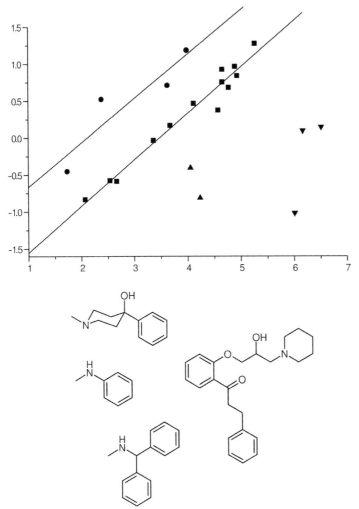

Figure 7.2 Plot of the calculated log P values versus $\log(1/EC_{50})$ values for a series of propafenones modified at the nitrogen atom: (■) N-alkyl derivatives; (▲) N-aryl derivatives; (▼) N-diphenylalkyl derivatives; (●) 4-hydroxy-4-phenylpiperidine derivatives.

analogous, conformationally restricted benzofurans [13]. Within both series, an excellent correlation is obtained, whereby benzofurans generally show a lower log potency/log P ratio than propafenones. Thus, for an equilipophilic pair of compounds, the benzofuran is generally an order of magnitude less active than the corresponding propafenone. This is further supported by Eq. (7.1) showing a coefficient of -1.16 for the indicator variable that encodes for the benzofuran scaffold (I_{bf}).

$$\log(1/EC_{50}) = 0.86 \log P - 1.16 I_{bf} - 3.33. \tag{7.1}$$

This relative loss of activity might be either due to a decrease in the flexibility of the molecules or due to the loss of the C=O group as H-bond acceptor. Systematic variations of the C=O group finally showed that the latter seems to be the case [14].

Figure 7.2 exemplifies the opposite case showing that a series of 4-hydroxy-4-phenylpiperidines generally exhibit higher $pIC_{50}/\log P$ ratios than expected according to their lipophilicity [15]. This indicates that the $-OH$ group utilizes an additional H-bond interaction with the protein. Figure 7.2 also outlines modifications in close vicinity of the basic nitrogen atom that can influence pharmacological activity independent of lipophilicity. Introduction of large groups such as diphenylalkyl gives rise to a dramatic relative loss of activity (relative to the log P of the moiety!). While the diphenylmethyl derivative showed a decrease of almost three orders of magnitude, which might be due to steric hindrance of the nitrogen atom, diphenylethyl-, -propyl-, and -isopropylamines exhibited almost identical activity values.

Projection of the hydrophobic potential onto the van der Waals surface of the molecules shows that the diphenyl group represents a huge hydrophobic moiety that might act like an anchor for the molecules in the lipid bilayer (Figure 7.3a). In this case, the phenone moiety, which has not been varied in this compound series, is supposed to interact with the protein. This possible change of the binding mode is also reflected by a different Hill coefficient of the dose–response curve (Figure 7.3b).

To further prove this hypothesis, a series of propafenone-type P-glycoprotein inhibitors were designed, synthesized, and tested in order to elucidate the influence of intermolecular hydrophobicity distribution. Results demonstrate that with increasing lipophilicity of the substituents on the amine moiety, the statistical significance of the indicator variables denoting the substitution pattern on the central aromatic ring system also increases [16]. This indicated that the distribution of hydrophobicity within the molecules influences the mode of interaction with P-gp. To further explore this hypothesis, we also implemented the concept of hydrophobic moments to use them as descriptors in multiple linear regression analysis [17]. Considering the zeroth moment as the sum of the atomic hydrophobicity coefficients (which is a measure for the total hydrophobicity of the molecule), the first moment (or hydrophobic dipole) is a measure for the asymmetry of the distribution of hydrophobicities and therefore is analogous to the electrostatic dipole (Figure 7.4). The use of these hydrophobic dipole moments as independent variables remarkably improved the predictive power of QSAR models obtained for this special set of propafenone-type inhibitors of P-gp.

The dramatic decrease in the diphenylmethylamine derivative already indicated the importance of the nitrogen atom. This is further strengthened by the very first pharmacophores published that mainly consisted of a positively charged nitrogen atom and two aromatic rings. However, systematic variation in the H-bond acceptor strength of the nitrogen atom in propafenone-type inhibitors revealed that H-bond acceptor strength in this region is quantitatively correlated with P-gp inhibitory activity (Figure 7.5) [18]. Thus, anilines, amides, and even esters show pharmacological activity, which rules out the hypothesis that the nitrogen atom interacts in positively charged form. The fundamental importance of H-bond acceptors has also been pointed out by several studies from Anna Seelig's group [19]. She described a

(a)

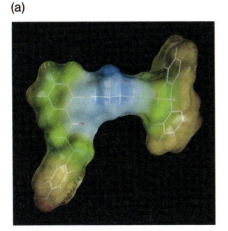

(b)

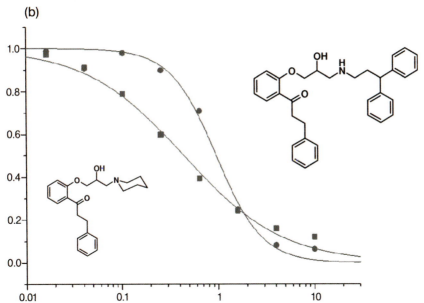

Figure 7.3 (a) Hydrophobic van der Waals surface of a diphenylalkylamine analogous propafenone derivative (GPV 0238). (b) Dose–response curves for the piperidine analogous propafenone GPV 0005 (■) and the diphenylalkylamine GPV 0238 (●).

scheme that comprises two H-bond acceptors at a spatial distance of either 2.5 or 4.6 Å. In the latter case, a third H-bond acceptor might be located in between the two primary electron donating groups.

Further proof of the negative influence of large substituents in close vicinity of the nitrogen atom has been obtained by CoMFA and CoMSIA analyses using a set of 131

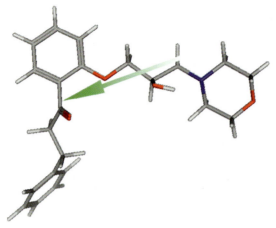

Figure 7.4 Hydrophobic moment (green arrow) for the morpholino analogue GPV 0057.

propafenone-type compounds. Analysis of the molecular interaction fields revealed an unfavorable steric interaction with compounds possessing a bulky substituent (e.g., diphenylmethyl) in close vicinity of the nitrogen atom (Figure 7.6). A favorable steric interaction was observed in the region of the phenyl ring of the phenylpropionyl moiety (i.e., more bulky substituents should improve activity). In the case of electrostatic interactions, both the carbonyl oxygen and the propanolamine nitrogen atoms are important for high activity. Analysis of the CoMSIA fields revealed favorable hydrophobic interactions along the propanolamine chain and in the vicinity of the phenyl ring of the arylpiperazine moiety [15].

This space-directed property of lipophilicity was first demonstrated by Pajeva and Wiese both for a series of phenothiazines and thioxanthenes [20] and for a subset of our propafenone-based library [21]. Addition of HINT-derived hydrophobic fields to

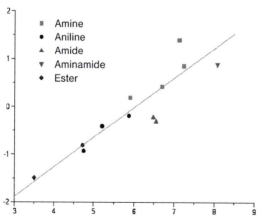

Figure 7.5 Plot of observed versus predicted $\log(1/IC_{50})$ values for a series of propafenone analogues with variations in the vicinity of the nitrogen atom.

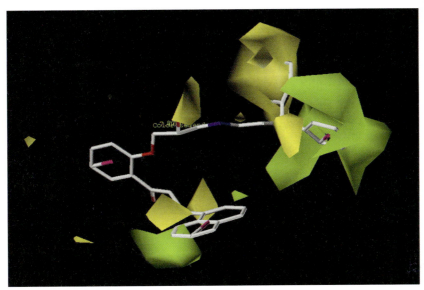

Figure 7.6 Steric favorable (green) and steric unfavorable (yellow) interaction fields for a series of 131 propafenone analogues as revealed in a CoMFA analysis.

the CoMFA input matrix remarkably improved the quality of the 3D-QSAR models. Recently, the same authors extended their studies to a series of 32 tariquidar analogues [22].

One of the major disadvantages of field-based 3D-QSAR methods is the need for a proper alignment of the molecules. This can be overcome by using descriptors derived from molecular interaction fields, such as VolSurf or GRIND. These are alignment free and thus allow the analysis of structurally diverse compound sets. Also, a model based on VolSurf descriptors has been successfully applied to identify new inhibitors in a virtual screening protocol [23]. Although these approaches do not enable one to rationalize the ligand–protein interaction, they might represent versatile tools for prefiltering large combinatorial libraries for compounds with P-gp activity.

Although all these QSAR studies give clear individual pictures and yield predictive models, the attempt to define distinct structural features necessary for high P-gp inhibitory activity leads to rather general features. Strong inhibitors are characterized by high lipophilicity (and/or molar refractivity) and possess at least two H-bond acceptors. Other features, such as H-bond donors, may act as additional interaction points. Furthermore, some steric constraints seem to apply in the vicinity of pharmacophoric structures. This is illustrated in Figure 7.7 taking propafenone derivatives as an example.

This picture has been supported by various pharmacophore modeling studies, particularly the most comprehensive studies by the group of Ekins [24, 25]. They used several different training sets, such as inhibitors of digoxin transport, inhibitors of

type inhibitors of P-gp

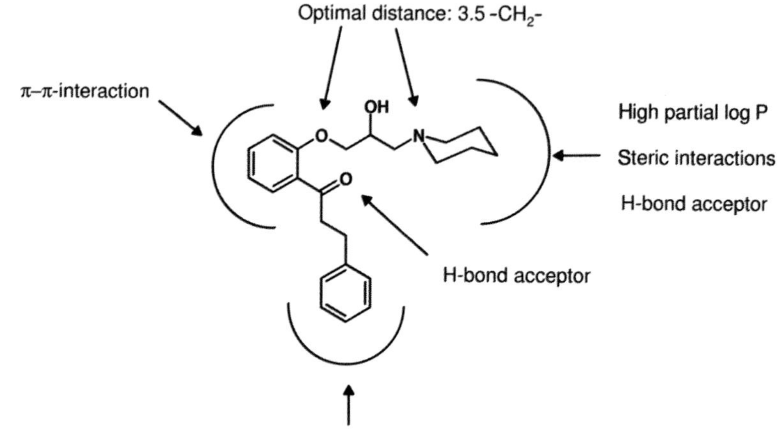

Figure 7.7 Summary of the results of structure–activity relationship studies on propafenone-type inhibitors of P-gp.

vinblastine binding, inhibitors of vinblastine accumulation, and inhibitors of calcein accumulation. Not really surprising, all four models retrieved showed differences both in the number and type of features involved and in the spatial arrangement of these features. A consensus model, which correctly ranked all four data sets, consisted of one H-bond acceptor, one aromatic feature, and two hydrophobic features. This further strengthens the hypothesis that toxins might bind to P-gp at different but overlapping sites. This was stressed also by Garrigues et al. who calculated the intramolecular distribution of polar and hydrophobic surfaces of a set of structurally diverse P-gp ligands and used the respective fields for superposition of the molecules. This led to the identification of two different but partially overlapping binding pharmacophores [26]. Using the genetic algorithm-based similarity program GASP, Pajeva and Wiese derived a general pharmacophore model for P-gp modulators using a diverse set of compounds binding to the verapamil site [27]. The final model consisted of two hydrophobic planes, three H-bond acceptors, and one H-bond donor. Penzotti et al. used pharmacophore sampling considering more than 3 million pharmacophores. The top 100 models, denoted as pharmacophore ensemble, contained 53 four-point pharmacophores, 39 three-point pharmacophores, and 8 two-point pharmacophores [28]. Roughly, half of the models included an H-bond acceptor, an H-bond donor, and hydrophobic areas.

We used a CATALYST model based on propafenone-type inhibitors for an *in silico* approach to identify new inhibitors of P-gp. The training set consisted of 27 propafenone-type inhibitors of daunorubicin efflux, and the model derived included one H-bond acceptor, two aromatic features, one hydrophobic area, and one positively charged group (Figure 7.8). The model was validated with additional 81

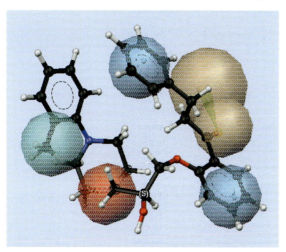

Figure 7.8 Pharmacophore model for propafenone-type P-glycoprotein inhibitors. Blue: aromatic; green: hydrophobic; brown: H-bond acceptor; red: positively ionizable.

compounds from our in-house data set and subsequently used to screen the World Drug Index. After applying an additional shape filter, 32 structurally diverse hits were retrieved. Nine out of these 32 compounds have already been described as P-gp inhibitors [29]. Thus, it is rather likely that the other compounds selected also bind to P-gp.

In recent years, algorithms based on machine learning have also been applied. Wang et al. used Bayesian-regularized neural networks to establish a model for a set of 57 flavonoids using molecular connectivity indices and electrotopological state values as descriptors. The Bayesian-regularized network performed slightly better than an analogous feedforward backpropagation network and far better than PLS [30]. In our studies on the use of artificial neural networks for drug discovery and design, we applied both supervised and unsupervised learning routines to create predictive models for P-gp inhibitors. Thus, a feedforward backpropagation network was trained to predict IC_{50} values of a series of propafenone-type derivatives. The final model obtained used log P values and six Free-Wilson-type indicator variables denoting the presence/absence of distinct substructures as input [16]. *In silico* screening of a small virtual library that consisted of all theoretically enumerable compounds using the 19 Free-Wilson indicator variables retrieved several compounds predicted to be active in the nanomolar range. Synthesis and pharmacological testing of selected derivatives proved the validity of the neural network model [31]. A completely different approach was used for identification of structurally new scaffolds. A set of 131 propafenone-type P-gp inhibitors was projected onto a self-organizing map. After testing several combinations of descriptors and network settings, a set of 30 2D autocorrelation vectors proved best for separating active and inactive compounds. Subsequently, the size of the map was enlarged and the propafenones were merged with the SPECS compound library (134 000 compounds).

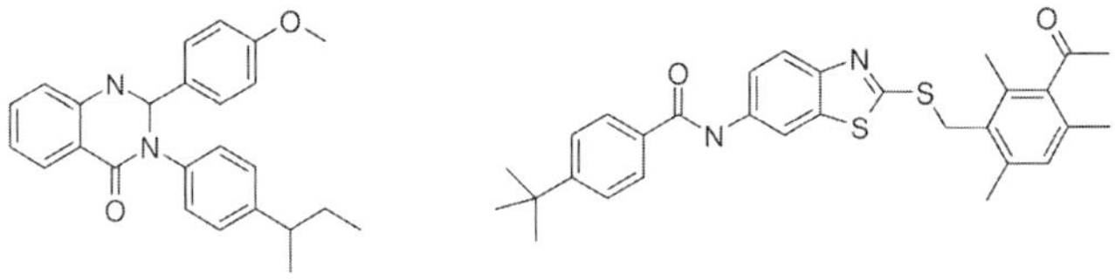

AG-690/11972772   AN-989/14669159

Figure 7.9 New hits for inhibitors of P-glycoprotein identified in an *in silico* screening of the SPECS compound library using a self-organizing map.

If the network indeed places highly active compounds in close vicinity with each other, compounds from the SPECS library colocalizing with highly active propafenone derivatives should also be active. Finally, seven compounds with completely different chemical scaffolds were retrieved and pharmacologically tested [32]. Two out of them showed inhibitory activity with IC_{50} values in the submicromolar range, which definitely renders them new lead compounds for P-gp (Figure 7.9).

7.3
Other ABC Transporter

In addition to ABCB1, inhibitors of the MDR-related proteins ABCC1 (multidrug resistance protein 1 (MRP1)) and ABCC2 (MRP2), the breast cancer resistance protein (BCRP) ABCG2, and the sister of P-gp ABCB11 (SPGP, bile salt export pump (BSEP)) have been published [33]. Further ABC proteins capable of transporting drugs comprise ABCC3 (MRP3), ABCC4 (MRP4), ABCC5 (MRP5), and ABCA2 [34]. These proteins are of increasing interest as targets and the above-mentioned computational methods might also be applied to these transporters both for identification of inhibitors and for selectivity profiling. However, only few *in vitro* data are available for these transporters, and QSAR studies with adequate validation sets are therefore rather rare.

7.3.1
ABCG2 (Breast Cancer Resistance Protein, MXR)

Analogous to ABCB1, ABCG2 also has a broad, partly overlapping, and diverse substrate specificity. It mainly transports mitoxantrone, methotrexate, camptothecins (topotecan, irinotecan), anthracyclines, etoposide, and flavonoids [35, 36]. Zhang *et al.* selected a panel of 25 flavonoids covering five different structural subclasses for QSAR studies. Using calculated log *P* values, molecular connectivity indices, Kappa shape indices, electrotopological state indices, information indices, subgraph count indices, molecular polarizability, weight and volume as input vector, and multiple linear regression analysis coupled with a genetic algorithm, a model with good predictive power ($q^2 = 0.78$) could be obtained. The most important descriptors were log P, count of all =C− groups, and the moment of the displacement between the center of mass and the center of dipole along the inertial Y-axis. Boumendjel et al. linked piperazines and phenylalkylamines to benzopyranones in order to obtain new inhibitors of ABCG2 [37]. The most active compounds had several structural features in common with the highly active ABCG2 inhibitors imatinib (STI 571) and the natural product fumitremorgin C (FTC), such as an alkylpiperazine moiety or methoxyphenylalkylamino groups. FTC also served as a starting point for the synthesis of a series of 42 structural analogous indolyl diketopiperazines. SAR studies demonstrated that lipophilic side chains in position 3 are important for high inhibition activity [38]. For a series of propafenone analogues, we could also demonstrate that for ABCG2 inhibitors, lipophilicity is a highly predictive descriptor. Both QSAR studies using a set of 10 ADME-related descriptors and qualitative pharmacophore feature modeling revealed that hydrophobicity, number of rotatable bonds, and number of H-bond acceptors are key features both for activity and for selectivity toward ABCB1 [39]. Results further indicate that for the class of propafenones, ABCG2 is more tolerant to structural modification than ABCB1. Selectivity is therefore mainly determined by the distinct QSAR pattern with respect to ABCB1 rather than a specific interaction with ABCG2. The main difference between selective ABCB1 inhibitors and rather selective ABCG2 inhibitors is the importance of the nitrogen atom as an H-bond acceptor for ABCB1 but not for ABCG2 (Figure 7.10a and b).

7.3.2
ABCC1 and ABCC2 (Multidrug Resistance-Related Proteins 1 and 2)

Multidrug resistance protein 1 (ABCC1) confers resistance toward vinca alkaloids, anthracyclines, epipodophyllotoxins, mitoxanthrone, and methotrexate, but not toward taxanes and bisantrene [40]. As for ABCB1 and ABCG2, a lot of structurally and functionally diverse inhibitors for ABCC1 have been identified and are summarized in a recent review [41]. These comprise verapamil, flavonoids, raloxifene, isoxazoles, quinazolinones, quinolines, pyrrolopyrimidines, and peptides. For the group of flavonoids, QSAR studies for both ABCC1 and ABCC2 have been performed. Results demonstrate three structural characteristics to be of major importance for ABCC1 inhibition: the total number of methoxy groups, the number of OH groups, and the dihedral angle between ring B and ring C. In parallel, the ABCC2 inhibitory potency was also investigated. For flavonoid-type inhibitors of ABCC2, the presence of a flavanol B-ring pyrogallol group seems to be an important structural characteristic. Only robinetin and myricetin were able to inhibit the activity by more than 50%. All other flavonoids did not reach 50% ABCC2 inhibition at concentrations up to 50 µm [42]. For a series of methotrexate analogues, octanol/water partition coefficient, hydrophobicity, and negative charge were identified as important features for high affinity to rat ABCC2. Furthermore, the addition of a benzoyl ornithine group at a distance of 9.3 Å from the negatively ionizable center gave rise to a 40-fold increase in affinity. These findings were supported by a pharmacophore model that consisted of two hydrophobic features, a negative ionizable feature, and two aromatic rings [43].

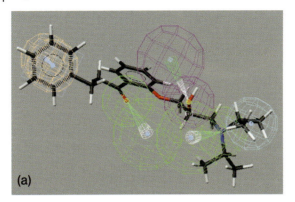

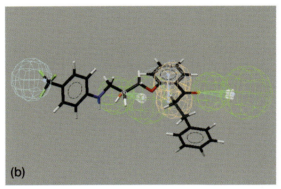

Figure 7.10 Pharmacophore model for propafenone analogues with selectivity toward (a) ABCB1 and (b) ABCG2.

Hirono and coworkers used 3D-QSAR-based receptor mapping of a series of 16 structurally diverse ABCC2 ligands in order to identify key functional groups for ligand binding. Molecular dynamics-based generation of conformers, superposition using the SUPERPOSE program [44], and subsequent CoMFA analysis gave a statistically significant model with a predictive power of $q^2 = 0.59$. This model consisted of two hydrophobic and two electrostatically positive sites as primary binding sites [45].

7.3.3
ABCB11 (Bile Salt Export Pump)

ABCB11 mainly eliminates bile salts from liver cells and thereby may be involved in several liver diseases. Hirano et al. used plasma vesicles prepared from insect cells to assess the ABCB11 inhibitory potency of a set of 40 structurally diverse compounds. The authors identified a set of chemical fragmentation codes generated with Markush TOPFRAG that are statistically significant and linked to the ABCB11 interaction [46]. For example, these comprise the descriptors M132 (ring-linking group containing one C atom), H181 (one amine bonded to aliphatic C), and ESTR (one ester group bonded to heterocyclic C via C=O).

7.4
Novel Methods

One of the problems when applying classical QSAR techniques is the right choice of the method and the descriptor combination. In principle, two general approaches might be undertaken to overcome this issue, which normally is pursued on a trial and error basis. One is to automatically combine feature selection algorithms with classification and regression tools and the other is to combinatorially explore the descriptor/method space. The latter was recently introduced by the group of Tropsha (combinatorial QSAR) [47].

We recently introduced the concept of using similarity values as independent variables in QSAR equations. Within this similarity-based SAR (SIBAR), similarity values between training set compounds and a set of reference compounds are calculated and subsequently used as molecular descriptors. The approach for calculating the SIBAR descriptors is outlined in Figure 7.11.

1. Selection of a reference compound set on the basis of maximum diversity and/or active/inactive.
2. Calculation of a set of descriptors for both the training set and the reference set.
3. Calculation of similarity values for each compound of the training set to each compound of the reference set; this leads to a given number of similarity values (equal to the number of reference compounds used) for each compound of the training set, which are assigned as SIBAR descriptors.
4. MLR, PLS, or SVM analysis of the training set data matrix.
5. Validation of the model using cross-validation procedures and external test sets.

So far the approach was successfully applied for a set of 131 propafenone-type inhibitors of P-gp. One hundred compounds were used in the training set. The 20 most diverse compounds of the SPECS library were used as reference set. SIBAR descriptors were calculated by using 39 physicochemical and topological descriptors as implemented in TSAR. Subsequent PLS analysis led to the models with a predictive power that is significantly higher than those obtained when using the descriptors alone [48]. Recent results obtained by Zdrazil further showed that among a panel of four different reference sets (A: highly diverse, drug-like compounds; B:

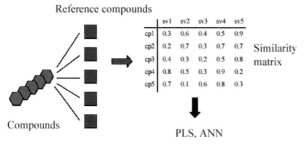

Figure 7.11 Workflow of the SIBAR approach.

P-gp inhibitors from our in-house library; C: P-gp substrates from the literature; D: chemicals), models derived by using reference sets related to the target P-gp (B, C) gave highest internal (leave-one-out cross-validation) and external (test set) predictivity [49].

7.5
Structural Basis for Polyspecificity

However, when comparing the 3D pharmacophore models, almost no overlap can be identified. Thus, a general applicable pharmacophoric pattern unifying the currently available hypotheses is still missing. Considering data from the very recently published X-ray structures from mouse P-glycoprotein in the apo form and with 2 enantiomeric inhibitors bound [50], this might even be an impossible task. The structure of the apo form of the protein shows an internal extremely large central cavity of 6000 Å^3. The two structures with cyclic peptide inhibitors bound reveal that binding is mainly driven by hydrophobic, aromatic, and van der Waals interactions and utilizes distinct, in part overlapping subsets of P-gp residues for the two enantiomeric ligands. Although one has to bear in mind that the resolution of this first X-ray structure of a mammalian ABC-transporter is quite low (3.8 Å), theses structures for the first time allow insights into ligand-protein interactions at P-gp.

7.6
Conclusions and Outlook

The success of traditional QSAR methods, such as Hansch analysis and CoMFA, heavily depends on the basic assumption that all compounds used bind to the same site and in the same mode to the target protein. In the case of polyspecific drug transport pumps such as those described in this chapter, there is experimental evidence that drug binding occurs at the interface of the two transmembrane domains and therefore the binding cavity is rather large accommodating simultaneously up to three ligands in the case of some transporters. Thus, conventional QSAR methods fail to decipher clear and distinct ligand–protein interaction patterns when structurally diverse compound sets are used. Success stories published so far mainly rely on VolSurf/GRIND descriptors, pharmacophore models, and machine learning methods. The last two approaches were also successfully applied for *in silico* screening of medium to large compound libraries in order to identify structurally new molecular scaffolds as ligands for P-gp. New approaches such as SVM and similarity-based descriptors may pave the way for the establishment of rapid *in silico* filters that are routinely applied in the early drug discovery phase. This will be of special importance in the field of predicting substrate properties of ABC transporters, as these are increasingly considered antitargets in the pharmaceutical industry.

P-glycoprotein, the paradigm transporter for the whole class of drug efflux pumps, has been known for over 30 years. However, both the molecular basis of the drug-protein interaction and the mechanism of transport still remain rather elusive. The recent X-ray structure of mouse P-gp as well as those of analogous bacterial transporter together with combined photoaffinity labeling/protein homology modeling approaches already have started to shed some light on the molecular basis of polyspecificity. Further in silico and in vitro studies and additional X-ray structures are expected soon and will help solve this amazingly complex biological puzzle.

References

1 Juliano, R.L. and Ling, V. (1976) A surface glycoprotein modulating drug permeability in Chinese hamster ovary cell mutants. *Biochimica et Biophysica Acta*, **455**, 152–162.

2 Gottesman, M.M. and Pastan, I. (1993) Biochemistry of multidrug resistance mediated by multidrug transporter. *Annual Review of Biochemistry*, **62**, 385–427.

3 Tsuruo, T., Iida, H., Tsukagoshi, S., and Sakurai, Y. (1981) Overcoming of vincristine resistance in P388 leukemia *in vivo* and *in vitro* through enhanced cytotoxicity on vincristine and vinblastine by verapamil. *Cancer Research*, **41**, 1967–1972.

4 Leonard, G.D., Polgar, O., and Bates, S.E. (2002) ABC transporters and inhibitors: new targets, new agents. *Current Opinion in Investigational Drugs*, **3**, 1652–1659.

5 Szakacs, G., Paterson, J.K., Ludwig, J.A., Booth-Genthe, C., and Gottesman, M.M. (2006) Targeting multidrug resistance in cancer. *Nature Reviews. Drug Discovery*, **5**, 219–234.

6 Raub, T.J. (2006) P-glycoprotein recognition of substrates and circumvention through rational drug design. *Molecular Pharmaceutics*, **1**, 3–25.

7 Pleban, K. and Ecker, G.F. (2005) Inhibitors of P-glycoprotein – lead identification and optimisation. *Mini Reviews in Medicinal Chemistry*, **5**, 153–163.

8 Wiese, M. and Pajeva, I.K. (2001) Structure–activity relationships of multidrug resistance reversers. *Current Medicinal Chemistry*, **8**, 685–713.

9 Tmej, C., Chiba, P., Huber, M., Richter, E., Hitzler, M., Schaper, K.J., and Ecker, G. (1998) A combined Hansch/Free-Wilson approach as predictive tool in QSAR studies on propafenone-type modulators of multidrug resistance. *Archiv der Pharmazie*, **331**, 233–240.

10 Kaiser, D., Smiesko, M., Kopp, S., Chiba, P., and Ecker, G.F. (2005) Interaction field based and hologram based QSAR analysis of propafenone-type modulators of multidrug resistance. *Medicinal Chemistry*, **1**, 431–444.

11 Tmej, C., Chiba, P., Schaper, K.J., Ecker, G., and Fleischhacker, W. (1999) Artificial neural networks as versatile tools for prediction of MDR-modulatory activity. *Advances in Experimental Medicine and Biology*, **457**, 95–105.

12 Klein, C., Kaiser, D., Kopp, S., Chiba, P., and Ecker, G.F. (2002) Similarity based SAR (SIBAR) as tool for early ADME profiling. *Journal of Computer-Aided Molecular Design*, **16**, 785–793.

13 Ecker, G., Chiba, P., Hitzler, M., Schmid, D., Visser, K., Cordes, H.P., Csöllei, J., Seydel, J.K., and Schaper, K.J. (1996) Structure–activity relationship studies on benzofurane analogs of propafenone-type modulators of tumor cell multidrug resistance. *Journal of Medicinal Chemistry*, **39**, 4767–4774.

14 Chiba, P., Ecker, G., Schmid, D., Drach, J., Tell, B., Goldenberg, S., and Gekeler, V. (1996) Structural requirements for activity of propafenone type modulators in PGP-mediated multidrug resistance. *Molecular Pharmacology*, **49**, 1122–1130.

15 Chiba, P., Hitzler, M., Richter, E., Huber, M., Tmej, C., and Ecker, G. (1997) Studies on propafenone-type modulators of multidrug resistance. III. Variations on the nitrogen. *Quantitative Structure–Activity Relationships*, **16**, 361–366.

16 Pleban, K., Hoffer, C., Kopp, S., Peer, M., Chiba, P., and Ecker, G.F. (2004) Intramolecular distribution of hydrophobicity influences pharmacological activity of propafenone-type MDR modulators. *Archiv der Pharmazie*, **337**, 328–334.

17 Koenig, G., Chiba, P., and Ecker, G.F. (2008) Hydrophobic moments as physicochemical descriptors in structure–activity relationship studies of P-glycoprotein inhibitors. *Monatsh Chemie*, **139**, 401–405.

18 Ecker, G., Huber, M., Schmid, D., and Chiba, P. (1999) The importance of a nitrogen atom in modulators of multidrug resistance. *Molecular Pharmacology*, **56**, 791–796.

19 Seelig, A. and Gatlik-Landwojtowicz, E. (2005) Inhibitors of multidrug efflux transporters: their membrane and protein interactions. *Mini Reviews in Medicinal Chemistry*, **5**, 135–151.

20 Pajeva, I.K. and Wiese, M. (1998) Molecular modelling of phenothiazines and related drugs as multidrug resistance modifiers: a comparative molecular field analysis study. *Journal of Medicinal Chemistry*, **41**, 1815–1826.

21 Pajeva, I.K. and Wiese, M. (1998) A comparative molecular field analysis of propafenone-type modulators of cancer multidrug resistance. *Quantitative Structure–Activity Relationships*, **17**, 301–312.

22 Globisch, C., Pajeva, I.K., and Wiese, M. (2006) Structure–activity relationships of a series of tariquidar analogs as multidrug resistance modulators. *Bioorganic and Medicinal Chemistry*, **14**, 1588–1598.

23 Kaiser, D., Bohl, M., Kopp, S., Chiba, P., and Ecker, G.F. (2004) A Volsurf model of propafenone-type inhibitors of P-glycoprotein for *in silico* screening of the Leadquest compound library. *Drugs Future*, **29** (Suppl A), 119.

24 Ekins, S., Kim, R.B., Leake, B.F., Dantzig, A.H., Schuetz, E., Lan, L.B. *et al.* (2002) Application of three-dimensional quantitative structure–activity relationships of P-glycoprotein inhibitors and substrates. *Molecular Pharmacology*, **61**, 974–981.

25 Ekins, S., Kim, R.B., Leake, B.F., Dantzig, A.H., Schuetz, E.G., Lan, L. *et al.* (2002) Three-dimensional quantitative structure–activity relationships of inhibitors of P-glycoprotein. *Molecular Pharmacology*, **61**, 964–973.

26 Garrigues, A., Loiseau, N., Delaforge, M., Ferte, J., Garrigos, M., Andre, F., and Orlowski, S. (2002) Characterization of two pharmacophores on the multidrug transporter P-glycoprotein. *Molecular Pharmacology*, **62**, 1288–1298.

27 Pajeva, I.K. and Wiese, M. (2002) Pharmacophore model of drugs involved in P-glycoprotein multidrug resistance: explanation of structural variety (hypothesis). *Journal of Medicinal Chemistry*, **45**, 5671–5686.

28 Penzotti, J.E., Lamb, M.L., Evensen, E., and Grootenhuis, P.D. (2002) A computational ensemble pharmacophore model for identifying substrates of P-glycoprotein. *Journal of Medicinal Chemistry*, **45**, 1737–1740.

29 Langer, T., Eder, M., Hoffmann, R.D., Chiba, P., and Ecker, G.F. (2004) Lead identification for modulators of multidrug resistance based on *in silico* screening with a pharmacophoric feature model. *Archiv der Pharmazie*, **337**, 317–327.

30 Wang, Y.H., Li, Y., Yang, S.L., and Yang, L. (2005) An *in silico* approach for screening flavonoids as P-glycoprotein inhibitors

based on a Bayesian-regularized neural network. *Journal of Computer-Aided Molecular Design*, **19**, 137–147.

31 Kaiser, D., Tmej, C., Chiba, P., Schaper, K.J., and Ecker, G. (2000) Artificial neural networks in drug design. II. Influence of learning rate and momentum factor on the predictive ability. *Scientia Pharmaceutica*, **68**, 57–64.

32 Kaiser, D., Terfloth, L., Kopp, S., Chiba, P., Gasteiger, J., and Ecker, G.F. (2006) Artificial neural networks for identification of new modulators of multidrug resistance, in *QSAR & Molecular Modelling in Rational Design of Bioactive Molecules* (eds E. Sener and I. Yalcin), pp. 274–277.

33 Chiba, P. and Ecker, G.F. (2004) Inhibitors of ABC-type drug efflux pumps – an overview on the actual patent situation. *Expert Opinion on Therapeutic Patents*, **14**, 499–508.

34 Gottesman, M.M., Fojo, T., and Bates, S.E. (2002) Multidrug resistance in cancer: role of ATP-dependent transporters. *Nature Reviews. Cancer*, **2**, 48–58.

35 van Herwaarden, A.E. and Schinkel, A.H. (2006) The function of breast cancer resistance protein in epithelial barriers, stem cells and milk secretion of drugs and xenotoxins. *Trends in Pharmacological Sciences*, **27**, 10–16.

36 Mao, Q. and Unadkat, J.D. (2005) Role of breast cancer resistance protein (ABCG2) in drug transport. *The AAPS Journal*, **7**, E118–E133.

37 Boumendjel, A., Nicolle, E., Moraux, T., Gerby, B., Blanc, M., Ronot, X., and Boutonnat, J. (2005) Piperazinobenzopyranones and phenylalkylaminobenzopyranones: potent inhibitors of breast cancer resistance protein (ABCG2). *Journal of Medicinal Chemistry*, **48**, 7275–7281.

38 van Loevezijn, A., Allen, J.D., Schinkel, A.H., and Koomen, G.J. (2001) Inhibition of BCRP-mediated drug efflux by fumitremorgin-type indolyl diketopiperazines. *Bioorganic & Medicinal Chemistry Letters*, **11**, 29–32.

39 Cramer, J., Kopp, S., Bates, S.E., Chiba, P., and Ecker, G.F. (2007) Multispecificity of drug transporters: probing inhibitor selectivity for the human drug efflux transporters ABCB1 and ABCG2. *ChemMedChem*, **2**, 1783–1788.

40 Schinkel, A.H. and Jonker, J.W. (2003) Mammalian drug efflux transporters of the ATP binding cassette (ABC) family: an overview. *Advanced Drug Delivery Reviews*, **55**, 3–29.

41 Boumendjel, A., Baubichon-Cortay, H., Trompier, D., Perrotton, T., and Di Pietro, A. (2005) Anticancer multidrug resistance mediated by MRP1: recent advances in the discovery of reversal agents. *Medicinal Research Reviews*, **25**, 453–472.

42 van Zanden, J.J., Wortelboer, H.M., Bijlsma, S., Punt, A., Usta, M., van Bladeren, P.J. *et al.* (2005) Quantitative structure–activity relationship studies on the flavonoid mediated inhibition of multidrug resistance proteins 1 and 2. *Biochemical Pharmacology*, **69**, 699–708.

43 Ng, C., Xiao, Y.D., Lum, B.L., and Han, Y.H. (2005) Quantitative structure–activity relationships of methotrexate and methotrexate analogues transported by the rat multispecific resistance-associated protein 2 (rMrp2). *European Journal of Pharmaceutical Sciences*, **26**, 405–413.

44 Iwase, K. and Hirono, S. (1999) Estimation of active conformation of drugs by a new molecular superposing procedure. *Journal of Computer-Aided Molecular Design*, **13**, 305–315.

45 Hirono, S., Nakagome, I., Imai, R., Maeda, K., Kusuhara, H., and Sugiyama, Y. (2005) Estimation of the three-dimensional pharmacophore of ligands for rat multidrug-resistance associated protein 2 using ligand-based drug design techniques. *Pharmaceutical Research*, **22**, 260–269.

46 Hirano, H., Kurata, A., Onishi, Y., Sakurai, A., Saito, H., Nakagawa, H. *et al.* (2006) High-speed screening and QSAR analysis of human ATP-binding cassette transporter ABCB11 (bile salt export pump) to predict drug-induced

intrahepatic cholestasis. *Molecular Pharmaceutics*, **3**, 252–265.

47 Kovatcheva, A., Golbraikh, A., Oloff, S., Xiao, Y.D., Zheng, W., Wolschann, P. et al. (2004) A combinatorial QSAR of ambergris fragrance compounds. *Journal of Chemical Information and Computer Sciences*, **44**, 582–595.

48 Klein, C., Kaiser, D., Kopp, S., Chiba, P., and Ecker, G.F. (2002) Similarity based SAR (SIBAR) as tool for early ADME profiling. *Journal of Computer-Aided Molecular Design*, **16**, 785–793.

49 Zdrazil, B., Kaiser, D., Kopp, S., Chiba, P., and Ecker, G.F. (2007) Similarity-based descriptors (SIBAR) as tool for QSAR-studies on P-glycoprotein inhibitors: influence of the reference set. *QSAR & Combinatorial Science*, **26**, 669–678.

50 Aller, S.G., Yu, J., Ward, A., Wenig, Y., Chittaboina, S., Zhuo, R., Harrell, P.M., Trinh, Y.T., Zhang, Q., Urbatsch, I.L., Chang, G. (2009) Structure of P-glycoprotein reveals a molecular basis for poly-specific drug binding. *Science*, **323**, 1718–1722.

8
Drug Transporter Pharmacophores

Sean Ekins

8.1
Introduction

The human genome contains nearly 900 genes that encode transporters, of which over 300 are intracellular transporters [1] responsible for transporting a wide range of molecules across the membrane [2]. Further classification of these transporters into families such as the solute carrier class (SLC) [3] and ATP binding cassette (ABC) family [4, 5] is possible. Transporters play a major role in clinical pharmacology as their adequate bioavailability determines the successful oral delivery of many therapeutics. Membrane transporter proteins are associated with drug absorption (uptake), tissue distribution (efflux and uptake), metabolism (hepatic efflux and uptake), and elimination (renal, biliary transporters, and breast milk efflux and uptake) [6, 7].

Hence, transporter proteins are being increasingly targeted to improve the oral bioavailability. Study of the structure, function, and regulation of such key transporters can enable the further development of drug molecules with optimal properties to improve bioavailability. A rational prediction of molecule interactions with transporters would therefore be useful to focus testing *in vitro* and *in vivo* and to enable prioritization of the most important molecules. As they are membrane-bound proteins, there is no high-resolution X-ray structure of a mammalian transporter available, and our insight into structure–transport relationships, therefore, can be derived only from computational and experimental methods. Such methods have been quite widely used, with a recent review [8] citing 27 homology models for 19 transporters and 38 predictive pharmacophore models for 15 drug transporters.

What is a pharmacophore? Paul Ehrlich defined a "pharmacophore" as a "molecular framework that carries (phoros) the essential features responsible for a drug's (pharmacon) biological activity" [9]. A pharmacophore model can be derived either from the protein binding site or by using a series of active molecules. To our knowledge, there are two key books on pharmacophores that are of general interest to the reader that contain further definitions and uses [10, 11]. A ligand-based transporter pharmacophore uses a series of known substrates or inhibitors to identify the key molecular features involved in ligand–transporter interactions such as recognition, binding, and transport. A recent review has described the various pharmacophore methods available and their use for database screening [12]. For the purpose of this chapter, a pharmacophore is the three-dimensional arrangement of minimum molecular features necessary for bioactivity/binding to the transporter [13, 14], and as an approach, it has already been widely applied to database screening for therapeutic targets [15–20].

Three-dimensional quantitative structure–activity relationships (3D-QSAR) and pharmacophore methods have become widely accepted methods for assessing the drug–transporter interactions [21]. The pharmacophore modeling approach is used when the data sets contain structurally diverse and conformationally flexible compounds [8]. Three popular pharmacophore modeling tools are DIStance COmparisons (DISCO) [14], Genetic Algorithm Similarity Program (GASP) [22], and Catalyst [8]. The Catalyst program used widely by our laboratory has two distinct modules, one using the common chemical features of a few active drug molecules (HIPHOP) [23] and the other is based on a series of molecules with varying structural activity and features [24]. This method and others have been used to model transporters by our group [25–31] and several other laboratories (Tables 8.1 and 8.2). We will describe the types of transporter pharmacophores that have been published to date while other QSAR models and protein-based models are detailed elsewhere [8, 12, 21, 32] and in other chapters in this book.

Table 8.1 SLC transporter pharmacophores.

SLC transporter	Acronym and gene	Pharmacophore references
Oligopeptide transporter 1	PEPT1 (SLC15A1)	[25, 73, 74]
Oligopeptide transporter 2	PEPT2 (SLC15A2)	[33, 34]
Organic anion transporter polypeptides	OATP (SLC21A family)	[30]
Organic anion transporters	OAT (SLC22A family)	[71]
Organic cation transporters	OCT (SLC22A family)	[29, 31, 75]
Sodium taurocholate cotransporting polypeptide	NTCP (SLC10A family)	[76]
Apical sodium-dependent bile acid transporter	ASBT (SLC10A2)	[77, 78]
Concentrative nucleoside transporter	CNT (SLC28A family)	[79–81]
Equilibrative nucleoside transporter	ENT (SLC29A family)	[79]

Table 8.2 ABC transporter pharmacophores.

ABC transporter	Acronym and gene	Pharmacophore references
P-glycoprotein	P-gp (MDR1, ABCB1)	[26–28, 37–39, 41, 82–88]
Breast cancer resistance protein	BCRP (ABCG2)	[12, 68, 86, 89]
Multidrug resistance protein superfamily	MRP1–9 (ABCC superfamily)	[53, 54]
Bile salt export pump	BSEP (ABCB11)	None[a]

[a]A recent study developed a QSAR model for this transporter [90].

8.2
Database Searching with Transporter Pharmacophores

Pharmacophores have been used as an *in silico* screening approach to select molecules for *in vitro* testing. This may assist in faster discovery of substrates and inhibitors for transporters by avoiding random or high-throughput screening of a large number of molecules *in vitro*. For example, the human peptide transporter (PEPT1, Table 8.1), a clinically relevant transporter of a broad range of substrates, is also an attractive prodrug target because of its high capacity and relatively broad substrate specificity. We have developed a HIPHOP pharmacophore model for human peptide transporter PEPT1 using three substrates Gly-Sar (dipeptide), bestatin (peptidomimetic), and enalapril (ACE inhibitor) [25]. The pharmacophore consisted of two hydrophobic features, a hydrogen bond donor, a hydrogen bond acceptor, and a negative ionizable group. The pharmacophore was used to search a database of over 8000 "drug-like" molecules in an attempt to identify other hPEPT1 ligands. One hundred forty-five virtual hits mapped to the pharmacophore features. Seven of the best scoring molecules with drug-like properties (i.e., MW < 500) were selected and purchased for *in vitro* testing to ascertain the predictability of the pharmacophore model. Two FDA-approved drug molecules, fluvastatin (antihyperlipidemic) and repaglinide (antidiabetic), and one component of the sugar substitute and pharmaceutical component, aspartame, were mapped to the pharmacophore features and were verified *in vitro* to be hPEPT1 inhibitors. This pharmacophore has also been used to assess the potential affinity of selected bacterial dipeptides in a recent study [33]. We observed that γ-iE-DAP scored highest mapping well to the pharmacophore features. This led to the development of a second pharmacophore using three high-affinity PEPT2 molecules [34], which contained two hydrogen bond acceptors, two hydrogen bond donors, and one hydrophobic feature, and γ-iE-DAP fitted well to all these features [33].

P-glycoprotein (P-gp, Table 8.2) plays an important role in determining drug distribution of many important drug candidates. P-gp substrates generally have reduced oral drug absorption and enhanced renal and biliary excretion. Limiting the exposure of xenobiotics to P-gp at the blood–brain barrier and placental barrier may also be important considerations. The rapid identification of P-gp substrates or inhibitors has resulted in many experimentally derived *in vitro* data sets that in turn has enabled extensive computational modeling [27, 28, 35, 36]. Several P-gp pharmacophores have been used for database screening and identification of potential substrates and inhibitors. Rebitzer and colleagues were first to develop a propafenone derivative MDR modulator-based pharmacophore model that was used to screen a database of molecules, namely, the Derwent World Drug Index. They selected from among the returned 28 hits 9 that were previously described MDR modulators [37], representing an initial validation of this approach. A more detailed discussion of this approach was later published [38]. We have recently compared different P-gp pharmacophore models for substrates and inhibitors, using them to search diverse molecule databases to identify new molecules and verify their potential to identify known P-gp inhibitors and substrates [26]. Using two-inhibitor and one-substrate pharmacophore model, we analyzed their predictive and enrichment capabilities with a database of 189 known P-gp substrates and nonsubstrates [39]. Following this quantitative validation, the pharmacophore models were applied to screen a database of over 500 commonly prescribed FDA-approved drugs and retrieved 7 molecules that had not been previously documented with P-gp affinities. These molecules were selected for purchase and *in vitro* testing using the (^{3}H)-digoxin transport assay and ATPase activation assay. All seven drugs were either micromolar inhibitors or substrates of P-gp, which also validated our approach [26]. The P-gp substrate pharmacophore was also used recently to map thiazolidinone derivatives that target the drug-resistant lung cancer cell line H460$_{taxR}$ that expresses large amounts of P-gp [40]. Nine out of 13 molecules (70%) were found to map to the P-gp substrate pharmacophore that were also found to be inactive in the cell lines likely due to P-gp-mediated export. In contrast, only 4 out of 11 molecules (36%) active against the cell line were predicted to map to the P-gp substrate pharmacophore. This indicates that the pharmacophore could discriminate the active thiazolidinone derivatives from nonactives based on whether they were likely to be P-gp substrates or not with a reasonable degree of accuracy. Such an approach may be useful for screening molecules in other areas where interaction with P-gp may be important for limiting bioactivity.

There have been several recently published P-gp pharmacophores. For example, Cianchetta *et al.* used GRID-alignment-independent descriptors (GRIND) for 129 substrates with a range of Calcein-AM assay data [41], to develop a pharmacophore that had some overlap in the features and distances with previously published models. A second GRIND 3D pharmacophore, containing multiple hydrophobic areas and two hydrogen bond acceptors, was also proposed to be similar to other published models by a second group [35]. Recently, Globisch *et al.* evaluated a series of 32 anthranilamide tariquidar analogues as P-gp inhibitors using the 3D-QSAR methods, CoMFA and CoMSIA [42]. Hydrogen bond acceptor, steric fields, and hydrophobic fields were found to be most important. A second group looked at 49 anthranilamide analogues using the same methods and had similar findings [43]. CoMFA and CoMSIA analyses had also been used with a series of 32 natural and synthetic coumarins to indicate the importance of the phenyl at position C4 as well as the α-(hydroisopropyl)dihydrofuran that possess a favorable electrostatic and steric volume [44]. To date there have been no additional examples of database searching for P-gp using pharmacophores, although QSAR methods have been used successfully [45, 46].

Another important member of the ABC transporter family is the multidrug resistance protein 1 (MRP1, Table 8.2) that transports a broad spectrum of substrates, ranging from anticancer drugs such as vincristine, mitoxantrone, and daunorubicin to organic anionic substrates such as the conjugates of glutathione, glucuronide, and sulfates [47, 48]. Due to its increasing significance in MDR, there is a major interest in the discovery of MRP1 inhibitors as MDR reversal agents [49]. We have previously selected five diverse and highly potent MRP1 inhibitors developed as MDR reversal agents (LY329146, LY402913, dehydrosilybin, indolopyrimidine, and phenoxymethyl quinoxalinone II) from a recent review [49] and generated a Catalyst HIPHOP model in the same manner as that previously described for P-gp and hPEPT1. The pharmacophore model contained three ring aromatic and three HBA features and was applied to screen the database of over 500 clinically used drugs [12]. Eight hits were retrieved including candesartan, eprosartan, fexofenadine, losartan, sulfasalazine, telmisartan, vancomycin, and zafirlukast. Among these, three drugs have been previously documented to have MRP1 affinity (losartan [50], sulfasalazine [51], and zafirlukast [52]). In addition, recent work [50] indicates that tetrazole compounds are particularly susceptible to P-gp- and MRP1-mediated efflux. This suggests that the tetrazole-containing candesartan could also inhibit MRP1. Further experimental verification of this preliminary model is certainly required to test the remaining selected compounds for activity. Interestingly, two groups have recently derived pharmacophores and QSAR models for the rat Mrp2 transporter. One of the groups used a narrow series of 25 methotrexate analogues and generated a Catalyst pharmacophore [53] containing three hydrophobic features, a negative ionizable, and a ring aromatic feature. An earlier study had also used 16 diverse molecules with SUPERPOSE and CoMFA and suggested a pharmacophore with two hydrophobic and two negative charged or hydrogen bond acceptor features in the compact pharmacophore [54]. To our knowledge, neither group evaluated their pharmacophores with large test sets or used them for database screening.

An additional ABC transporter is the breast cancer resistance protein (BCRP, Table 8.2) that is expressed on the apical side in placenta, breast, liver hepatocytes, and endothelium [55, 56]. BCRP expression confers resistance to several anticancer drugs such as mitoxantrone and anthracyclines [57–59], camptothecin-derived topoisomerase I inhibitors [60], methotrexate [61], and flavopiridol [62]. This half transporter, unlike P-gp and MRP1, has one ATP binding site and consumes one ATP molecule per substrate molecule transferred [63]. BCRP transports a broad range of drugs and has prompted the design of many high-affinity inhibitors [64]. This information has been used to generate a preliminary HIPHOP pharmacophore model using four BCRP inhibitors (GF120918, Ko143, nelfinavir, and nicardipine). This resulted in a pharmacophore containing three HBA and three hydrophobic features, which was then used to search the database of over 500 commercially available drugs. Among the 37 retrieved molecules, 6 were previously identified BCRP ligands, namely, digoxin [65], docetaxel [66], indinavir, lopinavir, ritonavir, saquinavir [67], and the training set compound nicardipine [12]. The remaining molecules represent a set that will be valuable for further testing *in vitro* to verify the role of BCRP and the utility of this model. We had earlier used a set of seven topoisomerase inhibitors [64] to generate a HIPHOP model for BCRP [68] that contained three hydrophobic features, two hydrogen bond donors, and a hydrogen bond acceptor feature. Our second pharmacophore for BCRP indicates that for this different set of structurally similar molecules, there may be an overlapping pharmacophore (three hydrophobic features and a hydrogen bond acceptor in common). A recent study has described a Catalyst HIPHOP pharmacophore for 28 BCRP inhibitors with far fewer molecular features, namely, two hydrophobic and one hydrogen bond acceptor feature. This did not appear to be used to screen a database of molecules or evaluated with a large test set of known BCRP inhibitors. It is likely based on our previous studies with BCRP inhibitors that the pharmacophores are likely to be large with many hydrophobic and hydrogen bonding features. It is also possible that, in an analogy to P-gp, there may be multiple overlapping pharmacophores for BCRP [27, 28].

8.3
Summary

The pharmacophore methods described above illustrate how they can be used to discover new inhibitors or substrates for transporters by first searching a database and then generating *in vitro* data (Figure 8.1). This is a reversal of the usual approach where computational models are often generated only after production of *in vitro* data and are rarely evaluated with new molecules or are used to prioritize what should be tested. Such an approach could be adopted in the drug discovery process to prioritize molecule testing for transporters in general and in drug design to avoid or focus on a

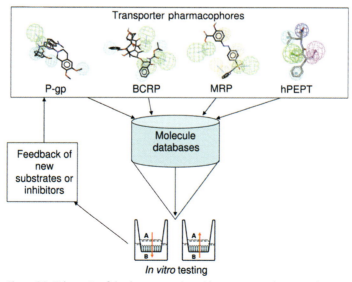

Figure 8.1 Schematic of database screening with transporter pharmacophores.

particular drug transporter. In our hands and those of others [38], the pharmacophore methods can also have a good hit rate that may be used alongside other QSAR methods for database screening. To date we have primarily limited our database searches to those of known FDA-approved molecules, but such approaches could also be applied to much larger databases of commercially available molecules. By focusing on FDA-approved molecules, our approach may have utility for drug repositioning efforts for transporters such as those described in Tables 8.1 and 8.2, to rapidly identify clinically useful molecules that could be valuable to improve bioavailability of other molecules that are transported by the same transporter. The discovery of potent BCRP inhibitors for use with anticancer compounds would be certainly valuable. It should be noted that the number of transporters addressed using pharmacophore methods in Tables 8.1 and 8.2 is very small while P-gp has been undoubtedly the most widely studied, enabling many pharmacophores (Table 8.1) and QSAR models to be generated [21]. There are examples of several drug transporters that have been studied *in vitro* with inhibitor and substrate data that as yet have rarely been used to develop pharmacophores (e.g., organic anion transporters [69, 70] (Table 8.1) from mouse, but not human, have been modeled [71]) or not at all as in the case of the bile salt export pump [72] (Table 8.2). The lack of pharmacophore availability may not be limited to the amount of available data, at least in the case of these two transporters. It is hoped that more pharmacophores for other SLC and ATP transporters would be added to these tables in future in order to provide a more complete picture of the molecular requirements. This will ultimately be of value to both scientists in the pharmaceutical industry and researchers involved in unraveling the physiological functions and regulation of the hundreds of transporters that are expressed in our cells.

Acknowledgments

S.E. gratefully acknowledges Drs Cheng Chang, Praveen M. Bahadduri, and Peter W. Swaan (The University of Maryland) for our collaboration on transporter modeling over the past several years and Accelrys are thanked for making Discovery Studio Catalyst available to us. Dr. Maggie A.Z. Hupcey is acknowledged for her encouragement.

References

1 Anderle, P., Huang, Y., and Sadee, W. (2004) Intestinal membrane transport of drugs and nutrients genomics of membrane transporters using expression microarrays. *European Journal of Pharmaceutical Sciences*, **21**, 17–24.

2 Huang, Y., Anderle, P., Bussey, K.J., Barbacioru, C., Shankavaram, U., Dai, Z., Reinhold, W.C., Papp, A., Weinstein, J.N., and Sadee, W. (2004) Membrane transporters and channels: role of the transportome in cancer chemosensitivity and chemoresistance *Cancer Research*, **64** (12), 4294–301.

3 Hediger, M.A., Romero, M.F., Peng, J.B., Rolfs, A., Takanaga, H., and Bruford, E.A.

(2004) The ABCs of solute carriers: physiological, pathological and therapeutic implications of human membrane transport proteins: introduction. *Pflugers Archives*, **447**, 465–468.

4 Zhang, E.Y., Phelps, M.A., Cheng, C., Ekins, S., and Swaan, P.W. (2002) Modeling of active transport systems. *Advanced Drug Delivery Reviews*, **54**, 329–354.

5 Zhang, E.Y., Knipp, G.T., Ekins, S., and Swaan, P.W. (2002) Structural biology and function of solute transporters: implications for identifying and designing substrates. *Drug Metabolism Reviews*, **34**, 709–750.

6 Ware, J.A. (2006) Membrane transporters in drug discovery and development: a new mechanistic ADME era. *Molecular Pharmaceutics*, **3**, 1–2.

7 Zhang, E.Y., Phelps, M.A., Cheng, C., Ekins, S., and Swaan, P.W. (2002) Modeling of active transport systems. *Advanced Drug Delivery Reviews*, **54**, 329–354.

8 Chang, C. and Swaan, P.W. (2006) Computational approaches to modeling drug transporters. *European Journal of Pharmaceutical Sciences*, **27**, 411–424.

9 Ehrlich, P. (1909) Present status of chemotherapy. *Berichte der Deutschen Chemischen Gesellschaft*, **42**, 17–47.

10 Guner, O.F. (ed.) (2000) *Pharmacophore, Perception, Development, and Use in Drug Design*, University International Line, San Diego, CA.

11 Langer, T. and Hoffman, R.D. (2006) *Pharmacophores and Pharmacophore Searches*, Wiley-VCH Verlag GmbH, Weinheim.

12 Chang, C., Ekins, S., Bahadduri, P., and Swaan, P.W. (2006) Pharmacophore-based discovery of ligands for drug transporters. *Advanced Drug Delivery Reviews*, **58**, 1431–1450.

13 Martin, Y.C. (1992) 3D database searching in drug design. *Journal of Medicinal Chemistry*, **35**, 2145–2154.

14 Martin, Y.C., Bures, M.G., Danaher, E.A., DeLazzer, J., Lico, I., and Pavlik, P.A. (1993) A fast new approach to pharmacophore mapping and its application to dopaminergic and benzodiazepine agonists. *Journal of Computer-Aided Molecular Design*, **7**, 83–102.

15 Barnum, D., Greene, J., Smellie, A., and Sprague, P. (1996) Identification of common functional configurations among molecules. *Journal of Chemical Information and Computer Sciences*, **36**, 563–571.

16 Sprague, P.W. (1995) Automated chemical hypothesis generation and database searching with catalyst. *Perspectives in Drug Discovery and Design*, **3**, 1–20.

17 Sprague, P.W. and Hoffman, R. (1997) CATALYST pharmacophore models and their utility as queries for searching 3D databases, in *Computer-Assisted Lead Finding and Optimization* (eds H. van der Waterbeemd, B. Testa, and G. Folkers), Verlag Helvetica Chimica Acta, Basel, pp. 225–240.

18 Kaminski, J.J., Rane, D.F., Snow, M.E., Weber, L., Rothofsky, M.L., Anderson, S.D. et al. (1997) Identification of novel farnesyl protein transferase inhibitors using three-dimensional searching methods. *Journal of Medicinal Chemistry*, **40**, 4103–4112.

19 Wang, S., Zaharevitz, D.W., Sharma, R., Marquez, V.E., Lewin, N.E., Du, L. et al. (1994) The discovery of novel, structurally diverse protein kinase C agonists through computer 3D-database pharmacophore search. Molecular modeling studies. *Journal of Medicinal Chemistry*, **37**, 4479–4489.

20 Nicklaus, M.C., Neamati, N., Hong, H., Mazumder, A., Sunder, S., Chen, J. et al. (1997) HIV-1 integrase pharmacophore: discovery of inhibitors through three-dimensional database searching. *The Journal of Medicinal Chemistry*, **40** (6), 920–929.

21 Ekins, S., Ecker, G.F., Chiba, P., and Swaan, P.W. (2007) Future directions for drug transporter modeling. *Xenobiotica*, **37**, 1152–1170.

22 Jones, G., Willett, P., and Glen, R.C. (1995) A genetic algorithm for flexible molecular

overlay and pharmacophore elucidation. *Journal of Computer-Aided Molecular Design*, **9**, 532–549.

23 Clement, O.O. and Mehl, A.T. (2000) *HipHop: Pharmacophore Based on Multiple Common-Feature Alignments*, IUL, San Diego, CA.

24 Evans, D.A., Doman, T.N., Thorner, D.A., and Bodkin, M.J. (2007) 3D QSAR methods: phase and catalyst compared. *Journal of Chemical Information and Modeling*, **47**, 1248–1257.

25 Ekins, S., Johnston, J.S., Bahadduri, P., D'Souza, V.M., Ray, A., Chang, C. et al. (2005) In vitro and pharmacophore-based discovery of novel hPEPT1 inhibitors. *Pharmaceutical Research*, **22**, 512–517.

26 Chang, C., Bahadduri, P.M., Polli, J.E., Swaan, P.W., and Ekins, S. (2006) Rapid identification of P-glycoprotein substrates and inhibitors. *Drug Metabolism and Disposition*, **34**, 1976–1984.

27 Ekins, S., Kim, R.B., Leake, B.F., Dantzig, A.H., Schuetz, E.G., Lan, L.B. et al. (2002) Application of three-dimensional quantitative structure–activity relationships of P-glycoprotein inhibitors and substrates. *Molecular Pharmacology*, **61**, 974–981.

28 Ekins, S., Kim, R.B., Leake, B.F., Dantzig, A.H., Schuetz, E.G., Lan, L.B. et al. (2002) Three-dimensional quantitative structure–activity relationships of inhibitors of P-glycoprotein. *Molecular Pharmacology*, **61**, 964–973.

29 Bednarczyk, D., Ekins, S., Wikel, J.H., and Wright, S.H. (2003) Influence of molecular structure on substrate binding to the human organic cation transporter, hOCT1. *Molecular Pharmacology*, **63**, 489–498.

30 Chang, C., Pang, K.S., Swaan, P.W., and Ekins, S. (2005) Comparative pharmacophore modeling of organic anion transporting polypeptides: a meta-analysis of rat Oatp1a1 and human OATP1B1. *The Journal of Pharmacology and Experimental Therapeutics*, **314**, 533–541.

31 Suhre, W.M., Ekins, S., Chang, C., Swaan, P.W., and Wright, S.H. (2005) Molecular determinants of substrate/inhibitor binding to the human and rabbit renal organic cation transporters hOCT2 and rbOCT2. *Molecular Pharmacology*, **67**, 1067–1077.

32 Chang, C. and Swaan, P.W. (2006) Computer optimization of biopharmaceutical properties, in *Computer Applications in Pharmaceutical Research and Development* (ed. S. Ekins), John Wiley & Sons, Inc., Hoboken, NJ, pp. 495–512.

33 Swaan, P.W., Bensman, T., Bahadduri, P., Hall, M.W., Sarker, A., Ekins, S. et al. (2008) Bacterial dipeptide recognition and immune activation facilitated by the human peptide transporter PEPT2. *American Journal of Respiratory Cell and Molecular Biology*, **39** (5), 536–542.

34 Biegel, A., Gebauer, S., Brandsch, M., Neubert, K., and Thondorf, I. (2006) Structural requirements for the substrates of the H^+/peptide cotransporter PEPT2 determined by three-dimensional quantitative structure–activity relationship analysis. *Journal of Medicinal Chemistry*, **49**, 4286–4296.

35 Crivori, P., Reinach, B., Pezzetta, D., and Poggesi, I. (2006) Computational models for identifying potential P-glycoprotein substrates and inhibitors. *Molecular Pharmaceutics*, **3**, 33–44.

36 Raub, T.J. (2006) P-glycoprotein recognition of substrates and circumvention through rational drug design. *Molecular Pharmaceutics*, **3**, 3–25.

37 Rebitzer, S., Annibali, D., Kopp, S., Eder, M., Langer, T., Chiba, P. et al. (2003) In silico screening with benzofurane- and benzopyrane-type MDR-modulators. *Farmaco*, **58**, 185–191.

38 Langer, T., Eder, M., Hoffmann, R.D., Chiba, P., and Ecker, G.F. (2004) Lead identification for modulators of multidrug resistance based on in silico screening with a pharmacophoric feature model. *Archiv der Pharmazie*, **337**, 317–327.

39 Penzotti, J.E., Lamb, M.L., Evensen, E., and Grootenhuis, P.D. (2002) A computational ensemble pharmacophore model for

identifying substrates of P-glycoprotein. *Journal of Medicinal Chemistry*, **45**, 1737–1740.

40 Zhou, H., Wu, S., Zhai, S., Liu, A., Sun, Y., Li, R. et al. (2008) Design, synthesis, cytoselective toxicity, structure–activity relationships and pharmacophores of thiazolidinone derivatives targeting drug-resistant lung cancer cells. *Journal of Medicinal Chemistry*, **51** (5), 1242–1251.

41 Cianchetta, G., Singleton, R.W., Zhang, M., Wildgoose, M., Giesing, D., Fravolini, A. et al. (2005) A pharmacophore hypothesis for P-glycoprotein substrate recognition using GRIND-based 3D-QSAR. *Journal of Medicinal Chemistry*, **48**, 2927–2935.

42 Globisch, C., Pajeva, I.K., and Wiese, M. (2006) Structure–activity relationships of a series of tariquidar analogs as multidrug resistance modulators. *Bioorganic & Medicinal Chemistry*, **14**, 1588–1598.

43 Labrie, P., Maddaford, S.P., Fortin, S., Rakhit, S., Kotra, L.P., and Gaudreault, R.C. (2006) A comparative molecular field analysis (CoMFA) and comparative molecular similarity indices analysis (CoMSIA) of anthranilamide derivatives that are multidrug resistance modulators. *Journal of Medicinal Chemistry*, **49**, 7646–7660.

44 Raad, I., Terreux, R., Richomme, P., Matera, E.L., Dumontet, C., Raynaud, J. et al. (2006) Structure–activity relationship of natural and synthetic coumarins inhibiting the multidrug transporter P-glycoprotein. *Bioorganic & Medicinal Chemistry*, **14**, 6979–6987.

45 Kaiser, D., Terfloth, L., Kopp, S., Schulz, J., de Laet, R., Chiba, P. et al. (2007) Self-organizing maps for identification of new inhibitors of P-glycoprotein. *Journal of Medicinal Chemistry*, **50**, 1698–1702.

46 Balakin, K.V., Ivanenkov, Y.A., Savchuk, N.P., Ivaschenko, A.A., and Ekins, S. (2005) Comprehensive computational assessment of ADME properties using mapping techniques. *Current Drug Discovery Technologies*, **2**, 99–113.

47 Jedlitschky, G., Hoffmann, U., and Kroemer, H.K. (2006) Structure and function of the MRP2 (ABCC2) protein and its role in drug disposition. *Expert Opinion on Drug Metabolism & Toxicology*, **2**, 351–366.

48 Morrow, C.S., Peklak-Scott, C., Bishwokarma, B., Kute, T.E., Smitherman, P.K., and Townsend, A.J. (2006) Multidrug resistance protein 1 (MRP1, ABCC1) mediates resistance to mitoxantrone via glutathione-dependent drug efflux. *Molecular Pharmacology*, **69**, 1499–1505.

49 Boumendjel, A., Baubichon-Cortay, H., Trompier, D., Perrotton, T., and Di Pietro, A. (2005) Anticancer multidrug resistance mediated by MRP1: recent advances in the discovery of reversal agents. *Medicinal Research Reviews*, **25**, 453–472.

50 Young, A.M., Audus, K.L., Proudfoot, J., and Yazdanian, M. (2006) Tetrazole compounds: the effect of structure and pH on Caco-2 cell permeability. *Journal of Pharmaceutical Sciences*, **95**, 717–725.

51 Mols, R., Deferme, S., and Augustijns, P. (2005) Sulfasalazine transport in *in-vitro*, *ex-vivo* and *in-vivo* absorption models: contribution of efflux carriers and their modulation by co-administration of synthetic nature-identical fruit extracts. *The Journal of Pharmacy and Pharmacology*, **57**, 1565–1573.

52 van Brussel, J.P., Oomen, M.A., Vossebeld, P.J., Wiemer, E.A., Sonneveld, P., and Mickisch, G.H. (2004) Identification of multidrug resistance-associated protein 1 and glutathione as multidrug resistance mechanisms in human prostate cancer cells: chemosensitization with leukotriene D4 antagonists and buthionine sulfoximine. *BJU International*, **93**, 1333–1338.

53 Ng, C., Xiao, Y.D., Lum, B.L., and Han, Y.H. (2005) Quantitative structure–activity relationships of methotrexate and methotrexate

analogues transported by the rat multispecific resistance-associated protein 2 (rMrp2). *European Journal of Pharmaceutical Sciences*, **26**, 405–413.

54 Hirono, S., Nakagome, I., Imai, R., Maeda, K., Kusuhara, H., and Sugiyama, Y. (2005) Estimation of the three-dimensional pharmacophore of ligands for rat multidrug-resistance-associated protein 2 using ligand-based drug design techniques. *Pharmaceutical Research*, **22**, 260–269.

55 Yeboah, D., Sun, M., Kingdom, J., Baczyk, D., Lye, S.J., Matthews, S.G. *et al.* (2006) Expression of breast cancer resistance protein (BCRP/ABCG2) in human placenta throughout gestation and at term before and after labor. *Canadian Journal of Physiology and Pharmacology*, **84**, 1251–1258.

56 Choudhuri, S. and Klaassen, C.D. (2006) Structure, function, expression, genomic organization, and single nucleotide polymorphisms of human ABCB1 (MDR1), ABCC (MRP), and ABCG2 (BCRP) efflux transporters. *International Journal of Toxicology*, **25**, 231–259.

57 Breuzard, G., Piot, O., Angiboust, J.F., Mantait, M., Candeil, L., Del Rio, M. *et al.* (2005) Changes in adsorption and permeability of mitoxantrone on plasma membrane of BCRP/MXR resistant cells. *Biochemical and Biophysical Research Communications*, **329**, 64–70.

58 Honjo, Y., Hrycyna, C.A., Yan, Q.W., Medina-Perez, W.Y., Robey, R.W., van de Laar, A. *et al.* (2001) Acquired mutations in the MXR/BCRP/ABCP gene alter substrate specificity in MXR/BCRP/ABCP-overexpressing cells. *Cancer Research*, **61**, 6635–6639.

59 Doyle, L.A. and Ross, D.D. (2003) Multidrug resistance mediated by the breast cancer resistance protein BCRP (ABCG2). *Oncogene*, **22**, 7340–7358.

60 Nagashima, S., Soda, H., Oka, M., Kitazaki, T., Shiozawa, K., Nakamura, Y. *et al.* (2006) BCRP/ABCG2 levels account for the resistance to topoisomerase I inhibitors and reversal effects by gefitinib in non-small cell lung cancer. *Cancer Chemotherapy and Pharmacology*, **58**, 594–600.

61 Volk, E.L. and Schneider, E. (2003) Wild-type breast cancer resistance protein (BCRP/ABCG2) is a methotrexate polyglutamate transporter. *Cancer Research*, **63**, 5538–5543.

62 Robey, R.W., Honjo, Y., van de Laar, A., Miyake, K., Regis, J.T., Litman, T. *et al.* (2001) A functional assay for detection of the mitoxantrone resistance protein, MXR (ABCG2). *Biochimica et Biophysica Acta*, **1512**, 171–182.

63 Rocchi, E., Khodjakov, A., Volk, E.L., Yang, C.H., Litman, T., Bates, S.E. *et al.* (2000) The product of the ABC half-transporter gene ABCG2 (BCRP/MXR/ABCP) is expressed in the plasma membrane. *Biochemical and Biophysical Research Communications*, **271**, 42–46.

64 Maliepaard, M., van Gastelen, M.A., Tohgo, A., Hausheer, F.H., van Waardenburg, R.C., de Jong, L.A. *et al.* (2001) Circumvention of breast cancer resistance protein (BCRP)-mediated resistance to camptothecins *in vitro* using non-substrate drugs or the BCRP inhibitor GF120918. *Clinical Cancer Research*, **7**, 935–941.

65 Pavek, P., Merino, G., Wagenaar, E., Bolscher, E., Novotna, M., Jonker, J.W. *et al.* (2005) Human breast cancer resistance protein: interactions with steroid drugs, hormones, the dietary carcinogen 2-amino-1-methyl-6-phenylimidazo[4,5-*b*]pyridine, and transport of cimetidine. *The Journal of Pharmacology and Experimental Therapeutics*, **312**, 144–152.

66 Huisman, M.T., Chhatta, A.A., van Tellingen, O., Beijnen, J.H., and Schinkel, A.H. (2005) MRP2 (ABCC2) transports taxanes and confers paclitaxel resistance and both processes are stimulated by probenecid. *International Journal of Cancer*, **116**, 824–829.

67 Gupta, A., Zhang, Y., Unadkat, J.D., and Mao, Q. (2004) HIV protease inhibitors

are inhibitors but not substrates of the human breast cancer resistance protein (BCRP/ABCG2). *The Journal of Pharmacology and Experimental Therapeutics*, **310**, 334–341.

68 Chang, C. and Ekins, S. (2006) Pharmacophores for human ADME/Tox-related proteins, in *Pharmacophores and Pharmacophore Searches* (eds T. Langer and R.D. Hoffmann), Wiley-VCH Verlag GmbH, Weinheim, pp. 299–324.

69 Nigam, S.K., Bush, K.T., and Bhatnagar, V. (2007) Drug and toxicant handling by the OAT organic anion transporters in the kidney and other tissues. *Nature Clinical Practice*, **3**, 443–448.

70 Rizwan, A.N. and Burckhardt, G. (2007) Organic anion transporters of the SLC22 family: biopharmaceutical, physiological, and pathological roles. *Pharmaceutical Research*, **24**, 450–470.

71 Kaler, G., Truong, D.M., Khandelwal, A., Nagle, M., Eraly, S.A., Swaan, P.W. et al. (2007) Structural variation governs substrate specificity for organic anion transporter (OAT) homologs. Potential remote sensing by OAT family members. *The Journal of Biological Chemistry*, **282**, 23841–23853.

72 Sakurai, A., Kurata, A., Onishi, Y., Hirano, H., and Ishikawa, T. (2007) Prediction of drug-induced intrahepatic cholestasis: in vitro screening and QSAR analysis of drugs inhibiting the human bile salt export pump. *Expert Opinion on Drug Safety*, **6**, 71–86.

73 Andersen, R., Jorgensen, F.S., Olsen, L., Vabeno, J., Thorn, K., Nielsen, C.U. et al. (2006) Development of a QSAR model for binding of tripeptides and tripeptidomimetics to the human intestinal di-/tripeptide transporter hPEPT1. *Pharmaceutical Research*, **23**, 483–492.

74 Ganapathy, M.E., Huang, W., Wang, H., Ganapathy, V., and Leibach, F.H. (1998) Valacyclovir: a substrate for the intestinal and renal peptide transporters PEPT1 and PEPT2. *Biochemical and Biophysical Research Communications*, **246**, 470–475.

75 Moaddel, R., Ravichandran, S., Bighi, F., Yamaguchi, R., and Wainer, I.W. (2007) Pharmacophore modelling of the stereoselective binding to the human organic cation transporter (hOCT1). *British Journal of Pharmacology*, **151**, 1305–1314.

76 Ekins, S., Mirny, L., and Schuetz, E.G. (2002) A ligand-based approach to understanding selectivity of nuclear hormone receptors PXR, CAR, FXR, LXRa and LXRb. *Pharmaceutical Research*, **19**, 1788–1800.

77 Baringhaus, K.H., Matter, H., Stengelin, S., and Kramer, W. (1999) Substrate specificity of the ileal and the hepatic Na^+/bile acid cotransporters of the rabbit. II. A reliable 3D QSAR pharmacophore model for the ileal Na^+/bile acid cotransporter. *Journal of Lipid Research*, **40**, 2158–2168.

78 Swaan, P.W., Szoka, F.C., Jr., and Oie, S. (1997) Molecular modeling of the intestinal bile acid carrier: a comparative molecular field analysis study. *Journal of Computer-Aided Molecular Design*, **11**, 581–588.

79 Chang, C., Swaan, P.W., Ngo, L.Y., Lum, P.Y., Patil, S.D., and Unadkat, J.D. (2004) Molecular requirements of the human nucleoside transporters hCNT1, hCNT2, and hENT1. *Molecular Pharmacology*, **65**, 558–570.

80 Hu, H., Endres, C.J., Chang, C., Umapathy, N.S., Lee, E.W., Fei, Y.J. et al. (2006) Electrophysiological characterization and modeling of the structure activity relationship of the human concentrative nucleoside transporter 3 (hCNT3). *Molecular Pharmacology*, **69**, 1542–1553.

81 Viswanadhan, V.N., Ghose, A.K., and Weinstein, J.N. (1990) Mapping the binding site of the nucleoside transporter protein: a 3D-OSAR study. *Biochimica et Biophysica Acta*, **1039**, 356–366.

82 Pearce, H.L., Safa, A.R., Bach, N.J., Winter, M.A., Cirtain, M.C., and Beck, W.T. (1989) Essential features of the P-glycoprotein pharmacophore as defined by a

series of reserpine analogs that modulate multidrug resistance. *Proceedings of the National Academy of Sciences of the United States of America*, **86**, 5128–5132.

83 Pearce, H.L., Winter, M.A., and Beck, W.T. (1990) Structural characteristics of compounds that modulate P-glycoprotein-associated multidrug resistance. *Advances in Enzyme Regulation*, **30**, 357–373.

84 Garrigues, A., Loiseau, N., Delaforge, M., Ferte, J., Garrigos, M., Andre, F. et al. (2002) Characterization of two pharmacophores on the multidrug transporter P-glycoprotein. *Molecular Pharmacology*, **62**, 1288–1298.

85 Yates, C.R., Chang, C., Kearbey, J.D., Yasuda, K., Schuetz, E.G., Miller, D.D. et al. (2003) Structural determinants of P-glycoprotein-mediated transport of glucocorticoids. *Pharmaceutical Research*, **20**, 1794–1803.

86 Cramer, J., Kopp, S., Bates, S.E., Chiba, P., and Ecker, G.F. (2007) Multispecificity of drug transporters: probing inhibitor selectivity for the human drug efflux transporters ABCB1 and ABCG2. *ChemMedChem*, **2**, 1783–1788.

87 Li, W.X., Li, L., Eksterowicz, J., Ling, X.B., and Cardozo, M. (2007) Significance analysis and multiple pharmacophore models for differentiating P-glycoprotein substrates. *Journal of Chemical Information and Modeling*, **47**, 2429–2438.

88 Muller, H., Pajeva, I.K., Globisch, C., and Wiese, M. (2008) Functional assay and structure–activity relationships of new third-generation P-glycoprotein inhibitors. *Bioorganic & Medicinal Chemistry*, **16** (5), 2448–2462.

89 Matsson, P., Englund, G., Ahlin, G., Bergstrom, C.A., Norinder, U., and Artursson, P. (2007) A global drug inhibition pattern for the human ATP-binding cassette transporter breast cancer resistance protein (ABCG2). *The Journal of Pharmacology and Experimental Therapeutics*, **323**, 19–30.

90 Hirano, H., Kurata, A., Onishi, Y., Sakurai, A., Saito, H., Nakagawa, H. et al. (2006) High-speed screening and QSAR analysis of human ATP-binding cassette transporter ABCB11 (bile salt export pump) to predict drug-induced intrahepatic cholestasis. *Molecular Pharmaceutics*, **3**, 252–265.

Part Four:
Transporters and ADME

9
Biological Membranes and Drug Transport
Gert Fricker

9.1
Biological Membranes

On their way from their site of absorption to their site of action, most drugs have to cross membranes. These may be plasma membranes of enterocytes at absorption in the gastrointestinal tract, sinusoidal and canalicular membranes during the passage through the liver, basolateral and apical membranes in the case of tubular secretion in the kidney, or membranes of brain capillary endothelial cells. Although they may vary with regard to structure and regulation, there are features common to all membranes. These characteristics will be summarized in this chapter including structure and function of membranes, which are relevant for absorption, disposition, and elimination of drugs.

9.1.1
Lipid Bilayer

The formation of plasma membranes is one of the essential evolutionary steps. It defines the size of a cell and forms the barrier between cellular fluids and external milieu. All membranes are composed of a 4–5 nm thick double phospholipid layer consisting of approximately 5×10^6 lipid molecules/mm^2 membrane and embedding proteins, cholesterol, and glycolipids. In general, both moieties of a double layer are organized in an asymmetric manner (Figure 9.1).

Phospholipids have one polar head group and two hydrophobic hydrocarbon chains ranging from 14 to 24 carbon atoms. In most cases, one of the chains has one or more double bonds (Figure 9.2). Using spin resonance spectroscopy, it can be demonstrated that a distinct lipid molecule changes its position with an adjacent molecule between 10^6 and 10^7 times/s. In contrast, a change with a molecule from the opposite membrane leaflet (flip-flop mechanisms) occurs only once per 1–2 weeks. Cholesterol has a significant impact on membrane fluidity; hydroxyl

---

Transporters as Drug Carriers: Structure, Function, Substrates. Edited by Gerhard Ecker and Peter Chiba
Copyright © 2009 WILEY-VCH Verlag GmbH & Co. KGaA, Weinheim
ISBN: 978-3-527-31661-8

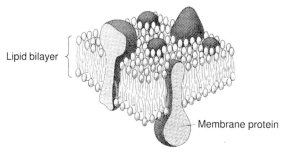

Figure 9.1 Simplified 3D image of a cell membrane with lipid bilayer and embedded membrane proteins.

groups of cholesterol are oriented toward the polar head groups of phospholipids, whereas the planar steroid ring system immobilizes the hydrocarbon chains below the head groups. Glycolipids containing an oligosaccharide are exclusively located in the outer leaflet of a bilayer. They include gangliosides with one or more sialic acid (*N*-acetyl neuraminic acid) residues in the polar head group.

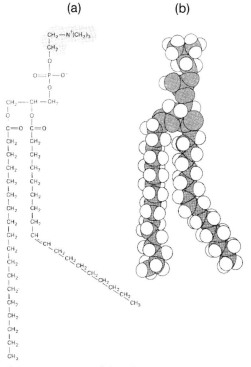

Figure 9.2 Structure of phosphatidylcholine, a typical phospholipid.

9.1.2
Membrane Proteins

Almost all plasma membranes also contain proteins amounting to 25–75% of a membrane. They participate in cell–cell connections, reception and release of signals, substrate transport, or enzymatic reactions. They may be embedded into both membrane leaflets (integral or transmembrane proteins) or only into one leaflet (peripheral proteins). Hydrophilic groups within a protein are oriented toward the liquid environment at the inner or outer surface.

9.1.3
Membrane Carbohydrates

Carbohydrates (2–10 wt% of a membrane) are bound to proteins or lipids, thus forming glycoproteins or glycolipids. Therefore, glycoproteins may have more than one carbohydrate chain, whereas glycolipids have only one residue. Oligosaccharides in outer cell membranes are exclusively oriented toward the outer membrane surface, whereas oligosaccharides of intracellular organelles are oriented toward the inner surface of these organelles. Oligosaccharides in the outer membrane belong to the so-called glycocalyx, which also contains glycoproteins and proteoglycans, which are only absorbed but are not covalently bound.

9.2
Membrane Transport

9.2.1
Mechanisms of Transport

For most polar molecules, the lipid bilayer of a cell membrane represents an impermeable membrane, which uncharged molecules can cross only by passive diffusion. During evolution, transport pathways also emerged allowing polar molecules, such as nutrients or metabolites, to cross a cell membrane. As a result, the transport of small molecules is mediated by transmembrane proteins, whereas macromolecules and small particles cross membranes by various cytotic mechanisms.

9.2.2
Transport Across Lipid Membranes

The velocity of diffusion across a membrane depends on the size of the respective molecule and its relative solubility within the lipid phase. Small nonpolar molecules exhibit good lipid solubility and have a rather high velocity of diffusion. Uncharged polar molecules such as H_2O or CO_2, which have a rather low lipid solubility, may also cross a membrane by passive diffusion. In contrast, charged molecules and ions do not diffuse across a membrane. The velocity of diffusion can be calculated according to Fick's law.

$$\left(\frac{dM}{dt}\right) = D \cdot P \cdot \frac{A}{d} \cdot (c_2 - c_1), \tag{9.1}$$

where dM/dt is the per time unit diffusion amount of compound, D is the diffusion coefficient, P is the distribution coefficient between membrane and outer medium, A is the area of diffusion, d is the thickness of membrane, and $(c_2 - c_1)$ is the concentration gradient.

In addition, the distribution coefficient significantly contributes to the formation of a concentration gradient (Figure 9.3). Compounds with a low distribution coefficient hardly enter a membrane, whereas compounds with a high distribution coefficient may accumulate within a membrane.

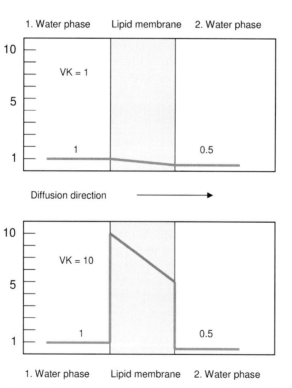

Figure 9.3 Concentration gradient during passage across membrane of two compounds with different distribution coefficients in the lipid bilayer. The compound with a distribution coefficient 10 has a steeper concentration gradient than the compound with a distribution coefficient 1. Since the rate of diffusion depends on the distribution coefficient, the compound with a coefficient 10 will pass the membrane 10 times faster (from Kurz and Neumann, 1977).

9.2.3
Protein-Coupled Membrane Transport

Transport of many compounds including drugs across cell membranes is mediated by membrane proteins called carrier proteins or channel proteins. Some of these proteins transport only one substrate molecule at a time across the membrane (uniport systems), while others act as cotransport systems (Figure 9.4). Depending on the direction of the second substrate, the proteins are also called symporters or antiporters, for example, Na^+/glucose cotransporter, H^+/peptide cotransporter, or Na^+/K^+ antiporter (=Na^+/K^+-ATPase).

Protein-coupled membrane transport may be an active, energy-dependent, or a passive process. The direction of transport may follow the concentration gradient of a compound. In case of charged molecules, both concentration and charge determine transport. Transport of positively charged molecules can be facilitated by an interior negative membrane potential, whereas the transport of negatively charged molecules can be hampered.

In contrast to pore forming channel proteins, carrier proteins bind their substrates at specific binding sites, resulting in the so-called "facilitated diffusion." Transport against an electrochemical gradient toward the higher concentration is coupled to energy consumption and is termed "active transport." In primary active systems, transport is directly coupled to the hydrolysis of ATP, and in secondary systems, transport is associated with concomitant transport of Na^+ or H^+ ions, the concentration of which depends on other transport proteins, such as the Na^+/K^+-ATPase. Most carrier proteins have a relatively narrow spectrum of substrate recognition. However, there are also transport proteins known to recognize several hundred different substrates, such as the multidrug resistance (MDR) protein P-glycoprotein [1–3] or the multidrug resistance-related proteins (MRP) [4, 5].

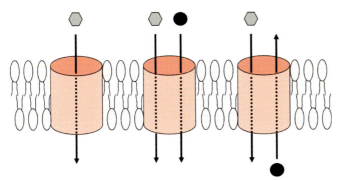

Figure 9.4 Transport proteins acting as uniport, symport, or antiport systems.

9.2.4
Kinetics of Carrier-Mediated Transport Processes

Kinetics of carrier-mediated transport processes is similar to enzyme–substrate reactions and can be described by the Michaelis–Menten equation (Eq. (9.2)), assuming that each transport system has one specific binding site for its substrates. Maximum transport velocity (V_{max}) is reached when all binding sites of the respective carrier proteins are occupied by substrate molecules. Substrate turnover can be delineated by the Michaelis constant K_M corresponding to the substrate concentration [S], at which half-maximum transport velocity has been reached (Figure 9.5). K_M also depends on pH and temperature. In cotransport systems transferring several substrates, the transport protein has a characteristic K_M for each molecule transported.

$$V = \frac{V_{max} \cdot [S]}{K_M + [S]}. \quad (9.2)$$

Calculation of V_{max} and K_M may occur by nonlinear regression or by linearization, such as the Lineweaver and Burk method. Thereafter, reciprocal values of V and [S] are plotted in a diagram and Eq. (9.2) is transformed to

$$\frac{1}{V} = \frac{1}{V_{max}} + \frac{K_M}{V_{max}} \cdot \frac{1}{[S]}. \quad (9.3)$$

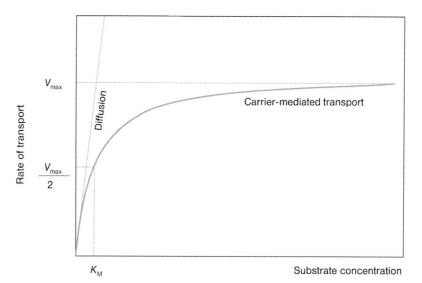

Figure 9.5 Kinetics of a carrier-mediated membrane transport processes and passive diffusion. At passive diffusion, the transport corresponds directly to the concentration of the diffusing compound. The transport rate of a carrier-mediated transport reaches a maximum. The rate at which half-maximum transport rate is reached is called Michaelis constant. It can be used as a measure of affinity of a substrate to the transport protein.

If several different substrates compete for a binding site, an inhibition may be observed following Eq. (9.4):

$$V = \frac{V_{max} \cdot [S]}{K_M[(1 + ([I]/K_I)] + [S]}, \tag{9.4}$$

with inhibitor concentration [I] and inhibition constant K_I.

9.2.5
Ion Gradient-Dependent Transport Processes

Many active transport processes are not directly driven by ATP hydrolysis but by ion gradients. For example, uptake of glucose or amino acids in enterocytes and kidney tubular cells is mediated by Na^+ cotransport systems [6–9]. Na^+ ions enter the cells along their electrochemical gradient, which is driven by the Na^+/K^+-ATPase (Figure 9.6).

The peptide transporter within the apical membrane of enterocytes represents an H^+ cotransport system, which transports 2–3-amino acid-long small peptides [10, 11]. This transport system is also responsible for uptake of drugs having a peptidomimetic structure, such as thrombin inhibitors, rennin inhibitors, angiotensin-converting enzyme inhibitors, or HIV-protease inhibitors. The underlying mechanisms of this transporter are subject of intensive research to clarify regulation and to identify requirements of substrate recognition. Peptide transporters of similar structure can also be found in other tissues, for example, the kidney [12]. Studies

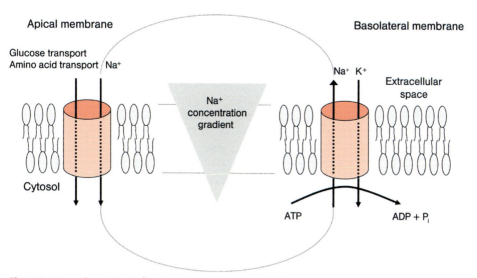

Figure 9.6 Secondary active Na^+ cotransport system in an apical cell membrane, which is driven by the Na^+/K^+-ATPase in the basolateral cell membrane.

concerning substrate structure showed that modification of the N-terminal amino residue of a substrate decreases its affinity to the carrier protein. Peptidomimetics containing a γ-amino acid are said not to be recognized by the transporter [13, 14].

Coupled transport systems frequently exhibit an asymmetric localization within plasma membranes. In enterocytes, the Na^+-dependent glucose transporter and the Na^+-dependent amino acid uptake systems are localized in apical (luminal) membrane, whereas the Na^+/K^+-ATPase is localized within the basolateral (blood-sided) membrane. Thus, the secondary active Na^+- or H^+-dependent transport systems are key elements for nutrient absorption, whereas subsequent transport across the basolateral membrane frequently follows the facilitated diffusion.

9.2.6
Cytotic Mechanisms: Transport of Macromolecules

Transport of macromolecules across cell membranes is basically different from transport of small molecules. While small molecules cross membranes by passive diffusion or protein-mediated mechanisms, macromolecules are transported by pinching-off of membrane vesicles (endocytosis or exocytosis). Unspecific uptake of liquids including dissolved compounds is termed pinocytosis and uptake of particles via large vesicles is called phagocytosis. Cytotic mechanisms may be of interest for targeted drug delivery, when substrates of surface receptors or antibodies versus such receptors are coupled to drug carrying entities, such as liposomes or nanoparticles. A well-known example is the use of the transferrin receptor at the blood–brain barrier: Antibodies versus the receptor can be covalently linked to liposomes, which subsequently exhibit selective binding and transcytosis across the blood–brain barrier. Recent *in vivo* experiments demonstrated significant transfer of daunomycin into the brain, which is normally not accessible to the drug. As a result, 30 vector molecules could be coupled to one liposome containing more than 30 000 drug molecules [15, 16].

Besides liposomes, polymeric nanoparticles may be used as effective drug carrier systems using cytotic pathways. Particle size and polymeric composition help control particle degradation and drug release. Recently, it was shown in a rat study that polybutylcyanoacrylate nanoparticles, which had been surface coated with polysorbate 80, exhibited a 20-fold higher uptake into brain capillary endothelial cells compared to noncoated nanoparticles [17]. It is assumed that association of lipoproteins at the surface triggers the endocytotic uptake of the nanoparticles.

9.2.7
Export Proteins

Export proteins that actively extrude substances out of a cell attract an increasing interest in the interpretation of drug pharmacokinetics and efficiency. Most important representatives of such proteins are members of the ABC protein superfamily, which comprises approximately 8000 proteins in bacteria, plants, animals, and man.

Typical characteristics of the proteins are ATP binding sites (ATP binding cassette proteins). With regard to drug pharmacokinetics, three protein families deserve special interest: the MDR proteins, MRP proteins, and breast cancer-related proteins (BCRP).

9.3
Pharmacokinetic-Relevant Membrane Barriers

During absorption, distribution, and elimination, drugs have to cross several epithelial and endothelial barriers. Cells of these barriers exhibit similar general characteristics, but each cell type also shows organ-specific distinctive features. In the following section, organs, cell types, and transport systems being relevant for the above-mentioned transport processes will be discussed.

9.3.1
Intestinal Drug Absorption

The intestinal tract consists of duodenum, jejunum, ileum, colon, and rectum. The total length of the small intestinal tract comprises 3–4 m in humans with duodenum having 12–15 cm, 150 cm jejunum and 70 cm ileum. Its apparent inner surface of approximately 3000 cm^2 is extended by microvilli to approximately 200 m^2.

The small intestinal tract is the major place of digestion and absorption. Protein digestion reaches approximately 15% in the stomach, and in the lower intestine, the hydrolyzed fraction amounts to about 60%. Seventy percent of fats, 60% of carbohydrates, and about 30% of protein and peptide are absorbed within the duodenum. Within enterocytes, triglycerides are resynthesized from long-chain fatty acids and transformed into chylomicrons, which are transported via the lymphatic system, whereas short- and medium-chain fatty acids are directly transported into mesenteric and portal vein system.

For the absorption of carbohydrates, amino acids, and peptides, a variety of transport systems following facilitated diffusion and active mechanisms have been identified on a molecular and functional level. D-Glucose is mainly absorbed via the Na$^+$-dependent transporter SGLT1 in the brush-border membrane of enterocytes [18–20]. It is transported across the basolateral membrane by facilitated diffusion via the hexose transporter GLUT-2. Besides SGLT1, the Na$^+$-independent transport protein GLUT-5 is localized in the apical enterocyte membrane, recognizing fructose as a substrate [21].

Amino acids are absorbed by Na$^+$-dependent transport proteins, with different proteins for acidic, neutral, and anionic amino acids. Small peptides are absorbed by the peptide transporter PEPT1 in the brush-border membrane [11]. The PEPT1 mRNA pattern exhibits regional differences with duodenum > jejunum > ileum [22]. The carrier works as an H$^+$ cotransport system and recognizes 2–3-amino acid-long small peptides as well as drugs with peptide-like structures such as renin inhibitors, ACE inhibitors [23], or β-lactam antibiotics (Figure 9.7). It also serves as a target for

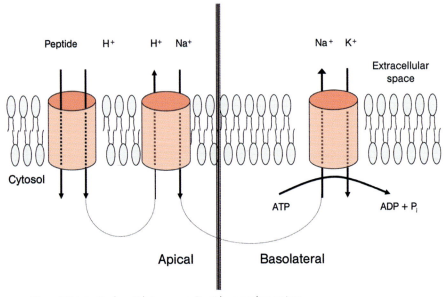

Figure 9.7 Intestinal peptide transport. Peptides are taken up into enterocytes together with H^+ ions. The proton gradient is maintained via an Na^+/H^+ antiport system in the apical cell membrane. The Na^+ gradient is guaranteed by the Na^+/K^+-ATPase in the basolateral cell membrane.

various prodrug approaches in order to improve absorption of otherwise poorly absorbable drugs, for example, L-α-methyldopa-L-phenylalanine or L-valine-acyclovir [24].

The ileal bile acid transporter (IBAT) transports conjugated bile acids in a Na^+-dependent manner. Like the peptide transporter, it serves as a target for prodrugs, which consist of drugs being coupled to the hydroxyl group at position 3 of the steroid ring system of a bile acid [25, 26].

The export protein P-glycoprotein (MDR1 gene product; GP170; ABCB1), which is relevant for the absorbed fraction of many drugs, exhibits an increasing expression from proximal to distal sections of the gastrointestinal tract [27]. It has an apparent molecular weight of 170 kDa and is located in the apical membrane of entrerocytes [28–32]. It consists of two units with 12 transmembrane domains, with each unit having an ATP binding site, which provides the energy necessary for transport by ATP hydrolysis (Figure 9.8). P-glycoprotein was originally discovered in drug-resistant cancer cells, but it is also expressed in healthy epithelial and endothelial tissues. It contributes to active secretion in gut–liver and kidney and diminishes passage of drugs across the blood–brain barrier by its excretory function [33, 34]. The protein transports predominantly lipophilic and cationic compounds, and recognizes an outstanding range of substrates including almost all classes of drugs and excipients

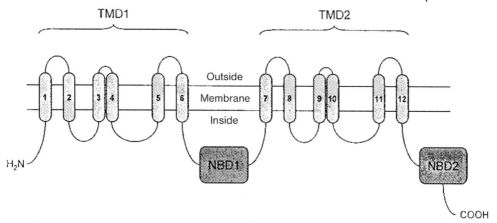

Figure 9.8 MDR1 gene product P-glycoprotein. The protein has 12 membrane spanning sequences; both ends of the protein are localized at the interior surface of the membrane. NBD, nucleotide-binding domain; TMD, transmembrane domain (from Ref. [37]).

(Table 9.1). Its mechanism of transport has not yet been fully elucidated, but it is assumed that the protein traps its substrates in the cell membrane and prevents further processing into the cells [35, 36].

P-glycoprotein shows a strong expression in cells with high metabolic activity and which also exhibit a high level of cytochrome P450 enzymes. It can be assumed that both systems represent elements of a general combined or complementary detoxification mechanism of the body [38–40]. Since P-glycoprotein decreases the concentration of many drugs and xenobiotics, the probability of an oversaturation of the enzymes is decreased and an efficient metabolism is guaranteed. The risk of a product inhibition may also be reduced by a P-glycoprotein-mediated excretion of metabolites. P-glycoprotein is also relevant in the context of adjunctive therapies, since several drugs exhibit competitive interactions at the transporter level and also up- or downregulate its expression resulting in a changed pharmacokinetic profile of concomitantly administered other drugs. Well-known examples are rifampicin or constituents of the extract of the herb St. John's wort, which may cause an induction of P-glycoprotein after long-term administration. For drugs such as cyclosporin A, digoxin, verapamil, or vinblastine, it could be shown that their absorption decreases with increasing expression of P-glycoprotein or that this absorption is improved after inhibition of the export pump [27, 41–44].

Several studies highlight the use of mdr1-knockout mice to investigate P-glycoprotein function *in vivo* [45, 46]. In bile duct-ligated mice, 16% of digoxin administered i.v. was secreted into the gut lumen of control animals, whereas only 2% was secreted in mdr1-knockout mice [47]. This finding supports the earlier observations

Table 9.1 Examples of compounds interacting with P-glycoprotein and Mrp1/2.

Drug category	Compound	Transporter
Cytostatics	Vinca alkaloids	p-GP; Mrp2
	Paclitaxel	p-GP
	Anthracyclins (doxorubicin, daunorubicin, epirubicin)	p-GP
Immunosuppressants	Cyclosporin A, rapamycin, tacrolimus (FK506)	p-GP
Alkaloids	Colchicine, reserpine	p-GP
Antiarrhythmics	Amiodarone, quinidine	p-GP
Antibiotics	Actinomycin D, puromycin, mitomycin C	p-GP
Lipid lowering agents	Atorvastatin, fluvastatin, pravastatin	p-GP
β-Adrenoceptor blocking agents	Cepiprolol, pafenolol, talinolol, acebutolol	p-GP
HIV-protease inhibitors	Ritonavir, saquinavir, indinavir, nelfinavir	p-GP; Mrp2
Ca^{2+} channel blockers	Bepridil, diltiazem, verapamil, nicardipine, nifedipine	p-GP
	Octreotide (somatostatin analogue)	p-GP, Mrp2
	Digoxin (cardiac glycoside)	p-GP
	Ivermectin (anthelminticum)	p-GP
	Morphine, morphine-6-glucuronide, loperamide (opiates)	p-GP
	Bilirubin diglucuronide	Mrp2
Pharmaceutical excipients	17β-Estradiol, 17β-D-glucuronide	Mrp2
	Cremophor EL	p-GP
	Pluronic F68, Pluronic L61	p-GP, Mrp2

of increased absorption and altered pharmacokinetics of digoxin during concomitant administration of quinidine, a P-glycoprotein blocker [48, 49]. For HIV-protease inhibitors, the influence of intestinal P-glycoprotein could also be demonstrated using mdr-knockout mice [50]. Compared to normal mice, plasma concentrations of the protease inhibitors were increased after oral administration in mdr-knockout animals. After i.v. administration, only small differences were seen, suggesting that intestinal P-glycoprotein activity reduced the extent of net absorption in normal animals. Finally, disruption of the murine mdr1a gene significantly reduced the intestinal transport of paclitaxel (Taxol) and altered drastically its biodistribution [51]. In another study, a novel taxane (IDN 5109) revealed an improved preclinical profile with respect to efficacy and tolerability [52]. Due to lack of interaction with P-glycoprotein, the new compound showed an intestinal absorption profile unique among taxanes.

The expression of another ABC transport protein, multidrug resistance-related protein 2 (MRP2) decreases from proximal to distal parts of the GI tract [53]. The protein recognizes mainly anionic metabolites such as glucuronides or glutathione conjugates (Table 9.1). Its function and substrate specificity will be further discussed in the context of hepatic transport systems.

The bidirectional cation transporter OCT1 (organic cation transporter 1) is localized in the basolateral membrane of enterocytes [54]. It belongs, as its analogues OCT2 and OCT3, to the transporter family SLC22 (SLC22A1–SLC22A3). The substrate specificities of these transport proteins exhibit a significant overlap, but there are species-dependent differences in affinity and maximum rates of transport. Table 9.2 shows cationic drugs interacting with OCT1 (from Ref. [55]).

Table 9.2 Drugs interacting with OCT1 (from Ref. [55]).

	Drug	IC_{50} or K_M (µM)
Receptor antagonists		
α-Adrenoceptor	Phenoxybenzamine	$IC_{50} = 2.7$
α-Adrenoceptor	Prazosin	$IC_{50} = 1.8$
β-Adrenoceptor	Acebutolol	$IC_{50} = 96$
Histamine H2 receptor	Cimetidine	$IC_{50} = 166$
Receptor agonists		
α-Adrenoceptor	Clonidine	$IC_{50} = 0.55$
β-Adrenoceptor	o-Methylisoprenaline	$IC_{50} > 100$
Ion channel blockers		
Na^+ channel	R-(−)-Disopyramide	$IC_{50} = 15$
Na^+ channel	Procainamide	$IC_{50} = 74$
Na^+ channel	Quinidine	$IC_{50} = 18$
Ca^{2+} channel	Verapamil	$IC_{50} = 2.9$
Psychoactive drugs		
Antidepressant	Desipramine	$IC_{50} = 5.4$
Antiviral drugs		
General	Acyclovir	$K_M = 151$
General	Ganciclovir	$K_M = 516$
HIV-protease inhibitor	Indinavir	$IC_{50} = 62$
HIV-protease inhibitor	Nelfinavir	$IC_{50} = 22$
HIV-protease inhibitor	Ritonavir	$IC_{50} = 5.2$
HIV-protease inhibitor	Aquinavir	$IC_{50} = 8.3$
Others		
Antidiabetic	Phenformin	$IC_{50} = 10$
Anesthetic	Midazolam	$IC_{50} = 3.7$
Antimalaric	Quinine	$IC_{50} = 23$
Muscle relaxant	Vecuronium	$IC_{50} = 232$

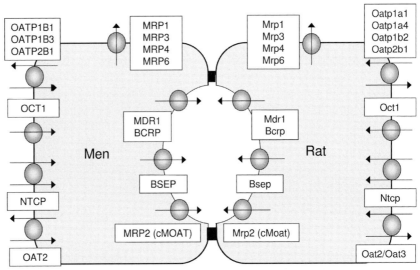

Figure 9.9 Transport systems participating in drug transport in hepatocyte membranes.

9.3.2
Liver

The tasks of the liver are manifold ranging from bile production, storage of carbohydrates, synthesis of plasma proteins, phase I and phase II metabolism, to formation of urea. It is the central organ of metabolism and elimination for a large variety of endobiotics and xenobiotics, and correspondingly hepatocytes express a multitude of transport proteins in their sinusoidal (basolateral) and canalicular plasma membrane (Figure 9.9).

The already mentioned proteins OCT1 and OCT3 transport small cationic substances, such as tetraalkyl ammonium compounds, polyamines such as spermine, monoamino-neurotransmitters, or N-methyl-nicotinamide across the basolateral plasma membrane [56]. OCTs play a key role in the distribution of cationic drugs and, therefore, drug interactions at the transporter level may become clinically relevant, as compounds with high affinity, such as prazosin or phenoxybenzamine, may affect the excretion of other substrates. Certain liver diseases or obstructive cholestasis may result in alterations of hepatic clearance via these transporters. In rats, a 7-day bile duct ligation resulted in a marked downregulation of Oct1 and an increased hepatic accumulation of the Oct1 substrate tetraethylammonium [57].

The Na^+/taurocholate cotransport system Ntcp belongs to the "solute carrier" superfamily (SLC) and represents one of the best-studied transport proteins in the basolateral membrane. It transports taurocholate and other conjugated bile acids from the blood into hepatocytes. In addition, an ontogenetically presumably older and Na^+-independent protein (Oatp1) is expressed in the basolateral membrane.

Oatps (organic anion transport proteins) also belong to the SLC superfamily. Hagenbuch and Meier [58] describe the phylogenetic relationship of these proteins, which allows the classification of distinct members in accordance with the guidelines of the human genome nomenclature committee. Organic anion transport protein 1 (Oatp1, Oatp1a1, Slc21a1), an 80 kDa protein, was the first member of this family identified in the basolateral hepatocyte membrane [59]. It has 12 transmembrane domains with an intracellular localization of both N- and C-terminal end of the protein.

Oatp1 exhibits a relatively wide pattern of substrate recognition and mediates the Na^+-independent uptake of bile acids, anionic conjugates of steroids (e.g., estrone-3-sulfate, estradiol-17β-glucuronide), uncharged steroids (aldosterone, cortisol, and ouabain), various peptides and hormones, and a large number of xenobiotics, including drugs such as enalapril or pravastatin [60, 61]. Oatp1 acts as an exchange protein with intracellular anions such as HCO_3^- or glutathione as counterions [62, 63].

Ntcp and Oatp1 are regulated by the nuclear receptor HNF4α (nuclear receptor 2A1), which is a key element of hepatocyte differentiation [64]. A functional downregulation occurs by phosphorylation by extracellular ATP and activation by protein kinase C, but not by protein kinase A [65, 66].

Oatp2 (Oatp1a4, Slc21a5), which is also localized in the basolateral hepatocyte membrane, shows about 77% structural homology with Oatp1 [67] and exhibits a large substrate overlap with that protein. The gene of Oapt2 contains pregnane X receptor response elements [68, 69], and activation of this nuclear receptor results in an upregulation of the transporter in parallel to cytochrome P450 enzymes [70].

Oatp3 has not unambiguously been identified in the liver, and Oatp4 (Oatp1b2, Slc21a10) was isolated from rat liver in two isoforms [71, 72]. One isoform, a shortened protein, appears to transport exclusively taurocholate, whereas the other subtype, the full-length protein, transports steroid hormones, prostaglandin E_2, leukotriene C4, and thyroidal hormones.

Another Oatp protein, the prostaglandin transporterPGT (Oatp2a1, Slc21a2), has been cloned from rat liver and has 37% structural homology to Oatp1 [73]. It recognizes prostanoids but not taurocholate or estradiol-17β-glucuronide [74].

Another organic anion transporter family in the hepatocyte membrane consists of polyspecific organic anion transporters, OAT, belonging to the SLC22A family. They transport a broad spectrum of small anions, including many drugs. In the rat, rOat2 (Slc2a7) has been identified and been localized in the basolateral membrane. The human analogue SLC22A7 was also identified in the liver. rOat2 recognizes substrates such as acetylsalicylic acid, salicylic acid, α-ketoglutarate, or prostaglandin E_2 by a Na^+-independent mechanism. In contrast to renal rOat1, it seems not to act as dicarboxylate/organic anion exchanger.

Northern blot analysis indicates a strong expression of another member of the OAT family, rOat3 [75]. It transports substrates such as *para*-aminohippuric acid, estrone-3-sulfate, or ochratoxin, not taurocholate or digoxin [75], and it also recognizes cationic cimetidine. Human hOAT3 seems to be expressed rather in the kidney than in the liver [76].

With respect to elimination of many drugs and their metabolites across the bile canalicular membrane, particular attention should be again on members of the ABC protein superfamily, which are directly driven by ATP hydrolysis [77]. P-glycoprotein and its pattern of substrate recognition have already been discussed in the context of intestinal drug absorption. Although MDR1 subtypes, which recognize preferably lipophilic cationic substrates, seem to be of secondary importance for the physiologic function of the liver, MDR3 (ABCB4) in the bile canalicular plasma membrane appears to contribute to the export of phospholipids out of the hepatocytes acting as the so-called phospholipid flippase [61, 78]. In addition, the protein accepts to a minor extent some drugs as substrates, such as paclitaxel or vinblastine [79].

A protein of similar structure, Spgp (sister of P-glycoprotein), is identical to the canalicular bile acid export protein Bsep (ABCB11), which has been identified in rodents and man [80]. Only few data are available concerning drug interactions with this transporter, but its importance is emphasized by the observation of a fatal peripheral cholangiocarcinoma, which developed in two girls with progressive familial intrahepatic cholestasis, ABCB11 mutations, and absent bile salt export pump (BSEP) expression. BSEP deficiency may have caused cholangiocarcinoma through bile composition shifts or bile acid damage within cells capable of hepatocytic/cholangiocytic differentiation [81].

Several members of the MRP family have been identified in the liver. Mrp1/MRP1 (ABCC1), Mrp3/MRP3 (ABCC3), and Mrp6/MRP6 are expressed at a relatively low level in the basolateral hepatocyte membrane [5, 82–84], whereas Mrp2/MRP2 (ABCC2) is expressed in the bile canalicular plasma membrane, where it acts as a multispecific organic anion transporter (cMOAT) recognizing a multiplicity of different substrates (e.g., glutathione conjugates including leukotriene C4, bilirubin, and estrogen glucuronides [85–88]). Interestingly, this protein also seems to interact with some cations, such as adriamycin [89, 90]. Mrp1, which is localized in the basolateral membrane, has a similar substrate recognition [91]. As a result of its substrate recognition, Mrp2 can be regarded as an active downstream elimination system for many phase II metabolites. A mutation of the protein and an aligned functional disturbance are of clinical relevance for the Dubin–Johnson syndrome, a hereditary hyperbilirubinemia. Functional impairment of Mrp2 can lead to an upregulation of Mrp3/MRP3 in order to compensate the decreased biliary excretion of organic anions [92, 93]. Table 9.3 gives an overview of transport systems identified in the liver (from Ref. [94]).

9.3.3
Kidney

The kidney regulates the concentration of metabolites and electrolytes in the extracellular space by elimination or by retention of water and solutes. The functional basic unit is a nephron, whereof a human kidney contains about 1.2 million. A nephron starts with a glomerular body, consisting of Bowman capsule and embedded

Table 9.3 Hepatic transport systems and their localization in hepatocyte membranes.

Transport system	Basolateral membrane	Bile canalicular membrane
Systems for ion homeostasis and pH regulation		
Na^+/K^+-ATPase	+	−
Ca^{2+}-ATPase	+	−
K^+ channel (several)	+	?
Cl^- channel (several)	+	+
Ca^{2+} channel (several)	+	?
Na^+/HCO_3^- cotransporter	+	−
Cl^-/HCO_3^- exchanger	−	+
OH^-/SO_4^{2-} exchanger	+	−
HCO_3^-/SO_4^{2-} exchanger (sat-1)	−	+
Systems for uptake of metabolic substrates		
Glucose transporter (GLUT-1, GLUT-2)	+	−
Amino acid transporter A	+	+
Amino acid transporter ASC	+	+
Amino acid transporter N	+	+
Amino acid transporter L	+	?
Na^+/adenosine cotransporter	?	+
Na^+/pyruvate, lactate cotransporter (MCT2)	+	−
Na^+/α-ketoglutarate cotransporter (Na DC-1)	+	−
Na^+/purine nucleoside transporter (SPNT)	−	+
Systems for directed transport of cholephilic substrates from blood to bile		
Na^+/bile acid cotransporter (Ntcp)	+	−
Organic anion transporter proteins (oatps)	+	−
Organic anion transporters (oats)	+	−
Prostaglandin transporter	(+)	−
Sinusoidal GSH transport (RsGshT)	+	−
Canalicular GSH transporter (RcGshT)	−	+
Organic cation transporter 1	+	−
ATP-dependent bile acid export protein (spgp or cBSEP)	−	−
ATP-dependent multidrug resistance-associated protein (mrp1)	+	−
ATP-dependent multidrug resistance-associated protein (mrp2, cMOAT)	−	+
ATP-dependent phospholipid translocator (mdr2)	−	+
ATP-dependent multidrug resistance protein (mdr1)	−	+
ATP-dependent breast cancer resistance protein	−	+

arterioles, and representing the place of renal filtration. The ultrafiltrate flows into the proximal tubules, where about two-thirds of the fluid is reabsorbed. Then, the fluid reaches the straight part of the tubule, followed by Henle's loop with a thin descending and a thicker ascending segment. Finally, the fluid flows through the distal tubule into the collecting duct. The total length of a nephron amounts to 45–65 mm or a total of 120 km in a whole human kidney.

The limit of size exclusion for glomerular filtration is in the range of 5000 Da, which means above the size of most therapeutically interesting drugs. Filtration out of arterioles occurs over a surface of about 1.5 m². In order to determine glomerular filtration, reference compounds are used, which exhibit no biotransformation, have no protein binding, do not accumulate in the kidney, and do not influence kidney function. Such a compound is inulin and to some extent also endogenous creatinine. From plasma concentration (C_p) at steady state and the amount excreted into urine per time ($V_u \times C_u$), the glomerular filtration rate (GFR) can be calculated (Eq. (9.5)).

$$\text{GFR} = \frac{V_u \cdot C_u}{C_p}. \tag{9.5}$$

In healthy adults, the value for inulin is in the range of 120–125 ml/day. Since inulin is neither absorbed nor secreted in the tubules, this value corresponds to the inulin clearance. Per day a total filtered volume of 180 l results, which is 60 times the plasma volume or 4 times the total body water. Because only 1–1.5 l urine is excreted per day the reabsorption of water is similar to the filtration rate.

The tubular walls reabsorb water and solutes and actively secrete a variety of compounds. Therefore, particularly in the proximal tubules different transport systems are expressed. The amount of a compound excreted into urine corresponds to its amount filtered in the glomeruli plus the net amount transported across the tubular walls. With tubular excretion the plasma clearance becomes larger than the glomerular filtration rate (up to 650 ml/min), and with tubular reabsorption it becomes smaller than the glomerular filtration rate (0–120 ml/min). After exclusively glomerular filtration, the ratio of drug clearance and inulin clearance is 1; with filtration and reabsorption, it is <1; and with filtration and active secretion, it is >1. Characteristic values are glucose 0 ml/min, diazepam 20–40 ml/min, urea 75 ml/min, inulin 125 ml/min, ampicillin 275 ml/min, p-aminohippuric acid 650 ml/min, acetylsalicylic acid 650 ml/min, and verapamil 1200 ml/min. As in all epithelial tissues, the passage across tubular walls occurs by passive diffusion, by active transport, or by cytotic mechanisms. A well-known marker compound to control the secretory capacity of tubules is p-aminohippuric acid, which is removed from plasma up to 91% during passage through the kidneys. Thus, the value for p-aminohippuric acid clearance reaches almost the total plasma perfusion rate of the kidney. A half-life of about 7 min results if a drug has the same rate of active tubular secretion as p-aminohippuric acid. Figure 9.10 shows some active transport systems that play a role in the renal clearance of drugs.

In proximal tubules, the organic anion transporters Oat1, Oat2, Oat3, Oat4, and Oat5 have been identified (Oat1–3 basolateral, Oat4 apical). In the kidney, Oat1 participates in the transport of many endogenous substrates such as α-ketoglutarate, cAMP, cGMP, folate, hippurate, prostaglandins, riboflavin, and different metabolites of neurotransmitters (e.g., 5-hydroxy-indolacetic acid, D,L-4-hydroxyl-3-methoxya-mygdalic acid). Oat2 transports, among others, cAMP, cholate, taurocholate,

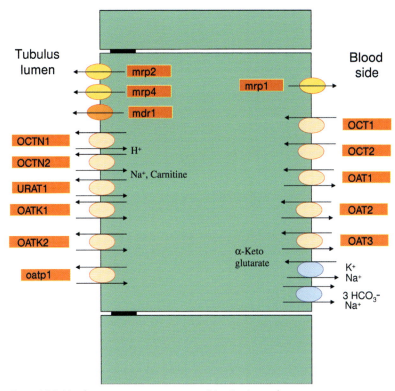

Figure 9.10 Membrane transport systems participating in renal secretion and reabsorption of drugs.

α-ketoglutarate, oxalacetate, and prostaglandin E_2. Estrone sulfate has a particular high affinity to Oat3. Table 9.4 shows examples of drugs transported by Oats.

It is assumed that the driving force for Oat1 follows a ternary mechanism, starting from Na^+/K^+-ATPase. The Na^+ gradient drives the Na^+/dicarboxylate transporter followed by a dicarboxylate/organic anion exchanger, by which organic anions are taken up into tubular cells. By this indirect way, negatively charged substrates can be transported against a concentration gradient and the electric potential. In contrast to Oat1, transport by Oat2/3 appears independent of Na^+ and glutarate [75]. Like many other transport proteins, Oats are regulated by multiple signaling pathways. For example, protein kinase C activation results in the downregulation of Oat1 activity. In addition to the regulation by phosphorylation and PKC, OAT activity can also be regulated by epidermal growth factor (EGF) through MAPK pathway [96].

Furthermore, in the apical tubular membrane, the transport proteins Oatp1, Oatk1, and Oatk2, the Na^+/phosphate cotransporter NPT1, and the primary active export proteins P-glycoprotein, Mrp2, Mrp4, and Bcrp have been identified [12, 95, 97–100]. Substrate requirements of the ABC transporters have already been discussed in the context of hepatic transport systems. Renal ABC transporters are well

Table 9.4 Substrates of Oat Transport Proteins.

Oat1	*Antibiotics:* amoxicillin, benzylpenicillin, carbenicillin, piperacillin, cephaloridine, cefadroxil, ceftazidime, levofloxacin, tetracycline *NSAIDs:* acetylsalicylic acid, antipyrine, paracetamol, ibuprofen, naproxen, phenacetin, piroxicam *Antiviral agents:* acyclovir, amantadine, azidothymidine *Diuretics:* furosemide, hydrochlorothiazide, ethacrynic acid *ACE inhibitors:* captopril, enalapril, ramipril, delapril, quinapril *Antineoplastics:* methotrexate, azathioprine, doxorubicin, 5-fluouracil *Antiepileptics:* valproic acid (from Ref. [95])
Oat2	Benzylpenicillin, erythromycin, tetracycline, rifampicin, glibenclamide, tolbutamide, zidovudine, ganciclovir, digoxin, enalapril, verapamil, methotrexate, acetylsalicylic acid
Oat3	Sitagliptin, cimetidine, ibuprofen, quinapril, indapamide, adefovir, cidofovir, tenofovir, cephalosporin, torsemide
Oat4	Candesartan, losartan, valsartan, torsemide, dehydroepiandrosterone sulfate

accessible for functional studies. Isolated kidney tubules can easily be incubated with fluorescent-labeled substrates, and subsequently, active secretion of the compounds into tubular lumens can be visualized by confocal laser scanning microscopy (Figure 9.11) [104–106].

Cation transporting membrane proteins in proximal tubules are Oct1 and Oct2, with a basolateral localization, as well as the apical protein OCTN1 and OCTN2. Similar to Oat proteins, these are involved in the renal secretion of a multitude of compounds (Oct1: choline, creatinine, guanidine, corticosterone, dopamine, histamine, and norepinephrine; Oct2: choline, creatinine, guanidine, corticosterone, dopamine, histamine, and epinephrine) and also interact with many drugs (Table 9.5)

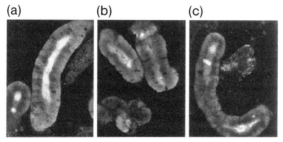

Figure 9.11 (a) Excretion of fluorescent-labeled anthelmintic drug ivermectin by P-glycoprotein in killifish proximal tubules. Tubules were incubated with 10 μM NBD-ivermectin; steady-state accumulation after 30 min of incubation. (b) Decreased excretion after inhibition of P-glycoprotein with 2 μM PSC-833, a potent P-glycoprotein blocking agent.

Table 9.5 Substrates of Oat Transport Proteins.

Oct1	Acyclovir, AZT, ganciclovir, indinavir, ritonavir, nelfinavir, saquinavir, amantandine, desipramine, nicotine, procainamide, vecuronium, cimetidine, clonidine, prazosin, reserpine, verapamil
Oct2	Amantandine, memantine, cocaine, desipramine, nicotine, procainamide, cimetidine, O-methylisoprenaline, prazosin, reserpine, verapamil
OCTN1	Cimetidine, procainamide, pyrilamine, quinidine, verapamil
OCTN2	Levofloxacin, grepafloxacin

9.3.4
Blood–Brain Barrier and Choroid Plexus

In order to ensure its complex sequences of information processing, the brain needs a constant ion homeostasis, which is guaranteed by the so-called blood–brain barrier. This barrier is formed by the endothelial cells of brain capillaries (Figure 9.12), which pervade the brain with a total length of 600 km and a mean distance of 40 μm to each other. They regulate passage of endogenous and exogenous compounds into the brain and allow only very limited access to the brain, thus minimizing fluctuation of solute concentrations as they occur in plasma. Brain capillaries are distinctively different from peripheral capillaries. Although the latter is fenestrated with relatively loose cell–cell contacts, endothelial cells of brain microvessels are connected by very dense tight junctions that exclude paracellular permeation. They produce extremely high transendothelial electrical resistances of up to $2000\,\Omega\,cm^2$. Permeation by pinocytosis across the blood–brain

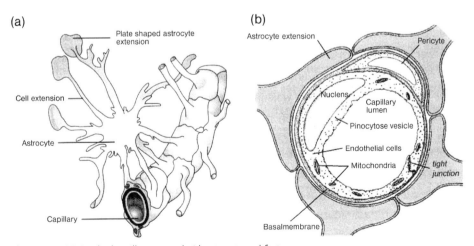

Figure 9.12 (a) Cerebral capillary covered with astrocyte end-feet. (b) Cross section of a brain microvessel formed by endothelial cells, pericytes, and astrocyte end-feet (from Ref. [104]).

barrier is also very limited, due to the very low pinocytotic activity of these endothelial cells. Besides the endothelial cells forming the actual barrier, the blood–brain barrier consists of additional structural elements, namely, astrocyte end-feet, pericytes, and basal membrane. Astrocytes belong to the so-called glia, which represents about 50% of all cells of the central nervous system (CNS). They are connected to neurons and capillaries via their end-feet, with a basal membrane separating astrocytes and capillary endothelial cells. About 80% of a capillary is covered with branches of the astrocytes. They are assumed to regulate junctional tightness as well as the expression of enzymes and transport systems in the endothelial cells.

Pericytes surround brain capillaries and are integrated into the basal membrane. They exhibit contractile properties, have phagocytic activity, and release growth hormones that are necessary for the proper assembly of capillaries.

The second barrier separating the central nervous system from blood circulation is the choroid plexus or plexus choroideus. It is formed by a vascular sponge, which is surrounded by epithelial cells (ECs) and which is located within the ventricles of the brain. The actual barrier is formed by the epithelial cells and not by the interior capillary. One of the major functions of the choroid plexus is the production of cerebrospinal fluid (liquor). In addition, the epithelial cells secrete ions, peptides, nutrients, and vitamins [105].

Brain capillary endothelial cells locate various transport proteins as well as receptor-mediated transport systems, which transport substrates from blood to brain and vice versa. Most of the transport proteins are isoforms of the proteins that have already been discussed in the context of other barrier tissues. For example, neutral, cationic, or anionic amino acids are transported from blood to brain by different Na^+-dependent and Na^+-independent amino acid transport proteins [106, 107]; glucose is taken up into endothelial cells by facilitated diffusion via the glucose transporter Glut1 [108, 109]. The pH-dependent transport protein Mct1, which has been detected in luminal and abluminal membranes, participates in the passage of monocarbonic acids, including pyruvate and lactic acid, and appears to play a role in the regulation of cerebral energy supply. Organic anions are transported by different proteins of the Oat and Oatp families [110]. Of particular importance for the barrier function are members of the ABC protein superfamily, which prevent entry of xenobiotics and toxic metabolites into the brain [111–115]. Best characterized is P-glycoprotein whose inhibition becomes clinically relevant for unwanted CNS side effects of P-glycoprotein substrates or for making cytostatics better accessible to the brain during the treatment of brain tumors [114]. Besides P-glycoprotein and several members of the Mrp family, which have already been discussed, the endothelial cells express the export proteins Abcg2 [116], which is also termed breast cancer resistance protein (Bcrp) due to its primary identification in breast cancer cell lines [117]. The spectrum of compounds interacting with Bcrp includes mitoxantrone, camptothecin-type topoisomerase I inhibitors, methotrexate, flavopiridol, quinazoline ErbB1 inhibitors, imatinib mesylate, herbal polyphenols, and flavonoids [118–121]. Its expression and function are regulated by estrogens [122–124].

Besides carrier-mediated transport processes, the blood–brain barrier features various receptor-mediated transport processes. Transferrin, insulin, and certain lipoproteins are guided through brain capillary endothelial cells by transcytosis. Especially, the transferring receptor gained interest in the development of drug delivery systems targeting the brain. Administration of "immuno"-liposomes, which had been linked to an antibody versus the transferrin receptor, resulted in elevated brain concentrations of otherwise poorly permeable compounds. For example, i.v. injection of daunomycin-loaded stealth liposomes led to a significant increase in daunomycin concentrations within the central nervous system. Thereafter, about 30 antibody molecules had been coupled to one liposome containing about 30 000 drug molecules. The transfer efficiency of the shuttle construct was several folds higher than that of a direct drug–antibody construct [15, 16].

Subtypes of the LDL receptor can also be used for an improved drug delivery to the brain: Coating of polymeric nanoparticles with surface-active compounds such as Tween 80 (polysorbate 80) leads to an association of plasma proteins such as apolipoprotein A1 or apolipoprotein E with the particles and subsequently leads to an improved CNS efficiency of incorporated drugs. It has been suggested that coated particles are taken into endothelial cells via an LDL receptor [17, 125].

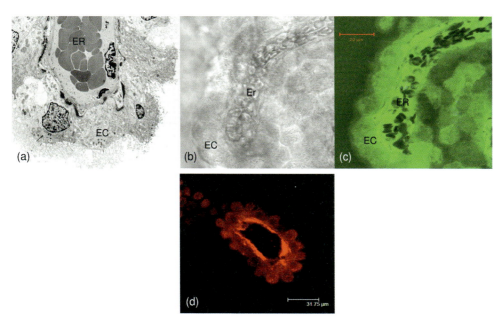

Figure 9.13 (a) Electron micrograph of the plexus choroideus. It contains a blood capillary filled with erythrocytes (ER) and surrounded by epithelial cell with microvilli oriented toward the liquor space of the ventricle. (b) Phase contrast image of a plexus blood capillary and epithelial cells. (c) Fluorescent microscopic image of (b). Fluorescein is actively transported from the liquor space to the blood capillary (with permission of D.S. Miller, NIEHS). (d) Active transport of Texas Red from the liquor space into the central capillary of choroids plexus tissue.

Active transport processes in the choroid plexus (Figure 9.13) have been less well characterized than those in the blood–brain barrier, although different transporters have been identified in the abluminal (blood-oriented) and luminal (liquor-oriented) membranes of plexus epithelial cells. The localization of ABC proteins deserves special attention: Although P-glycoprotein is located in the blood-oriented membrane of brain capillary endothelial cells, its localization in plexus epithelial cells seems to differ. Immunostaining suggests that P-glycoprotein in the choroid plexus is rather located in subapical vesicles than in the luminal membrane [126]. It has not yet been clarified whether the protein has any secretory function in the choroid plexus. Secretion into the cerebrospinal fluid would be detrimental, as potentially toxic compounds might reach the brain. In contrast, Mrp1 is highly expressed in the blood-sided membrane of plexus epithelial cells. Different from renal tubular cells, where it is expressed in the luminal membrane, Mrp4 also exhibits high expression levels in the blood-sided abluminal membrane of the plexus epithelium (Leggas et al., 2004). Its substrates comprise cyclic nucleotides [54, 127] (Schuetz et al., 1999) including antiviral drugs such as adefovir or tenofovir (Imaoka et al., 2007) and estradiol-17β-glucuronide. Furthermore, it interacts with cephalosporin antibiotics such as ceftizoxime or cefazolin (Ci et al., 2007) or diuretics such as hydrochlorothiazide or furosemide (Hasegawa et al., 2007). Recently, in an elegant approach, fluorescent-labeled compounds have been used in order to visualize transport of substrates from cerebrospinal fluid into blood circulation (Figure 9.13) [126, 127].

References

1 Seelig, A. (1998) A general pattern for substrate recognition by P-glycoprotein. *European Journal of Biochemistry*, **251**, 252–261.

2 Borst, P., Evers, R., Kool, M., and Wijnholds, F. J. (1999) The multidrug resistance protein family. *Biochimica et Biophysica Acta*, **1461**, 347–357.

3 Chen, Y. and Simon, S.M. (2000) In situ biochemical demonstration that P-glycoprotein is a drug efflux pump with broad specificity. *The Journal of Cell Biology*, **148**, 863–870.

4 Seelig, A., Blatter, X.L., and Wohnsland, F. (2000) Substrate recognition by P-glycoprotein and the multidrug resistance-associated protein MRP1: a comparison. *International Journal of Clinical Pharmacology and Therapeutics*, **38**, 111–121.

5 König, J., Rost, D., Cui, Y., and Keppler, D. (1999) Characterization of the human multidrug resistance protein isoform MRP3 localized to the basolateral hepatocyte membrane. *Hepatology*, **29**, 1156–1163.

6 Murer, H. and Hopfer, U. (1974) Demonstration of electrogenic Na^+-dependent D-glucose transport in intestinal brush border membranes. *Proceedings of the National Academy of Sciences of the United States of America*, **71**, 484–488.

7 Hopfer, U. (1976) Sugar and amino acid transport in animal cells. *Horizons in Biochemistry and Biophysics*, **2**, 106–133.

8 Silverman, M. (1976) Glucose transport in the kidney. *Biochimica et Biophysica Acta*, **457**, 303–351.

9 Kinne, R.K. (1995) Amino acid transporters. *Current Opinion in Nephrology and Hypertension*, **4**, 412–415.

10 Ganapathy, V. and Leibach, F.H. (1991) Proton-coupled solute transport in the animal cell plasma membrane. *Current Opinion in Cell Biology*, **3**, 695–701.

11 Daniel, H. (1996) Function and molecular structure of brush border membrane peptide/H^+ symporters. *The Journal of Membrane Biology*, **154**, 197–203.

12 Inui, K.I., Masuda, S., and Saito, H. (2000) Cellular and molecular aspects of drug transport in the kidney. *Kidney International*, **58**, 944–958.

13 Addison, J.M., Burton, J.A., Dalrymple, J.A., Matthews, D.D., Payner, J.W., Sleisinger, M.H., and Wilkinson, S. (1975) A common mechanism for transport of di- and tripeptides by hamster jejunum *in vitro*. *Clinical Science & Molecular Medicine*, **49**, 313–322.

14 Bai, J.P.F. and Amidon, G.L. (1992) Structural specificity of mucosal-cell transport and metabolism of peptide drugs: implication for oral peptide delivery. *Pharmaceutical Research*, **9**, 969–978.

15 Huwyler, J., Wu, D., and Pardridge, W.M. (1996) Brain drug delivery of small molecules using immunoliposomes. *Proceedings of the National Academy of Sciences of the United States of America*, **93**, 14164–14169.

16 Huwyler, J., Yang, J., and Pardridge, W.M. (1997) Receptor mediated delivery of daunomycin using immunoliposomes: pharmacokinetics and tissue distribution in the rat. *The Journal of Pharmacology and Experimental Therapeutics*, **282**, 1541–1546.

17 Alyaudtin, R.N., Reichel, A., Löbenberg, R., Ramge, P., Kreuter, J., and Begley, D.J. (2001) Interaction of poly (butylcyanoacrylate) nanoparticles with the blood–brain barrier *in vivo* and *in vitro*. *Journal of Drug Targeting*, **9**, 209–221.

18 Crane, R.K. (1960) Intestinal absorption of sugars. *Physiological Reviews*, **40**, 789–825.

19 Wright, E.M., Hirayama, B., Hazama, A., Loo, D.D., Supplisson, S., Turk, E., and Hager, K.M. (1993) The sodium/glucose cotransporter (SGLT1). *Society of General Physiologists Series*, **48**, 229–241.

20 Wright, E.M., Loo, D.D., Panayotova-Heiermann, M., Hirayama, B.A., Turk, E., Eskandari, S., and Lam, T.J. (1998) Structure and function of the Na^+/glucose cotransporter. *Acta Physiologica Scandinavica Supplement*, **643**, 257–264.

21 Takata, K. (1996) Glucose transporters in the transepithelial transport of glucose. *Journal of Electron Microscopy*, **45**, 275–284.

22 Terada, T., Shimada, Y., Pan, X., Kishimoto, K., Sakurai, T., and Doi, R. *et al.* (2005) Expression profiles of various transporters for oligopeptides, amino acids and organic ions along the human digestive tract. *Biochemical Pharmacology*, **70**, 1756–1763.

23 Lee, V.H. (2000) Membrane transporters. *European Journal of Pharmaceutical Sciences*, **11** (Suppl 2), S41–S50.

24 Sugawara, M., Huang, W., Fei, Y.J., Leibach, F.H., Ganapathy, V., and Ganapathy, M.E. (2000) Transport of valganciclovir, a ganciclovir prodrug, via peptide transporters PEPT1 and PEPT2. *Journal of Pharmaceutical Sciences*, **89**, 781–789.

25 Petzinger, E., Wickboldt, A., Pagels, P., Starke, D., and Kramer, W. (1999) Hepatobiliary transport of bile acid amino acid, bile acid peptide, and bile acid oligonucleotide conjugates in rats. *Hepatology*, **30**, 1257–1268.

26 Baringhaus, K.H., Matter, H., Stengelin, S., and Kramer, W. (1999) Substrate specificity of the ileal and the hepatic Na^+/bile acid cotransporters of the rabbit. II. A reliable 3D QSAR pharmacophore model for the ileal Na^+/bile acid

cotransporter. *Journal of Lipid Research*, **40**, 2158–2168.

27 Fricker, G., Drewe, J., Huwyler, J., Gutmann, H., and Beglinger, C. (1996) Relevance of P-glycoprotein for the enteral absorption of cyclosporin A: *in vitro–in vivo* correlation. *British Journal of Pharmacology*, **118**, 1841–1847.

28 Juliano, R.L. and Ling, V. (1976) A surface glycoprotein modulating drug permeability in Chinese hamster ovary cell mutants. *Biochimica et Biophysica Acta*, **455**, 152–162.

29 Gottesman, M.M. and Pastan, I. (1988) The multidrug transporter, a double-edged sword. *The Journal of Biological Chemistry*, **263**, 12163–12166.

30 Endicott, J.A. and Ling, V. (1989) The biochemistry of P-glycoprotein-mediated multidrug resistance. *Annual Review of Biochemistry*, **58**, 137–171.

31 Borst, P. (1997) Multidrug resistant proteins. *Seminars in Cancer Biology*, **8**, 131–134.

32 Schinkel, A.H. and Borst, P. (1991) Multidrug resistance mediated by P-glycoproteins. *Seminars in Cancer Biology*, **2**, 213–226.

33 Schinkel, A.H. (1997) The physiological function of drug-transporting P-glycoproteins. *Seminars in Cancer Biology*, **8**, 161–170.

34 Fromm, M.F. (2000) P-glycoprotein: a defense mechanism limiting oral bioavailability and CNS accumulation of drugs. *International Journal of Clinical zPharmacology and Therapeutics*, **38**, 69–74.

35 Sharom, F.J. (1997) The P-glycoprotein efflux pump: how does it transport drugs? *The Journal of Membrane Biology*, **160**, 61–175.

36 Ueda, K., Yoshida, A., and Amachi, T. (1999) Recent progress in P-glycoprotein research. *Anti-Cancer Drug Design*, **14**, 115–121.

37 Jones, P.M. and George, A.M. (2000) Symmetry and structure in P-glycoprotein and ABC transporters. What goes around comes around. *European Journal of Biochemistry*, **267**, 5298–5305.

38 Schuetz, E.G. and Schinkel, A.H. (1999) Drug disposition as determined by the interplay between drug-transporting and drug-metabolizing systems. *Journal of Biochemical and Molecular Toxicology*, **13**, 219–222.

39 Bennet, L.Z. and Cummings, C.L. (2001) The drug efflux–metabolism alliance: biochemical aspects. *Advanced Drug Delivery Reviews*, **50** (Suppl 1), S3–S11.

40 Zhang, Y. and Bennet, L.Z. (2001) The gut as a barrier to drug absorption: combined role of cytochrome P450 3A and P-glycoprotein. *Clinical Pharmacokinetics*, **40**, 159–168.

41 Terao, T., Hisanaga, E., Sai, Y., Tamai, I., and Tsuji, A. (1996) Active secretion of drugs from the small intestinal epithelium in rats by P-glycoprotein functioning as an absorption barrier. *The Journal of Pharmacy and Pharmacology*, **48**, 1083–1089.

42 van Asperen, J., van Tellingen, O., and Beijnen, J.H. (1998) The pharmacological role of P-glycoprotein in the intestinal epithelium. *Pharmacological Research*, **37**, 429–435.

43 Sababi, M., Borga, O., and Hultkvist-Bengtsson, U. (2001) The role of P-glycoprotein in limiting intestinal regional absorption of digoxin in rats. *European Journal of Pharmaceutical Sciences*, **14**, 21–27.

44 Fricker, G. and Miller, D.S. (2002) Relevance of multidrug resistance proteins for intestinal drug absorption *in vitro* and *in vivo*. *Pharmacology & Toxicology*, **90**, 5–13.

45 Schinkel, A.H., Smit, J.J., van Tellingen, O., Beijnen, J.H., Wagenaar, E., van Deemter, L., Mol, C.A., van der Valk, M.A., Robanus-Maandag, E.C., te Riele, H.P. *et al.* (1994) Disruption of the mouse mdr1a P-glycoprotein gene leads to a deficiency in the blood–brain barrier and

to increased sensitivity to drugs. *Cell*, **77**, 491–502.

46 Schinkel, A.H., Mol, C.A., Wagenaar, E., van Deemter, L., Smit, J.J., and Borst, P. (1995) Multidrug resistance and the role of P-glycoprotein knockout mice. *European Journal of Cancer*, **31**, 1295–1298.

47 Mayer, U., Wagenaaar, E., Beijnen, J.H., Smit, J.H., Meijer, D.K., van Asperen, J., Borst, P., and Schinkel, A.H. (1996) Substantial excretion of digoxin via the intestinal mucosa and prevention of long-term digoxin accumulation in the brain by the mdr1a P-glycoprotein. *British Journal of Pharmacology*, **119**, 1038–1044.

48 Leahey, E.B., Jr., Reiffel, J.A., Drusin, R.E., Heissenbuttel, R.H., Lovejoy, W.P., and Bigger, J.T., Jr. (1978) Interaction between quinidine and digoxin. *The Journal of the American Medical Association*, **240**, 533–534.

49 Pedersen, K.E., Christiansen, B.D., Klitgaard, N.A., and Nielsen-Kudsk, F. (1983) Effect of quinidine on digoxin bioavailability. *European Journal of Clinical Pharmacology*, **24**, 41–47.

50 Kim, R.B., Fromm, M.F., Wandel, C., Leake, B., Wood, A.J., Roden, D.M., and Wilkinson, G.R. (1998) The drug transporter P-glycoprotein limits oral absorption and brain entry of HIV-1 protease inhibitors. *The Journal of Clinical Investigation*, **101**, 289–294.

51 Sparreboom, A., van Asperen, J., Mayer, U., Schinkel, A.H., Smit, J.W., Meijer, D.K., Borst, P., Nooijen, W.J., Beijnen, J.H., and van Tellingen, O. (1997) Limited oral bioavailability and active epithelial excretion of paclitaxel (Taxol) caused by P-glycoprotein in the intestine. *Proceedings of the National Academy of Sciences of the United States of America*, **94**, 2031–2035.

52 Polizzi, D., Pratesi, G., Monestiroli, S., Tortoreto, M., Zunino, F., Bombardelli, E., Riva, A., Morazzoni, P., Colombo, T., D'Incalci, M., and Zucchetti, M. (2000) Oral efficacy and bioavailability of a novel taxane. *Clinical Cancer Research*, **6**, 2070–2074.

53 Berggren, S., Gall, C., Wollnitz, N., Ekelund, M., Karlbom, U., Hoogstraate, J., Schrenk, D., and Lennernas, H. (2007) Gene and protein expression of P-glycoprotein, MRP1, MRP2, and CYP3A4 in the small and large human intestine. *Molecular Pharmaceutics*, **4**, 252–257.

54 Chen, J.J. *et al.* (2001) Maintenance of serotonin in the intestinal mucosa and ganglia of mice that lack the high-affinity serotonin transporter: abnormal intestinal motility and the expression of cation transporters. *The Journal of Neuroscience*, **21**, 6348–6361.

55 Koepsell, H. (2004) Polyspecific organic cation transporters: their functions and interactions with drugs. *Trends in Pharmacological Sciences*, **25**, 375–382.

56 Martel, F., Vetter, T., Russ, H., Gründemann, D., Azevedo, I., Koepsell, H., and Schomig, E. (1996) Transport of small organic cations in the rat liver. The role of the organic cation transporter OCT1. *Naunyn-Schmiedeberg's Archives of Pharmacology*, **354**, 320–326.

57 Denk, U.G. *et al.* (2004) Down-regulation of the organic cation transporter 1 of rat liver in obstructive cholestasis. *Hepatology*, **39**, 1382–1389.

58 Hagenbuch, B. and Meier, P.J. (2003) Organic anion transporting polypeptides of the OATP/SLC21 family: phylogenetic classification as OATP/SLCO superfamily, new nomenclature and molecular/functional properties. *Pflugers Archives*, **447**, 653–665.

59 Jacquemin, E., Hagenbuch, B., Stieger, B., Wolkoff, A.W., and Meier, P.J. (1994) Expression cloning of a rat liver Na^+-independent organic anion transporter. *Proceedings of the National Academy of Sciences of the United States of America*, **91** (1), 133–137.

60 Meier, P.J., Eckhardt, U., Schroeder, A., Hagenbuch, B., and Stieger, B. (1997)

Substrate specificity of sinusoidal bile acid and organic anion uptake systems in rat and human liver. *Hepatology*, **26**, 1667–1677.

61 Müller, M. and Jansen, P.L. (1997) Molecular aspects of hepatobiliary transport. *The American Journal of Physiology*, **272**, G1285–G1303.

62 Shi, X., Bai, S., Ford, A.C., Burk, R.D., Jacquemin, E., Hagenbuch, B., Meier, P.J., and Wolkoff, A.W. (1995) Stable inducible expression of a functional rat liver organic anion transport protein in HeLa cells. *The Journal of Biological Chemistry*, **270** (43), 25591–25595.

63 Li, L., Lee, T.K., Meier, P.J., and Ballatori, N. (1998) Identification of glutathione as a driving force and leukotriene C4 as a substrate for oatp1, the hepatic sinusoidal organic solute transporter. *The Journal of Biological Chemistry*, **273**, 16184–16191.

64 Hayhurst, G.P., Lee, Y.H., Lambert, G., Ward, J.M., and Gonzalez, F.J. (2001) Hepatocyte nuclear factor 4alpha (nuclear receptor 2A1) is essential for maintenance of hepatic gene expression and lipid homeostasis. *Molecular and Cellular Biology*, **21** (4), 1393–1403.

65 Glavy, J.S., Wu, S.M., Wang, P.J., Orr, G.A., and Wolkoff, A.W. (2000) Down-regulation by extracellular ATP of rat hepatocyte organic anion transport is mediated by serine phosphorylation of oatp1. *The Journal of Biological Chemistry*, **275**, 1479–1484.

66 Guo, G.L. and Klaassen, C.D. (2001) Protein kinase C suppresses rat organic anion transporting polypeptide 1- and 2-mediated uptake. *The Journal of Pharmacology and Experimental Therapeutics*, **299** (2), 551–557.

67 Noe, B., Hagenbuch, B., Stieger, B., and Meier, P.J. (1997) Isolation of a multispecific organic anion and cardiac glycoside transporter from rat brain. *Proceedings of the National Academy of Sciences of the United States of America*, **94**, 10346–10350.

68 Staudinger, J.L., Goodwin, B., Jones, S.A., Hawkins-Brown, D., MacKenzie, K.I., LaTour, A., Liu, Y., Klaassen, C.D., Brown, K.K., Reinhard, J., Willson, T.M., Koller, B.H., and Kliewer, S.A. (2001) The nuclear receptor PXR is a lithocholic acid sensor that protects against liver toxicity. *Proceedings of the National Academy of Sciences of the United States of America*, **98** (6), 3369–3374.

69 Guo, G.L., Staudinger, J., Ogura, K., and Klaassen, C.D. (2002) Induction of rat organic anion transporting polypeptide 2 by pregnenolone-16alpha-carbonitrile is via interaction with pregnane X receptor. *Molecular Pharmacology*, **61**, 832–839.

70 Jones, S.A., Moore, L.B., Shenk, J.L., Wisely, G.B., Hamilton, G.A., McKee, D.D., Tomkinson, N.C., LeCluyse, E.L., Lambert, M.H., Willson, T.M., Kliewer, S.A., and Moore, J.T. (2000) The pregnane X receptor: a promiscuous xenobiotic receptor that has diverged during evolution. *Molecular Endocrinology*, **14**, 27–39.

71 Kakyo, M., Unno, M., Tokui, T., Nakagomi, R., Nishio, T., Iwasashi, H., Nakai, D., Seki, M., Suzuki, M., Naitoh, T., Matsuno, S., Yawo, H., and Abe, T. (1999) Molecular characterization and functional regulation of a novel rat liver-specific organic anion transporter rlst-1. *Gastroenterology*, **117**, 770–775.

72 Cattori, V., Hagenbuch, B., Hagenbuch, N., Stieger, B., Ha, R., Winterhalter, K.E., and Meier, P.J. (2000) Identification of organic anion transporting polypeptide 4 (Oatp4) as a major full-length isoform of the liver-specific transporter-1 (rlst-1) in rat liver. *FEBS Letters*, **474**, 242–245.

73 Nishio, T., Adachi, H., Nakagomi, R., Tokui, T., Sato, E., Tanemoto, M., Fujiwara, K., Okabe, M., Onogawa, T., Suzuki, T., Nakai, D., Shiiba, K.,

Suzuki, M., Ohtani, H., Kondo, Y., Michiaki, U., Sadayoshi, I., Iinuma, K., Nunoki, K., Matsuno, S., and Abe, T. (2000) Molecular identification of a rat novel organic anion transporter moat1, which transports prostaglandin D_2, leukotriene C_4, and taurocholate. *Biochemistry and Biophysics Research Communications*, **275** (3), 831–838.

74 Kanai, N., Lu, R., Satriano, J.A., Bao, Y., Wolkoff, A.W., and Schuster, V.L. (1995) Identification and characterization of a prostaglandin transporter. *Science*, **268** (5212), 866–869.

75 Kusuhara, H., Sekine, T., Utsunomiya-Tate, N., Tsuda, M., Kojima, R., Cha, S.H., Sugiyama, Y., Kanai, Y., and Endou, H. (1999) Molecular cloning and characterization of a new multispecific organic anion transporter from rat brain. *The Journal of Biological Chemistry*, **274**, 13675–13680.

76 Cha, S.H., Sekine, T., Fukushima, J.I., Kanai, Y., Kobayashi, Y., Goya, T., and Endou, H. (2001) Identification and characterization of human organic anion transporter 3 expressing predominantly in the kidney. *Molecular Pharmacology*, **59**, 1277–1286.

77 Hooiveld, G.J., van Montfoort, J.E., Meijer, D.K., and Muller, M. (2001) Function and regulation of ATP-binding cassette transport proteins involved in hepatobiliary transport. *European Journal of Pharmaceutical Sciences*, **12**, 525–543.

78 Borst, P., Zelcer, N., and van Helvoort, A. (2000) ABC transporters in lipid transport. *Biochimica et Biophysica Acta*, **1486**, 128–144.

79 Smith, A.J., van Helvoort, A., van Meer, G., Szabo, K., Welker, E., Szakacs, G., Varadi, A., Sarkadi, B., and Borst, P. (2000) MDR3 P-glycoprotein, a phosphatidylcholine translocase, transports several cytotoxic drugs and directly interacts with drugs as judged by interference with nucleotide trapping. *The Journal of Biological Chemistry*, **275**, 23530–23539.

80 Kullak-Ublick, G.A., Stieger, B., Hagenbuch, B., and Meier, P.J. (2000) Hepatic transport of bile salts. *Seminars in Liver Disease*, **20**, 273–292.

81 Scheimann, A.O., Strautnieks, S.S., Knisely, A.S., Byrne, J.A., Thompson, R.J., and Finegold, M.J. (2007) Mutations in bile salt export pump (ABCB11) in two children with progressive familial intrahepatic cholestasis and cholangiocarcinoma. *The Journal of Pediatrics*, **150**, 556–559.

82 Hirohashi, T., Suzuki, H., Ito, K., Ogawa, K., Kume, K., Shimizu, T., and Sugiyama, Y. (1998) Hepatic expression of multidrug resistance-associated protein-like proteins maintained in Eisai hyperbilirubinemic rats. *Molecular Pharmacology*, **53**, 1068–1075.

83 Madon, J., Hagenbuch, B., Landmann, L., Meier, P.J., and Stieger, B. (2000) Transport function and hepatocellular localization of mrp6 in rat liver. *Molecular Pharmacology*, **57**, 634–641.

84 Kool, M., van der Linden, M., de Haas, M., Baas, F., and Borst, P. (1999) Expression of human MRP6, a homologue of the multidrug resistance protein gene MRP1, in tissues and cancer cells. *Cancer Research*, **59**, 175–182.

85 Müller, M., Roelofsen, H., and Jansen, P.L. (1996) Secretion of organic anions by hepatocytes: involvement of homologues of the multidrug resistance protein. *Seminars in Liver Disease*, **16**, 211–220.

86 Keppler, D., Jedlitschky, G., and Leier, I. (1998) Transport function and substrate specificity of multidrug resistance protein. *Methods in Enzymology*, **292**, 607–616.

87 Koenig, J., Nies, A.T., Cui, Y., Leier, I., and Keppler, D. (1999) Conjugate export pumps of the multidrug resistance protein (MRP) family: localization, substrate specificity, and MRP2-mediated drug resistance. *Biochimica et Biophysica Acta*, **1461**, 377–394.

88 Keppler, D. and Koenig, J. (2000) Hepatic secretion of conjugated drugs and endogenous substances. *Seminars in Liver Disease*, **20**, 265–272.

89 van Aubel, R.A., Koenderink, J.B., Peters, J.G., van Os, C.H., and Russel, F.G. (1999) Mechanisms and interaction of vinblastine and reduced glutathione transport in membrane vesicles by the rabbit multidrug resistance protein Mrp2 expressed in insect cells. *Molecular Pharmacology*, **56**, 714–719.

90 Renes, J., de Vries, E.G., Nienhuis, E.F., Jansen, P.L., and Muller, M. (1999) ATP- and glutathione-dependent transport of chemotherapeutic drugs by the multidrug resistance protein MRP1. *British Journal of Pharmacology*, **126**, 681–688.

91 Leslie, E.M., Deeley, R.G., and Cole, S.P. (2001) Toxicological relevance of the multidrug resistance protein 1, MRP1 (ABCC1) and related transporters. *Toxicology*, **167**, 3–23.

92 Keppler, D. and König, J. (2000) Hepatic secretion of conjugated drugs and endogenous substances. *Seminars in Liver Disease*, **20**, 265–272.

93 Ogawa, K., Suzuki, H., Hirohashi, T., Ishikawa, T., Meier, P.J., Hirose, K., Akizawa, T., Yoshioka, M., and Sugiyama, Y. (2000) Characterization of inducible nature of MRP3 in rat liver. *American Journal of Physiology. Gastrointestinal and Liver Physiology*, **278**, G438–G446.

94 Wehner, F., Kinne, R.K., and Petzinger, E. (1996) Second International Ringberg Conference: "Cell Biology and Molecular Basis of Liver Transport". *Hepatology*, **24**, 259–267.

95 Sekine, T., Cha, S.H., and Endou, H. (2000) The multispecific organic anion transporter (OAT) family. *Pflugers Archives*, **440**, 337–350.

96 Zhou, F. and You, G. (2007) Molecular insights into the structure–function relationship of organic anion transporters OATs. *Pharmaceutical Research*, **24**, 28–36.

97 van Aubel, R.A., Masereeuw, R., and Russel, F.G. (2000) Molecular pharmacology of renal organic anion transporters. *The American Journal of Physiology*, **279**, F216–F232.

98 Uchino, H., Tamai, K., Yamashita, K., Minemoto, Y., Sai, Y., Yabuuchi, H., Miyamoto, K., Takeda, E., and Tsuji, A. (2000) p-Aminohippuric acid transport at renal apical membrane mediated by human inorganic phosphate transporter NPT1. *Biochemical and Biophysical Research Communications*, **270**, 254–259.

99 Schaub, T.P., Kartenbeck, J., Koenig, J., Spring, H., Dorsam, J., Staehler, G., Storkel, S., Thon, W.F., and Keppler, D. (1999) Expression of the MRP2 gene-encoded conjugate export pump in human kidney proximal tubules and in renal cell carcinoma. *Journal of the American Society of Nephrology*, **10**, 1159–1169.

100 Hilgendorf, C., Ahlin, G., Seithel, A., Artursson, P., Ungell, A.L., and Karlsson, J. (2007) Expression of thirty-six drug transporter genes in human intestine, liver, kidney, and organotypic cell lines. *Drug Metabolism and Disposition*, **35**, 1333–1340.

101 Miller, D., Fricker, G., and Drewe, J. (1997). p-Glycoprotein mediated transport of a fluorescent rapamycin derivative in renal proximal tubule. *Journal of Pharmacology and Experimental Therapeutics*, **282**, 440–444.

102 Fricker, G., Gutmann, H., Droulle, A., Drewe, J., and Miller, D.S. (1999) Epithelial transport of anthelmintic ivermectin in a novel model of isolated proximal kidney tubules. *Pharmaceutical Research*, **16**, 1570–1575.

103 Gutmann, H., Miller, D.S., Droulle, A., Drewe, J., Fahr, A., and Fricker, G. (2000). P-glycoprotein and mrp2-mediated octreotide transport in renal proximal tubule. *British Journal of Pharmacology*, **129**, 251–256.

104 Goldstein, G.W. and Betz, A.L. (1986) The blood–brain barrier. *Scientific American*, **255**, 70–79.

105 Spector, R. and Johanson, C.E. (1989) The mammalian choroid plexus. *Scientific American*, **261**, 48–53.
106 Smith, Q.R. (2000) Transport of glutamate and other amino acids at the blood–brain barrier. *The Journal of Nutrition*, **130**, S1016–S1022.
107 Pardridge, W.M. (1998) Blood–brain barrier carrier-mediated transport and brain metabolism of amino acids. *Neurochemical Research*, **23**, 635–644.
108 Olson, A.L. and Pessin, J.E. (1996) Structure, function, and regulation of the mammalian facilitative glucose transporter gene family. *Annual Review of Nutrition*, **16**, 235–256.
109 Kumagai, A.K. (1999) Glucose transport in brain and retina: implications in the management and complications of diabetes. *Diabetes/Metabolism Research and Reviews*, **15**, 261–273.
110 Bart, J., Groen, H.J., Hendrikse, N.H., van der Graaf, W.T., Vaalburg, W., and de Vries, E.G. (2000) The blood–brain barrier and oncology: new insights into function and modulation. *Cancer Treatment Reviews*, **26**, 449–462.
111 Miller, D.S., Nobmann, S.N., Gutmann, H., Toeroek, M., Drewe, J., and Fricker, G. (2000) Xenobiotic transport across isolated brain microvessels studied by confocal microscopy. *Molecular Pharmacology*, **58**, 1357–1367.
112 Fricker, G. and Miller, D.S. (2004) Modulation of drug transporters at the blood–brain barrier. *Pharmacology*, **70**, 169–176.
113 Bauer, B., Hartz, A.M., Fricker, G., and Miller, D.S. (2005) Modulation of P-glycoprotein transport function at the blood–brain barrier. *Experimental Biology and Medicine*, **230**, 118–127.
114 Fellner, S., Schaffrik, M., Spruß, T., Färber, L., Gschaidhammer, H., Bernhardt, G., Buschauer, A., Miller, D.S., Bauer, B., Graeff, C., and Fricker, G. (2002) Transport of Paclitaxel (Taxol) across the blood brain barrier in vitro and in vivo. *The Journal of Clinical Investigation* **110**, 1309–1318.
115 Zhang, Y., Han, H., Elmquist, W.F., and Miller, D.W. (2000) Expression of various multidrug resistance-associated protein (MRP) homologues in brain microvessel endothelial cells. *Brain Research*, **8**, 148–153.
116 Eisenblätter, T., Huwel, S., and Galla, H.J. (2003) Characterisation of the brain multidrug resistance protein (BMDP/ABCG2/BCRP) expressed at the blood–brain barrier. *Brain Research*, **971**, 221–231.
117 Doyle, L.A., Yang, W., Abruzzo, L.V., Krogmann, T., Gao, Y., Rishi, A.K., and Ross, D.D. (1998) A multidrug resistance transporter from human MCF-7 breast cancer cells. *Proceedings of the National Academy of Sciences of the United States of America*, **95** (26), 15665–15670.
118 Doyle, L.A. and Ross, D.D. (2003) Multidrug resistance mediated by the breast cancer resistance protein BCRP (ABCG2). *Oncogene*, **22**, 7340–7358.
119 Cooray, H.C., Janvilisri, T., van Veen, H.W., Hladky, S.B., and Barrand, M.A. (2004) Interaction of the breast cancer resistance protein with plant polyphenols. *Biochemical and Biophysical Research Communications*, **317**, 269–275.
120 Zhang, S., Yang, X., and Morris, M.E. (2004) Flavonoids are inhibitors of breast cancer resistance protein (ABCG2)-mediated transport. *Molecular Pharmacology*, **65**, 1208–1216.
121 Burger, H., van Tol, H., Boersma, A.W., Brok, M., Wiemer, E.A., Stoter, G., and Nooter, K. (2004) Imatinib mesylate (STI571) is a substrate for the breast cancer resistance protein (BCRP)/ABCG2 drug pump. *Blood*, **104**, 2940–2942.
122 Ee, P.L.R., Kamalakaran, S., Tonetti, D., He, X., Ross, D.D., and Beck, W.T. (2004) Identification of a novel estrogen response element in the breast cancer resistance protein (ABCG2) gene. *Cancer Research*, **64**, 1247–1251.

123 Imai, Y., Ishikawa, E., Asada, S., and Sugimoto, Y. (2005) Estrogen-mediated post transcriptional down-regulation of breast cancer resistance protein/ABCG2. *Cancer Research*, **65**, 596–604.

124 Wang, H., Zhou, L., Gupta, A., Vethanayagam, R.R., Zhang, Y., Unadkat, J.D., and Mao, Q. (2006) Regulation of BCRP/ABCG2 expression by progesterone and 17beta-estradiol in human placental BeWo cells. *American Journal of Physiology. Endocrinology and Metabolism*, **290**, E798–807.

125 Ramge, P., Unger, R.E., Oltrogge, J.B., Zenker, D., Begley, D., Kreuter, J., and Von Briesen, H. (2000) Polysorbate-80 coating enhances uptake of polybutylcyanoacrylate (PBCA)-nanoparticles by human and bovine primary brain capillary endothelial cells. *The European Journal of Neuroscience*, **12**, 1931–1940.

126 Baehr, C., Reichel, V., and Fricker, G. (2006) Choroid plexus epithelial monolayers – a cell culture model from porcine brain. *Cerebrospinal Fluid Research*, **3**, 13.

127 Reichel, V., Masereeuw, R., van den Heuvel, J.J.M.W., Miller, D.S., and Fricker, G. (2007) Transport of a fluorescent cAMP analog in teleost proximal tubules. *The American Journal of Physiology*, **293**, R2382–R2389.

128 Breen, C.M., Sykes, D.B., Baehr, C., Fricker, G., and Miller, D.S. (2004) Fluoresceinmethotrexate transport in intact rat choroid plexus. *American Journal of Physiology*, **287**, F562–F569.

10
Transport at the Blood–Brain Barrier
Winfried Neuhaus and Christian R. Noe

10.1
The Blood–Brain Barrier

The blood–brain barrier (BBB) is a selective barrier formed by endothelial cells that line the cerebral microvessels and is present in all vertebrate animals [1]. It consists of a network of capillaries in the human brain with an approximately total length of 600 km and an average distance of 40 µm from each capillary [2].

The concept of a barrier preventing the movement of certain materials between blood and the adult brain originates from studies of dye injections made into the circulation. In 1885, the German scientist Paul Ehrlich reported that after parenteral injection in adult animals of a variety of vital dyes, practically all animal organs were stained, except the brain and spinal cord. Although Ehrlich himself described the observation that after the intravenous application of some aniline dyes, most of the animal tissues were stained with the exception of the central nervous system (CNS), he thought that this difference was due to different binding affinities [3, 4]. In 1900, Lewandowsky introduced the term "blood–brain barrier" to describe the phenomenon that intravenously injected cholic acids or sodium ferrocyanide had no pharmacological effects on the CNS, whereas neurological symptoms occurred after intraventricular application of the same substances [3]. Further experiments by Goldmann, an associate of Ehrlich, however, indicated that after injection of the acidic dye Trypan blue into the cerebrospinal fluid (CSF) of dogs and rabbits, the brain, but not the bloodstream and other organs, was stained [5]. Therefore, it was hypothesized that a barrier between the blood and the brain exists, which was termed the blood–brain barrier, that serves as a barrier to the free entry of molecules into the brain from the blood circulation. Walter and Spatz were the first in the early 1930s who differentiated between the blood–brain barrier and blood–CSF barrier (BCSFB) [6, 7]. They assumed that gas exchange in the CNS is supported by the CSF flow in an insufficient manner. In 1946, Krogh [8] thought that active transport systems may play an important role in the delivery of nutrients into the CNS. However, it was still believed that glia cells were the main components of the blood–brain barrier [9]. One of the most important key steps to learn about the anatomy of the blood–brain barrier was the use of electron scanning microscopy. This novel technique enabled to confirm the anatomical evidence of this barrier in the late 1960s. In 1967, Reese and Karnovsky showed for the first time at ultrastructural level that the endothelium of mouse cerebral capillaries constitutes a structural barrier to the horseradish peroxidase (HRP) [10]. They found that HRP was able to enter the interendothelial spaces only up to, but not beyond, the first luminal interendothelial tight junctions (TJ) in cerebral capillaries [3]. Furthermore, Brightman demonstrated with his studies that on injecting HRP or ferritin intraventricularly, the anatomical site of the BBB was neither the astrocytic end-feet nor the basement membrane, but rather the endothelium itself [11, 12]. Later on, it could be shown by freeze fracture that the tight junctions between endothelial cells of CNS capillaries and venules are arranged in six to eight parallel strands with complex net-like anastomoses all along the upper circumference of the endothelial cell [13, 14]. The historical data were mainly extracted from a review from Ribatti et al. [3].

The BBB serves two important functions. First, it protects the brain from xenobiotics, and second, it maintains an ideal environment for the brain [15].

In addition to the blood–brain barrier, two other barrier layers limit and regulate molecular exchange at the interface between the blood and the neural tissue and its fluid spaces: the choroid plexus epithelium between blood and ventricular CSF and the arachnoid epithelium between blood and subarachnoid CSF. These CNS barriers perform a number of functions such as the ionic homeostasis, the restriction of small molecule permeation, the specific transport of small molecules required to enter or leave the brain, the restriction and regulation of large molecule traffic by reducing the fluid-phase endocytosis via pinocytotic vesicles, the separation of peripheral and central neurotransmitter pools, and the immune privilege [16].

The extracellular space of the brain can be divided into two major compartments, the CSF and the interstitial fluid (ISF). The CSF and the ISF are separated from the blood by the choroid plexus or the BCSFB and the brain capillary or BBB, respectively. No anatomical barrier exists between the CSF and the ISF; a functional barrier is built up by the flow of CSF from its formation site (choroid plexus) to its absorption site (arachnoid villi) [15]. In the case of a human brain, 20 ml CSF is produced per hour and the complete turnover of the total 100 ml CSF occurs approximately within 4–5 h, whereas only 2 ml ISF is renewed per hour compared to the total amount of 300 ml ISF [17, 18]. Neurons are bathed by the extracellular (or interstitial) fluid of the brain (ECF = ISF) that forms the microenvironment of the CNS [19]. ISF and CSF are low-protein fluids (plasma:CSF ratio $\sim$260) due to the tightness of the CNS barrier layers [20]; furthermore, the brain has no true lymph or lymphatics.

Both the BBB and the BCSFB actively regulate the type and concentration of molecules transported to and from the extracellular fluid, CSF, and intracellular fluid [21]. Because of its large surface area ($\sim$20 m^2/1.3 kg brain) and the short diffusion distance between neurons and capillaries (8–20 µm), the endothelium plays a predominant role in regulating the brain environment [1]. The BCSFB also contributes to this process besides playing other roles [22]. It was proposed that the surface area of the BBB is several thousand times larger than the one of the B-CSF [23, 24]. In this context, it has to be mentioned that Keep and Jones suggested that the choroid plexus may play a more important role in the regulation of the brain microenvironment than previously thought, since they showed that the total apical surface area of the choroid plexus (75 cm^2) was almost half of the corresponding BBB (155 cm^2) of rats studied with stereological techniques, when the apical microvilli were taken into account [25].

The main component of the blood–brain barrier is the brain endothelium, which exhibits a physical, an efflux and a metabolic barrier for the transport of drugs into the CNS. The physical barrier, an efflux, is a result of the tight junctions between adjacent endothelial cells, which are around 50–100 times tighter than in the peripheral endothelium, so that penetration across the endothelium is effectively confined to transcellular mechanisms [26, 27]. These junctions significantly restrict even the movement of small ions such as Na$^+$ and Cl$^-$, so that the transendothelial electrical resistance (TEER), which is typically 2–20 Ω cm^2 in peripheral capillaries, can be over 1000 Ω cm^2 in brain endothelium [28].

Specific transport systems present on the luminal and the abluminal membranes regulate the transcellular traffic of small hydrophilic molecules, which provides a selective transport barrier, permitting or facilitating the entry of required nutrients and excluding or effluxing potentially harmful compounds [29]. Minimal pinocytotic activity and the absence of aqueous pores in the endothelium also lead to restricted passage of most molecules from the cerebrovascular circulation into the CNS [30]. The metabolic barrier consists of a combination of intracellular and extracellular enzymes as ectoenzymes, for example, peptidases and nucleotidases, which are capable of metabolizing peptides and ATP, and intracellular enzymes, for example, monoamine oxidase and cytochrome P450, which inactivate many neuroactive and toxic compounds [31]. In summary, the term BBB covers a number of static and dynamic properties that enable the endothelium to protect and regulate the brain microenvironment [32].

The endothelial cells themselves are rather flat, their luminal and abluminal membranes are separated by a 300 nm thick endothelial cytoplasm [33]. Several BBB properties are defined as typical BBB markers, although a number of these properties are also expressed to some degree in peripheral capillary endothelium. However, most of them are upregulated in brain endothelium to such an extent that they can be identified as "markers" of BBB phenotype and function [26]. Von Willebrand factor (vWF), apolipoprotein A1, lectins (e.g., UEA-1), and the uptake of acetylated LDL are often determined as general endothelial markers, whereas enzymes such as γ-glutamyltranspeptidase and alkaline phosphatase, the glucose carrier GLUT-1, the efflux transport system P-glycoprotein (P-gp), and tight junctional proteins are used as typical BBB markers.

Brain endothelial cells form the BBB. Several recent studies have highlighted the importance of the environmental conditions to induce and maintain BBB properties in brain endothelial cells. Neighboring astrocytes, pericytes, neurons, and even the basal lamina are able to excite BBB properties. A scheme of the localization and the surrounding components of the BBB is shown in Figure 10.1.

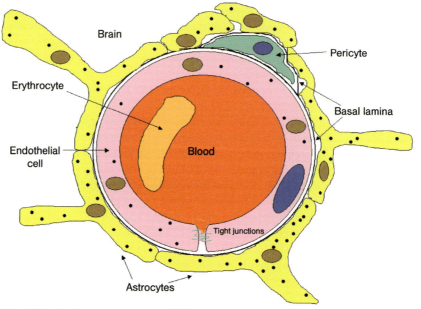

Figure 10.1 A schematic cross section of a brain microvessel.

The brain endothelial cells are distributed along the length of the vessel and completely encircle the lumen. The thin basement membrane supports the abluminal surface of the endothelium. The basal lamina surrounds the endothelial cells and pericytes, the region in between is known as the Virchow–Robin space. Astrocytes are adjacent to the endothelial cells, with astrocytic end-feet sharing the basal lamina. The association of pericytes with blood vessels has been suggested to regulate endothelial cell proliferation, survival, migration differentiation, and vascular branching [30, 34].

Astrocytes are glial cells as well as oligodendrocytes, microglia cells (mononulcear macrophages), and ependymal cells. In general, glial cells are mainly responsible for the mechanical support, neuronal nutrition, and phagocytosis in the brain. Astrocytes show a number of different morphologies, depending on their localization and association with other cell types [1]. Of the approximately 11 distinct phenotypes that can be readily distinguished, 8 involve specific interactions with blood vessels [35]. End-feet of astrocytic glia are closely apposed to the brain endothelium and are important in the induction and maintenance of the BBB [26, 27]. There is now strong evidence, particularly from studies on cell culture, that astrocytes can upregulate many BBB features, leading to tighter junctions (physical barrier), the expression and polarized localization of transporters, including P-gp and GLUT-1 (transport barrier), and specialized enzyme systems (metabolic barrier) [1, 36–38]. Furthermore, astrocytes were able to modulate the cellular physiology both in longer and shorter terms [26, 39].

Especially, *in vitro* studies helped to understand the different ways of induction by astrocytes. These are able to secrete a range of chemical agents including transforming growth factor-β (TGF-β), glial-derived neurotrophic factor (GDNF), basic fibroblast growth factor (bFGF), and angiopoetin-1 (ANG-1), which can induce aspects of the BBB phenotype in endothelial cells *in vitro* [1]. Furthermore, induction of BBB features was proved *in vitro* using nonbrain endothelial cells as human umbilical vein and aortic endothelial cells [40] regardless of whether cells were grown in coculture with astrocytes, C6 glioma cells, or even only with astrocyte-conditioned medium (ACM). Ramsohoye and Fritz [41] suggested that nonproteinaceous substances derived from ACM were also able to increase the tightness of endothelial cell layers *in vitro*. Conversely, endothelium-derived leukemia inhibitory factor (LIF) has been shown to induce astrocytic differentiation [42].

In addition to the ability of astrocytes to induce BBB in endothelial cells, it was shown that pericytes and neurons can influence the layer's integrity. Pericytes in the brain occur every two to four endothelial cells. As fourth important component, the basement membrane has to be introduced. The basal lamina is the extracellular matrix layer produced by the basal cell membrane used as an anchoring and signaling site for cell–cell interactions. It is a thin basement membrane, comprising laminin, fibronectin, collagens, and other proteins, that surrounds the endothelial cells and associated pericytes and provides both mechanical support and barrier function. Cell adhesion to the basal lamina involves the integrins. Integrins are transmembrane receptors that bridge the cytoskeletal elements of a cell to the extracellular matrix and are heterodimers of α and β subunits [30].

Based on this knowledge, the concept of the neurovascular unit was developed [1]. It is a functional unit composed of groups of neurons and their associated astrocytes, interacting with smooth muscle cells and endothelial cells on the microvessels (arterioles) responsible for their blood supply, and capable of regulating the local blood flow. Within this organization, further modular structures can be detected. In particular, the proposed gliovascular units, in which individual astrocytic glia support the function of particular neuronal populations and territories, communicate with associated segments of the microvasculature [43, 44]. This novel concept of single functional units over the total BBB structure is in concordance with Ge *et al.* [45], who summarized many data supporting differences in the expression levels of several BBB markers in arterioles, capillaries, and venules and asked finally "where is the blood–brain barrier … really?"

Taken together, Abbott suggested that the BBB should not be seen as a static entity, but as one that can alter its function according to local needs [46]. The hypothesis about a dynamic BBB was supported by the results of animal studies and confocal imaging of postmortem human brain material, which provide some evidence for a small percentage ($<5\%$) of BBB tight junctions being open at any time under physiological conditions.

10.2
Transport Mechanisms Across the Blood–Brain Barrier

To maintain brain homeostasis, the blood–brain barrier selectively transports nutrients into the brain via the expression of a number of surface transporters.

An overview of several possible transport routes is given in Figure 10.2. During various disease states, alterations in the levels or distribution of transporters can be seen. Tight junctions limit the paracellular diffusion of molecules and the formation of extracellular fluid [47].

Small lipophilic molecules such as oxygen, CO_2, and ethanol can freely diffuse across the lipid membranes of the endothelium. In the case of passive transcellular diffusion, higher lipid solubility favors this process. Hence, compounds of high lipophilicity generally show higher permeability [16]. In addition to passive transport, which is driven by a concentration gradient, the transport of substances can also be catalyzed by carrier- or receptor-mediated processes [48]. In general, carrier-mediated transport is nonenergy dependent and transported down a concentration gradient. Contrary to this, the active transport process is energy dependent. At least 10 different transport systems have been identified [49]. The BBB also regulates the ion balance of the brain. Therefore, ions can be transported by several carriers as ion channels, ion symporters, and ion antiporters. In fact, brain edema formation after a stroke has been linked to the inability of the BBB to maintain necessary ion gradients [50]. In addition, ion transport can be energy dependent as well, for example, to create an electro/osmotic Na^+ gradient by the basolaterally positioned antiport transport system Na^+/K^+ ATPase. Small polar solutes needed for brain function are transported by a number of specific carriers (e.g., GLUT-1 for glucose, L-system carrier L1 for large neutral amino acids such as leucine), and specific carriers mediate the efflux of potentially toxic metabolites (e.g., glutamate) from the CNS [26, 30]. Glucose is transported by means of the carrier GLUT-1 due to facilitated diffusion. In this case, glucose is bound to the carrier on one side of the membrane that triggers a conformational change in the protein. As a result, the substance is carried through to the other side of the membrane, from high to low concentration. Facilitated diffusion is passive and contributes to the transport of many substances at the BBB such as monocarboxylates, hexoses, amino acids, nulceosides, glutathione, small peptides, and so on [30]. GLUT-1 can be present on both luminal and abluminal membranes [1]. On the contrary, Na^+-dependent transporters are generally abluminal, specialized for moving solutes out of the brain [51, 52]. They include several Na^+-dependent glutamate transporters (excitatory amino acid transporters 1–3, EAAT1–3) [53], which transport glutamate out of the brain against the large opposing concentration gradient ($<1\,\mu M$ in ISF compared to $\sim 100\,\mu M$ in plasma).

In addition, energy-dependent transporters of the BBB efflux waste products and exogenous compounds of potential toxicity. The most prominent efflux transporter is P-gp [54], which possesses broad substrate specificity that keeps out more hydrophobic molecules. It has been localized to the luminal brain endothelial membrane and plays a major role in protecting the brain from xenobiotics. In the context of the phenomenon "multidrug resistance," other known drug carriers also play an important role, such as the family of multidrug resistance associated proteins (MRPs), breast cancer resistance protein (BCRP), organic anion transporters (OATs), and the organic anion transporting polypeptides (OATPs).

In general, most transporters are members of the solute carrier family (SLC) or of the active and energy-consuming ATP binding cassette (ABC) transporter family.

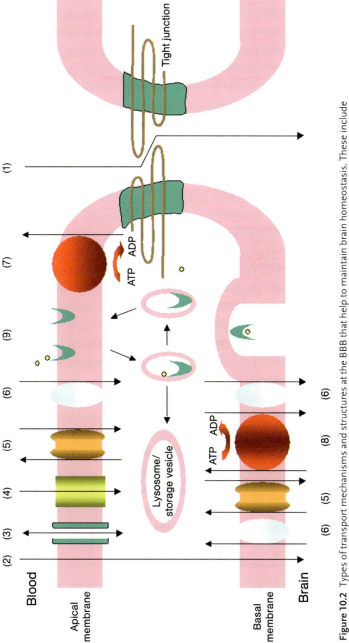

Figure 10.2 Types of transport mechanisms and structures at the BBB that help to maintain brain homeostasis. These include (1) paracellular transport, (2) transcellular diffusion (EtOH), (3) cation channels (K^+ ions), (4) ion symports ($Na^+/K^+/Cl^-$ cotransporter), (5) ion antiports (Na^+/H^+ exchange), (6) facilitated diffusion (glucose via GLUT-1), (7) active transport (P-gp), (8) active antiport transport (Na^+/K^+ ATPase), and (9) endocytosis (receptor as insulin or transferrin or adsorption mediated). Adapted from Huber et al. [47].

At last, substances can be transported across the BBB by endocytosis. Bulk-phase endocytosis (pinocytosis) is a nonspecific uptake of extracellular fluids and occurs at a constitutive level within the cell through mechanisms that depend on ligand binding. Bulk-phase endocytosis is temperature and energy dependent, noncompetitive, and nonsaturable. The brain endothelium has lower levels of endocytosis/transcytosis than peripheral capillaries [26]. However, molecules too large for carrier-mediated entry, such as peptides and proteins, may be able to cross the endothelium to a limited degree via a vesicular route, either by specific receptor-mediated transcytosis (RMT) or by following nonspecific adsorption of cationic molecules to the membrane surface (adsorption-mediated transcytosis (AMT)) [55]. A number of receptors are expressed on both the luminal and the abluminal surfaces of the endothelial cells [15]. RMT occurs in the brain for substances such as transferrin, insulin, leptin, and insulin-like growth factors (IGF-I, IGF-II), and it is a highly specific type of energy-dependent transport. Substances that enter a cell by means of RMT are bound to receptors present in specialized areas of the plasma membrane known as coated pits. The coated pits contain the electron-dense clathrin protein and other proteins [56]. After a compound binding to the receptor and cellular uptake of the coated vesicles, the clathrin vesicle coat is rapidly removed to form smooth-coated endosomes. Within the "compartment of uncoupling receptor and ligand (CURL)," the ligand dissociates from the receptor due to an acidification within the endosome caused by an endosomal membranal proton ATPases [57]. The endosome with the receptor can be reincorporated into the plasma membrane for further endocytosis.

Adsorption-mediated transport is triggered by an electrostatic interaction between a positively charged substance, usually a charge moiety of a peptide, and the negatively charged plasma membrane surface (i.e., glycocalyx) [58].

10.3
The Physical Barrier: Paracellular Transport and Its Characterization

The endothelial monolayer forms the main physical barrier within the brain capillaries for the transport of hydrophilic, polar substances. Furthermore, the network of the basal lamina can contribute to the prevention of the permeation of macromolecules. The restriction of the unregulated paracellular transport is a prerequisite to enable homeostasis of the brain microenvironment and to control the influx as well as the efflux of nutrients, waste products, and xenobiotics. In contrast to the peripheral endothelium, adjacent endothelial cells of the blood–brain barrier are connected to each other by intercellular tight junctions. They are the most apical element of the junctional complex, which includes both tight and adherens junctions [47]. Tight junctions define apical and basal membrane polarity by limiting the exchange of membrane lipids between the two and regulate the paracellular transport of water, solutes, and immune cells [59]. In the BBB, they are composed of an intricate combination of transmembrane and cytoplasmic proteins linked to an actin-based cytoskeleton that allows the tight junctions to form a seal while remaining capable of rapid modulation and regulation. These tight junctions consist of three kinds of integral proteins: claudins, occludin, and junctional adhesion molecules. The transmembrane proteins are connected on the cytoplasmic side to a complex array of peripheral membrane proteins that form large protein complexes: the cytoplasmic plaques. Within the plaques are adaptor proteins with many protein–protein interaction domains, including ZO-1, ZO-2, and ZO-3. In the BBB, the presence of claudin-1 and claudin-5 was reported and identified as major component of tight junction strands [60–62]. In contrast, Abbott et al. [1] summarized recent findings and concluded that claudin-3 (not claudin-1) and claudin-5 and possibly claudin-12 appeared to contribute to the high TEER at the BBB.

The paracellular component of the transport is distinct from the transcellular transport in several ways. It is a completely passive transport, resulting from paracellular dissipation of the electro/osmotic gradients established by the transcellular transport. Paracellular transport has two basic characteristics: permeability (the magnitude of the barrier) and permselectivity (ability to discriminate molecular size and ionic charge). In practice, most investigators quantify permeability by two complementary techniques. The first is by measuring TEER, which is the determination of the barrier to small ions (predominantly Na^+ and Cl^-) in an experimentally applied electrical field in the bathing media. The second is by measuring the flux of tracer solutes, such as radiolabeled inulin or mannitol, which only traverse the endothelium through the intercellular space [63].

A disadvantage of the measurement of TEER is that it is excessively sensitive to small regions with low resistance such as crushed cells near the edge of a mounting apparatus or a patch of dead cells within the monolayer. Another problem is that the lateral interspace contributes to the paracellular resistance directly by its length and inversely by its width. Both of these factors can be affected by seemingly trivial factors, such as changing culture media or the time a grown monolayer has been in culture. Electrical resistance measurement methods model a cell monolayer as a circuit of parallel resistors composed of all the individual transcellular and paracellular elements. In this case, the total resistance is a function of the inverse sums of individual resistances ($1/R_{total} = 1/R_1 + 1/R_2 + \cdots$). Thus, TEER is dominated by elements with the lowest resistance [63].

Permeability, measured by tracer flux, is linearly proportional to the area across which diffusion occurs. Thus, it is theoretically less sensitive to trivial defects in the monolayer, and in comparison to the measurement of TEER, it more reliably reports changes induced in the junction by experimental manipulations. For the characterization of the paracellular route, tracers should be hydrophilic, polar, and should not be substrates for transport systems, brain endothelial receptors, or an endothelial enzyme to minimize contribution of transcellular transport. For smaller molecular weight range, sucrose, mannitol, fluorescein, and carboxyfluorescein are often applied, whereas for higher ranges inulin and dextrans are markers of choice. Horseradish peroxidase and labeled albumin are often used to describe the permeability of serum proteins [64, 65]. The tracers are then often labeled radioactively or with a fluorescent dye. In the case of low molecular weight markers, the usage of radiolabeled mannitol seems to be favorable since Garberg et al. [66] suggested that sucrose could be cleaved to 14C-labeled monosaccharides in the cell and pretend sucrose permeability. Also, fluorescein and carboxyfluorescein were suggested as substrates for some efflux transporter systems [67–69]. Until now, it was indispensable to carry out several transport studies with different paracellular markers of different molecular weight to obtain information about paracellular permeability over a wider and pharmaceutically relevant molecular size range. For this reason, we have developed a novel paracellular marker, the so-called APTS–dextran ladder [65]. Normally, fluorescent-labeled dextrans such as FITC–dextran consist of a mixture of several dextrans. It can be used to determine significant changes in tightness by measuring total fluorescence of the dextran mixtures. However, labeling at the terminal carbonyl moiety of a chosen dextran with fluorescent APTS by reductive amination enables the separation of the single-labeled dextran fractions by capillary electrophoresis with a resolution of one glucose unit (Figure 10.3a). Analysis of each single fluorescent fraction in samples after transport studies provides the possibility to generate molecular size-dependent permeability patterns from free APTS and APTS–glucose to APTS–dextran consisting of up to 35 glucose units (Figure 10.3b). These patterns may allow a more detailed correlation of paracellular permeability to the permeability of drugs in the corresponding molecular weight range. The applicability of this dextran ladder was already proven for *in vitro* test systems, but still has to be evaluated for *in vivo* studies.

10.4
The Efflux Barrier: Transport Proteins at the Blood–Brain Barrier

In general, most transporters are members of the SLC or of the active and energy-consuming ABC transporter family. The transporters at the blood–brain barrier are of great importance for the medicinal chemists since these mechanisms can influence the efficacy of brain-targeted drugs on the one hand and minimize side effects of compounds that should primarily act at the periphery on the other hand. As an example, the development of the different generations of H1 antihistaminic drugs should be mentioned here. Antihistamines of the first generation such as diphenhydramine and mepyramine exhibited strong CNS side effects, whereas antihistamines of the second generation such as fexofenadine did not. First suggestions attributed these changes in BBB permeation to different physicochemical properties, for example, higher lipophilicity of antihistaminics of the first generation [70]. Studies reported by Chishty *et al.* [71] showed that antihistamines of the second generation inhibited the efflux of colchicine from an immortalized rat cerebral endothelial cell line named RBE4, while older antihistamines did not. They concluded that the antihistaminics of the second generation might be substrates of the efflux pump P-gp. With regard to this, we accomplished transport studies with several antihistamines across a BBB model with the immortalized porcine brain microvascular endothelial cell line PBMEC/C1–2. High expression of P-gp was proven for this cell line [72]. The ranking of the permeability coefficients after normalizing to the internal standard diazepam showed that loratadine, astemizole, fexofenadine, and cetirizine (second generation) migrated significantly slower than

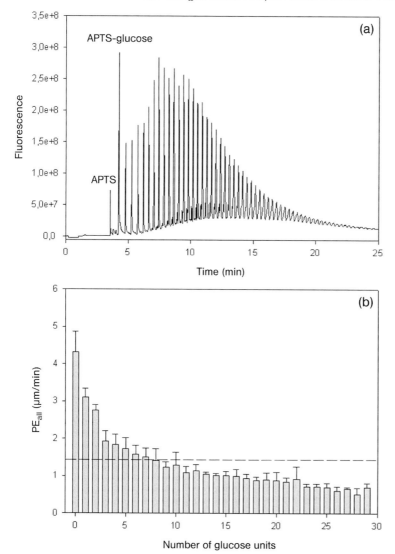

Figure 10.3 A novel tool to characterize paracellular transport: the APTS–dextran ladder. Due to an optimized labeling strategy, the single fractions of the dextran mixture could be separated by capillary electrophoresis (a) with a resolution of one glucose unit between each fraction. This analytical technique enabled to generate molecular size-dependent permeability patterns of a paracellular marker that was applied at once across cell layers. In (b), a typical permeability pattern across an *in vitro* transwell model of the blood–brain barrier using human cell line ECV304 is displayed. The dashed line indicates the permeability coefficient determined for the total APTS–dextran mixture.

diazepam, whereas diphenhydramine and pheniramine (first generation) crossed our model distinctly faster than diazepam [73]. Since Chen *et al.* [74] proposed that antihistaminics of the newer generations were P-gp substrates, it was concluded that the PBMEC/C1–2 model was able to display the *in vivo* conditions in this case due to its high expression of P-gp. The case of the drug development of the antihistamines was an impressive example that the traditional method of medicinal chemists to increase the lipophilicity resulted in the construction of new P-gp substrates [75].

Consequently, the knowledge about possibly involved transport systems and their substrate specificities has to be considered when someone wants to design a new drug. Furthermore, it is vital to know that interspecies differences in the presence of transporters and in their substrate specificities can occur when interpreting literature data correctly.

In this context, studies on transformants of LLC-PK$_1$ cells that expressed P-gp derived from human, monkey, canine, rat, and mouse impressively showed altered efflux activities and rankings depended on the species for substances such as clarithromycin, daunorubicin, digoxin, etoposide, paclitaxel, quinidine, ritonavir, saquinavir, verapamil, and vinblastine [76]. Subsequent experiments confirmed different inhibitory effects of verapamil and quinidine on the transport of daunorubicin, digoxin, and cyclosporin A across LLC-PK$_1$ cells with P-gp from different species [77]. These reports clearly pointed out that qualitative statements, whether a substance is a transporter substrate or not, are possible. But it was also underlined that one has to be really careful when applying permeability data of *in vitro* experiments or *in vivo* animal studies to human conditions. In general, the functional consequences of species variation may vary from compound to compound, and further studies are needed on this aspect [78].

In the following section, we have tried to summarize and to give an overview of the transporter systems that are believed to be relevant for drug transport, in particular for drug export, across the BBB. In Figure 10.4, the most important drug transporters at the BBB are displayed.

In addition to these transporters, specific transporters for hexoses (GLUT-1), amino acids (EAAT1–3, LAT1), and nucleosides (concentrative nucleoside transporters – CNT (SLC28 family) and equilibrative nucleoside transporters – ENT (SLC29 family) were reported as being present at the BBB.

10.5
Transporter of the SLC Transporter Family

10.5.1
Monocarboxylate Transporters

The discovery of monocarboxylate transporters (MCTs), their characterization, and the study of their distribution has been a major landmark, and it provides interesting clues about their possible roles in regulating monocarboxylate fluxes not only between the blood and the brain but also between the different intraparenchymal cell types. Monocarboxylate transporters may represent an important site for both entry and exit of numerous substances, including several pharmaceuticals, in the brain [79].

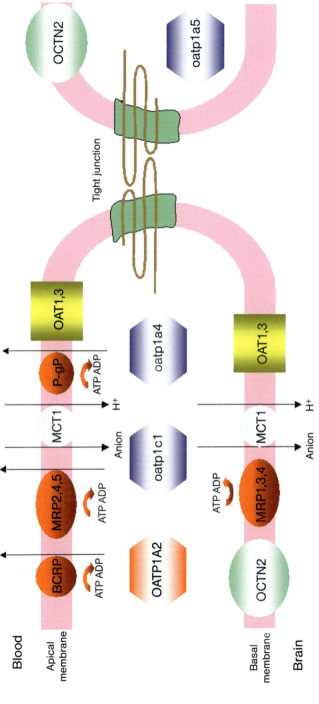

Figure 10.4 Scheme of proposed localization of important drug transporters at the blood–brain barrier. Adapted from Miecz et al. [121], Dallas et al. [184], and Miller [198].

The important role of MCTs in the brain was proven by the transport of monocarboxylates such as lactate, pyruvate, and ketone bodies acetoacetate and β-hydroxybutyrate, which are potential energy substrates for the brain [80]. These substrates can serve for energy production instead of glucose during pathological conditions such as diabetes and prolonged starvation [81, 82]. Several NMR studies as well as investigations at the cellular level showed that lactate can be used by brain cells, in particular, by neurons as a preferred oxidative substrate. Moreover, other cells in the brain such as oligodendrocytes and astrocytes are able to utilize lactate as energy substrate or for neoglucogenesis and glycogen synthesis [83]. In contrast, lactate efflux from the brain into the bloodstream probably mediated by brain endothelial cells was also observed during, for example, ischemia or brain injury [84, 85]. Saturable transport of monocarboxylates was shown for several brain cell types, and it was proposed that each cell type possesses different major MCT systems, which are summarized in the SLC16 gene family [79].

In 2004, 14 members of this transporter family were identified as MCT1–9, MCT11–14, and TAT-1 based on their sequence homologies [86]. MCT1 is currently the most studied one and is also believed to be the most important one acting at the blood–brain barrier. MCT1 was detected by Western blotting with a molecular weight of approximately 43–45 kDa, both in brain microvessels and in the murine cell line b.End5 [87, 88]. It is distributed at the luminal as well as at the abluminal side of brain blood vessels. The presence and the role of MCT2 at the BBB are still under a controversial discussion [89]. Pellerin and Pierre [79] stated that MCT2 is much less abundant than MCT1 at the BBB. Furthermore, several data also supported species differences [90]. MCT3–7 were found in different tissues but not at the BBB to a significant extent. Friesema et al. [91] proved the presence of MCT8 in the brain by Western blotting of total brain extracts, but did not specify local distribution.

The transport of monocarboxylates mediated by MCT1–4 was shown to be a symport accompanied by one proton per anion of the corresponding monocarboxylate following a sequential mechanism [92–104]. For the other MCTs, no proton dependence was shown, which suggested other mechanisms. For example, TAT1 is most probably a transport system for aromatic amino acids, whereas MCT8 transports thyroid hormones [91, 105].

Lactate and pyruvate seemed to be the endogenous substrates with the highest affinity to MCTs at the BBB, whereas other monocarboxylates such as acetate, propionate, and butyrate were transported to a lower extent [93, 106]. This is consistent with BBB uptake studies published by Oldendorf [107]. Moreover, stereoselectivity of MCT1 for L-lactate transport was shown toward the D-isomer in studies carried out by Bröer et al. [93, 94].

In Table 10.1 exogenous substrates are listed for which MCT influx and/or efflux across the BBB were suggested (summarized by Pellerin and Pierre [79]).

10.5.2
Organic Ion Transporters and Transporting Peptides

The organic anion transporters (human OATs, rodent oats) are classified within the SLC22A solute carrier family, and the transporting peptides (human OATPs, rodent

Table 10.1 Suggested exogenous substrates for MCT transport at the blood–brain barrier.

Substance	Drug class	Comment	Test system	References
Salicylate	NSAID		Brain uptake	[108]
Benzoate	Antibacterial	Preservative		[109, 110]
Valproate	Antiepileptic drug	Probable MCT1	Isolated neurons	[111]
Nicotinic acid	Vitamin B3	Used in hyperlipidemia	Rat astrocytes	[112]
β-Lactam antibiotics	Antibiotic drugs			
Foscarnet	Antiviral	Herpes viruses	In vitro brain capillary cells	[113]
R- and S-mandelic acid	Antibacterial			
Simvastatin	HMG-CoA reductase inhibitors	Acetate as inhibitor	In vivo; in vitro across bovine brain endothelial cells	[114, 115]
Lovastatin	HMG-CoA reductase inhibitors			
Ifosfamide and its metabolites SCMC and TDGA[a]	Anticancer	Alkylating agent		[116]
D-Lactate		Usage as MCT inhibitor for L-lactate transport		[117]

[a]SCMC, S-carboxymethylcysteine; TDGA, thiodiglycolic acid.

oatps) are summarized in the SLC21A solute carrier family. Substrates for these transporters are drugs and xenobiotics, neuroactive peptides, thyroid hormones, bile salts, and steroid conjugates [118]. Since the nomenclature for the SLC family was reorganized recently, it is important to know the older names of several transporters such as OATP-A, OATP for OATP1A2, oatp2 for oatp1a4, oatp3 for oatp1a5, and oapt2 or 14 for oatp1c1 to understand and classify previous data. OATP1A2 was found in brain capillaries. In general, it is believed that OATPs and OATs are responsible for transport in both directions – on the one hand, for the efflux from the brain through the endothelium into the bloodstream, and on the other hand, for the transport from the endothelium into the brain.

OAT1 (SLC22A6) is polyspecific and was reported to transport organic anions similar to OAT3 (SLC22A8), which prefers dicarboxylates [119]; bile acids, organic anions and cations, and steroids are supposed to be substrates for human OATP1A2 (SLC21A3), digoxin, bile acids, organic anions, and cations for rodent oatp1a4 (Slco21a5), bile acids and organic anions for rodent oatp1a5 (Slco21a7) and T4, rT3, and BSP for oatp1c1 (Slco21a14) [120].

OCTN2 (SLC22A3) is a transport protein for organic cations at the blood–brain barrier. Typical substrates are L-carnitine (Na^+ symport) and organic cations (H^+ antiport). It is believed that OCTN2 acts in a polyspecific manner [119]; Na^+-dependent transport of antibiotic cephaloridine, L-lysine, and L-methionine and Na^+-independent transport of TEA, pyrilamine, verapamil, choline, and quinidine have been published. Recently, studies of Miecz et al. [121] proposed that OCTN2 is localized in the basolateral membrane and in the vicinity of the nuclei in endothelial cells of an in vitro BBB model. This suggested that OCTN2 may also transport carnitine from the brain into the endothelial cell, which means that OCTN2 plays an important role in the removal of certain acyl esters.

10.6
Transporter of the ABC Transporter Family

10.6.1
P-Glycoprotein

P-glycoprotein is considered as the most important ABC efflux transporter (ABCB1, MDR1) at the blood–brain barrier [54]. It is a member of the ABC family and consumes ATP during its active efflux transport mechanisms. The major role of P-gp in the phenomenon of "multidrug resistance" is attributed to its broad substrate specificity.

Several species express multidrug resistance proteins and each species possesses a different number of gene products. In addition to MDR1, a second gene product called MDR3 is found in humans. On the contrary, three different genes, namely, mdr1a, mdr1b, and mdr2 were identified in rodents [122], three in hamsters (pgp1, pgp2, pgp3), and two in dogs (MDR1, MDR2), sheep (MDR1, MDR2), pig (pgp1: A–D), and bovine (MDR1, MDR2). However, it seems that only the gene products of class I MDRs (MDR1, mdr1a, mdr1b) contribute to the multidrug resistance in the brain, whereas it was published that MDR2 and mdr2 gene products mainly secrete phosphatidylcholine from the liver into the bile [123].

P-gp (MDR1, mdr1a) is found at the luminal side of brain capillary endothelial cells in human, rats, and mouse [75]. Pardridge et al. [124] proposed that P-gp is mainly distributed in the cell membranes of astrocytic foot processes. This is contrary to other observations, and the general view nowadays is that P-gp is expressed as a kind of first line defense at the brain capillary endothelial cells. P-gp transports a very broad range of compounds that are mainly lipophilic, planar, and either neutral or cationic. It recognizes and transports an impressive array of substrates ranging in size from approximately 250 Da (cimetidine) to more than 1850 Da (gramicidin D) [125]. Examples are listed in Table 10.2 and the column about the specific drug class highlights the wide substrate specificity of P-gp. Many chemotherapeutical compounds of natural origin such as anthracyclines, vinca alkaloids, and taxanes, some immunosuppressants, cardiac glycosides, antipsychotics, antidepressants, antihistaminics, and HIV protease inhibitors are substrates and/or modulators of P-gp.

Table 10.2 Suggested substrates for P-glycoprotein transport.

Substance	Drug class	Comment	Test system	References
Aldosterone	Endogenous substrate	mdr1a, MDR1	KO mice, LLC-GA5-COL150 cells	[130]
Amiodarone	Antiarrhythmic drug		KO mice	[131]
Amitriptyline	Tricyclic antidepressant	mdr1a	KO mice	[132]
Asasetron	Antiemetic, 5-HT$_3$ receptor antagonist	mdr1a		[133]
Asimadoline	Antianalgesic, κ-opioid receptor agonist	mdr1a, MDR1	KO mice	[134]
Astemizole	H1-antihistamine		Cell lines RBE4, PBMEC/C1-2	[71, 73]
Carebastine	H1-antihistamine, metabolite of ebastine	mdr1a	Rats, cell lines	[135]
Cetirizine	H1-antihistamine	mdr1a	KO mice, MDCK cells	[74]
Cimetidine	H2-antihistamine		Caco-2 cells	[125, 136]
Citalopram	Psychotropic drug, SSRI	mdr1a	KO mice	[137]
Chlorpromazine	Antipsychotic drug		ATPase assay	[78]
Colchicine	CT antimicrotubulus drug	mdr1a	rats	[138]
Cortisol	Endogenous substrate	mdr1a, MDR1	KO mice	[130]
Cyclosporin A	Immunosuppressant	mdr1a, MDR1		[139]
Daunorubicin^a	CT anthracylines, DNA intercalating	mdr1a, MDR1		[75]
Dexamethasone	Glucocorticoid	mdr1a	Comment: modulates P-gp	[75, 140]
Digoxin	Cardiac glycoside	mdr1a, MDR1	KO mice	[141]
Diltiazem	Ca^{2+} channel blocker	MDR1	Cell lines	[140]
Doxepin	Psychotropic drug	mdr1a	KO mice	[142]
				(Continued)

Table 10.2 (Continued)

Substance	Drug class	Comment	Test system	References
Doxorubicin	CT anthracyclines, DNA intercalating	mdr1a, MDR1		[75]
Ebastine	H1-antihistamine	mdr1a	Cell lines	[135]
Etoposide	CT epipodophyllotoxin	mdr1a, MDR1	LLC-PK1 cells	[76]
Fexofenadine	H1-antihistamine	MDR1	KO mice	[126]
FK506[a]	Immunosuppressant	MDR1	LLC-GA5-COL300 cells	[139]
Fluoxetine	Antidepressant drug, SSRI	mdr1a	KO mice	[143]
Glucuronides	Conjugates of phase II metabolized drugs	MDR1		[75]
Gramicidin-D	Antibacterial	MDR1		[75]
Grepafloxacin	Antibacterial	mdr1a	MDCKII-MDR1 cells	[144]
HSR-903[a]	Antibacterial	mdr1a	KO mice	[145]
Indinavir	HIV protease inhibitor	mdr1a	BBMEC cells, ATPase assay	[146]
Ivermectin	Antihelmintic drug	mdr1a, MDR1	KO mice	[147]
Loperamide	Antidiarrheal agent, opioid	mdr1a, MDR1	KO mice	[125]
Loratadine	H1-antihistamine	mdr1a, MDR1	MDCK-MDR	[74]
Methadone	Opioid		KO mice	[78]
Methotrexate	Dihydrofolate reductase inhibitor	MDR1		[75]
Morphine	Analgesic drug	mdr1a, MDR1		[78]
Morphine-6-glucuronide	Phase II conjugate	Increased uptake by PSC833		[148]
Nelfinavir	HIV protease inhibitor	mdr1a	KO mice, monkeys, BBMEC	[146, 149]
Nifedipine[b]	Ca^{2+} channel blocker	MDR1	Controversial data	[75]
Nortriptyline	Psychotropic drug	mdr1a	KO mice	[132]
Olanzapine	Antipsychotic drug	MDR1	ATPase assay, LLC-PK1 cells	[78, 150]

Ondansetron	5-HT$_3$ receptor antagonist	mdr1a, MDR1	Rat brain capillaries	[75]
Paclitaxel (Taxol®)	CT taxane	Increased uptake by PSC833		[151]
Paroxetine	Psychotropic drug	mdr1a	KO mice	[142]
Phenytoin	Antiepileptic drug	mdr1a, MDR1	KO mice	[147, 152]
Quetiapine	Antipsychotic drug	MDR1	ATPase assay, LLC-PK1 cells	[78, 150]
Quinidine	Antiarrhythmic drug	mdr1	KO mice	[78]
Ranitidine	H2-antihistamine	MDR1	Caco-2 cells	[125, 136]
Rapamycin	Immunosuppressant	MDR1		[75]
Rhodamine 123	Laser dye	mdr1a, MDR1		[146]
Risperidone	Antipsychotic drug	MDR1	ATPase assay, LLC-PK1 cells	[78, 150]
9-OH risperidone	Antipsychotic drug, metabolite	MDR1	ATPase assay, LLC-PK1 cells	[78, 150]
Saquinavir	HIV protease inhibitor	mdr1a		[146]
Trimipramine	Tricyclic antidepressant	mdr1a, MDR1	KO mice	[137]
Desmethyltrimipramine	Tricyclic antidepressant	mdr1a, MDR1	KO mice	[137]
Valinomycin	Peptide ionophore	mdr1a		[153]
Vecuronium	Muscle relaxant	mdr1a	LLC-PK1 cells	[154]
Venlafaxine	Antidepressant, SNRI	mdr1a, MDR1	KO mice	[142]
Vinblastine	CT vinca alkaloid, antimitotic drug	mdr1a, MDR1	KO mice	[155]
Vincristine[a,b]	CT vinca alkaloid, antimitotic drug	mdr1a, MDR1	KO mice, controversial data	[155]

CT, chemotherapeutic.
[a] Daunorubicin: daunomycin; FK506: tacrolismus, fujimycin; HSR-903: olamufloxacin; vincristine: leurocristine.
[b] Controversial data, whether this substance is really a P-gp substrate.

Furthermore, there is a strong evidence that the question of stereoselectivity is of certain relevance for the P-gp-mediated transport [126]. Since many drugs are administered as racemats, these facts certainly complicate the appraisal during combination therapies. Especially, the involvement of P-gp on the effectivity of psychotropic drugs may influence the therapeutical outcome in patients concerned. Recently, Linnet and Ejsing [78] summarized the impact of P-gp on the penetration of some of these drugs into the brain. They reviewed the evidence of drug–drug interactions involving primarily psychotropic drugs and P-gp at the BBB on the basis of *in vitro* studies, animal experiments, and observations in humans. In some cases, results were controversial depending on the different test systems used, such as ATPase activity, Caco-2 cell permeation, knockout mice, or drug–drug interaction studies with P-gp inhibiting substances. Using *in vitro* tests, it was pointed out that the membrane environment may influence P-gp activity. The experimental findings of the interaction studies support the notion that P-gp plays an important role in brain uptake of psychotropic drugs and underline the potential risk of neurotoxicity when potent P-gp inhibitors are coadministered. For example, serious respiratory depression occurred when loperamide was coadministered with quinidine, whereas when loperamide was applied alone no adverse effects were observed [127].

In the context of P-gp substrates and ADME, it has to be mentioned that P-gp and CYP3A4 show a striking substrate overlap [128]. Thus, CNS side effects could arise from the combination of drug–drug interactions in relation to CYP enzymes and with regard to P-gp at the BBB. Since few plasma-level data in humans are known, many adverse effects were attributed to CYP activities. However, CNS side effects may occur not only due to conversion of compounds by CYP3A4 to active agents but also due to coadministration of substances that are P-gp substrates, which can result in competitive P-gp inhibition and consequently in enhanced BBB permeation of one of the compounds. For example, grapefruit juice components are well known to inhibit CYP3A4, but recently it was also shown that they are able to inhibit P-gp transport activities [129].

In general, the influence of P-gp has to be evaluated for each case and care should be taken not to overrate the importance of P-gp in the clinical context. Finding out that a substance is a P-gp substrate does not necessarily coincide with a clinical effect. For example, Linnet and Ejsing [78] concluded that the possible clinical effects of most of the psychotropic drugs caused by interaction with P-gp – although many of them are believed to be P-gp substrates – are rather limited when considering the relatively low impact of P-gp absence in KO mice.

10.6.2
Multidrug Resistance-Associated Proteins

One of the major tasks of the blood–brain barrier in addition to the maintenance of CNS homeostasis is the efflux of xenobiotics and waste products by an array of efflux pumps. The most prominent one is the P-glycoprotein, also called ABCB1, which is encoded by the gene named multidrug resistance 1 (MDR1). The active efflux of drugs mediated by this protein has already been described. In addition to P-gp, another group of efflux pumps, which are also members of the ATP binding cassette family, is involved in the phenomenon called "multidrug resistance." At least nine of these MRPs are known and denoted as MRP1–9 or equally as ABCC1–6 and ABCC10–12. It was reported that MRP2, MRP4, and MRP5 occur in the brain endothelium, whereas MRP1 expression was low in brain capillaries but high at the choroid plexus [156]. Lee et al. [157] found low expression of MRP3 and MRP6 in the brain suggesting a minor role at the BBB for these transport mechanisms. In the case of epilepsy, Dombrowsky et al. [158] showed that P-gp, MRP2, and MRP5 were upregulated, whereas MDR3 and MRP1 were unchanged in comparison to control tissues. MRP1 and MRP3 were found at the basolateral side, whereas MRP2 and MRP5 were localized at the apical side, and MRP4 was detected at both sides [156].

MRPs are multispecific and transport different kinds of drugs. Mainly negatively charged acidic anions (purine- and pyrimidine-based nucleotide analogues), natural compounds, and drugs bound to glutathione, glucuronate, and sulfate are substrates for MRPs, so are neutral drugs if they are cotransported with glutathione [159–161]. In the case of cancer radiation therapy that affects the glutathione cycle and causes oxidation to the disulfide GSSG, radiation may increase effectivity of, for example, simultaneously applied chemotherapeutical substances such as nucleotide analogues due to prolonged maintenance of the compounds into the cancer cell. It was reported that MRP1 transports the endogenous substance leukotriene C4 indicating involvement in inflammation. However, Wijnholds [156] suggested no major role for MRP1 at the BBB due to its low presence there, whereas MRP2 presence at the BBB was significantly high indicating an important role at the BBB. As typical inhibitors, probenecid, MK571, and sulfinpyrazone may be used for the investigation of MRP transport. In Table 10.3, several substrates of possible BBB-relevant MRP transporters are listed.

Table 10.3 Suggested substrates for MRP transporters [159, 162–166].

MRP	Substance	Comment
MRP1	Estrone sulfate+ glutathione, indinavir, methotrexate, MDR drugs	Symport with glutathione; MRP1 plays a major role in astrocytes [167]
MRP2	Similar to MRP1 and MDR drugs, nonconjugated amphiphatic organic anions, phenytoin, cyclic peptides (BQ-123), cisplatin	Phenytoin plus probenecid resulted in increased uptake in rats [165]
MRP3	Substrates conjugated with glucuronides or sulfates, methotrexaten	
MRP4	Prostaglandins, folate nucleotide and nucleoside analogues, methotrexate, E$_2$17βG^a	Indometacin inhibited transport of prostaglandins PGE1, PGE2 (ATPase assay) [166]
MRP5	cAMP, cGMP, fluorescein diacetate	

^{a}E$_2$17βG: estradiol-17betaglucuronide.

10.6.3
Breast Cancer Resistance Protein

The BCRP belongs to the group of ABC transporters and is classified as ABCG2. It was recognized in tumor cell lines (e.g., MCF/AdrVp, SI-MI-80), which did not overexpress P-gp or MRPs but were still resistant to several cytostatic drugs. BCRP was identified at the luminal side of human endothelial cells that line brain capillaries [168, 169]. However, the role of BCRP has not been elucidated until now. It is believed that BCRP has to dimerize to be functionally active. Many substrates, both exogenous and endogenous, have been identified. In comparison to P-gp and MRP substrates, selectivity of BCRP overlaps with these transporters suggesting its involvement in the multidrug resistance machinery. However, significant differences in the wide substrate range of P-gp was shown, since substances such as vinca alkaloids, verapamil, or paclitaxel have not been transported by BCRP [170]. A comprehensive list of identified substrates for BCRP is provided in Table 10.4. In addition to several anticancer and HIV drugs, some dietary agents were found to affect BCRP activity. For example, plant polyphenols such as resveratrol and quercetin influenced the uptake of the BCRP substrate mitoxantrone. Furthermore, the chlorophyll metabolite pheophorbide was transported in cell lines overexpressing murine Bcrp1 and human BCRP [171]. In addition to cytostatics and dietary drugs, endogenous substances are also thought to be substrates of BCRP. Imai et al. [172] supposed estrogens and their derivatives to interact with BCRP, which was confirmed by Suzuki et al. [173] who showed that the presence of estrogen agonists and antagonists increased drug accumulation of, for example, mitoxantrone in mammalian cell lines. Furthermore, based on the data of the transport properties of other ABCG family members such as ABCG1, ABCG5, and ABCG8, Barrand [174] concluded a possible transport of phospholipids such as phosphatidylserine by BCRP, thus maintaining the asymmetry of lipids in the cell membrane. Several substances such as Ko-143, tryptostatin A, fumitremorgin C (too toxic for *in vivo*

Table 10.4 Suggested substrates for BCRP transport [170, 176–178].

Substance	Drug class	Comment
Doxorubicin	CT anthracyclin	DNA intercalating
Daunorubicin	CT anthracyclin	DNA intercalating
Etoposide (VP-16)	CT topoisomerase II inhibitor	
Topotecan [176]	CT topoisomerase I inhibitor	
Mitoxantrone	CT anthracycline	Topoisomerase II inhibitor
Methotrexate	Dihydrofolate reductase inhibitor	
SN-38 [177]	CT topoisomerase I inhibitor	Metabolite of CPT-11
Irinotecan (CPT-11)		
Prazosin	α-Adrenergic blocker	
Azidothymidine (AZT)	Reverse transcriptase inhibitor	HIV treatment
Lamivudine [178]	Reverse transcriptase inhibitor	HIV treatment

CT, chemotherapeutic.

studies), and GF120918 (also inhibits P-gp) are recommended for use as BCRP activity inhibitors [175].

Recently, a novel transporter called RFLIP76 (RALBP-1) was reported to play a major role in drug resistance in epilepsy [179]. It is a non-ABC transporter that is found ubiquitously from *Drosophila* to humans and displays inhibitory GTPase activity toward Rho/Rac class G-protein cdc42. Awasthi *et al.* [180] claimed that RFLIP76 is a multispecific transporter of chemotherapeutic agents and glutathione conjugates. RFLIP76 was found luminally colocalized with P-gp in endothelial cells of the brain vasculature and showed broad substrate specificity, including anthracyclines, vinca alkaloids, and antiepileptic drugs as phenytoin and carbamazepine. It was concluded that RFLIP76 is involved in major drug resistance, especially in the case of epilepsy [179, 180]. However, data still need to be confirmed and the role of this transporter has to be clarified. In summary, RFLIP76 could be another interesting member of the transporter machinery that is responsible for the phenomenon of multidrug resistance.

10.7
The Metabolic Barrier: Enzymes at the Blood–Brain Barrier

In addition to being a physical and a transport barrier, the blood–brain barrier also represents a metabolic barrier for drugs as well as for endogenous substances. In comparison to peripheral endothelial cells, brain capillary endothelial cells possess a 5–10-fold higher density of mitochondria, which results in a high metabolic activity [181]. The presence and activity of several enzymes for phase I and phase II biotransformations as well as for the endogenous metabolism have been found in brain capillary endothelial cells. Modifications as hydroxylation or dealkylation (phase I) or conjugation (phase II) can convert substances to BBB impermeable compounds. On the one hand, this will prevent the efflux of neurotransmitters and neurohormones into the circulation, and on the other hand, toxins, drugs, and peripheral neuroactive substances as well as circulating neurotransmitters will not cross the BBB after being metabolized [182]. However, it should be mentioned that the conversion of exogenous substances by BBB enzymes can also result in the formation of pharmacologically active or neurotoxic compounds [31, 183]. Bauer [184] summarized enzymes that have been identified in brain endothelial cells and may contribute to the metabolic blood–brain barrier. Cytochrome P450-monooxygenase, NADPH-cytochrome P450 reductase, alcohol dehdyrogenase, aldehyde dehydrogenase, ketone dehydrogenase, epoxide hydrolase, L-amino acid decarboxylase, and alkaline phosphatase may act in phase I biotransformations, whereas glutathione S-transferase, enzymes of the UDP-glucuronosyltransferase family, catechol O-methyltransferase, and phenol sulfotransferase support biotransformations in phase II [185, 186]. Furthermore, other enzymes such as angiotensin-converting enzyme (ACE), γ-glutamyltranspeptidase, monoaminoxidases (A and B), choline esterases, carboanhydrase, aminopeptidases (A and M), and enkephalinase were reported to play important roles in the metabolism of endogenous substrates in brain endothelial cells [187, 188]. Moreover, highly active enzymes in brain endothelial cells are used as markers for the BBB. For example, the enzyme γ-glutamyltranspeptidase was found to be overexpressed at the BBB compared to at the peripheral endothelium and pial microvessels [189–191].

10.8
How to Overcome the Blood–Brain Barrier

There are several strategies to overcome the BBB with regard to its different barrier functionalities. For example, to open experimentally the tight junctions, EDTA can be added to withdraw Ca^{2+} ions. Furthermore, mannitol can be applied in high concentrations (e.g., 1.6 M) to build up high osmotic pressure and, consequently, to open the tight junctions due to shrunken cells [21, 192]. The application of inhibitors targeted at the efflux transport systems was recommended as a possibility to circumvent the export barrier. For example, verapamil or cyclosporin A could be used to reduce effects of P-gp. Furthermore, the coadministration of probenecid to increase the effects of valproate for the treatment of epilepsy is discussed as an alternative [193, 194]. However, it has to be clearly stated that opening of the tight junctions and inhibition of transport systems could cause severe side effects since these treatments are not specific. Therefore, these rough methods should be used only when there is no other alternative. In this context, it was discussed that the focus has to be on the cell biological influence of the expression of the transporters as future targets rather than to use inhibitors [193]. Another possibility to import hydrophilic drugs in the brain is to develop prodrugs that may be recognized by transport systems as MCTs [195]. Finally, there have been several attempts aimed at drug delivery using receptor-mediated or adsorption-mediated endocytosis, although pinocytotic activity is reduced at the BBB. However, recently, the receptors of transferrin, insulin, heparin binding epidermal growth factor, and low-density apolipoproteins have successfully been targeted for a brain drug delivery [196–199].

Note: Parts of this book chapter had been extracted from the Dissertation of Winfried Neuhaus [201].

References

1 Abbott, N.J., Rönnbäck, L., and Hansson, E. (2006) Astrocyte–endothelial interactions at the blood–brain barrier. *Nature Reviews. Neuroscience*, **7**, 41–53.

2 Pardrigde, W.M. (1996) Brain drug delivery and blood–brain barrier transport. *Drug Delivery*, **3**, 99–115.

3 Ribatti, D., Nico, B., Crivellato, E., and Artico, M. (2006) Development of the blood–brain barrier: a historical point of view. *Anatomical Record. Part B. New Anatomist*, **2898**, 3–8.

4 Joo, F. (1993) The blood–brain barrier *in vitro*: the second decade. *Neurochemistry International*, **23**, 499–521.

5 Goldmann, E. (1913) Vitalfärbung am Zentralnervensystem: Beitrag zur Physiopathologie des

plexus choroideus der Hirnhäute. *Abhandlungen der Preußischen Akademie der Wissenschaften, Physkalisch-Mathematische Klasse*, **1**, 1–60.

6 Walter, F.K. (1930) Die allgemeinen Grundlagen des Stoffaustausches zwischen dem Zentralnervensystems und dem übrigen Körper. *Archiv für Psychiatrie und Nervenkrankheiten*, **101**, 195–230.

7 Spatz, H. (1933) Die Bedeutung der vitalen Färbung für die Lehre vom Stoffaustausch zwischen dem Zentralnervensystem und dem übrigen Körper. *Archiv für Psychiatrie und Nervenkrankheiten*, **101**, 267–358.

8 Krogh, A. (1946) The active and passive exchange of inorganic ions through the surface of living cells and through living membranes generally. *Proceedings of the Royal Society of London: Biological Sciences*, **133**, 140–200.

9 Dempsey, E.W. and Wislocky, G.B. (1955) An electron microscopic study of the blood–brain barrier in the rat, employing silver nitrate as a vital stain. *Journal of Biophysical and Biochemical Cytology*, **1**, 245–256.

10 Reese, T.S. and Karnovsky, M.J. (1967) Fine structural localisation of a blood–brain barrier to exogenous peroxidase. *The Journal of Cell Biology*, **34**, 207–217.

11 Brightman, M.W. (1965) The distribution within the brain of ferritin injected into cerebrospinal fluid compartments. II. Parenchymal distribution. *The Journal of Cell Biology*, **26**, 99–123.

12 Brightman, M.W. (1968) The intracerebral movement of proteins injected into blood and cerebrospinal fluid of mice. *Progress in Brain Research*, **29**, 19–40.

13 Nagy, Z., Peters, H., and Huttner, I. (1984) Fracture faces of cell junctions in cerebral endothelium during normal and hyperosmotic conditions. *Laboratory Investigation*, **50**, 313–322.

14 Shivers, R.R., Betz, A.L., and Goldstein, G.W. (1984) Isolated rat brain capillaries possess intact, structurally complex, interendothelial tight junctions: freeze-fracture verification of tight junction integrity. *Brain Research*, **324**, 313–322.

15 Dash, A.K. and Elmquist, W.F. (2003) Separation methods that are capable of revealing blood–brain barrier permeability. *Journal of Chromatography B*, **797**, 241–254.

16 Abbott, N.J. (2004) Prediction of blood–brain barrier permeation in drug discovery from *in vivo*, *in vitro* and *in silico* models. *Drug Discovery Today*, **1**, 407–416.

17 Davson, H., Welch, K., and Segal, M.B. (1987) Secretion of the cerebrospinal fluid, in *The Physiology and Pathophysiology of the Cerebrospinal Fluid* (ed. H. Davson), Churchill Livingstone, London.

18 Huwyler, J. (2007) *In silico* prediction of brain and CSF permeation of small molecules using PLS regression models. Oral lecture at the 9th Blood–Brain Barrier Expert Meeting, May 22, 2007, Bad Herrenalb, Germany.

19 Abbott, N.J. (2004) Evidence for bulk flow of brain interstitial fluid: significance for physiology and pathology. *Neurochemistry International*, **45**, 545–552.

20 Davson, H. and Segal, M.B. (1995) *Physiology of the CNS and Blood–Brain Barriers*, CRC Press, Boca Raton, FL.

21 Baehr, C., Reichel, V., and Fricker, G. (2006) Choroid plexus epithelial monolayers – a cell culture model from porcine brain. *Cerebrospinal Fluid Research*, **3**, 1–15.

22 Chodobski, A. and Szmydynger-Chodobski, J. (2001) Choroid plexus: target for polypeptides and site of their synthesis. *Microscopy Research and Technique*, **52**, 65–82.

23 Pardridge, W.M. (2004) Log(BB), PS products and *in silico* models of drug

brain penetration. *Drug Discovery Today*, **9**, 392–393.

24 Graff, C.L. and Pollack, G.M. (2004) Drug transport at the blood–brain barrier and the chroroid plexus. *Current Drug Metabolism*, **5**, 95–108.

25 Keep, R.F. and Jones, H.C. (1990) A morphometric study on the development of the lateral ventricle choroid plexus, choroid plexus capillaries and ventricular ependyma in the rat. *Brain Research. Developmental Brain Research*, **56**, 47–53.

26 Abbott, N.J. (2002) Astrocyte–endothelial interactions and blood–brain barrier permeability. *Journal of Anatomy*, **200**, 629–638.

27 Abbott, N.J. (2005) Dynamics of CNS barriers: evolution, differentiation and modulation. *Cellular and Molecular Neurobiology*, **25**, 5–23.

28 Butt, A.M., Jones, H.C., and Abbott, N.J. (1990) Electrical resistance across the blood–brain barrier in anaesthetised rats: a developmental study. *The Journal of Physiology*, **429**, 47–62.

29 Begley, D. and Brightman, M.W. (2003) Structural and functional aspects of the blood–brain barrier. *Progress in Drug Research*, **61**, 40–78.

30 Lauer, R. (2003) Development and evaluation of an *in vitro* model of the blood–brain barrier. PhD thesis. Department of Medicinal Chemistry, University of Vienna, Austria.

31 El-Bacha, R.S. and Minn, A. (1999) Drug metabolizing enzymes in cerebrovascular endothelial cells afford a metabolic protection to the brain. *Cellular and Molecular Biology (Noisy-le-Grand, France)*, **45**, 15–23.

32 Abbott, N.J. and Romero, I.A. (1996) Transporting therapeutics across the blood–brain barrier. *Molecular Medicine Today*, **2**, 106–113.

33 Brightman, M.W. and Reese, T.S. (1969) Junctions between initimately apposed cell membranes in the vertebrate brain. *The Journal of Cell Biology*, **40**, 648–677.

34 Dohgu, S., Takata, F., Yamauchi, A., Nakagawa, S., Egawa, T., Naito, M., Tsuruo, T., Sawada, Y. et al. (2005) Brain pericytes contribute to the induction and up-regulation of blood–brain barrier functions through transforming growth factor-β production. *Brain Research*, **1038**, 208–215.

35 Reichenbach, A. and Wolburg, H. (2004) Astrocytes and ependymal glia, in *Neuroglia*, 2nd edn (eds H. Kettenmann and B.R. Ransom), Oxford University Press, New York.

36 Dehouck, M.P., Meresse, S., Delorme, P., Fruchart, J.S., and Cecchelli, R. (1990) An easier reproducible and mass-production method to study the blood–brain barrier in vitro. *Journal of Neurochemistry*, **54**, 1798–1801.

37 Sobue, K., Yamamoto, N., Yoneda, K., Hodgson, M.E., Yamashiro, K., Tsuruoka, N., Tsuda, T., Katsuya, H. et al. (1999) Induction of blood–brain barrier properties in immortalized bovine brain endothelial cells by astrocytic factors. *Neuroscience Research*, **35**, 155–164.

38 Haseloff, R.F., Blasig, I.E., Bauer, H.-C., and Bauer, H. (2005) In search of the astrocytic factor(s) modulating blood–brain barrier function in brain capillary endothelial cells *in vitro*. *Cellular and Molecular Neurobiology*, **25**, 25–39.

39 Hartz, A.M.S., Bauer, B., Fricker, G., and Miller, D.S. (2004) Rapid regulation of P-glycoprotein at the blood–brain barrier by endothelin-1. *Molecular Pharmacology*, **66**, 387–394.

40 Stanness, K.A., Guatteo, E., and Janigro, D. (1996) A dynamic model of the blood–brain barrier "*in vitro*". *Neurotoxicology*, **17**, 481–496.

41 Ramsohoye, P.V. and Fritz, I.B. (1998) Preliminary characterization of glial-secreted factors responsible for the induction of high electrical resistances across endothelial monolayers in a blood–brain barrier

model. *Neurochemical Research*, **23**, 1545–1551.

42 Mi, H., Haeberle, H., and Barres, B.A. (2001) Induction of astrocyte differentiation by endothelial cells. *The Journal of Neuroscience*, **21**, 1538–1547.

43 Anderson, C.M. and Nedergaard, M. (2003) Astrocyte-mediated control of cerebral microcirculation. *Trends in Neurosciences*, **26**, 340–344.

44 Nedergaard, M., Ransom, B., and Goldman, S.A. (2003) New roles for astrocytes: rendering the functional architecture of the brain. *Trends in Neurosciences*, **26**, 523–530.

45 Ge, S., Song, L., and Pachter, J.S. (2005) Where is the blood–brain barrier ... really? *Journal of Neuroscience Research*, **79**, 421–427.

46 Abbott, N.J. (2004) The ABCs of the BBB. Abstract at the Peripheral Markers of Blood–Brain Barrier Failure III Symposium, November 4, 2004, Cleveland, OH.

47 Huber, J.D., Egleton, R.D., and Davis, T. (2001) Molecular physiology and pathophysiology of tight junctions in the blood–brain barrier. *Trends in Neurosciences*, **24**, 719–725.

48 Pardridge, W.M. (1999) Blood–brain barrier biology and methodology. *Journal of Neurovirology*, **5**, 556–569.

49 Van Bree, J.B., de Boer, A.G., Danhof, M., and Breimer, M.M. (1992) Drug transport across the blood–brain barrier. I. Anatomical and physiological aspects. *Pharmaceutisch weekblad. Scientific edition*, **14**, 305–310.

50 Betz, A.L., Keep, R.F., Beer, M.E., and Ren, X.D. (1994) Blood–brain barrier permeability and brain concentration of sodium, potassium, and chloride during focal ischemia. *Journal of Cerebral Blood Flow and Metabolism*, **14**, 29–37.

51 Hawkins, R.A., Peterson, D.R., and Vina, J.R. (2002) The complementary membranes forming the blood–brain barrier. *IUMB Life*, **54**, 101–107.

52 O'Kane, R. and Hawkins, R.A. (2003) Na^+-dependent transport of large neutral amino acids occurs at the abluminal membrane of the blood–brain barrier. *American Journal of Physiology. Endocrinology and Metabolism*, **285**, E1167–E1173.

53 O'Kane, R.L., Martinez-Lopez, I., DeJoseph, M.R., Vina, J.R., and Hawkins, R.A. (1999) Na^+-dependent glutamate transporters (EAAT1, EAAT2, and EAAT3) of the blood–brain barrier. *The Journal of Biological Chemistry*, **274**, 31891–31895.

54 Loscher, W. and Potschka, H. (2005) Blood–brain barrier active efflux transporters: ATP-binding cassette gene family. *NeuroRx*, **2**, 86–98.

55 Pardridge, W.M. (2003) Blood–brain barrier drug targeting: the future of brain drug development. *Molecular Interventions*, **3**, 90–105.

56 Moore, G.R. and Raine, C.S. (1988) Immunogold localization and analysis of IgG during immune-mediated demyelination. *Laboratory Investigation*, **59**, 641–648.

57 Stahl, P. and Schwartz, A.L. (1986) Receptor-mediated endocytosis. *The Journal of Clinical Investigation*, **77**, 657–662.

58 Gonatas, N.K., Stieber, A., Hickey, W.F., Herbert, S.H., and Gonatas, J.O. (1984) Endosomes and Golgi vesicles in adsorptive and fluid phase endocytosis. *The Journal of Cell Biology*, **99**, 1379–1390.

59 Heiskala, M., Peterson, P.A., and Yang, Y. (2001) The roles of claudin superfamily proteins in paracellular transport. *Traffic*, **2**, 92–98.

60 Huber, J.D., Witt, K.A., Hom, S., Egleton, R.D., Mark, K.S., and Davis, T.P. (2001) Inflammatory pain alters blood–brain barrier permeability and tight junctional protein expression. *American Journal of Physiology. Heart and Circulatory Physiology*, **280**, H1241–H1248.

61 Lippoldt, A., Kniesel, U., Liebner, S., Kalbacher, H., Kirsch, T., Wolburg, H.,

and Haller, H. (2000) Structural alterations of tight junctions are associated with loss of polarity in stroke-prone spontaneously hypertensive rat blood–brain barrier endothelial cells. *Brain Research*, **885**, 251–261.

62 Vorbrodt, A.W. and Dobrogowska, D.H. (2003) Molecular anatomy of intercellular junctions in brain endothelial and epithelial barriers: electron microscopist's view. *Brain Research Reviews*, **42**, 221–242.

63 Fanning, A.S., Mitic, L.L., and Anderson, J.M. (1999) Transmembrane proteins in the tight junction barrier. *Journal of the American Society of Nephrology*, **10**, 1337–1345.

64 Deli, M.A., Abraham, C.S., Kataoka, Y., and Niwa, M. (2005) Permeability studies on *in vitro* blood–brain barrier models: physiology, pathology, and pharmacology. *Cellular and Molecular Neurobiology*, **25**, 59–127.

65 Neuhaus, W., Bogner, E., Wirth, M., Trzeciak, J., Lachmann, B., Gabor, F., and Noe, C.R. (2006) A novel tool to characterize paracellular transport: the APTS–dextran ladder. *Pharmaceutical Research*, **23**, 1491–1501.

66 Garberg, P., Ball, M., Borg, N., Cecchelli, R., Fenart, L., Hurst, R.D., Lindmark, T., Mabondzo, A. et al. (2005) In vitro models for the blood–brain barrier. *Toxicology In Vitro*, **19**, 299–334.

67 Sun, H., Johnson, D.R., Finch, R.A., Sartorelli, A.C., Miller, D.W., and Elmquist, W.F. (2001) Transport of fluorescein in MDCKII-MRP1 transfected cells and mrp1-knockout mice. *Biochemical and Biophysical Research Communications*, **284**, 863–869.

68 Cantz, T., Nies, A.T., Brom, M., Hofmann, A.F., and Keppler, D. (2000) MRP2, a human conjugate export pump, is present and transports fluo 3 into apical vacuoles of Hep G2 cells. *American Journal of Physiology. Gastrointestinal and Liver Physiology*, **278**, G522–G531.

69 Cihlar, T. and Ho, E.S. (2000) Fluorescence-based assay for the interaction of small molecules with the human renal organic anion transporter 1. *Analytical Biochemistry*, **283**, 49–55.

70 ter Laak, A.M., Tsai, R.S., Donne-Op den Kelder, G.M., Carrupt, P.-A., Testa, B., and Timmermann, H. (1994) Lipophilicity and hydrogen-bonding capacity of H_1-antihistaminic agents in relation to their central side effects. *European Journal of Pharmaceutical Sciences*, **2**, 373–384.

71 Chishty, M., Reichel, A., Abbott, N.J., and Begley, D. (2001) Affinity for the P-glycoprotein efflux pump at the blood–brain barrier may explain the lack of CNS side effects of modern antihistamines. *Journal of Drug Targeting*, **9**, 223–228.

72 Suda, K., Rothen-Rutishauser, B., Günthert, M., and Wunderli-Allenspach, H. (2001) Phenotypic characterization of human umbilical vein endothelial (ECV304) and urinary carcinoma (T24) cells: endothelial versus epithelial features. *In Vitro Cellular & Developmental Biology – Animal*, **37**, 505–514.

73 Pawlowitsch, R., Neuhaus, W., Lauer, R., Linz, B., Lachmann, B., Ecker, G.F., and Noe, C.R. (2004) Permeation studies of a set of antihistaminic drugs with an *in vitro* blood–brain barrier model. *European Journal of Pharmaceutical Sciences*, **23**, PO-74, S60 (Suppl 1).

74 Chen, C., Hanson, E., Watson, J.W., and Lee, J.S. (2003) P-glycoprotein limits the brain penetration of nonsedating but not sedating H1-antagonists. *Drug Metabolism and Disposition*, **31**, 312–318.

75 Begley, D.J. (2005) P-glycoprotein: the prototypical BBB efflux transporter, in *Efflux Transporters and the Blood–Brain Barrier* (ed. E.M. Taylor), Nova Science Publishers, Inc., New York.

76 Takeuchi, T., Yoshitomi, S., Higuchi, T., Ikemoto, K., Niwa, S.I., Ebihara, T., Katoh, M., Yokoi, T. et al. (2006) Establishment and characterization of

the transformants stably-expressing MDR1 derived from various animal species in LLC-PK$_1$. *Pharmaceutical Research*, **23**, 1460–1472.

77 Suzuyama, N., Katoh, M., Takeuchi, T., Yoshitomi, S., Higuchi, T., Asahi, S., and Yokoi, T. (2007) Species differences of inhibitory effects on P-glycoprotein-mediated drug transport. *Journal of Pharmaceutical Sciences*, **96**, 1609–1618.

78 Linnet, K. and Ejsing, T.B. (2008) A review on the impact of P-glycoprotein on the penetration of drugs into the brain. Focus on psychotropic drugs. *European Neuropsychopharmacology*, **18** (3), 157–169.

79 Pellerin, L. and Pierre, K. (2005) Monocarboxylate transporters and their dual role in efflux and influx across the blood–brain barrier and beyond, in *Efflux Transporters and the Blood–Brain Barrier* (ed. E.M. Taylor), Nova Science Publishers, Inc., New York.

80 Nehlig, A. and Pereira de Vasconcelos, A. (1993) Glucose and ketone body utilization by the brain of neonatal rats. *Progress in Neurobiology*, **40**, 163–221.

81 Gjedde, A. and Crone, C. (1975) Induction processes in blood–brain transfer of ketone bodies during starvation. *The American Journal of Physiology*, **229**, 1165–1169.

82 Hawkins, R.A., Mans, A.M., and Davis, D.W. (1986) Regional ketone body utilization by rat brain in starvation and diabetes. *The American Journal of Physiology*, **250**, E169–E178.

83 Dringen, R., Schmoll, D., Cesar, M., and Hamprecht, B. (1993) Incorporation of radioactivity from [^{14}C]lactate into the glycogen of cultured mouse astroglial cells. Evidence for gluconeogenesis in brain cells. *Biological Chemistry Hoppe-Seyler*, **374**, 343–434.

84 Frerichs, K.U., Lindsberg, P.J., Hallenbeck, J.M., and Feuerstein, G.Z. (1990) Increased cerebral lactate output to cerebral venous blood after forebrain ischemia in rats. *Stroke*, **21**, 614–617.

85 Inao, S., Marmarou, A., Clarke, G.D., Andersen, B.J., Fatouros, P.P., and Young, H.F. (1988) Production and clearance of lactate from brain tissue, cerebrospinal fluid, and serum following experimental brain injury. *Journal of Neurosurgery*, **69**, 736–744.

86 Halestrap, A.P. and Meredith, D. (2004) The SLC16 gene family – from monocarboxylate transporters (MCTs) to aromatic amino acid transports and beyond. *Pflugers Archives*, **447**, 619–628.

87 Matson, C.T. and Drewes, L.R. (2003) Immunoblot detection of brain vascular proteins. *Methods in Molecular Medicine*, **89**, 479–487.

88 Yokel, R.A., Wilson, M., Harris, W.R., and Halestrap, A.P. (2002) Aluminium citrate uptake by immortalized brain endothelial cells: implications for its blood–brain barrier transport. *Brain Research*, **930**, 101–110.

89 Gerhart, D.Z., Enerson, B.E., Zhdankina, O.Y., Leino, R.L., and Drewes, L.R. (1998) Expression of the monocarboxylate transporter MCT2 by rat brain glia. *Glia*, **22**, 272–281.

90 Pellerin, L., Bergersen, L.H., Halestrap, A.P., and Pierre, K. (2005) Cellular and subcellular distribution of monocarboxylate transporters in cultured brain cells and in the adult brain. *Journal of Neuroscience Research*, **79**, 55–64.

91 Friesema, E.C., Ganguly, S., Abdalla, A., Manning Fox, J.E., Halestrap, A.P., and Visser, T.J. (2003) Identification of monocarboxylate transporter 8 as a specific thyroid hormone transporter. *The Journal of Biological Chemistry*, **278**, 40128–40135.

92 Bröer, S., Rahman, B., Pellegri, G., Pellerin, L., Martin, J.L., Verleysdonk, S., Hamprecht, B., and Magistretti, P.J. (1997) Comparison of lactate transport in astroglial cells and monocarboxylate transporter 1 (MCT1) expressing *Xenopus laevis* oocytes expression of two different monocarboxylate transporters in

astroglial cells and neurons. *The Journal of Biological Chemistry*, **272**, 30096–30102.

93 Bröer, S., Schneider, H.P., Bröer, A., Rahman, B., Hamprecht, B., and Deitmer, J.W. (1998) Characterization of the monocarboxylate transporter MCT1 in *Xenopus laevis* oocytes by changes in cytosolic pH. *The Biochemical Journal*, **333**, 167–174.

94 Bröer, S., Bröer, A., Schneider, H.P., Stegen, C., Halestrap, A.P., and Deitmer, J.W. (1999) Characterization of the high-affinity monocarboxylate transporter MCT2 in *Xenopus laevis* oocytes. *The Biochemical Journal*, **341**, 529–535.

95 Dimmer, K.S., Friedrich, B., Lang, F., Deimer, J.W., and Bröer, S. (2000) The low-affinity monocarboxylate transporter MCT4 is adapted to the export of lactate in highly glycolytic cells. *The Biochemical Journal*, **350**, 219–227.

96 Garcia, C.K., Goldstein, J.L., Pathak, R.K., Anderson, R.G., and Brown, M.S. (1994) Molecular characterization of a membrane transporter for lactate, pyruvate, and other monocarboxylates: implications for the Cori cycle. *Cell*, **76**, 865–873.

97 Garcia, C.K., Brown, M.S., Pathak, R.K., and Goldstein, J.L. (1995) cDNA cloning of MCT2, a second monocarboxylate transporter expressed in different cells than MCT1. *The Journal of Biological Chemistry*, **270**, 1843–1849.

98 Lin, R.Y., Vera, J.C., Chaganti, R.S., and Golde, D.W. (1998) Human monocarboxylate transporter 2 (MCT2) is a high affinity pyruvate transporter. *The Journal of Biological Chemistry*, **273**, 28959–28965.

99 Takanaga, H., Tamai, I., Inaba, S., Sai, Y., Higashida, H., Yamamoto, H., and Tsuji, A. (1995) cDNA cloning and functional characterization of rat intestinal monocarboxylate transporter. *Biochemical and Biophysical Research Communications*, **217**, 370–377.

100 Tamai, I., Takanaga, H., Maeda, H., Sai, Y., Ogihara, T., Higashida, H., and Tsuji, A. (1995) Participation of a proton-cotransporter, MCT1, in the intestinal transport of monocarboxylic acids. *Biochemical and Biophysical Research Communications*, **214**, 482–489.

101 Wilson, M.C., Jackson, V.N., Heddle, C., Price, N.T., Pilegaard, H., Juel, C., Bonen, A., Montgomery, I. et al. (1998) Lactic acid efflux from white skeletal muscle is catalyzed by the monocarboxylate transporter isoform MCT3. *The Journal of Biological Chemistry*, **273**, 15920–15926.

102 Yoon, H., Fanelli, A., Grollman, E.F., and Philp, N.J. (1997) Identification of a unique monocarboxylate transporter (MCT3) in retinal pigment epithelium. *Biochemical and Biophysical Research Communications*, **234**, 90–94.

103 De Bruijne, A.W., Vreeburg, H., and Van Steveninck, J. (1983) Kinetic analysis of L-lactate transport in human erythrocytes via the monocarboxylate-specific carrier system. *Biochimica et Biophysica Acta*, **732**, 562–568.

104 De Bruijne, A.W., Vreeburg, H., and Van Steveninck, J. (1985) Alternative-substrate inhibition of L-lactate transport via the monocarboxylate-specific carrier system in human erythrocytes. *Biochimica et Biophysica Acta*, **812**, 841–844.

105 Kim, D.K., Kanai, Y., Chairoungdua, A., Matsuo, H., Cha, S.H., and Endou, H. (2001) Expression cloning of a Na^+-independent aromatic amino acid transporter with structural similarity to H^+-/monocarboxylate transporters. *The Journal of Biological Chemistry*, **276**, 17221–17228.

106 Jackson, V.N. and Halestrap, A.P. (1996) The kinetics, substrate, and inhibitor specificity of the monocarboxylate (lactate) transporter of rat liver cells determined using the fluorescent intracellular pH indicator, 2′,7′-bis(carboxyethyl)-5(6)-carboxyfluorescein. *The Journal of Biological Chemistry*, **271**, 861–868.

107 Oldendorf, W.H. (1973) Carrier-mediated blood–brain barrier transport of short-

chain monocarboxylic organic acids. *The American Journal of Physiology*, **224**, 1450–1453.

108 Kang, Y.S., Terasaki, T., and Tsuji, A. (1990) Acidic drug transport *in vivo* through the blood–brain barrier: a role of the transport carrier for monocarboxylic acids. *Journal of Pharmacobio-Dynamics*, **13**, 158–163.

109 Terasaki, T., Kang, Y.S., Ohnishi, T., and Tsuji, A. (1991) *In-vitro* evidence for carrier-mediated uptake of acidic drugs by isolated bovine brain capillaries. *The Journal of Pharmacy and Pharmacology*, **43**, 172–176.

110 Terasaki, T., Takakuwa, S., Moritani, S., and Tsuji, A. (1991) Transport of monocarboxylic acids at the blood–brain barrier: studies with monolayers of primary cultured bovine brain capillary endothelial cells. *Journal of Pharmacology and Experimental Therapeutics*, **258**, 932–937.

111 Mac, M., Nehlig, A., Nalecz, M.J., and Nalecz, K.A. (2000) Transport of alpha-ketoisocaproate in rat cerebral cortical neurons. *Archives of Biochemistry and Biophysics*, **376**, 347–355.

112 Shimada, A., Nakagawa, Y., Morishige, H., Yamamoto, A., and Fujita, T. (2006) Functional characteristics of H^+-dependent nicotinate transport in primary cultures of astrocytes from rat cerebral cortex. *Neuroscience Letters*, **392**, 207–212.

113 Kido, Y., Tamai, I., Okamoto, M., Suzuki, F., and Tsuji, A. (2000) Functional clarification of MCT1-mediated transport of monocarboxylic acid at the blood–brain barrier using *in vitro* cultured cells and *in vivo* BUI studies. *Pharmaceutical Research*, **17**, 55–62.

114 Saheki, A., Terasaki, T., Tamai, I., and Tsuji, A. (1994) *In vivo* and *in vitro* blood–brain barrier transport of 3-hydroxy-3-methylglutaryl coenzyme A (HMG-CoA) reductase inhibitors. *Pharmaceutical Research*, **11**, 305–311.

115 Tsuji, A., Saheki, A., Tamai, I., and Terasaki, T. (1993) Transport mechanism of 3-hydroxy-5-methylglutaryl coenzyme A reductase inhibitors at the blood–brain barrier. *Journal of Pharmacology and Experimental Therapeutics*, **267**, 1085–1090.

116 Chatton, J.Y., Idle, J.R., Vagbo, C.B., and Magistretti, P.J. (2001) Insights into the mechanisms of ifosfamide encephalopathy: drug metabolites have agonistic effects on alpha-amino-3-hydroxy-5-methyl-4-isoxazolepropionic acid (AMPA)/kainate receptors and induce cellular acidification in mouse cortical neurons. *Journal of Pharmacology and Experimental Therapeutics*, **299**, 1161–1168.

117 Cassidy, C.J., Phillis, J.W., and O'Regan, M.H. (2001) Further studies on the effects of topical lactate on amino acid efflux from the ischemic rat cortex. *Brain Research*, **901**, 30–37.

118 Meier, P.J., Eckhardt, U., Schroeder, A., Hagenbuch, B., and Stieger, B. (1997) Substrate specificity of sinusoidal bile acid and organic anion uptake systems in rat and human liver. *Hepatology*, **26**, 1667–1677.

119 Koepsell, H. and Endou, H. (2004) The SLC22 drug transporter family. *Pflugers Archives*, **447**, 666–676.

120 Hagenbuch, B. and Meier, P.J. (2004) Organic anion transporting polypeptides of the OATP/SLC21 family: phylogenetic classification as OATP/SLCO superfamily, new nomenclature and molecular/functional properties. *Pflugers Archives*, **447**, 653–665.

121 Miecz, D., Januszewicz, E., Czeredys, M., Hinton, B.T., Berezowski, V., Cecchelli, R., and Nalecz, K.A. (2008) Localization of organic/carnitine transporter (OCTN2) in cells forming the blood–brain barrier. *Journal of Neurochemistry*, **104**, 113–123.

122 Chin, J.E., Soffir, R., Noonan, K.E., Choi, K.Y., and Roninson, I.B. (1989) Structure and expression of the human MDR (P-glycoprotein) gene family.

123 Smit, J.J.M., Schinkel, A.H., Oude Elferinck, R.P.J., Groen, A.K., Wagenaar, E., van Deemter, L., Mol, C., Ottenhoff, R. et al. (1993) Homozygous disruption of the murine mdr2 P-glycoprotein gene leads to a complete absence of phospholipid from bile and to liver disease. *Cell*, **75**, 451–462.

124 Pardridge, W.M., Golden, P.L., Kang, Y.-S., and Bickel, U. (1997) Brain microvascular and astrocyte localisation of P-glycoprotein. *Journal of Neurochemistry*, **68**, 1278–1285.

125 Schinkel, A.H. (1999) P-glycoprotein, a gatekeeper in the blood–brain barrier (review). *Advanced Drug Delivery Reviews*, **36**, 179–194.

126 Miura, M., Uno, T., Tateishi, T., and Suzuki, T. (2007) Pharmacokinetics of fexofenadine enantiomers in healthy subjects. *Chirality*, **19**, 223–227.

127 Sadeque, A.J.M., Wandel, C., He, H., Shah, S., and Wood, A.J.J. (2000) Increased drug delivery to the brain by P-glycoprotein inhibition. *Clinical Pharmacology and Therapeutics*, **68**, 231–237.

128 Fromm, M.F. (2004) Importance of P-glycoprotein at blood–tissue barriers. *Trends in Pharmacological Sciences*, **25**, 423–429.

129 Wang, E.J., Casciano, C.N., Clement, R.P., and Johnson, W.W. (2001) Inhibition of P-glycoprotein transport function by grapefruit juice psoralen. *Pharmaceutical Research*, **18**, 432–438.

130 Uhr, M., Holsboer, F., and Müller, M.B. (2002) Penetration of endogenous steroid hormones corticosterone, cortisol, aldosterone and progesterone into the brain is enhanced in mice deficient for both mdr1a and mdr1b P-glycoproteins. *Journal of Neuroendocrinology*, **14**, 753–759.

131 Katoh, M., Nakajima, M., Yamazaki, H., and Yokoi, T. (2001) Inhibitory effects of CYP3A4 substrates and their metabolites on P-glycoprotein-mediated transport. *European Journal of Pharmaceutical Sciences*, **12**, 505–513.

132 Uhr, M., Steckler, T., Yassouridis, A., and Holsboer, F. (2000) Penetration of amitriptyline, but not of fluoxetine, into brain is enhanced in mice with blood–brain barrier deficiency due to Mdr1a P-glycoprotein gene disruption. *Neuropsychopharmacology*, **22**, 380–387.

133 Tamai, I. (1997) Molecular characterization of intestinal absorption of drugs by carrier-mediated transport mechanisms. *Yakugaku Zasshi*, **117**, 415–434.

134 Jonker, J.W., Wagenaar, E., van Deemter, L., Gottschlich, R., Bender, H.M., Drasenbrock, J., and Schinkel, A.H. (1999) Role of blood–brain barrier P-glycoprotein in limiting brain accumulation and sedative side effects of asimadoline, a peripherally acting analgesic drug. *British Journal of Pharmacology*, **127**, 43–50.

135 Tamai, I., Kido, Y., Yamashita, J., Sai, Y., and Tsuji, A. (2000) Blood–brain barrier transport of H1-antagonist ebastine and its metabolite carebastin. *Journal of Drug Targeting*, **8**, 383–393.

136 Collett, A., Higgs, N.B., Sims, E., Rowland, M., and Warhurst, G. (1999) Modulation of the permeability of H2 receptor antagonists cimetidine and ranitidine by P-glycoprotein in rat intestine and the human colonic cell line Caco-2. *Pharmacology and Experimental Therapeutics*, **288**, 171–178.

137 Uhr, M. and Grauer, M.T. (2003) abcb1ab P-glycoprotein is involved in uptake of citalopram and trimipramine into the brain of mice. *Journal of Psychiatric Research*, **37**, 179–185.

138 Cisternino, S., Rousselle, C., Debray, M., and Scherrmann, J.C. (2003) In vivo saturation of the transport of vinblastine and colchicine by P-glycoprotein at the rat blood–brain barrier. *Pharmaceutical Research*, **20**, 1607–1611.

139 Saeki, T., Ueda, K., Tanigawara, Y., Hori, R., and Komano, T. (1993) P-glycoprotein

transports cyclosporine A and FK506. *The Journal of Biological Chemistry*, **268**, 6077–6080.

140 Katoh, M., Suzuyama, M., Takeuchi, E., Yoshitomi, S., Asahi, S., and Yokoi, T. (2006) Kinetic analyses for species differences in P-glycoprotein-mediated drug transport. *Journal of Pharmaceutical Sciences*, **95**, 2673–2683.

141 Begley, D. (2004) ABC transporters and the blood–brain barrier. *Current Pharmaceutical Design*, **10**, 1295–1312.

142 Uhr, M., Grauer, M.T., and Holsboer, F. (2003) Differential enhancement of antidepressant penetration into the brain of mice with abcb1ab (mdr1ab) P-glycoprotein gene disruption. *Biological Psychiatry*, **54**, 840–846.

143 Doran, A., Obach, R.S., Smith, B.J., Hosea, N.A., Becker, S., Callegari, E., Chen, C., Chen, X. et al. (2005) The impact of P-glycoprotein on the disposition of drugs targeted for indications of the central nervous system: evaluation using mdr1a/1b knockout mouse model. *Drug Metabolism and Disposition*, **33**, 165–174.

144 Pal, D. and Mitra, A.K. (2006) MDR- and CYP3A4-mediated drug–drug interactions. *Journal of Neuroimmune Pharmacology*, **1**, 323–329.

145 Murata, M., Tamai, I., Kato, H., Nagata, O., and Tsuji, A. (1999) Efflux transport of a new quinolone antibacterial agent, HSR-903, across the blood–brain barrier. *The Journal of Pharmacology and Experimental Therapeutics*, **290**, 51–57.

146 Bachmeier, C.J., Spitzenberger, T.J., Elmquist, W.F., and Miller, D.W. (2005) Quantitative assessment of HIV-1 protease inhibitor interactions with drug efflux transporters in the blood–brain barrier. *Pharmaceutical Research*, **22**, 1259–1268.

147 Schinkel, A.H., Wagenaar, E., Mol, C.A., and van Deemter, M. (1996) P-glycoprotein in the blood–brain barrier of mice influences the brain penetration and pharmacological activity of many drugs. *The Journal of Clinical Investigation*, **97**, 2517–2524.

148 Lötsch, J., Schmidt, R., Vetter, G., Schmidt, H., Neiderberger, E., Geisslinger, G., and Tegeder, I. (2002) Increased CNS uptake and enhanced antinociception of morphine-6-glucuronide in rats after inhibition of P-glycoprotein. *Journal of Neurochemistry*, **83**, 241–248.

149 Kaddoumi, A., Choi, S.U., Kinman, L., Whittington, E., Tsai, C.C., Ho, R.J., Anderson, B.D., and Unadkat, J.D. (2007) Inhibition of P-glycoprotein activity at the primate blood–brain barrier increases the distribution of nelfinavir into the brain but not into the cerebrospinal fluid. *Drug Metabolism and Disposition*, **35**, 1459–1462.

150 Wang, J.S., Zhu, H.J., Markowitz, J.S., Donovan, J.L., and De Vane, C.L. (2006) Evaluation of antipsychotic drugs as inhibitors of multidrug resistance transporter P-glycoprotein. *Psychopharmacology*, **187**, 415–423.

151 Fellner, S., Bauer, B., Miller, D.S., Schaffrik, M., Fankhänel, M., Spruss, T., Bernhardt, G., Graeff, C. et al. (2002) Transport of paclitaxel (Taxol) across the blood–brain barrier in vitro and in vivo. *The Journal of Clinical Investigation*, **110**, 1309–1318.

152 Potschka, H. and Löscher, W. (2001) In vivo evidence for P-glycoprotein-mediated transport of phenytoin at the blood–brain barrier of rats. *Epilepsia*, **42**, 1231–1240.

153 Kwan, T. and Gros, P. (1998) Mutational analysis of the P-glycoprotein first intracellular loop and flanking transmembrane domains. *Biochemistry*, **37**, 3337–3350.

154 Smit, J.W., Weert, B., Schinkel, A.H., and Meijer, D.K. (1998) Heterologous expression of various P-glycoproteins in polarized epithelial cells induces directional transport of small (type 1) and bulky (type 2) cationic drugs. *The Journal of Pharmacology and*

Experimental Therapeutics, **286**, 321–327.

155 Cisternino, S., Rousselle, C., Dagenais, C., and Scherrmann, J.C. (2001) Screening of multidrug-resistance sensitive drugs by *in situ* brain perfusion in P-glycoprotein-deficient mice. *Pharmaceutical Research*, **18**, 183–190.

156 Wijnholds, J. (2005) Multidrug resistance-associated proteins and efflux of organic anions at the blood–brain and blood cerebrospinal fluid barrier, in *Efflux Transporters and the Blood–Brain Barrier* (ed. E.M. Taylor), Nova Science Publishers, Inc., New York.

157 Lee, Y.J., Kusuhara, H., and Sugiyama, Y. (2004) Do multidrug resistance-associated protein-1 and -2 play any role in the elimination of estradiol-17beta-glucuronide and 2,4-dinitrophenyl-S-glutathione across the blood–cerebrospinal fluid barrier? *Journal of Pharmaceutical Sciences*, **93**, 99–107.

158 Dombrowski, S.M., Desai, S.Y., Marroni, M., Cucullo, L., Goodrich, K., Bingaman, W., Mayberg, M.R., Bengez, L. et al. (2001) Overexpression of multiple drug resistance genes in endothelial cells from patients with refractory epilepsy. *Epilepsia*, **42**, 1501–1506.

159 Borst, P., Evers, R., Kool, M., and Wijnholds, J. (2000) A family of drug transporters: the multidrug resistance-associated proteins. *Journal of the National Cancer Institute*, **92**, 1295–1302.

160 Hipfner, D.R., Deeley, R.G., and Cole, S.P. (1999) Structural, mechanistic and clinical aspects of MRP1. *Biochimica et Biophysica Acta*, **1609**, 1–18.

161 Kruh, G.D., Zeng, H., Rea, P.A., Liu, G., Chen, Z.S., Lee, K., and Belinsky, M.G. (2001) MRP subfamily transporters and resistance to anticancer drugs. *Journal of Bioenergetics and Biomembranes*, **33**, 493–501.

162 Bodo, A., Bakos, E., Szeri, F., Varadi, A., and Sarkadi, B. (2003) The role of multidrug transporters in drug availability, metabolism and toxicity. *Toxicology Letters*, **140–141**, 133–143.

163 Sun, H., Dai, H., Shaik, N., and Elmquist, W.F. (2003) Drug efflux transporters in the CNS. *Advanced Drug Delivery Reviews*, **55**, 83–105.

164 Miller, D.S., Nobmann, S.N., Gutmann, H., Toeroek, M., Trewe, J., and Fricker, G. (2004) Xenobiotic transport across isolated brain microvessels studied by confocal microscopy. *Molecular Pharmacology*, **58**, 1357–1367.

165 Potschka, H. and Löscher, W. (2001) Multidrug resistance-associated protein is involved in the regulation of extracellular levels of phenytoin in the brain. *Neuroreport*, **12**, 2387–2389.

166 Reid, G., Wielinga, P., Zelcer, N., van der Heijden, I., Kuil, A., de Haas, M., Wijnholds, J., and Borst, P. (2003) The human multidrug resistance protein MRP4 functions as a prostaglandin efflux transporter and is inhibited by nonsteroidal antiinflammatory drugs. *Proceedings of the National Academy of Sciences of the United States of America*, **100**, 9244–9249.

167 Minich, T., Riemer, J., Schulz, J.B., Wielinga, P., Wijnholds, J., and Dringen, R. (2006) The multidrug resistance protein 1 (Mrp1), but not Mrp5, mediates export of glutathione and glutathione disulfide from brain astrocytes. *Journal of Neurochemistry*, **97**, 373–384.

168 Eisenblätter, T. and Galla, H.J. (2002) A new multidrug resistance protein at the blood-brain barrier. *Biochemical and Biophysical Research Communications*, **293**, 1273–1278.

169 Eisenblätter, T., Hüwel, S. and Galla, H.J. (2003) Characterisation of the brain multidrug resistance protein (BMDP/ABCG2/BCRP) expressed at the blood-brain barrier. *Brain Research*, **971**, 221–231.

170 Doyle, L.A. and Ross, D.D. (2003) Multidrug resistance mediated by the breast cancer resistance protein BCRP (ABCG2). *Oncogene*, **22**, 7340–7358.

171 Jonker, J.W., Buitelaar, M., Wagenaar, E., Van Der Valk, M.A., Scheffer, G.L., Scheper, R.J., Plosch, T., Kuipers, F. et al. (2002) The breast cancer resistance protein protects against a major chlorophyll-derived dietary phototoxin and protoporphyria. *Proceedings of the National Academy of Sciences of the United States of America*, **99**, 15649–15654.

172 Imai, Y., Tsukahara, S., Ishikawa, E., Tsuruo, T., and Sugimoto, Y. (2002) Estrone and 17beta-estradiol reverse breast cancer resistance protein-mediated multidrug resistance. *Japanese Journal of Cancer Research: Gann*, **93**, 231–235.

173 Suzuki, M., Suzuki, H., Sugimoto, Y., and Sugiyama, Y. (2003) ABCG2 transports sulfated conjugates of steroids and xenobiotics. *The Journal of Biological Chemistry*, **278**, 22644–22649.

174 Barrand, M.A. (2005) The potential role of breast cancer resistance protein in the blood–brain barrier, in *Efflux Transporters and the Blood–Brain Barrier* (ed. E.M. Taylor), Nova Science Publishers, Inc., New York.

175 van Loevezijn, A., Allen, J.D., Schinkel, A.H., and Koomen, G.J. (2001) Inhibition of BCRP-mediated drug efflux by fumitremorgin-type indolyl diketopiperazines. *Bioorganic & Medicinal Chemistry Letters*, **11**, 29–32.

176 Schellens, J.H., Maliepaard, M., Scheper, R.J., Scheffer, G.L., Jonker, J.W., Smit, J.W., Beijnen, J.H., and Schinkel, A.H. (2000) Transport of topoisomerase I inhibitors by the breast cancer resistance protein. Potential clinical implications. *Annals of the New York Academy of Sciences*, **922**, 188–194.

177 Kawabata, S., Oka, M., Shiozawa, K., Tsukamoto, K., Nakatomi, K., Soda, H., Fukuda, M., Ikegami, Y. et al. (2001) Breast cancer resistance protein directly confers SN-38 resistance of lung cancer cells. *Biochemical and Biophysical Research Communications*, **280**, 1216–1223.

178 Wang, X., Furukawa, T., Nitanda, T., Okamoto, M., Sugimoto, Y., Akiyama, S., and Baba, M. (2003) Breast cancer resistance protein (BCRP/ABCG2) induces cellular resistance to HIV-1 nucleoside reverse transcriptase inhibitors. *Molecular Pharmacology*, **63**, 65–72.

179 Awasthi, S., Hallene, K.L., Fazio, V., Singhal, S.S., Cucullo, L., Awasthi, Y.C., Dini, G., and Janigro, D. (2005) RLIP76, a non-ABC transporter, and drug resistance in epilepsy. *BMC Neuroscience*, **6**, 61.

180 Awasthi, Y.C., Sharma, R., Yadav, S., Dwivedi, S., Sharma, A., and Awasthi, S. (2007) The non-ABC drug transporter RLIP76 (RALBP-1) plays a major role in the mechanisms of drug resistance. *Current Drug Metabolism*, **8**, 315–323.

181 Bradbury, M.W.B. (1985) The blood–brain barrier. Transport across the cerebral endothelium. *Circulation Research*, **57**, 213–222.

182 Ghersi-Egea, J.F., Leininger-Muller, B., Cecchelli, R., and Fenstermacher, J.D. (1995) Blood–brain interfaces: relevance to cerebral drug metabolism. *Toxicology Letters*, **82–83**, 645–653.

183 Minn, A., Ghersi-Egea, J.F., Perrin, R., Leininger, B., and Siest, G. (1991) Drug metabolizing enzymes in the brain and cerebral microvessels. *Brain Research Reviews*, **16**, 65–82.

184 Bauer, B. (2002) In vitro Zellkulturmodelle der Blut-Hirn Schranke zur Untersuchung der Permeation und P-Glycoprotein-Interaktion von Arzneistoffen. PhD thesis. Ruprecht-Karls-University, Heidelberg, Germany.

185 Ghersi-Egea, J.F., Minn, A., and Siest, G. (1988) A new aspect of the protective functions of the blood–brain barrier: activities of four drug-metabolizing enzymes in isolated rat brain microvessels. *Life Sciences*, **42**, 2515–2523.

186 Dallas, S., Miller, D.S., and Bendayan, R. (2006) Multidrug resistance-associated proteins: expression and function in the

central nervous system. *Pharmacological Reviews*, **58**, 140–161.
187 DeBault, L.E. and Cancilla, P.A. (1980) Gamma-glutamyl transpeptidase in isolated brain endothelial cells: induction by glial cells *in vitro*. *Science*, **207**, 653–655.
188 Guillot, F.L. and Audus, K.L. (1990) Angiotensin peptide regulation of fluid-phase endocytosis in brain microvessel endothelial cell monolayers. *Journal of Cerebral Blood Flow and Metabolism*, **10**, 827–834.
189 Lawrenson, J.G., Reid, A.R., Finn, T.M., Orte, C., and Allt, G. (1999) Cerebral and pial microvessels: differential expression of gamma-glutamyl transpeptidase and alkaline phosphatase. *Anatomy and Embryology*, **199**, 29–34.
190 Orte, C., Lawrenson, J.G., Finn, T.M., Reid, A.R., and Allt, G. (1999) A comparison of blood–brain barrier and blood–nerve barrier endothelial cell markers. *Anatomy and Embryology*, **199**, 509–517.
191 Roux, F., Durieu-Trautmann, O., Chaverot, N., Claire, M., Mailly, P., Bourre, J.M., Strosberg, A.D., and Couraud, P.O. (1994) Regulation of gamma-glutamyl transpeptidase and alkaline phosphatase activities in immortalized rat brain microvessel endothelial cells. *Journal of Cellular Physiology*, **159**, 101–113.
192 Cucullo, L., Hossain, M., Rapp, E., Manders, T., Marchi, N., and Janigro, D. (2007) Development of a humanized *in vitro* blood–brain barrier model to screen for brain penetration of antiepileptic drugs. *Epilepsia*, **48**, 505–516.
193 Löscher, W. (2007) Überexpression von Multi-drug-Transportern und Pharmkoresistenz bei Epilepsien: Epiphenomen oder kausaler Zusammenhang? Oral lecture at the 9th Blood–Brain Barrier Expert Meeting, May 22, 2007, Bad Herrenalb, Germany.
194 Hermann, D. (2007) Rolle von ABC-Transportern beim ischämischen Schlaganfall. Oral lecture at the 9th Blood–Brain Barrier Expert Meeting, May 22, 2007, Bad Herrenalb, Germany.
195 Noe, C.R., Kruse, J., Lachmann, B., and Lauer, R. (2003) Compounds containing lactic acid elements, method for the production and use thereof as pharmaceutically active substances. Patent WO03016259.
196 Boado, R.J., Zhang, Y., Zhang, Y., Wang, Y., and Pardridge, W.M. (2008) GDNF fusion protein for targeted-drug delivery across the human blood–brain barrier. *Biotechnology and Bioengineering*, **100** (2), 387–396.
197 Gaillard, P.J., Brink, A., and deBoer, A.G. (2005) Diphtheria toxin receptor-targeted brain drug delivery. *International Congress Series*, **1277**, 209–214.
198 Kreuter, J., Hekmatara, T., Dreis, S., Vogel, T., Gelperina, S., and Langer, K. (2007) Covalent attachment of apolipoprotein A-1 and apolipoprotein B-100 to albumin nanoparticles enables drug transport into the brain. *Journal of Controlled Release*, **118**, 54–58.
199 Neuwelt, E., Abbott, N.J., Abrey, L., Banks, W.A., Blakley, B., Davis, T., Engelhardt, B., Grammas, P., Nedergaard, M., Nutt, J., Pardridge, W., Rosenberg, G.A., Smith, Q., and Drewes, L.R. (2008) Strategies to advance translational research into brain barriers. *The Lancet*, **7**, 84–98.
200 Miller, D.S. (2005) The future of efflux transporters at the blood–brain barrier: lessons from the periphery, in *Efflux Transporters and the Blood–Brain Barrier* (ed. E.M. Taylor), Nova Science Publishers, Inc., New York.
201 Neuhaus, W. (2007) Development and Validation of *in Vitro* Models of the Blood-Brain Barrier. *Dissertation*, Department of Medicinal Chemistry, University of Vienna, Austria.

11
Bile Canalicular Transporters

Dirk R. de Waart and Ronald P.J. Oude Elferink

Abbreviations

2-AAF	2-acetylaminofluorene
ABC	ATP binding cassette
ABCP	placenta ABC protein
BCRP1	breast cancer resistance protein 1
BRIC	benign recurrent intrahepatic cholestasis
BSEP	bile salt export pump
CAR	constitutive androstane receptor
CYP	cytochrome P450
DBSP	dibromosulfophthalein
DNP-SG	dinitrophenyl glutathione
E_1S	estrone-3-sulfate
$E_217\beta G$	estradiol-17-β-D-glucuronide
E3040G	E3040-glucuronide
E3040S	E3040-sulfate
EHBR	mutant Eisai hyperbilirubinemic (Sprague Dawley) rat
5-FU	5-fluorouracil
FXR	farnesoid X activated receptor
GP170	*MDR1* gene product P-glycoprotein
GSH	reduced glutathione
ICG	indocyanine green
ICP	intrahepatic cholestasis of pregnancy
IR-1	inverted repeat-1
LRH-1	liver receptor homologue-1
LTC4	leukotriene C4
MDR1	multidrug resistance protein 1
MDR3	multidrug resistance protein 3
MRP2	multidrug resistance-associated protein 2
4-MUG	4-methylumbelliferone glucuronide

Transporters as Drug Carriers: Structure, Function, Substrates. Edited by Gerhard Ecker and Peter Chiba
Copyright © 2009 WILEY-VCH Verlag GmbH & Co. KGaA, Weinheim
ISBN: 978-3-527-31661-8

4-MUS	4-methylumbelliferone sulfate
MXR1	mitoxantrone resistance protein 1
NBD	nucleotide binding domains
NPC1L1	Niemann-Pick C1-like 1 protein
PB	phenobarbital
PC	phosphatidylcholine
PCN	pregnenolone 16α-carbonitrile
PFIC	progressive familial intrahepatic cholestasis
P-gp	P-glycoprotein
PGE$_2$	prostaglandin E$_2$
PhIP	2-amino-1-methyl-6-phenylimidazo[4,5-*b*]pyridine
PXR	pregnane X receptor
RXRα	9-*cis*-retinoic acid receptor α
SHP	small heterodimeric partner
TCPOBOP	1,4-bis[2-(3,5-dichloropyridyloxy)]benzene
TLC-S	taurolithocholate sulfate
TMD	transmembrane domain
TR$^-$	Abcc2-deficient (Wistar) rat strain
TUDC	taurodeoxycholate
UGT2B4	uridine 5-diphosphate-glucuronosyltransferase 2B4
VDR	vitamin D receptor

11.1
Introduction

Bile formation is an osmotic process that depends on the transport of bile salts and other compounds across the apical (canalicular) membrane of the liver cell (hepatocyte). The concentration of these substances exceeds the concentration in the liver and blood, thereby creating an osmotic gradient that attracts water. As a consequence, transporters important in the formation of bile excrete substrates against a steep gradient. For bile salts, the concentration factor can go up to a 1000-fold. The transporters involved require ATP as an energy source, they are all ATP binding cassette (ABC) transporters. In this chapter, the ABC transporters are discussed that reside in the canalicular membrane (Figure 11.1 and Table 11.1). In general, the important ABC transporters for bile formation are ABCB11, ABCB4, and ABCG5/G8. For elimination of drugs and metabolites of endogenous and exogenous compounds are ABCC2, ABCG2, and ABCB1. In this review, we discuss for every transporter the following subjects: size, diseases associated with mutations in the corresponding gene, examples of endogenous and exogenous substrates that have been recognized, and the regulation of transporter gene expression by nuclear receptor(s) and their ligands. Thereafter, relations between the different transporters in relation to homeostasis of endogenous compounds and elimination of exogenous compounds are discussed. Also, the question is discussed whether these transporters are potential (anti-)targets in drug delivery.

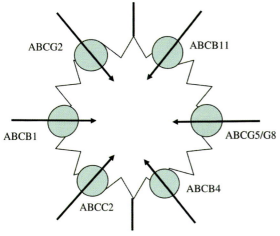

Figure 11.1 ABC transporters located at the canalicular membrane.

11.2
ABCC2

ABCC2 (MRP2) is a 190-kDa integral membrane glycoprotein. It consists of 17 transmembrane helices and 2 nucleotide binding domains (NBDs) [1]. Mutations in the *ABCC2* gene cause the Dubin–Johnson syndrome [2] (Table 11.2). Patients afflicted by this syndrome suffer from an inherited conjugated hyperbilirubinemia, which means that in these patients bilirubin is taken up by the liver from the blood and can be conjugated with glucuronic acid but that it cannot be excreted into bile via ABCC2, which is defective. Instead, it is secreted back into blood most likely via ABCC3 [2, 3]. ABCC3 shares some substrates with ABCC2 but is situated on the opposite side of the hepatocyte in the basolateral membrane. Other endogenous substrates of ABCC2 are reduced and oxidized glutathione [4, 5], leukotriene C4 (LTC4) [6], glucuronic acid conjugates of estradiol ($E_2 17\beta G$) [6] and hyodeoxycholate [7], tauroursodeoxycholate [172], taurolithocholate sulfate (TLC-S) [8], cholecystokinin-8-sulfate [9], prostaglandin E_2 [10], and estrone-3-sulfate ($E_1 S$) [11] (Table 11.3).

Table 11.1 Canalicular transporters.

Gene code	Trivial names		
ABCB1	P-gp	MDR1	GP170
ABCB4	MDR3		
ABCB11	BSEP	sPGP	
ABCC2	MRP2		
ABCG2	BCRP1	MXR1	ABCP
ABCG5	Sterolin-1		
ABCG8	Sterolin-2		

Table 11.2 Transporters and related diseases.

Transporter	Disease
ABCB11	Progressive familial intrahepatic cholestasis type 2
ABCB4	Progressive familial intrahepatic cholestasis type 3 Benign recurrent intrahepatic cholestasis type 2 Intrahepatic cholestasis of pregnancy Cholesterol gallstone disease
ABCC2	Dubin–Johnson syndrome
ABCG5/G8	Sitosterolemia

Many exogenous drugs were also found to be substrates of ABCC2: dibromosulfophthalein (DBSP) [12], indocyanine green (ICG) [13], ampicillin [14], ceftriaxone [15], carboxydichlorofluorescein [16], pravastatin [17], methotrexate [171], and probenecid [18] (Table 11.4). Some anions are transported by ABCC2 after the formation of a complex with GSH: α-naphthylisothiocyanate [19] and arsenite [20]. Furthermore, transport of uncharged compounds such as the food-derived carcinogen 2-amino-1-methyl-6-phenylimidazo[4,5-b]pyridine (PhIP), vinblastine, and sulfinpyrazone can be mediated by ABCC2 in cotransport with GSH [21, 22]. Also, metal cations such as Cd^{2+} and Zn^{2+} can be transported by ABCC2 after forming a complex with GSH [23, 24]. In case of transport of uncharged and positively charged molecules, complex formation gives the molecule a negative charge receiving the requirement to be transported by ABCC2. Furthermore, exogenous compounds that are conjugated can also be substrates for ABCC2/Abcc2. Glutathione conjugates are bromosulfophthalein glutathione [25], dinitrophenyl glutathione (DNP-SG) [26], and acetaminophen glutathione [27]. Glucuronic acid conjugates are acetaminophen glucuronide [28], mycophenolic acid glucuronide [29], indomethacin glucuronide [30], morphine glucuronide

Table 11.3 Endogenous substrates of the transporters ABCC2, ABCG2, and ABCB1.

ABCC2	ABCG2	ABCB1
Bilirubin monoglucuronide	Estrone-3-sulfate	Cortisol
Bilirubin diglucuronide	Taurolithocholate sulfate	Aldosterone
GSH	Estradiol-17β-glucuronide	Ethynylestradiol
GSSG	Protoporphyrin IX	Estrone
Leukotriene C4		Estriol
Estradiol-17β-glucuronide		Estradiol-17β-glucuronide
Hyodeoxycholate glucuronide		
Taurolithocholate sulfate		
Cholecystokinin-8-sulfate		
Prostaglandin E_2		
Estrone-3-sulfate		
Tauroursodeoxycholate		

[170], and phenobarbital glucuronide [31]. Sulfuric acid conjugates are acetaminophen sulfate [32], resveratrol sulfate [33], and phenolphthalein sulfate [34].

As mentioned above, bile formation is an osmotic process that depends largely on canalicular excretion of bile salts and glutathione. Lack of functional ABCC2 protein (in the animal models, the TR$^-$ rat, the EHBR rat, and the Abcc2$^{-/-}$ mice) diminishes bile flow roughly by one-third due to almost complete abrogation of biliary glutathione excretion. Since there is no apparent pathology in livers of ABCC2-deficient Dubin–Johnson patients, it may be suggested that alternative elimination routes are at least partially capable of lowering hepatic levels of these endogenous substrates to nontoxic levels. ABCC3 in the basolateral membrane has a preference for glucuronides and is therefore probably the most important transporter responsible for the extrusion back into blood of the glucuronic acid conjugates such as those of hyodeoxycholate and estradiol [3, 35]. Furthermore, the canalicular transporter ABCG2 (see below) is also capable of transporting organic anions such as estradiol-17 β-glucuronide into bile [36, 37]. The elimination of LTC4 and PGE$_2$ will take place via peptidolysis and thereafter via ω- and β-oxidation for the former and β-oxidation for the latter. Hepatic taurolithocholate sulfate and E$_1$S [37] can be excreted into bile via ABCG2 and the former can be alternatively eliminated via another basolateral transporter, most likely MRP4 [38], into the blood compartment. For the hepatic sulfate of cholecystokinin, the fate is not known in case of a nonfunctional ABCC2 protein.

The general picture that emerges from the substrate spectrum of ABCC2 is that it transports anions with one or, preferably, two negative charges. These can be parent compounds, such as glutathione and glucuronic and sulfuric acid, or glutathione conjugates.

How ABCC2 transports substrates is not known in detail. The mechanism of transport has been described in a model based on transport studies. First, Bakos et al. demonstrated that the transport of the GSH conjugate of N-ethylmaleimide by ABCC2 is stimulated by several other organic anions [18]. Experiments with polarized cells led to a model in which ABCC2 cotransports drugs from two distinct binding sites [22]. Zelcer et al. studied drug interactions with ABCC2 using transport assays with membrane vesicles from *Spodoptera frugiperda* insect cells that were infected with a baculovirus construct containing *ABCC2*. They proposed that ABCC2 contains two distinguishable binding sites: one site from which drug is transported and the second site that allosterically regulates the former [39, 40]. On the basis of their own data, Bodo et al. proposed a similar model [40]. Apart from the stimulation of ABCC2 activity by drugs, the expression of the *ABCC2* gene can also be upregulated under the influence of drugs. Hormones such as the glucocorticoid dexamethasone and structurally unrelated drugs such as 2-acetylaminofluorene, phenobarbital (PB), rifampicin, pregnenolone 16α-carbonitrile (PCN), 1,4-bis[2-(3,5-dichloropyridyloxy)]benzene (TCPOBOP), clotrimazole, ursodeoxycholate, chenodeoxycholate, arsenite, and hyperforin can provoke such induction [41–45]. These molecules induce expression directly via nuclear receptors. Three different nuclear receptors are thought to be involved in the regulation of *ABCC2* transcription: farnesoid X activated receptor (FXR), pregnane X receptor (PXR), and constitutive androstane receptor (CAR). All three receptors form a heterodimer with the 9-*cis*-retinoic acid receptor α (RXRα) and are thought to bind to an

26 bp sequence within the *ABCC2* promoter [41]. Ligands for FXR are bile salts, like chenodeoxycholate. Ligands for PXR are clorimazole, PCN, rifampicin, dexamethasone and many other xenobiotics.

Ligands for CAR are phenobarbital and TCPOBOP. All can induce the expression of human *ABCC2* or mouse *Abcc2* [41, 45–48]. These nuclear receptors also regulate the expression of drug metabolizing enzymes such as different cytochrome P450 (CYP) proteins, UDP-glucuronosyltransferases, and glutathione sulfotransferases [49]. In that way, coordinated upregulation of the so-called phase I (CYPs), phase II (conjugation), and phase III (transport) genes enhances the metabolism and elimination of compounds.

11.3
ABCG2

ABCG2 is a 70-kDa integral membrane protein. The functional transporter is a homodimer of 140 kDa, each monomer consists of 6 transmembrane helices and 1 nucleotide binding domain [50]. Endogenous substrates of ABCG2 are estrone-3-sulfate, taurolithocholate sulfate, $E_2 17\beta G$, and protoporphyrin IX [36, 37, 51] (Table 11.3). These substrates (except for protoporphyrin IX) are also substrates for ABCC2. Examples of exogenous substrates for ABCG2/Abcg2 are food-derived carcinogen such as the dietary carcinogen PhIP, pheophorbide α, anthracyclines, anthracenes, camptothecin derivates, methotrexate, nucleoside analogues, rhodamine 123, Hoechst 33342, lysotracker green, topotecan, imatinib, albendazole sulphoxide, and pitavastatin [36, 51–59, 173–175] (Table 11.4). ABCG2/Abcg2 shares with ABCC2/Abcc2 the possibility of transporting conjugated drugs. Glucuronic acid conjugates are $E_2 17\beta G$, 4-methylumbelliferone glucuronide (4-MUG), and E3040-glucuronide (E3040G) [37]. Sulfuric acid conjugates are $E_1 S$, TLC-S, 4-methylumbelliferone sulfate (4-MUS), and E3040-sulfate (E3040S) [37]. Glutathione conjugate is DNP-SG [37]. Indeed, ABCG2 shares with ABCC2 glutathione, glucuronic acid, and sulfuric acid conjugates as substrates, but it is not a general rule that when a drug conjugate is a substrate for ABCC2, it is also a substrate for ABCG2 and vice versa. Certain neutral amphipathic drugs (e.g., PhIP, vinblastine, and sulfinpyrazone) are transported by ABCC2 in cotransport with GSH, but the transport of neutral amphipathic drugs (e.g., PhIP and mitoxantrone) by ABCG2 does not require GSH [21, 60]. Furthermore, data show that ABCG2 consists of two identical units with two symmetric binding sides [50]. One site on each unit exclusively interacts with rhodamine 123 and the other site interacts with daunomycin, doxorubicin, prazosin, mitoxantrone, and Hoechst 33342. Moreover, binding of daunomycin, doxorubicin, and prazosin has a negative allosteric interaction and binding of mitoxantrone and Hoechst 33342 has a positive allosteric interaction with the doxorubicin binding site on the other unit. For the rhodamine 123 site, such a mechanism was not possible to prove [61]. So far, there is no disease known that is linked to mutations within the *ABCG2* gene, and no polymorphisms are known that lead to altered transporter expression, stability, or function [62].

Table 11.4 Exogenous substrates of the transporters ABCC2, ABCG2, and ABCB1.

ABCC2	ABCG2	ABCB1
Dibromosulfophthalein	PhIP	Doxorubicin
Indocyanine green	Pheophorbide α	Daunorubicin
Ampicillin	Anthracyclines	Vincristine
Ceftriaxone	Anthracenes	Vinblastine
Carboxydichlorofluorescein-diacetate	Camptothecin derivates	Paclitaxel
Pravastatin	Methotrexate	Etoposide
Probenecid	Nucleoside analogues	Teniposide
α-Naphthylisothiocyanate	Rhodamine 123	Topotecan
Arsenite	Hoechst 33342	Mitomycin C
PhIP	Lysotracker green	Colchicines
Vinblastine	Pitavastatin	Ethidium bromide
Sulfinpyrazone	4-Methylumbelliferone glucuronide	Gramicidin D
Cadmium	E3040-glucuronide	Valinomycin
Zinc	4-Methylumbelliferone sulfate	Opioid peptides
Bromosulfophthalein glutathione	Dinitrophenyl glutathione	Ritonavir
Dinitrophenyl glutathione	Mitoxantrone	Indinavir
Acetaminophen glutathione	Daunomycin	Saquinavir
Acetaminophen glucuronide	Doxorubicin	Hoechst 33342
Mycophenolic acid glucuronide	Prazosin	Rhodamine 123
Indomethacin glucuronide	Topotecan	Calcein-AM
Phenobarbital glucuronide	Imatinib	Ivermectin
Acetaminophen sulfate	Albendazole sulphoxide	
Resveratrol sulfate		
Phenolphthalein sulfate		
Paclitaxel		
Morphine glucuronide		
Methotrexate		

Like *ABCC2* expression, *ABCG2* expression is also upregulated via CAR and PXR after treatment with the CAR ligand PB, with the PXR ligand rifampicin in primary human hepatocytes [63], and with the PXR ligand 2-acetylaminofluorene (2-AAF) in murine liver [46].

11.4
ABCB1

ABCB1 is a 170-kDa integral membrane glycoprotein. It consists of 12 transmembrane helices and 2 nucleotide binding domains. There is no disease known that is linked to a nonfunctional protein due to mutations within the *ABCB1* gene, so far. In recent years, many screening studies with human individuals have been performed on the association of *ABCB1* polymorphisms with *ABCB1* expression and function in tissues and with the pharmacokinetics and pharmacodynamics of drugs [64]. However, still there are discrepancies in the results, and furthermore, no firm conclusions can be drawn to relate *ABCB1* genotypes to altered pharmacokinetics of drugs [65].

Potential physiological substrates include steroids such as cortisol, aldosterone, ethynylestradiol, estrone, and estriol [66, 67] (Table 11.3). Exogenous substrates include the pesticide ivermectin [176]; chemotherapeutic drugs such as doxorubicin, daunorubicin, vincristine, vinblastine, paclitaxel, etoposide, teniposide, topotecan, and mitomycin C; cytotoxic agents such as colchicines and ethidium bromide; cyclic and linear peptides such as gramicidin D, valinomycin, a number of other biologically active amidated peptides including opioid peptides [68] and also HIV-protease inhibitors such as ritonavir, indinavir, and saquinavir; and other compounds such as Hoechst 33342, rhodamine 123, and calcein-AM [69] (Table 11.4). In general, ABCB1 substrates are neutral or positively charged amphipathic molecules. However, there is also an example of a negatively charged substrate, namely, estradiol-17β-glucuronide [70]. ABCB1 is organized as two homologous halves that are joined by a linker region (like all other ABC transporters) [71]. The drug binding pocket is at the interface between TMDs [71–74]. How ABCB1 is able to recognize so many structurally diverse compounds is still a matter of debate, as is the case of all drug transporting ABC transporters. One suggestion is that ABCB1 contains up to four drug binding sites [75–78]. Another model proposes a common drug binding pocket in which drugs bind through a "substrate-induced fit" mechanism [74]. The common drug binding pocket is thought to be relatively large and can accommodate different substrates simultaneously [73, 78–80]. For both models holds that binding of one substrate affects the binding of another [74, 77, 81, 82].

Expression of *ABCB1* can be strongly upregulated via PXR after binding with ligands such as HIV-protease inhibitors [83] and hyperforin (the active compound of St. John's wort) [84].

11.5
ABCB4

ABCB4 is a 170-kDa integral membrane glycoprotein. It consists of 12 transmembrane helices and 2 nucleotide binding domains. Mutations in the *ABCB4* gene lead to progressive familial intrahepatic cholestasis type 3 (PFIC3), intrahepatic cholestasis of pregnancy, and cholesterol gallstone disease [85–88] (Table 11.2). Mutations in this gene have also been found in patients with symptoms of primary biliary cirrhosis [89]. PFIC3 patients (almost) completely lack functional ABCB4 activity and therefore are unable to translocate phosphatidylcholine (PC) into bile causing very severe damage to hepatocytes and cholangiocytes [90]. In other diseases, milder mutations or heterozygosity gives rise to the less severe phenotype. No other canalicular ABC transporter can transport PC at a similar rate and therefore the concentration of PC in bile of these patients is very low. In PFIC3, liver histology reveals fibrosis (progressing into cirrhosis) with portal inflammation and strong bile duct proliferation. No other endogenous compounds, except PC, are reported to be substrates for ABCB4. Since ABCB4 and ABCB1 are 77% identical at the amino acid level, attempts were made to examine if exogenous ABCB1 substrates could also be ABCB4 substrates. Using polarized monolayers of ABCB4-transfected cells Smith *et al.* found that the ABCB1 substrates digoxin, paclitaxel, and vinblastine are also ABCB4 substrates [91]. Furthermore, analysis of B-cell leukemias showed a correlation between *ABCB4* overexpression and daunorubicin transport [92, 93]. Kino *et al.* provided results to suggest that ABCB4 transports aureobasidin A [94]. It is accepted that ABCB4 is a "floppase" that translocates PC from the inner to the outer leaflet of the canalicular membrane. How PC subsequently leaves the plasma membrane is still a matter of debate. Two models exist how PC ends up in bile: The first model describes the direct extraction of PC from the outer membrane by bile salts. The second model is more complex. First, translocated PC remains in microdomains. Further active transport causes a phospholipid excess in the outer leaflet that destabilizes these microdomains. Bile salts have an increasing destabilizing effect on these domains, and ongoing translocation of PC will result in vesicular structures that pinch off to yield biliary vesicles. For more detailed reviews, see [90, 95]. Expression of *ABCB4* is upregulated via FXR by the FXR ligands chenodeoxycholate and GW4064 [96].

11.6
ABCB11

ABCB11 (BSEP or bile export pump) is a 160-kDa integral membrane glycoprotein. It consists of 12 transmembrane helices and 2 nucleotide binding domains. Mutations in the *ABCB11* gene lead to progressive familial intrahepatic cholestasis type 2 (PFIC2) [97, 98] (Table 11.2). The patients lack a functional ABCB11 protein and therefore are unable to transport bile salt into bile. Biliary bile salt concentrations in these patients are less than 1% of normal [98]. In liver biopsies, there is a prominent giant cell transformation of hepatocytes, chronic inflammation, and fibrosis. Endogenous substrates are bile salts such as taurocholate, glycocholate, taurochenodeoxycholate, glycochenodeoxycholate, taurodeoxycholate (TUDC), glycodeoxycholate, and tauroursodeoxycholate [99–102] (Table 11.5). The transport of bile salts in humans largely depends on ABCB11. In mice, however, other canalicular transporters can take over the transport of bile salt to a considerable extent, as it was observed that ABCB11-knockout animals suffer from mild and not severe cholestasis [103]. It has been suggested that Abcb1 may fulfill this function [104]. However, this difference in phenotype is also, at least partially, caused by the different bile salt composition in these two species.

There are some reports that suggest that ABCB11/Abcb11 also transport nonbile salts such as taxol and vinblastine, but otherwise ABCB11 seems to be a bile salt-specific transporter [105, 106]. There are no reports describing (the) ligand binding site(s) on ABCB11 and whether drugs can modulate the activity of the protein. Expression of *ABCB11* is tightly regulated by the nuclear receptor FXR that is activated by bile salts. After binding of bile salts to FXR, a complex with RXR is formed that can bind to an inverted repeat-1 (IR-1) element on the ABCB11 promoter.

Table 11.5 Endogenous substrates of the canalicular transporters ABCB4, ABCB11, and ABCG5/G8.

ABCB4	ABCB11	ABCG5/G8
Phosphatidylcholine	Taurocholate	Cholesterol
	Glycocholate	Sitosterol
	Taurochenodeoxycholate	Stigmasterol
	Glycochenodeoxycholate	Campesterol
	Taurodeoxycholate	5α-Cholestanol
	Glycodeoxycholate	5α-Campestanol
	Tauroursodeoxycholate	5α-Sitostanol
		22-Dehydrocholesterol
		Brassicasterol
		24-Methylene cholesterol

As a consequence, transcription and expression of *ABCB11* is induced [107, 108]. This induction can be counteracted by the vitamin D receptor (VDR) after binding of the ligand 1,25-dihydroxyvitamin D3 [109]. Furthermore, VDR is also activated by bile salts such as lithocholate and its metabolites, as is the nuclear receptor PXR [110–112]. FXR not only regulates the expression of *ABCB11* but also the synthesis of bile salts from cholesterol, through reduction of expression of the rate limiting enzymes *CYP7A1* and *CYP8B1* via induction of expression of the *small heterodimeric partner* (SHP) [113, 114]. Also, binding of lithocholate to PXR results in downregulation of the *CYP7A1* expression [112]. Bile salt metabolizing enzymes can also be induced. Chenodeoxycholate can upregulate the expression of *dehydroepiandrosterone sulfotransferase* after binding to FXR [115]. Furthermore, chenodeoxycholate can increase expression levels, after binding to FXR, of *uridine 5-diphosphate-glucuronosyltransferase 2B4* (UGT2B4), an enzyme responsible for glucuronidation of bile salts [116].

11.7
ABCG5 and ABCG8

ABCG5 and ABCG8 are two half transporters that form a heterodimer. The size of the monomers is 65–70 kDa [117]. Each monomer consists of 6 transmembrane helices and 1 nucleotide binding domain. Mutations in the *ABCG5* or *ABCG8* gene lead to sitosterolemia [118, 119] (Table 11.2). Patients with sitosterolemia lack a functional ABCG5/G8 protein and are characterized by increased intestinal absorption and decreased biliary excretion of not only dietary sterols, particularly plant sterols, but also cholesterol. As a consequence, they suffer from hypercholesterolemia and premature coronary atherosclerosis [118]. Upon liver biopsy, deposition of electron dense pigment in the vicinity of the bile canaliculi was observed. The Golgi cisterns looked markedly dilated and the Golgi vesicles were loaded with electron dense material. The rough endoplasmic reticulum profiles appeared somewhat dilated and lost some of their ribosomes [120]. Patients with sitosterolemia have elevated levels of cholesterol and plant sterols such as sitosterol, stigmasterol, campesterol, 5α-cholestanol, 5α-campestanol, and 5α-sitostanol, 22-dehydrocholesterol, brassicasterol, and 24-methylene cholesterol (Table 11.5). Since sitosterolemia is caused by mutations in the *ABCG5* or *ABCG8* gene, it was concluded that cholesterol and the above-mentioned plant sterols are substrates for ABCG5/G8. Since only the heterodimer shows transport of substrates, the protein complex will be named hereafter as ABCG5/G8. There are no reports describing (the) ligand binding site(s) on ABCG5/G8, and it is also not known whether drugs can modulate the activity of the protein. Expression of *ABCG5/G8*, however, is regulated by the nuclear receptor LXR that binds oxysterols such as 24(*S*),25-epoxycholesterol, 22(*R*)-hydroxycholesterol, and 24(*S*)-hydroxycholesterol [121–123]. Furthermore, deoxycholate represses the expression of *ABCG5/G8* indirectly via the nuclear receptor FXR. FXR induces the expression of the *SHP* and subsequently SHP reduces the expression of the *liver receptor homologue-1* (LRH-1). LRH-1 is a regulator of *ABCG5/G8* expression, and therefore reduction of *LRH-1* expression results in the reduction of *ABCG5/G8* expression [124]. FXR not only indirectly regulates the expression of the export pump of cholesterol but also the metabolism of cholesterol to bile salts, through reduction of expression of the bile salt synthesis rate limiting enzymes *CYP7A1* and *CYP8B1* via induction of *SHP* expression [113, 114]. Furthermore, LXR upregulates the expression not only of *ABCG5/G8* but also of *CYP7A1* [125].

11.8
Canalicular Transporters as Targets for Drug Delivery

For several therapeutic purposes, it is desirable to aim for excretion of drugs via canalicular transporters. First, delivery of drugs to bile can promote drug action in the biliary tree. Second, it may be attractive in terms of pharmacokinetics to have a drug that undergoes enterohepatic circulation that involves transport via canalicular transporters. Examples of these applications will be discussed hereafter.

Bacterial infections of the biliary tract can be treated with antibiotics delivered in bile. Particularly, cephalosporins have good activity against organisms involved in these infections [126]. Treatment of these infections requires a sufficient concentration at the site of action at nontoxic doses. Ceftriaxone fits these requirements as it is transported into the bile when applied at a single dose of 1 g per day resulting in an average concentration in the bile of 250 µg/ml, which substantially exceeds the minimum effective concentration of 10 µg/ml [127]. In Abcc2-deficient rats, no ceftriaxone is seen in bile; hence, Abcc2 is the specific target for ceftriaxone delivery in bile [15]. Probably, ABCC2 is also the specific target for biliary delivery of ceftriaxone in humans. Similarly, specific antibiotic targeting to the biliary tract can be performed with another Abcc2 substrate, namely, ampicillin [14].

Surgery is currently the only treatment option for biliary tract cancer. Patients with an unresectable advanced stage of disease often receive palliative systemic chemotherapy. Some combination therapies have moderate efficacy against biliary tract cancers such as the FAM regimen (5-fluorouracil (5-FU), doxorubicin, and mitomycin) and the CEF regimen (5-FU, cisplatin, and epirubicin) [128–131]. In case of the FAM regimen, multiple ABC transporters may be involved in biliary excretion: ABCB1 is known to transport doxorubicin and mitomycin [69] and ABCG2 is known to transport doxorubicin [132]. For the CEF regimen, biliary transport of cisplatin is mediated by ABCC2 [133] and biliary transport of epirubicin is mediated by ABCB1 [69] and ABCG2 [134]. In case of ceftriaxone, levels of ceftriaxone in bile were high enough to be effective, but in case of doxorubicin, epirubicin, mitomycin, and cisplatin, this is questionable. Cholangiocarcinoma cells also express the transporters involved (namely, ABCC2, ABCG2, and ABCB1) and therefore are able to excrete these anticancer drugs. Moreover, these transporters are also expressed in the intestine [135] and therefore are able to transport drugs, absorbed by the enterocytes, back into the lumen of the intestine.

Inhibiting these transporters in the intestine could be an attractive way of increasing the plasma concentration of these anticancer drugs, if they are orally administered. Drugs such as atorvastatin, tamoxifen, cyclosporin, elacridar (GF120918), zosuquidar (LY335979), oc144093 (ONT-093), tariquidar (XR9576), and laniquidar (R101933) are inhibitors of ABCB1 [136] and could be used for this purpose. However, these drugs also inhibit ABCB1 in the canalicular membrane of the liver. Therefore, in an ideal situation, a dose should be used that is high enough to inhibit ABCB1 in the intestine but low enough to obtain minimal inhibition in the liver, which can be achieved only when such an inhibitor stays in the intestine to a large extent and/or is hardly taken up by the liver. Furthermore, ABCB1 inhibitors can also be inhibitors of other ABC transporters, which can have major disadvantages. For instance, cyclosporin also inhibits ABCC2, ABCG2, and ABCB11. As a consequence, inhibition of ABCB11 by cyclosporin causes cholestasis that is an unwanted side effect. Elacridar and tamoxifen are inhibitors of both ABCB1 and ABCG2, which could be an advantage in the FAM regimen, since excretion of doxorubicin by ABCG2 would also be inhibited. Zosuquidar has been extensively characterized in terms of specificity for ABC transporters but was found not to inhibit ABCC2 and ABCG2 [137, 138]. Clinical trials have been performed to test if the clearance of anticancer drugs could be decreased through inhibiting ABCB1 by specific inhibitors, such as zosuquidar. The results are, however, at best only modest [139]. Therefore, no applicable regimen is yet available to achieve satisfying results.

In cases of cancers of the biliary tract, when surgery is not applicable, the treatment is not curative but palliative. Therefore, long-term unwanted side effects will be considered as less important. A complicating factor is that an inhibitor that achieves strong or total inhibition of ABCB1 may have considerable side effects. An example of this was observed in $Abcb1^{-/-}$ mice that were exposed to ivermectin, an acaricide, and anthelmintic drug. These knockout mice died after being sprayed with a dilute solution of ivermectin, a standard treatment against worms that is often applied to (wild-type) laboratory mice [140]. The toxicity is thought to result from an interaction of ivermectin with a neurotransmitter system in the central nervous system. The penetration of the Abcb1 substrate ivermectin into the brain is normally very low due to the presence of Abcb1 in the blood–brain barrier. The knockout animals, which lack the ability to transport ivermectin back to the blood compartment, showed as a consequence a 87-fold higher level of ivermectin in the brain. Thus, treatment of cancer (e.g., in the biliary tract) with anticancer drugs in combination with such a putative ABCB1 inhibitor would give undesirable brain damage due to high levels of these anticancer drugs in the brain, and if ABCB1 is inhibited, also in the blood–brain barrier. Therefore, total inhibition of ABCB1 by a specific inhibitor may not be a clinically acceptable option since a major drug–drug interaction can be expected. Chemoradiotherapy with doxofluridine and paclitaxel seemed to produce a good clinical response without severe toxicity and improve survival rates in patients with extrahepatic bile duct cancer [141]. Paclitaxel is eliminated from enterocytes by ABCB1. The survival rates could be improved by increasing the oral availability by using an ABCB1 inhibitor. Interestingly, biliary excretion is maintained, when hepatic ABCB1 is also inhibited, because Abcc2 is also able to excrete paclitaxel into bile [142]. ABCC2 is then the target for biliary delivery and ABCB1 the antitarget.

Not only can ABCB1 inhibition have unwanted (severe) side effects but also ABCG2 inhibition is at risk. This was shown in $Abcg2^{-/-}$ mice that lack a functional Abcg2 protein. These $Abcg2^{-/-}$ mice display a type of genetic porphyria characterized by increased levels of protoporphyrin IX in erythrocytes [51]. Furthermore, these $Abcg2^{-/-}$ mice became extremely sensitive to the dietary chlorophyll breakdown product pheophorbide α, resulting in severe, sometimes lethal phototoxic lesions on light-exposed skin. Abcg2 transports pheophorbide α and is highly efficient in limiting its uptake from ingested food [51]. In case of total inhibition of ABCG2 in the intestine, as part of a therapeutic regimen, such toxic diet components could be increasingly absorbed because of diminished excretion from the enterocyte by ABCG2. An example is the food-derived carcinogen PhIP. $Abcg2^{-/-}$ mice showed a higher plasma concentration of PhIP after oral feeding. Furthermore, liver, brain, and kidney concentrations of PhIP were increased in $Abcg2^{-/-}$ mice [54]. As PhIP is a mutagenic and carcinogenic compound, this represents a highly undesirable side effect [143]. Besides Abcg2, Abcc2 is also able to transport PhIP. $Abcc2^{-/-}$ rats showed reduced elimination of PhIP and metabolites. After administration with PhIP, elevated blood concentrations and liver and kidney content of PhIP and its metabolites were seen in Abcc2-deficient rats in comparison with the wild type [21, 144]. Therefore, total inhibition of ABCC2 in humans could also lead to increased absorbance of PhIP and its metabolites.

Obstetric cholestasis or intrahepatic cholestasis of pregnancy is an important clinical problem, giving rise not only to pruritis, sometimes gallstones, but also to risks for the baby. Causes for this cholestasis are multifactorial, with genetic, environmental, and hormonal factors being involved. Evidence was presented that the steroid profiles in serum are profoundly altered [145]. Furthermore, ICP has been associated with an abnormal reaction of the maternal liver to endogenous sex steroids and their metabolites [146, 147]. Certain steroids such as estradiol-17β-glucuronide induce cholestasis by transinhibition of ABCB11 [99, 148]. To restore the activity of this transporter, patients can be treated with ursodeoxycholate. This compound is effectively conjugated in the liver with taurine forming TUDC. TUDC is a potent intracellular signaling agent that induces stimulation of impaired hepatocellular secretion. Vesicular exocytosis is increased, and an increased number of transport proteins is mobilized to the canalicular membrane; as a consequence, transport systems involved in the biliary secretion of steroid mono- and disulfates as well as bile salts are stimulated [149, 150]. Moreover, TUDC may also directly activate canalicular transporters through modification of their phosphorylation status [151]. Other characteristics of TUDC are at the level of cell survival. Toxic bile salts such as glycochenodeoxycholate and glycodeoxycholate can induce apoptosis in hepatocytes at concentrations comparable to those found in chronic cholestasis. TUDC can block apoptosis *in vitro* and *in vivo* in the rat and in human hepatocytes by interrupting classic pathways of apoptosis [152–154]. Hydrophobic bile salts are also capable of disrupting plasma membranes *in vitro*. TUDC has been shown to counteract this disruption of membranes probably by alteration of the structure and composition of micelles rather than by direct membrane interactions [155, 156]. To achieve this effect of TUDC, a high concentration in the millimolar range is needed. These concentrations are reached within the biliary tree after transport of TUDC by the target ABCB11. High therapeutic doses of up to 15 mg/kg per day are needed. It is not exactly known which mechanism will be the most important for the treatment of ICP. Stimulation of transport systems involved in the biliary secretion of steroid mono- and disulfates may be beneficial [157]. Stimulation of bile salt transport will most likely reduce cholestatic symptoms. For other chronic cholestatic diseases such as primary biliary cirrhosis and primary sclerosing cholangitis, ursodeoxycholate treatment is also proven to be beneficial [158–161].

Treatment of hypercholesterolemia can be performed by inhibiting the uptake of cholesterol in the intestine. This can be achieved by treatment with ezetimibe [162, 163]. The target of ezetimibe is Niemann-Pick C1-like 1 protein (NPC1L1), which is involved in cholesterol uptake [164, 165]. Ezetimibe itself is also taken up by enterocytes and extensively glucuronidated [163]. Conjugation in general makes drugs more hydrophilic and elimination is in that way facilitated. However, clearance of ezetimibe glucuronide from blood is not straightforward; the concentration–time profiles exhibit multiple peaks [166, 167]. This is caused by enterohepatic circulation: after being glucuronidated in the enterocyte, ezetimibe travels via portal blood to the liver, is then extruded into bile, presumably by MRP2 [168, 169], and enters the intestine again, the site where inhibition of cholesterol uptake takes place. In this way, enterohepatic circulation is necessary for prolonged inhibition of NPC1L1, with MRP2 being the target for biliary excretion of ezetimibe glucuronide.

In conclusion, canalicular transporters play a crucial role in pharmacokinetics and pharmacodynamics of many drugs; for example, delivery of ceftriaxone mediated by ABCC2 to the biliary tract after a bacterial infection is applicable. On the other hand, delivery of anticancer drugs to cells of the biliary tract is limited by the expression of the same transporters in cholangiocarcinoma cells. The use of inhibitors of ABC transporters to increase intestinal drug uptake may have serious potential side effects. Furthermore, patients with chronic intrahepatic cholestasis can be treated with ursodeoxycholate with the canalicular transporter ABCB11 being the specific target for biliary delivery.

References

1 Borst, P., Zelcer, N., and van de Wetering, K. (2006) MRP2 and 3 in health and disease. *Cancer Letters*, **234** (1), 51–61.

2 Dubin, I.N. and Johnson, F.B. (1954) Chronic idiopathic jaundice with unidentified pigment in liver cells: a new clinicopathologic entity with a report of 12 cases. *Medicine*, **33** (3), 155–197.

3 Zelcer, N., van de Wetering, K., de Waart, R., Scheffer, G.L., Marschall, H.U., Wielinga, P.R. *et al.* (2006) Mice lacking Mrp3 (Abcc3) have normal bile salt transport, but altered hepatic transport of endogenous glucuronides. *Journal of Hepatology*, **44** (4), 768–775.

4 Oude Elferink, R.P., Ottenhoff, R., Liefting, W.G., Schoemaker, B., Groen, A.K., and Jansen, P.L. (1990) ATP-dependent efflux of GSSG and GS-conjugate from isolated rat hepatocytes. *The American Journal of Physiology*, **258** (5 Pt 1), G699–G706.

5 Paulusma, C.C., van Geer, M.A., Evers, R., Heijn, M., Ottenhoff, R. Borst, P. *et al.* (1999) Canalicular multispecific organic anion transporter/multidrug resistance protein 2 mediates low-affinity transport of reduced glutathione. *The Biochemical Journal*, **338** (Pt 2), 393–401.

6 Cui, Y., Konig, J., Buchholz, J.K., Spring, H., Leier, I., and Keppler, D. (1999) Drug resistance and ATP-dependent conjugate transport mediated by the apical multidrug resistance protein, MRP2, permanently expressed in human and canine cells. *Molecular Pharmacology*, **55** (5), 929–937.

7 Keppler, D., Leier, I., and Jedlitschky, G. (1997) Transport of glutathione conjugates and glucuronides by the multidrug resistance proteins MRP1 and MRP2. *Biological Chemistry*, **378** (8), 787–791.

8 Kuipers, F., Enserink, M., Havinga, R., van der Steen, A.B., Hardonk, M.J., Fevery, J. *et al.* (1988) Separate transport systems for biliary secretion of sulfated and unsulfated bile acids in the rat. *The Journal of Clinical Investigation*, **81** (5), 1593–1599.

9 Letschert, K., Komatsu, M., Hummel-Eisenbeiss, J., and Keppler, D. (2005) Vectorial transport of the peptide CCK-8 by double-transfected MDCKII cells stably expressing the organic anion transporter OATP1B3 (OATP8) and the export pump ABCC2. *The Journal of Pharmacology and Experimental Therapeutics*, **313** (2), 549–556.

10 de Waart, D.R., Paulusma, C.C., Kunne, C., and Oude Elferink, R.P. (2006) Multidrug resistance associated protein 2 mediates transport of prostaglandin E2. *Liver International*, **26** (3), 362–368.

11 Kopplow, K., Letschert, K., Konig, J., Walter, B., and Keppler, D. (2005) Human hepatobiliary transport of organic anions analyzed by quadruple-transfected cells. *Molecular Pharmacology*, **68** (4), 1031–1038.

12 Cui, Y., Konig, J., and Keppler, D. (2001) Vectorial transport by double-transfected cells expressing the human uptake transporter SLC21A8 and the apical export pump ABCC2. *Molecular Pharmacology*, **60** (5), 934–943.

13 Jansen, P.L., van Klinken, J.W., van Gelder, M., Ottenhoff, R., and Elferink, R.P. (1993) Preserved organic anion transport in mutant TR⁻ rats with a hepatobiliary secretion defect. *The American Journal of Physiology*, **265** (3 Pt 1), G445–G452.

14 Verkade, H.J., Wolbers, M.J., Havinga, R., Uges, D.R., Vonk, R.J., and Kuipers, F. (1990) The uncoupling of biliary lipid from bile acid secretion by organic anions in the rat. *Gastroenterology*, **99** (5), 1485–1492.

15 Oude Elferink, R.P. and Jansen, P.L. (1994) The role of the canalicular multispecific organic anion transporter in the disposal of endo- and xenobiotics. *Pharmacology & Therapeutics*, **64** (1), 77–97.

16 Kitamura, T., Jansen, P., Hardenbrook, C., Kamimoto, Y., Gatmaitan, Z., and Arias, I.M. (1990) Defective ATP-dependent bile canalicular transport of organic anions in mutant (TR⁻) rats with conjugated hyperbilirubinemia. *Proceedings of the National Academy of Sciences of the United States of America*, **87** (9), 3557–3561.

17 Yamazaki, M., Akiyama, S., Ni'inuma, K., Nishigaki, R., and Sugiyama, Y. (1997) Biliary excretion of pravastatin in rats: contribution of the excretion pathway mediated by canalicular multispecific organic anion transporter. *Drug Metabolism and Disposition*, **25** (10), 1123–1129.

18 Bakos, E., Evers, R., Sinko, E., Varadi, A., Borst, P., and Sarkadi, B. (2000) Interactions of the human multidrug resistance proteins MRP1 and MRP2 with organic anions. *Molecular Pharmacology*, **57** (4), 760–768.

19 Dietrich, C.G., Ottenhoff, R., de Waart, D.R., and Oude Elferink, R.P. (2001) Role of MRP2 and GSH in intrahepatic cycling of toxins. *Toxicology*, **167** (1), 73–81.

20 Kala, S.V., Kala, G., Prater, C.I., Sartorelli, A.C., and Lieberman, M.W. (2004) Formation and urinary excretion of arsenic triglutathione and methylarsenic diglutathione. *Chemical Research in Toxicology*, **17** (2), 243–249.

21 Dietrich, C.G., de Waart, D.R., Ottenhoff, R., Bootsma, A.H., van Gennip, A.H., and Elferink, R.P. (2001) Mrp2-deficiency in the rat impairs biliary and intestinal excretion and influences metabolism and disposition of the food-derived carcinogen 2-amino-1-methyl-6-phenylimidazo. *Carcinogenesis*, **22** (5), 805–811.

22 Evers, R., de Haas, M., Sparidans, R., Beijnen, J., Wielinga, P.R., Lankelma, J. et al. (2000) Vinblastine and sulfinpyrazone export by the multidrug resistance protein MRP2 is associated with glutathione export. *British Journal of Cancer*, **83** (3), 375–383.

23 Dijkstra, M., Havinga, R., Vonk, R.J., and Kuipers, F. (1996) Bile secretion of cadmium, silver, zinc and copper in the rat. Involvement of various transport systems. *Life Sciences*, **59** (15), 1237–1246.

24 Houwen, R., Dijkstra, M., Kuipers, F., Smit, E.P., Havinga, R., and Vonk, R.J. (1990) Two pathways for biliary copper excretion in the rat. The role of glutathione. *Biochemical Pharmacology*, **39** (6), 1039–1044.

25 Jansen, P.L., Groothuis, G.M., Peters, W.H., and Meijer, D.F. (1987) Selective hepatobiliary transport defect for organic anions and neutral steroids in mutant rats with hereditary-conjugated hyperbilirubinemia. *Hepatology*, **7** (1), 71–76.

26 Elferink, R.P., Ottenhoff, R., Liefting, W., de Haan, J., and Jansen, P.L. (1989) Hepatobiliary transport of glutathione and glutathione conjugate in rats with hereditary hyperbilirubinemia. *The Journal of Clinical Investigation*, **84** (2), 476–483.

27 Chen, C., Hennig, G.E., and Manautou, J.E. (2003) Hepatobiliary excretion of acetaminophen glutathione conjugate and its derivatives in transport-deficient (TR⁻) hyperbilirubinemic rats. *Drug Metabolism and Disposition*, **31** (6), 798–804.

28 Xiong, H., Turner, K.C., Ward, E.S., Jansen, P.L., and Brouwer, K.L. (2000) Altered hepatobiliary disposition of acetaminophen glucuronide in isolated perfused livers from multidrug resistance-associated protein 2-deficient TR(−) rats. *The Journal of Pharmacology and Experimental Therapeutics*, **295** (2), 512–518.

29 Westley, I.S., Brogan, L.R., Morris, R.G., Evans, A.M., and Sallustio, B.C. (2006) Role of Mrp2 in the hepatic disposition of mycophenolic acid and its glucuronide metabolites: effect of cyclosporine. *Drug Metabolism and Disposition*, **34** (2), 261–266.

30 Kouzuki, H., Suzuki, H., and Sugiyama, Y. (2000) Pharmacokinetic study of the hepatobiliary transport of indomethacin. *Pharmaceutical Research*, **17** (4), 432–438.

31 Patel, N.J., Zamek-Gliszczynski, M.J., Zhang, P., Han, Y.H., Jansen, P.L., Meier, P.J. et al. (2003) Phenobarbital alters hepatic Mrp2 function by direct and indirect interactions. *Molecular Pharmacology*, **64** (1), 154–159.

32 Zamek-Gliszczynski, M.J., Hoffmaster, K.A., Tian, X., Zhao, R., Polli, J.W., Humphreys, J.E. et al. (2005) Multiple mechanisms are involved in the biliary excretion of acetaminophen sulfate in the rat: role of Mrp2 and Bcrp1. *Drug Metabolism and Disposition*, **33** (8), 1158–1165.

33 Kaldas, M.I., Walle, U.K., and Walle, T. (2003) Resveratrol transport and metabolism by human intestinal Caco-2 cells. *The Journal of Pharmacy and Pharmacology*, **55** (3), 307–312.

34 Tanaka, H., Sano, N., and Takikawa, H. (2003) Biliary excretion of phenolphthalein sulfate in rats. *Pharmacology*, **68** (4), 177–182.

35 Hirohashi, T., Suzuki, H., and Sugiyama, Y. (1999) Characterization of the transport properties of cloned rat multidrug resistance-associated protein 3 (MRP3). *The Journal of Biological Chemistry*, **274** (21), 15181–15185.

36 Chen, Z.S., Robey, R.W., Belinsky, M.G., Shchaveleva, I., Ren, X.Q., Sugimoto, Y. et al. (2003) Transport of methotrexate, methotrexate polyglutamates, and 17beta-estradiol 17-(beta-D-glucuronide) by ABCG2: effects of acquired mutations at R482 on methotrexate transport. *Cancer Research*, **63** (14), 4048–4054.

37 Suzuki, M., Suzuki, H., Sugimoto, Y., and Sugiyama, Y. (2003) ABCG2 transports sulfated conjugates of steroids and xenobiotics. *The Journal of Biological Chemistry*, **278** (25), 22644–22649.

38 Zelcer, N., Reid, G., Wielinga, P., Kuil, A., van der Heijden, I., Schuetz, J.D. et al. (2003) Steroid and bile acid conjugates are substrates of human multidrug-resistance protein (MRP) 4 (ATP-binding cassette C4). *The Biochemical Journal*, **371** (Pt 2), 361–367.

39 Zelcer, N., Huisman, M.T., Reid, G., Wielinga, P., Breedveld, P., Kuil, A. et al. (2003) Evidence for two interacting ligand binding sites in human multidrug resistance protein 2 (ATP binding cassette C2). *The Journal of Biological Chemistry*, **278** (26), 23538–23544.

40 Bodo, A., Bakos, E., Szeri, F., Varadi, A., and Sarkadi, B. (2003) Differential modulation of the human liver conjugate transporters MRP2 and MRP3 by bile acids and organic anions. *The Journal of Biological Chemistry*, **278** (26), 23529–23537.

41 Kast, H.R., Goodwin, B., Tarr, P.T., Jones, S.A., Anisfeld, A.M., Stoltz, C.M. et al. (2002) Regulation of multidrug resistance-associated protein 2 (ABCC2) by the nuclear receptors pregnane X receptor, farnesoid X-activated receptor, and constitutive androstane receptor. *The Journal of Biological Chemistry*, **277** (4), 2908–2915.

42 Schrenk, D., Baus, P.R., Ermel, N., Klein, C., Vorderstemann, B., and Kauffmann, H.M. (2001) Up-regulation of transporters of the MRP family by drugs and toxins. *Toxicology Letters*, **120** (1–3), 51–57.

43 Kauffmann, H.M., Pfannschmidt, S., Zoller, H., Benz, A., Vorderstemann, B., Webster, J.I. et al. (2002) Influence of redox-active compounds and PXR-activators on human MRP1 and MRP2 gene expression. *Toxicology*, **171** (2–3), 137–146.

44 Fickert, P., Zollner, G., Fuchsbichler, A., Stumptner, C., Pojer, C., Zenz, R. et al. (2001) Effects of ursodeoxycholic and cholic acid feeding on hepatocellular transporter expression in mouse liver. *Gastroenterology*, **121** (1), 170–183.

45 Tzameli, I., Pissios, P., Schuetz, E.G., and Moore, D.D. (2000) The xenobiotic compound 1,4-bis[2-(3,5-dichloropyridyloxy)]benzene is an agonist ligand for the nuclear receptor CAR. *Molecular and Cellular Biology*, **20** (9), 2951–2958.

46 Anapolsky, A., Teng, S., Dixit, S., and Piquette-Miller, M. (2006) The role of pregnane X receptor in 2-acetylaminofluorene-mediated induction of drug transport and metabolizing enzymes in mice. *Drug Metabolism and Disposition*, **34** (3), 405–409.

47 Guo, G.L., Lambert, G., Negishi, M., Ward, J.M., Brewer, H.B., Jr., Kliewer, S.A. et al. (2003) Complementary roles of farnesoid X receptor, pregnane X receptor, and constitutive androstane receptor in protection against bile acid toxicity. *The Journal of Biological Chemistry*, **278** (46), 45062–45071.

48 Johnson, D.R. and Klaassen, C.D. (2002) Regulation of rat multidrug resistance protein 2 by classes of prototypical microsomal enzyme inducers that activate distinct transcription pathways. *Toxicological Sciences*, **67** (2), 182–189.

49 Xu, C., Li, C.Y., and Kong, A.N. (2005) Induction of phase I, II and III drug metabolism/transport by xenobiotics. *Archives of Pharmacal Research*, **28** (3), 249–268.

50 Mitomo, H., Kato, R., Ito, A., Kasamatsu, S., Ikegami, Y., Kii, I. et al. (2003) A functional study on polymorphism of the ATP-binding cassette transporter ABCG2: critical role of arginine-482 in methotrexate transport. *The Biochemical Journal*, **373** (Pt 3), 767–774.

51 Jonker, J.W., Buitelaar, M., Wagenaar, E., van der Valk, M.A., Scheffer, G.L., Scheper, R.J. et al. (2002) The breast cancer resistance protein protects against a major chlorophyll-derived dietary phototoxin and protoporphyria. *Proceedings of the National Academy of Sciences of the United States of America*, **99** (24), 15649–15654.

52 Hirano, M., Maeda, K., Matsushima, S., Nozaki, Y., Kusuhara, H., and Sugiyama, Y. (2005) Involvement of BCRP (ABCG2) in the biliary excretion of pitavastatin. *Molecular Pharmacology*, **68** (3), 800–807.

53 Litman, T., Brangi, M., Hudson, E., Fetsch, P., Abati, A., Ross, D.D. et al. (2000) The multidrug-resistant phenotype associated with overexpression of the new ABC half-transporter, MXR (ABCG2). *Journal of Cell Science*, **113** (Pt 11), 2011–2021.

54 van Herwaarden, A.E., Jonker, J.W., Wagenaar, E., Brinkhuis, R.F., Schellens, J.H., Beijnen, J.H. et al. (2003) The breast cancer resistance protein (Bcrp1/Abcg2) restricts exposure to the dietary carcinogen 2-amino-1-methyl-6-phenylimidazo[4,5-b]pyridine. *Cancer Research*, **63** (19), 6447–6452.

55 Nakatomi, K., Yoshikawa, M., Oka, M., Ikegami, Y., Hayasaka, S., Sano, K. et al. (2001) Transport of 7-ethyl-10-hydroxycamptothecin (SN-38) by breast cancer resistance protein ABCG2 in human lung cancer cells. *Biochemical and Biophysical Research Communications*, **288** (4), 827–832.

56 Rajendra, R., Gounder, M.K., Saleem, A., Schellens, J.H., Ross, D.D., Bates, S.E. et al. (2003) Differential effects of the breast cancer resistance protein on the cellular accumulation and cytotoxicity of 9-aminocamptothecin and 9-nitrocamptothecin. *Cancer Research*, **63** (12), 3228–3233.

57 Wang, X., Furukawa, T., Nitanda, T., Okamoto, M., Sugimoto, Y., Akiyama, S. et al. (2003) Breast cancer resistance protein (BCRP/ABCG2) induces cellular resistance to HIV-1 nucleoside reverse transcriptase inhibitors. *Molecular Pharmacology*, **63** (1), 65–72.

58 Robey, R.W., Honjo, Y., van de Laar, A., Miyake, K., Regis, J.T., Litman, T. et al. (2001) A functional assay for detection of the mitoxantrone resistance protein, MXR (ABCG2). *Biochimica et Biophysica Acta*, **1512** (2), 171–182.

59 Scharenberg, C.W., Harkey, M.A., and Torok-Storb, B. (2002) The ABCG2 transporter is an efficient Hoechst 33342 efflux pump and is preferentially expressed by immature human hematopoietic progenitors. *Blood*, **99** (2), 507–512.

60 Diah, S.K., Smitherman, P.K., Aldridge, J., Volk, E.L., Schneider, E., Townsend, A.J. et al. (2001) Resistance to mitoxantrone in multidrug-resistant MCF7 breast cancer cells: evaluation of mitoxantrone transport and the role of multidrug resistance protein family proteins. *Cancer Research*, **61** (14), 5461–5467.

61 Clark, R., Kerr, I.D., and Callaghan, R. (2006) Multiple drugbinding sites on the R482G isoform of the ABCG2 transporter. *British Journal of Pharmacology*, **149** (5), 506–515.

62 Honjo, Y., Morisaki, K., Huff, L.M., Robey, R.W., Hung, J., Dean, M. et al. (2002) Single-nucleotide polymorphism (SNP) analysis in the ABC half-transporter ABCG2 (MXR/BCRP/ABCP1). *Cancer Biology & Therapy*, **1** (6), 696–702.

63 Jigorel, E., Le, V.M., Boursier-Neyret, C., Parmentier, Y., and Fardel, O. (2006) Differential regulation of sinusoidal and canalicular hepatic drug transporter expression by xenobiotics activating drug-sensing receptors in primary human hepatocytes. *Drug Metabolism and Disposition*, **34** (10), 1756–1763.

64 Yates, C.R., Zhang, W., Song, P., Li, S., Gaber, A.O., Kotb, M. et al. (2003) The effect of CYP3A5 and MDR1 polymorphic expression on cyclosporine oral disposition in renal transplant patients. *Journal of Clinical Pharmacology*, **43** (6), 555–564.

65 Sakaeda, T. (2005) MDR1 genotype-related pharmacokinetics: fact or fiction? *Drug Metabolism and Pharmacokinetics*, **20** (6), 391–414.

66 Ueda, K., Okamura, N., Hirai, M., Tanigawara, Y., Saeki, T., Kioka, N. et al. (1992) Human P-glycoprotein transports cortisol, aldosterone, and dexamethasone, but not progesterone. *The Journal of Biological Chemistry*, **267** (34), 24248–24252.

67 Kim, W.Y. and Benet, L.Z. (2004) P-glycoprotein (P-gp/MDR1)-mediated efflux of sex-steroid hormones and modulation of P-gp expression *in vitro*. *Pharmaceutical Research*, **21** (7), 1284–1293.

68 Oude Elferink, R.P. and Zadina, J. (2001) MDR1 P-glycoprotein transports endogenous opioid peptides. *Peptides*, **22** (12), 2015–2020.

69 Ambudkar, S.V., Dey, S., Hrycyna, C.A., Ramachandra, M., Pastan, I., and Gottesman, M.M. (1999) Biochemical, cellular, and pharmacological aspects of the multidrug transporter. *Annual Review of Pharmacology and Toxicology*, **39**, 361–398.

70 Huang, L., Hoffman, T., and Vore, M. (1998) Adenosine triphosphate-dependent transport of estradiol-17beta (beta-D-glucuronide) in membrane vesicles by MDR1 expressed in insect cells. *Hepatology*, **28** (5), 1371–1377.

71 Chen, C.J., Chin, J.E., Ueda, K., Clark, D.P., Pastan, I., Gottesman, M.M. et al. (1986) Internal duplication and homology with bacterial transport proteins in the mdr1 (P-glycoprotein) gene from multidrug-resistant human cells. *Cell*, **47** (3), 381–389.

72 Loo, T.W. and Clarke, D.M. (2001) Defining the drug-binding site in the human multidrug resistance P-glycoprotein using a methanethiosulfonate analog of verapamil, MTS-verapamil. *The Journal of Biological Chemistry*, **276** (18), 14972–14979.

73 Loo, T.W. and Clarke, D.M. (2001) Determining the dimensions of the drug-binding domain of human P-glycoprotein using thiol cross-linking compounds as molecular rulers. *The Journal of Biological Chemistry*, **276** (40), 36877–36880.

74 Loo, T.W. and Clarke, D.M. (2002) Location of the rhodamine-binding site in the human multidrug resistance P-glycoprotein. *The Journal of Biological Chemistry*, **277** (46), 44332–44338.

75 Dey, S., Ramachandra, M., Pastan, I., Gottesman, M.M., and Ambudkar, S.V.

(1997) Evidence for two nonidentical drug-interaction sites in the human P-glycoprotein. *Proceedings of the National Academy of Sciences of the United States of America*, **94** (20), 10594–10599.

76 Pascaud, C., Garrigos, M., and Orlowski, S. (1998) Multidrug resistance transporter P-glycoprotein has distinct but interacting binding sites for cytotoxic drugs and reversing agents. *The Biochemical Journal*, **333** (Pt 2), 351–358.

77 Shapiro, A.B., Fox, K., Lam, P., and Ling, V. (1999) Stimulation of P-glycoprotein-mediated drug transport by prazosin and progesterone. Evidence for a third drug-binding site. *European Journal of Biochemistry*, **259** (3), 841–850.

78 Lugo, M.R. and Sharom, F.J. (2005) Interaction of LDS-751 with P-glycoprotein and mapping of the location of the R drug binding site. *Biochemistry*, **44** (2), 643–655.

79 Loo, T.W., Bartlett, M.C., and Clarke, D.M. (2003) Simultaneous binding of two different drugs in the binding pocket of the human multidrug resistance P-glycoprotein. *The Journal of Biological Chemistry*, **278** (41), 39706–39710.

80 Sauna, Z.E., Andrus, M.B., Turner, T.M., and Ambudkar, S.V. (2004) Biochemical basis of polyvalency as a strategy for enhancing the efficacy of P-glycoprotein (ABCB1) modulators: stipiamide homodimers separated with defined-length spacers reverse drug efflux with greater efficacy. *Biochemistry*, **43** (8), 2262–2271.

81 Martin, C., Berridge, G., Higgins, C.F., Mistry, P., Charlton, P., and Callaghan, R. (2000) Communication between multiple drug binding sites on P-glycoprotein. *Molecular Pharmacology*, **58** (3), 624–632.

82 Kondratov, R.V., Komarov, P.G., Becker, Y., Ewenson, A., and Gudkov, A.V. (2001) Small molecules that dramatically alter multidrug resistance phenotype by modulating the substrate specificity of P-glycoprotein. *Proceedings of the National Academy of Sciences of the United States of America*, **98** (24), 14078–14083.

83 Owen, A., Chandler, B., Back, D.J., and Khoo, S.H. (2004) Expression of pregnane-X-receptor transcript in peripheral blood mononuclear cells and correlation with MDR1 mRNA. *Antiviral Therapy*, **9** (5), 819–821.

84 Watkins, R.E., Maglich, J.M., Moore, L.B., Wisely, G.B., Noble, S.M., Davis-Searles, P.R. et al. (2003) 2.1 Å crystal structure of human PXR in complex with the St. John's Wort compound hyperforin. *Biochemistry*, **42** (6), 1430–1438.

85 de Vree, J.M., Jacquemin, E., Sturm, E., Cresteil, D., Bosma, P.J., Aten, J. et al. (1998) Mutations in the MDR3 gene cause progressive familial intrahepatic cholestasis. *Proceedings of the National Academy of Sciences of the United States of America*, **95** (1), 282–287.

86 Dixon, P.H., Weerasekera, N., Linton, K.J., Donaldson, O., Chambers, J., Egginton, E. et al. (2000) Heterozygous MDR3 missense mutation associated with intrahepatic cholestasis of pregnancy: evidence for a defect in protein trafficking. *Human Molecular Genetics*, **9** (8), 1209–1217.

87 Jacquemin, E. (2001) Role of multidrug resistance 3 deficiency in pediatric and adult liver disease: one gene for three diseases. *Seminars in Liver Disease*, **21** (4), 551–562.

88 Rosmorduc, O., Hermelin, B., and Poupon, R. (2001) MDR3 gene defect in adults with symptomatic intrahepatic and gallbladder cholesterol cholelithiasis. *Gastroenterology*, **120** (6), 1459–1467.

89 Lucena, J.F., Herrero, J.I., Quiroga, J., Sangro, B., Garcia-Foncillas, J., Zabalegui, N. et al. (2003) A multidrug resistance 3 gene mutation causing cholelithiasis, cholestasis of pregnancy, and adulthood biliary cirrhosis. *Gastroenterology*, **124** (4), 1037–1042.

90 Oude Elferink, R.P., Paulusma, C.C., and Groen, A.K. (2006) Hepatocanalicular transport defects: pathophysiologic

mechanisms of rare diseases. *Gastroenterology*, **130** (3), 908–925.

91 Smith, A.J., van Helvoort, A., van Meer, G., Szabo, K., Welker, E., Szakacs, G. *et al.* (2000) MDR3 P-glycoprotein, a phosphatidylcholine translocase, transports several cytotoxic drugs and directly interacts with drugs as judged by interference with nucleotide trapping. *The Journal of Biological Chemistry*, **275** (31), 23530–23539.

92 Nooter, K., Sonneveld, P., Janssen, A., Oostrum, R., Boersma, T., Herweijer, H. *et al.* (1990) Expression of the mdr3 gene in prolymphocytic leukemia: association with cyclosporin-A-induced increase in drug accumulation. *International Journal of Cancer*, **45** (4), 626–631.

93 Herweijer, H., Sonneveld, P., Baas, F., and Nooter, K. (1990) Expression of mdr1 and mdr3 multidrug-resistance genes in human acute and chronic leukemias and association with stimulation of drug accumulation by cyclosporine. *Journal of the National Cancer Institute*, **82** (13), 1133–1140.

94 Kino, K., Taguchi, Y., Yamada, K., Komano, T., and Ueda, K. (1996) Aureobasidin A, an antifungal cyclic depsipeptide antibiotic, is a substrate for both human MDR1 and MDR2/P-glycoproteins. *FEBS Letters*, **399** (1–2), 29–32.

95 Elferink, R.P., Tytgat, G.N., and Groen, A.K. (1997) Hepatic canalicular membrane 1: the role of mdr2 P-glycoprotein in hepatobiliary lipid transport. *The FASEB Journal*, **11** (1), 19–28.

96 Huang, L., Zhao, A., Lew, J.L., Zhang, T., Hrywna, Y., Thompson, J.R. *et al.* (2003) Farnesoid X receptor activates transcription of the phospholipid pump MDR3. *The Journal of Biological Chemistry*, **278** (51), 51085–51090.

97 Strautnieks, S.S., Bull, L.N., Knisely, A.S., Kocoshis, S.A., Dahl, N., Arnell, H. *et al.* (1998) A gene encoding a liver-specific ABC transporter is mutated in progressive familial intrahepatic cholestasis. *Nature Genetics*, **20** (3), 233–238.

98 Jansen, P.L., Strautnieks, S.S., Jacquemin, E., Hadchouel, M., Sokal, E.M., Hooiveld, G.J. *et al.* (1999) Hepatocanalicular bile salt export pump deficiency in patients with progressive familial intrahepatic cholestasis. *Gastroenterology*, **117** (6), 1370–1379.

99 Byrne, J.A., Strautnieks, S.S., Mieli-Vergani, G., Higgins, C.F., Linton, K.J., and Thompson, R.J. (2002) The human bile salt export pump: characterization of substrate specificity and identification of inhibitors. *Gastroenterology*, **123** (5), 1649–1658.

100 Mita, S., Suzuki, H., Akita, H., Hayashi, H., Onuki, R., Hofmann, A.F. *et al.* (2006) Vectorial transport of unconjugated and conjugated bile salts by monolayers of LLC-PK1 cells doubly transfected with human NTCP and BSEP or with rat Ntcp and Bsep. *American Journal of Physiology. Gastrointestinal and Liver Physiology*, **290** (3), G550–G556.

101 Noe, J., Stieger, B., and Meier, P.J. (2002) Functional expression of the canalicular bile salt export pump of human liver. *Gastroenterology*, **123** (5), 1659–1666.

102 Gerloff, T., Stieger, B., Hagenbuch, B., Madon, J., Landmann, L., Roth, J. *et al.* (1998) The sister of P-glycoprotein represents the canalicular bile salt export pump of mammalian liver. *The Journal of Biological Chemistry*, **273** (16), 10046–10050.

103 Wang, R., Salem, M., Yousef, I.M., Tuchweber, B., Lam, P., Childs, S.J. *et al.* (2001) Targeted inactivation of sister of P-glycoprotein gene (spgp) in mice results in nonprogressive but persistent intrahepatic cholestasis. *Proceedings of the National Academy of Sciences of the United States of America*, **98** (4), 2011–2016.

104 Lam, P., Wang, R., and Ling, V. (2005) Bile acid transport in sister of P-glycoprotein (ABCB11) knockout mice. *Biochemistry*, **44** (37), 12598–12605.

105 Childs, S., Yeh, R.L., Hui, D., and Ling, V. (1998) Taxol resistance mediated by transfection of the liver-specific sister

gene of P-glycoprotein. *Cancer Research*, **58** (18), 4160–4167.

106 Lecureur, V., Sun, D., Hargrove, P., Schuetz, E.G., Kim, R.B., Lan, L.B. et al. (2000) Cloning and expression of murine sister of P-glycoprotein reveals a more discriminating transporter than MDR1/P-glycoprotein. *Molecular Pharmacology*, **57** (1), 24–35.

107 Ananthanarayanan, M., Balasubramanian, N., Makishima, M., Mangelsdorf, D.J., and Suchy, F.J. (2001) Human bile salt export pump promoter is transactivated by the farnesoid X receptor/bile acid receptor. *The Journal of Biological Chemistry*, **276** (31), 28857–28865.

108 Schuetz, E.G., Strom, S., Yasuda, K., Lecureur, V., Assem, M., Brimer, C. et al. (2001) Disrupted bile acid homeostasis reveals an unexpected interaction among nuclear hormone receptors, transporters, and cytochrome P450. *The Journal of Biological Chemistry*, **276** (42), 39411–39418.

109 Honjo, Y., Sasaki, S., Kobayashi, Y., Misawa, H., and Nakamura, H. (2006) 1,25-Dihydroxyvitamin D3 and its receptor inhibit the chenodeoxycholic acid-dependent transactivation by farnesoid X receptor. *The Journal of Endocrinology*, **188** (3), 635–643.

110 Makishima, M., Lu, T.T., Xie, W., Whitfield, G.K., Domoto, H., Evans, R.M. et al. (2002) Vitamin D receptor as an intestinal bile acid sensor. *Science*, **296** (5571), 1313–1316.

111 Xie, W., Radominska-Pandya, A., Shi, Y., Simon, C.M., Nelson, M.C., Ong, E.S. et al. (2001) An essential role for nuclear receptors SXR/PXR in detoxification of cholestatic bile acids. *Proceedings of the National Academy of Sciences of the United States of America*, **98** (6), 3375–3380.

112 Staudinger, J.L., Goodwin, B., Jones, S.A., Hawkins-Brown, D., MacKenzie, K.I., LaTour, A. et al. (2001) The nuclear receptor PXR is a lithocholic acid sensor that protects against liver toxicity. *Proceedings of the National Academy of Sciences of the United States of America*, **98** (6), 3369–3374.

113 Lu, T.T., Makishima, M., Repa, J.J., Schoonjans, K., Kerr, T.A., Auwerx, J. et al. (2000) Molecular basis for feedback regulation of bile acid synthesis by nuclear receptors. *Molecular Cell*, **6** (3), 507–515.

114 del Castillo-Olivares, A. and Gil, G. (2001) Suppression of sterol 12alpha-hydroxylase transcription by the short heterodimer partner: insights into the repression mechanism. *Nucleic Acids Research*, **29** (19), 4035–4042.

115 Song, C.S., Echchgadda, I., Baek, B.S., Ahn, S.C., Oh, T., Roy, A.K. et al. (2001) Dehydroepiandrosterone sulfotransferase gene induction by bile acid activated farnesoid X receptor. *The Journal of Biological Chemistry*, **276** (45), 42549–42556.

116 Barbier, O., Torra, I.P., Sirvent, A., Claudel, T., Blanquart, C., Duran-Sandoval, D. et al. (2003) FXR induces the UGT2B4 enzyme in hepatocytes: a potential mechanism of negative feedback control of FXR activity. *Gastroenterology*, **124** (7), 1926–1940.

117 Graf, G.A., Li, W.P., Gerard, R.D., Gelissen, I., White, A., Cohen, J.C. et al. (2002) Coexpression of ATP-binding cassette proteins ABCG5 and ABCG8 permits their transport to the apical surface. *The Journal of Clinical Investigation*, **110** (5), 659–669.

118 Berge, K.E., Tian, H., Graf, G.A., Yu, L., Grishin, N.V., Schultz, J. et al. (2000) Accumulation of dietary cholesterol in sitosterolemia caused by mutations in adjacent ABC transporters. *Science*, **290** (5497), 1771–1775.

119 Lee, M.H., Lu, K., Hazard, S., Yu, H., Shulenin, S., Hidaka, H. et al. (2001) Identification of a gene, ABCG5, important in the regulation of dietary cholesterol absorption. *Nature Genetics*, **27** (1), 79–83.

120 Nguyen, L.B., Shefer, S., Salen, G., Ness, G.C., Tint, G.S., Zaki, F.G. et al. (1990) A

molecular defect in hepatic cholesterol biosynthesis in sitosterolemia with xanthomatosis. *The Journal of Clinical Investigation*, **86** (3), 923–931.

121 Janowski, B.A., Willy, P.J., Devi, T.R., Falck, J.R., and Mangelsdorf, D.J. (1996) An oxysterol signalling pathway mediated by the nuclear receptor LXR alpha. *Nature*, **383** (6602), 728–731.

122 Lehmann, J.M., Kliewer, S.A., Moore, L.B., Smith-Oliver, T.A., Oliver, B.B., Su, J.L. et al. (1997) Activation of the nuclear receptor LXR by oxysterols defines a new hormone response pathway. *The Journal of Biological Chemistry*, **272** (6), 3137–3140.

123 Janowski, B.A., Grogan, M.J., Jones, S.A., Wisely, G.B., Kliewer, S.A., Corey, E.J. et al. (1999) Structural requirements of ligands for the oxysterol liver X receptors LXRalpha and LXRbeta. *Proceedings of the National Academy of Sciences of the United States of America*, **96** (1), 266–271.

124 Freeman, L.A., Kennedy, A., Wu, J., Bark, S., Remaley, A.T., Santamarina-Fojo, S. et al. (2004) The orphan nuclear receptor LRH-1 activates the ABCG5/ABCG8 intergenic promoter. *Journal of Lipid Research*, **45** (7), 1197–1206.

125 Gupta, S., Pandak, W.M., and Hylemon, P.B. (2002) LXR alpha is the dominant regulator of CYP7A1 transcription. *Biochemical and Biophysical Research Communications*, **293** (1), 338–343.

126 Richards, D.M., Heel, R.C., Brogden, R.N., Speight, T.M., and Avery, G.S. (1984) Ceftriaxone. A review of its antibacterial activity, pharmacological properties and therapeutic use. *Drugs*, **27** (6), 469–527.

127 Hayton, W.L., Schandlik, R., and Stoeckel, K. (1986) Biliary excretion and pharmacokinetics of ceftriaxone after cholecystectomy. *European Journal of Clinical Pharmacology*, **30** (4), 445–451.

128 Harvey, J.H., Smith, F.P., and Schein, P.S. (1984) 5-Fluorouracil, mitomycin, and doxorubicin (FAM) in carcinoma of the biliary tract. *Journal of Clinical Oncology*, **2** (11), 1245–1248.

129 Ellis, P.A., Norman, A., Hill, A., O'Brien, M.E., Nicolson, M., Hickish, T. et al. (1995) Epirubicin, cisplatin and infusional 5-fluorouracil (5-FU) (ECF) in hepatobiliary tumours. *European Journal of Cancer*, **31** (10), 1594–1598.

130 Morizane, C., Okada, S., Okusaka, T., Ueno, H., and Saisho, T. (2003) Phase II study of cisplatin, epirubicin, and continuous-infusion 5-fluorouracil for advanced biliary tract cancer. *Oncology*, **64** (4), 475–476.

131 Ishii, H., Furuse, J., Yonemoto, N., Nagase, M., Yoshino, M., and Sato, T. (2004) Chemotherapy in the treatment of advanced gallbladder cancer. *Oncology*, **66** (2), 138–142.

132 Allen, J.D., Brinkhuis, R.F., Wijnholds, J., and Schinkel, A.H. (1999) The mouse Bcrp1/Mxr/Abcp gene: amplification and overexpression in cell lines selected for resistance to topotecan, mitoxantrone, or doxorubicin. *Cancer Research*, **59** (17), 4237–4241.

133 Kawabe, T., Chen, Z.S., Wada, M., Uchiumi, T., Ono, M., Akiyama, S. et al. (1999) Enhanced transport of anticancer agents and leukotriene C4 by the human canalicular multispecific organic anion transporter (cMOAT/MRP2). *FEBS Letters*, **456** (2), 327–331.

134 Han, B. and Zhang, J.T. (2004) Multidrug resistance in cancer chemotherapy and xenobiotic protection mediated by the half ATP-binding cassette transporter ABCG2. *Current Medicinal Chemistry. Anticancer Agents*, **4** (1), 31–42.

135 Borst, P. and Elferink, R.O. (2002) Mammalian ABC transporters in health and disease. *Annual Review of Biochemistry*, **71**, 537–592.

136 Kim, R.B. (2002) Drugs as P-glycoprotein substrates, inhibitors, and inducers. *Drug Metabolism Reviews*, **34** (1–2), 47–54.

137 Shepard, R.L., Cao, J., Starling, J.J., and Dantzig, A.H. (2003) Modulation of P-glycoprotein but not MRP1- or BCRP-mediated drug resistance by LY335979. *International Journal of Cancer*, **103** (1), 121–125.

138 Dantzig, A.H., Shepard, R.L., Law, K.L., Tabas, L., Pratt, S., Gillespie, J.S. et al. (1999) Selectivity of the multidrug resistance modulator, LY335979, for P-glycoprotein and effect on cytochrome P-450 activities. *The Journal of Pharmacology and Experimental Therapeutics*, **290** (2), 854–862.

139 Le, L.H., Moore, M.J., Siu, L.L., Oza, A.M., MacLean, M., Fisher, B. et al. (2005) Phase I study of the multidrug resistance inhibitor zosuquidar administered in combination with vinorelbine in patients with advanced solid tumours. *Cancer Chemotherapy and Pharmacology*, **56** (2), 154–160.

140 Schinkel, A.H., Smit, J.J., van Tellingen, O., Beijnen, J.H., Wagenaar, E., van Deemter, L. et al. (1994) Disruption of the mouse mdr1a P-glycoprotein gene leads to a deficiency in the blood–brain barrier and to increased sensitivity to drugs. *Cell*, **77** (4), 491–502.

141 Park, J.Y., Park, S.W., Chung, J.B., Seong, J., Kim, K.S., Lee, W.J. et al. (2006) Concurrent chemoradiotherapy with doxifluridine and paclitaxel for extrahepatic bile duct cancer. *American Journal of Clinical Oncology*, **29** (3), 240–245.

142 Lagas, J.S., Vlaming, M.L., van Tellingen, O., Wagenaar, E., Jansen, R.S., Rosing, H. et al. (2006) Multidrug resistance protein 2 is an important determinant of paclitaxel pharmacokinetics. *Clinical Cancer Research*, **12** (20 Pt 1), 6125–6132.

143 Felton, J.S., Knize, M.G., Shen, N.H., Lewis, P.R., Andresen, B.D., Happe, J. et al. (1986) The isolation and identification of a new mutagen from fried ground beef: 2-amino-1-methyl-6-phenylimidazo[4,5-*b*]pyridine (PhIP). *Carcinogenesis*, **7** (7), 1081–1086.

144 Dietrich, C.G., de Waart, D.R., Ottenhoff, R., Schoots, I.G., and Elferink, R.P. (2001) Increased bioavailability of the food-derived carcinogen 2-amino-1-methyl-6-phenylimidazo[4,5-*b*]pyridine in MRP2-deficient rats. *Molecular Pharmacology*, **59** (5), 974–980.

145 Sjovall, J. and Sjovall, K. (1970) Steroid sulphates in plasma from pregnant women with pruritus and elevated plasma bile acid levels. *Annals of Clinical Research*, **2** (4), 321–337.

146 Reyes, H., Ribalta, J., Gonzalez, M.C., Segovia, N., and Oberhauser, E. (1981) Sulfobromophthalein clearance tests before and after ethinyl estradiol administration, in women and men with familial history of intrahepatic cholestasis of pregnancy. *Gastroenterology*, **81** (2), 226–231.

147 Vore, M. (1987) Estrogen cholestasis. Membranes, metabolites, or receptors? *Gastroenterology*, **93** (3), 643–649.

148 Stieger, B., Fattinger, K., Madon, J., Kullak-Ublick, G.A., and Meier, P.J. (2000) Drug- and estrogen-induced cholestasis through inhibition of the hepatocellular bile salt export pump (Bsep) of rat liver. *Gastroenterology*, **118** (2), 422–430.

149 Beuers, U., Bilzer, M., Chittattu, A., Kullak-Ublick, G.A., Keppler, D., Paumgartner, G. et al. (2001) Tauroursodeoxycholic acid inserts the apical conjugate export pump, Mrp2, into canalicular membranes and stimulates organic anion secretion by protein kinase C-dependent mechanisms in cholestatic rat liver. *Hepatology*, **33** (5), 1206–1216.

150 Galan, A.I., Jimenez, R., Munoz, M.E., and Gonzalez, J. (1990) Effects of ursodeoxycholate on maximal biliary secretion of bilirubin in the rat. *Biochemical Pharmacology*, **39** (7), 1175–1180.

151 Noe, J., Hagenbuch, B., Meier, P.J., and St-Pierre, M.V. (2001) Characterization of the mouse bile salt export pump overexpressed in the baculovirus system. *Hepatology*, **33** (5), 1223–1231.

152 Benz, C., Angermuller, S., Otto, G., Sauer, P., Stremmel, W., and Stiehl, A. (2000) Effect of tauroursodeoxycholic acid on bile acid-induced apoptosis in primary human hepatocytes. *European Journal of Clinical Investigation*, **30** (3), 203–209.

153 Benz, C., Angermuller, S., Tox, U., Kloters-Plachky, P., Riedel, H.D., Sauer,

P. et al. (1998) Effect of tauroursodeoxycholic acid on bile-acid-induced apoptosis and cytolysis in rat hepatocytes. *Journal of Hepatology*, **28** (1), 99–106.

154 Rodrigues, C.M., Fan, G., Wong, P.Y., Kren, B.T., and Steer, C.J. (1998) Ursodeoxycholic acid may inhibit deoxycholic acid-induced apoptosis by modulating mitochondrial transmembrane potential and reactive oxygen species production. *Molecular Medicine*, **4** (3), 165–178.

155 Heuman, D.M. and Bajaj, R. (1994) Ursodeoxycholate conjugates protect against disruption of cholesterol-rich membranes by bile salts. *Gastroenterology*, **106** (5), 1333–1341.

156 Heuman, D.M., Bajaj, R.S., and Lin, Q. (1996) Adsorption of mixtures of bile salt taurine conjugates to lecithin-cholesterol membranes: implications for bile salt toxicity and cytoprotection. *Journal of Lipid Research*, **37** (3), 562–573.

157 Pusl, T. and Beuers, U. (2006) Ursodeoxycholic acid treatment of vanishing bile duct syndromes. *World Journal of Gastroenterology*, **12** (22), 3487–3495.

158 Poupon, R., Chretien, Y., Poupon, R.E., Ballet, F., Calmus, Y., and Darnis, F. (1987) Is ursodeoxycholic acid an effective treatment for primary biliary cirrhosis? *Lancet*, **1** (8537), 834–836.

159 Beuers, U., Spengler, U., Kruis, W., Aydemir, U., Wiebecke, B., Heldwein, W. et al. (1992) Ursodeoxycholic acid for treatment of primary sclerosing cholangitis: a placebo-controlled trial. *Hepatology*, **16** (3), 707–714.

160 Stiehl, A. (1994) Ursodeoxycholic acid therapy in treatment of primary sclerosing cholangitis. *Scandinavian Journal of Gastroenterology. Supplement*, **204**, 59–61.

161 O'Brien, C.B., Senior, J.R., Arora-Mirchandani, R., Batta, A.K., and Salen, G. (1991) Ursodeoxycholic acid for the treatment of primary sclerosing cholangitis: a 30-month pilot study. *Hepatology*, **14** (5), 838–847.

162 van Heek, M., France, C.F., Compton, D.S., McLeod, R.L., Yumibe, N.P., Alton, K.B. et al. (1997) In vivo metabolism-based discovery of a potent cholesterol absorption inhibitor, SCH58235, in the rat and rhesus monkey through the identification of the active metabolites of SCH48461. *The Journal of Pharmacology and Experimental Therapeutics*, **283** (1), 157–163.

163 van Heek, M., Farley, C., Compton, D.S., Hoos, L., Alton, K.B., Sybertz, E.J. et al. (2000) Comparison of the activity and disposition of the novel cholesterol absorption inhibitor, SCH58235, and its glucuronide, SCH60663. *British Journal of Pharmacology*, **129** (8), 1748–1754.

164 Altmann, S.W., Davis, H.R., Jr., Zhu, L.J., Yao, X., Hoos, L.M., Tetzloff, G. et al. (2004) Niemann-Pick C1 Like 1 protein is critical for intestinal cholesterol absorption. *Science*, **303** (5661), 1201–1204.

165 Davis, H.R., Jr., Zhu, L.J., Hoos, L.M., Tetzloff, G., Maguire, M., Liu, J. et al. (2004) Niemann-Pick C1 Like 1 (NPC1L1) is the intestinal phytosterol and cholesterol transporter and a key modulator of whole-body cholesterol homeostasis. *The Journal of Biological Chemistry*, **279** (32), 33586–33592.

166 Ezzet, F., Krishna, G., Wexler, D.B., Statkevich, P., Kosoglou, T., and Batra, V.K. (2001) A population pharmacokinetic model that describes multiple peaks due to enterohepatic recirculation of ezetimibe. *Clinical Therapeutics*, **23** (6), 871–885.

167 Patrick, J.E., Kosoglou, T., Stauber, K.L., Alton, K.B., Maxwell, S.E., Zhu, Y. et al. (2002) Disposition of the selective cholesterol absorption inhibitor ezetimibe in healthy male subjects. *Drug Metabolism and Disposition*, **30** (4), 430–437.

168 Oswald, S., Westrup, S., Grube, M., Kroemer, H.K., Weitschies, W., and Siegmund, W. (2006) Disposition and

sterol-lowering effect of ezetimibe in multidrug resistance-associated protein 2-deficient rats. *The Journal of Pharmacology and Experimental Therapeutics*, **318** (3), 1293–1299.
169 Oswald, S., Haenisch, S., Fricke, C., Sudhop, T., Remmler, C., Giessmann, T. et al. (2006) Intestinal expression of P-glycoprotein (ABCB1), multidrug resistance associated protein 2 (ABCC2), and uridine diphosphate-glucuronosyltransferase 1A1 predicts the disposition and modulates the effects of the cholesterol absorption inhibitor ezetimibe in humans. *Clinical Pharmacology and Therapeutics*, **79** (3), 206–217.
170 van de Wetering, K., Zelcer, N., Kuil, A., Feddema, W., Hillebrand, M., Vlaming, M.L. et al. (2007) Multidrug resistance proteins 2 and 3 provide alternative routes for hepatic excretion of morphine-glucuronides. *Molecular Pharmacology*, **72** (2), 387–394.
171 Hooijberg, J.H., Broxterman, H.J., Kool, M., Assaraf, Y.G., Peters, G.J., Noordhuis, P. et al. (1999) Antifolate resistance mediated by the multidrug resistance proteins MRP1 and MRP2. *Cancer Research*, **59** (11), 2532–2535.
172 Gerk, P.M., Li, W., Megaraj, V., and Vore, M. (2007) Human multidrug resistance protein 2 transports the therapeutic bile salt tauroursodeoxycholate. *The Journal of Pharmacology and Experimental Therapeutics*, **320** (2), 893–899.
173 Jonker, J.W., Smit, J.W., Brinkhuis, R.F., Maliepaard, M., Beijnen, J.H., Schellens, J.H. et al. (2000) Role of breast cancer resistance protein in the bioavailability and fetal penetration of topotecan. *Journal of the National Cancer Institute*, **92** (20), 1651–1656.
174 Breedveld, P., Pluim, D., Cipriani, G., Wielinga, P., van Tellingen, O., Schinkel, A.H. et al. (2005) The effect of Bcrp1 (Abcg2) on the *in vivo* pharmacokinetics and brain penetration of imatinib mesylate (Gleevec): implications for the use of breast cancer resistance protein and P-glycoprotein inhibitors to enable the brain penetration of imatinib in patients. *Cancer Research*, **65** (7), 2577–2582.
175 Merino, G., Alvarez, A.I., Redondo, P.A., Garcia, J.L., Larrode, O.M., and Prieto, J.G. (1999) Bioavailability of albendazole sulphoxide after netobimin administration in sheep: effects of fenbendazole coadministration. *Research in Veterinary Science*, **66** (3), 281–283.
176 Schinkel, A.H., Smit, J.J., van Tellingen, O., Beijnen, J.H., Wagenaar, E., van Deemter, L. et al. (1994) Disruption of the mouse mdr1a P-glycoprotein gene leads to a deficiency in the blood–brain barrier and to increased sensitivity to drugs. *Cell*, **77** (4), 491–502.

12
Interplay of Drug Metabolizing Enzymes and ABC Transporter
Walter Jäger

12.1
Combined Role of Cytochrome P450 3A and ABCB1

Cytochrome P450 3A (CYP3A) is the major drug metabolizing subfamily in humans. Among the CYP3A subfamily, CYP3A4 is the predominant and important enzyme responsible for the biotransformation of about 55% of all prescribed drugs [1]. CYP3A4 is found throughout the body but is most highly expressed in the liver and the intestinal epithelium making these organs to predominate sites for drug elimination. Besides CYP3A4, CYP3A5 is also expressed in 10–30% of human adult livers, where it accounts for about 25% of the hepatic CYP3A content. CYP3A5 is commonly expressed in the intestine at much lower levels than CYP3A4 and represents the major CYP3A isoform in extrahepatic tissues including the blood, kidney, and lung [2]. CYP3A5 has complete overlapping substrate specificity and almost equal enzyme activity with CYP3A4. However, comparative analysis of CYP3A expression in human liver indicates that the contribution of CYP3A5 to hepatic drug metabolism in Caucasians is insignificant [2]. A third CYP3A isozyme, namely, CYP3A7 is expressed only in the fetus and diminishes during infancy [3].

P-glycoprotein (P-gp, ABCB1) is a cellular membrane glycoprotein functioning as an energy-dependent drug efflux pump that lowers intercellular drug concentrations. In addition to the expression in tumor cells, ABCB1 is also localized on the apical surfaces of epithelial cells in the small biliary ductules of the liver, small intestine, colon mucosa cells, the proximal tubule of the kidney, the adrenal gland, and the pancreatic duct [4, 5]. Furthermore, it is also highly expressed in the capillary endothelium of the brain and testes. ABCB1 therefore plays a key role as a defense mechanism against drugs and xenotoxins by preventing their gastrointestinal absorption or penetration through the blood–brain barrier [4, 5]. In addition, ABCB1 also facilitates the elimination of already absorbed xenobiotics and toxins into bile and urine.

In the small intestine, ABCB1 and CYP3A4 form a cooperative barrier against the oral absorption of drugs and xenobiotics. While ABCB1 expression increased longitudinally along the intestine with lowest levels in the stomach and highest levels in the colon [6], CYP3A4 protein and catalytic activity decreased longitudinally

Transporters as Drug Carriers: Structure, Function, Substrates. Edited by Gerhard Ecker and Peter Chiba
Copyright © 2009 WILEY-VCH Verlag GmbH & Co. KGaA, Weinheim
ISBN: 978-3-527-31661-8

along the small intestine [7]. ABCB1 and CYP3A are highly variable between individuals with 2–8-fold variations for ABCB1 and up to 30-fold for CYP3A4 in the human intestine [8, 9]. CYP3A4 variations reported for human liver are even higher and up to 100-fold [10]. Total CYP3A (CYP3A4 and CYP3A5) levels in the small intestine are lower and about 10–50% of those found in the liver. However, recent data demonstrated that CYP3A concentrations might be equal to or even exceed the concentration in the liver [10].

Based on a significant overlap between substrate specificity and tissue, a functional interaction between CYP3A and ABCB1 has been observed in the small intestine and in the liver [11]. However, based on data from healthy volunteers and kidney transplant patients, intersubject ABCB1 levels seem not to be correlated either with CYP3A concentration in the enterocyte or with liver CYP3A activities, ruling out any coordinate regulation of these two proteins [9]. ABCB1 may rather act to regulate the exposure of drugs to metabolism by CYP3A. Drugs taken up into the enterocytes may be pumped out by ABCB1 and taken up again. Repeated exposure to CYP3A4 and CYP3A5 isoenzymes may increase the probability of drugs being metabolized [5, 11, 12].

Cummings and coworkers were among the first to demonstrate that for compounds that were substrates for ABCB1 and CYP3A4, inhibition of ABCB1 in the intestine would increase absorption by blocking efflux transport and metabolism, resulting in a significantly enhanced intestinal bioavailability [13, 14]. Using four CYP3A4 substrates, namely, K77, an investigational cyteine protease inhibitor, sirolimus, an immunosuppressive agent, midazolam, an anesthetic drug, and felodipine, a calcium channel blocker Benet and coworkers could show that K77 and sirolimus, but not midazolam and felodipine, were good ABCB1 substrates, as basolateral to apical efflux, using CYP3A4-transfected Caco-2 cells, was 9-fold and 2.5-fold greater than their apical to basolateral one (see Table 12.1). When added to the

Table 12.1 Effect of P-glycoprotein inhibition on the extraction ratios of drugs across CYP3A4-overexpressing Caco-2 monolayers [13, 14].

Drug	Substrate for		Efflux ratio	Extraction ratio (%) ± SD		
	CYP3A4	ABCB1	B to A/ A to B	Drug alone	Drug + CsA	Drug + GG918
K77 (10 µM)	Yes	Yes	9	33 ± 3 (356 ± 26)	5.7 ± 0.3 (1830 ± 40)	14 ± 1 (1600 ± 60)
Sirolimus (1 µM)	Yes	Yes	2.5	60 ± 5 (56 ± 10)	15 ± 1 (212 ± 19)	45 ± 1 (73 ± 3)
Midazolam (3 µM)	Yes	No	1	25 ± 2 (282 ± 50)	10 ± 1 (354 ± 15)	23 ± 2 (349 ± 30)
Felodipine (10 µM)	Yes	No	1	26 ± 1 (3750 ± 130)	14 ± 1 (4030 ± 90)	24 ± 2 (3710 ± 220)

A to B: ratio apical to basolateral; B to A: ratio basolateral to apical. Intracellular drug amounts (picomole) are shown in parentheses. Cyclosporin is a known CYP3A4 and ABCB1 substrate and inhibitor. GG918 is a specific ABCB1 inhibitor.

Caco-2 cells, all four compounds were significantly metabolized. As expected, addition of the CYP3A inhibitor cyclosporin significantly inhibited the extraction ratios for all drugs. Surprisingly, the decrease in the extraction ratio for midazolam and felodipine, which are not substrates for ABCB1, was 46–60% while these values for the CYP3A and ABCB1 substrates K77 and sirolimus were found to be 74–83%. As cyclosporin is also an ABCB1 inhibitor, the greater reduction in the extraction ratio for K77 and sirolimus might be explained by an additional contribution of ABCB1 in their transcellular transport. The involvement of ABCB1 was confirmed when K77 and sirolimus were added to the Caco-2 cells in combination with the ABCB1-specific inhibitor GG918, a compound that does not inhibit CYP3A. Specific inhibition of ABCB1 was more pronounced for the better ABCB1 substrate K77 than for sirolimus (58 and 25% decrease of extraction ratios, respectively). As expected, incubation of midazolam and felodipine with the ABCB1-specific inhibitor GG918 did not alter the extraction ratios for these compounds. In order to investigate whether modulation of ABCB1 may affect intracellular drug concentration, Cummings and coworkers also quantified intracellular drug levels [13, 14]. When ABCB1 was inhibited by GG918 or both CYP3A4 and ABCB1 were inhibited by cyclosporin, a significant increase of intracellular K77 and sirolimus concentration was observed. However, GG918 and cyclosporin increased only marginally intracellular amounts of midazolam and felodipine (see Table 12.1).

In a perfused rat liver model, Wu and Benet showed different functional interactions between CYP3A4 and ABCB1 in the liver and small intestine [15]. Specific inhibition of ABCB1 significantly decreased the concentration of tacrolimus, a CYP3A4 and ABCB1 substrate, in the perfusate. However, when felodipine, a CYP3A4 but not an ABCB1 substrate, was studied in the perfused rat liver, no difference in the perfusate concentration of felodipine was observed [12, 15]. This indicates that inhibition of ABCB1 in the liver increases CYP3A4 dependent drug metabolism whereas in the intestine it reduces biotransformation of CYP3A4 substrates [12]. These findings can be explained by major differences between intestinal mucosa cells and hepatocytes. Expression and activity of CYP3A enzymes in the hepatocytes are significantly higher than in mucosa cells leading to higher rates of CYP3A-catalyzed metabolism. In the mucosa cells of the small intestine, however, CYP3A expression is much lower and may be saturated by higher substrate concentration leading to a higher transmembrane penetration of nonmetabolized drug. Local differences in the expression of ABCB1 in the intestine and liver may also strongly affect metabolism. In the intestine, drugs enter the mucosa cells through the apical membrane [11–13]. As ABCB1 is located at the apical membrane, this efflux transporter will pump a drug back into the gut lumen where it is reabsorbed (see Figure 12.1), thereby ABCB1 regulating access to CYP3A-dependent metabolism. After CYP3-catalyzed biotransformation, metabolites and nonmetabolized drugs may again come in contact with ABCB1 or finally reach the portal vein [11–13]. Contrary to mucosa cells, drugs in the liver enter the hepatocytes from the blood through the basolateral membrane first coming into contact with CYP3A enzymes before reaching the portal vein, which is located in the apical biliary membrane (see Figure 12.1). This simplified model may reflect the gut- and

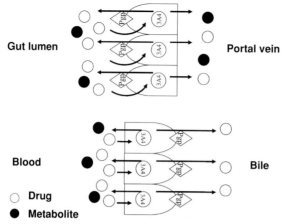

Figure 12.1 Sequential differences of efflux transporter and drug metabolism in the gut mucosa (top) and hepatocytes (bottom). After Refs [1, 11].

liver-specific metabolism of various clinically important CYP3A- and ABCB1-specific drugs. However, it may be more complex for compounds that are substrates for uptake transporters [11–13].

As seen in Table 12.2, in both liver and intestine, inhibition of CYP3A will decrease metabolism leading to increased bioavailability. Inhibition of ABCB1 in the intestinal mucosa, however, will reduce biotransformation of drugs whereas in the liver inhibition of ABCB1 will increase metabolism [11–13]. Dual inhibition of CYP3A and ABCB1 in the intestine will result in an even more potent inhibition of drug metabolism and a higher drug concentration in the portal vein [11–13]. Contrary to the gut, inhibition of CYP3A and ABCB1 in the liver can be predicted as metabolism may be increased, decreased, or not affected depending on the affinity of substrates and inhibitors to CYP3a and ABCB1 [11–13]. *In vitro* it is relatively easy to approximately estimate the relative contribution of CYP3A4 and ABCB1 to drug metabolism and interaction. Because of overlapping substrate specificity between CYP3A4 and ABCB1 and also because of the similarities in the inhibitors and inducers between these two proteins, it is very difficult to quantitatively differentiate the relative contribution of ABCB1 and CYP3A to the overall drug interactions *in vivo* [11–13]. Therefore, care should be taken in interpreting data for drug/transporter-mediated interactions in patients, particularly in terms of the underlying mechanisms.

Table 12.2 Predicted direction of metabolic change in the intestine and liver for dual CYP3A and ABCB1 substrates when coincubated with inhibitors [11].

	Intestine	Liver
Inhibit ABCB1	↓	↑
Inhibit CYP3A	↓	↓
Inhibit ABCB1 + CYP3A	↓↓	↔ ↑ ↓

In addition to the functional interplay between drug metabolizing enzymes and transporters, a coordinate regulation of these proteins may have even a greater impact on drug disposition as demonstrated for the concomitant induction of ABCB1 and CYP3A4 by the nuclear hormone receptor PXR (pregnane X receptor) [16]. PXR (also termed SXR in humans) seems to mediate a general protective response against various xenobiotics via activation of several detoxification/elimination pathways [16]. PXR is a promiscuous receptor that is activated by a wide variety of xenobiotics and endogenous compounds such as progesterone, phytoestrogens, dexamethasone, bile acids, and drugs such as rimpamin, peptide mimetic protease inhibitors, and paclitaxel [12]. Several studies have shown that in the human liver and intestine, PXR can induce CYP3A4 and CYP3A5 [16]. Coregulation of drug metabolism and efflux via CYP3A and ABCB1 in the liver and intestine by PXR has been shown for the anticancer drug paclitaxel. By activating the PXR, paclitaxel reduces its own oral bioavailability, metabolism, and biliary elimination. Hydroxylated paclitaxel metabolites and the structurally similar docetaxel did not interact with PXR [16].

PXR was also shown to be activated by herbal dietary supplements such as St John's wort (*Hypericum perforatum*), a most popular herbal remedy for treating depression available without a prescription [12]. Results from clinical studies and case reports indicate that self-administered St John's wort reduces steady-state plasma concentrations of amitriptyline, cyclosporin, digoxin, fexofenadine, amprenavir, indonavir, lopinavir, ritonavir, saquinavir, bezodiazepines, theophyline, irinotecan, midazolam, and warfarin [17]. This herbal agent has also been reported to cause bleeding and unwanted pregnancies when concomitantly administered with oral contraceptives [12]. Most of these drugs are substrates for CYP3A4 and ABCB1. The effects of a 12-day pretreatment with St John's wort on the disposition of selected *in vivo* probe drugs were determined in 21 young healthy subjects [18]. Midazolam after oral administration was used to assess CYP3A activity in both intestinal epithelium and liver, whereas the disposition of cyclosporin after an oral dose was assumed to reflect both CYP34 and ABCB1 activities [18]. Pretreatment with St John's wort resulted in a 53% reduction in maximal plasma concentration (C_{max}) whereby these changes were reflected in midazolam's oral bioavailability being reduced by almost 50%. After oral administration of cyclosporin, the consumption of St John's wort produced a 63% increase in oral clearance, with a corresponding reduction in C_{max} of 28%. Although the disposition of both drugs was altered by St John's wort, the extent of induction was more pronounced for the CYP3A4 substrate midazolam than for cyclosporin [18]. Quantitative aspects of inductions seem to be complex and strongly dependent on drug and relative contribution of CYP3A and ABCB1 in its disposition [19].

12.2
Combined Role of Cytochrome P450 3A and OATPs

An interplay between metabolic enzymes may also occur with uptake transporters that later control the access of drug molecules to the enzymes. Therefore, any change in the function of transporters can modulate intestinal and hepatic metabolism without directly changing enzyme activity.

Organic anion transporting polypeptides (OATPs) form a superfamily of sodium-independent transport systems and mediate the cellular uptake of many endogenous and exogenous chemicals including drugs in clinical use. Eleven members of the OATP family have so far been identified in humans [20, 21]. They are expressed in a variety of tissues including intestine, liver, kidney, and brain, and they play a critical role in drug absorption, distribution, and excretion. Although multispecificity and wide tissue distribution are common characteristics of many OATPs, some members have a high substrate specificity and exhibit unique cellular expression in distinct organs [20, 21].

OATP1B1, 1B3, and OATP2B1, in particular, are highly expressed in the human liver and are involved in the active hepatic uptake of various drugs. This active hepatic drug uptake process can be subject to inhibition resulting in higher blood levels and in a range of severe side effects such as those reported for statins after coadministration with the OATP inhibitor cyclosporin [22].

In a clinical study, plasma concentrations of the cholesterol-lowering drug cerivastatin were determined after oral administration of a 0.2 mg single dose of cerivastatin to 12 kidney transplant recipients and healthy volunteers [22]. The mean AUC value of cerivastatin (36.2 ng/ml h) in the kidney transplant recipients with cyclosporin was approximately fourfold higher than that in healthy volunteers (9.5 ng/ml h) who received the same oral dose of cerivastatin without cyclosporin treatment [22]. Due to the almost complete absorption of cerivastatin after oral administration and the fact that cyclosporin does not affect the elimination half-life, the authors suggested that the increased AUC observed in transplant patients cannot be explained only by cyclosporin-based CYP3A inhibition [23].

Further *in vitro* studies by Shitara *et al.* [24] using human liver microsomes and hepatocytes indeed revealed that cyclosporin was only a weak inhibitor of cerivastatin metabolism with an $IC_{50} < 50\,\mu M$. In contrast, cyclosporin was a potent inhibitor of cerivastatin OATB1B1 hepatic uptake with a K_i value of $0.3\,\mu M$ strongly indicating that the cerivastatin–cyclosporin interaction was mainly due to the inhibition of the hepatic influx transporter OATP1B1. Other cholesterol-lowering drugs, such as atorvastatin (ATV), also showed significant interactions with CYP3A4 inhibitors such as itraconazole [25, 26] and erythromycin [27] leading to increased AUC values not only for the parent compound but also for its active metabolites 2-OH atorvastatin (2-OH ATV) and 4-OH atorvastatin (4-OH ATV). In a recent study on rats, Lau *et al.* [28] could also show that oral coadministration of rifampicin (RIF), an inducer of CYP3A4 and a potent inhibitor of several OATPs, markedly increased (up to 3.5-fold) the plasma concentrations of atorvastatin and its two metabolites (see Figure 12.2). As

Figure 12.2 Mean ($\pm$) SD plasma concentrations of (a) atorvastatin, (b) 2-OH ATV, and (c) 4-OH ATV in rats ($n=5$) after a single oral dose of 10 mg/kg atorvastatin with and without rifampicin given as a bolus intravenous dose (20 mg/kg). Solid circles indicate the ATV-alone control group; open circles indicate the RIF-treatment group. Data are depicted on a semilogarithmic scale. After Ref. [28].

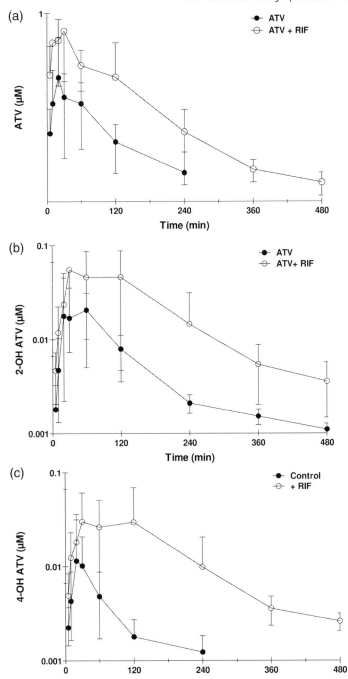

12 Interplay of Drug Metabolizing Enzymes and ABC Transporter

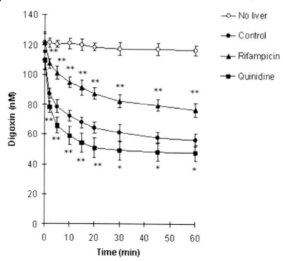

Figure 12.3 Influence of rifampicin and quinidine on concentrations of digoxin in perfusate after addition of 10 µg of digoxin to the perfusate. Values are mean ± SD, $n = 6$ per group; no liver, $n = 3$. $*p < 0.05$; $**p < 0.01$ for values compared to control. From Ref. [29].

both atorvastatin and hydroxylated metabolites are predominantly excreted into bile and only marginally into urine, the increased plasma levels of atorvastatin and its hydroxylated metabolites might be due to an inhibition of hepatic uptake by rifampicin. Besides OATPs in the liver, rifampicin also induces CYP3A4 in the intestine explaining the elevated atorvastatin metabolite formation. OATPs may therefore significantly alter biliary drug elimination and metabolism.

A similar type of interplay was observed in rat liver for the CYP3A/ABCB1 substrate digoxin. In rat, digoxin is extensively metabolized by CYP3A4 into Dg2. As investigated in an isolated perfused rat liver model, Lau et al. [29] could show that digoxin concentrations were significantly reduced during coadministration of the Oatp2 inhibitor rifampicin, suggesting that rifampicin limits the hepatic uptake of this drug. Rifampicin also inhibits CYP3A4-mediated formation to Dg2 by preventing digoxin from entering hepatocytes. This leads to decreased perfusate concentration of this biotransformation product (Figure 12.3).

Digoxin, however, is also a good substrate of ABCB1 that is located on the canalicular membrane of hepatocytes. As seen in Figure 12.4, coadministration of quinidine increased digoxin and Dg2 perfusate concentrations by inhibiting ABCB1 from pumping digoxin into the bile [29]. More digoxin now has contact with CYP3A enzymes that consequently leads to an increased Dg2 formation.

12.3
Combined Role of UDP-Glucuronosyltransferases and ABCC2

UDP-glucuronosyltransferases (UGTs) represent one of the major classes of enzymes involved in phase II conjugative metabolism. These enzymes catalyze the

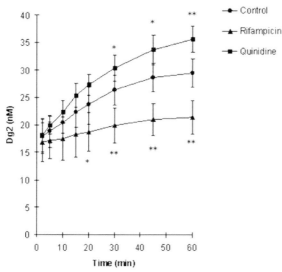

Figure 12.4 Influence of rifampicin and quinidine on concentrations of Dg2 in perfusate after addition of 10 µg of digoxin to the perfusate. Values are mean ± SD, $n = 6$ per group; no liver, $n = 3$. *$p < 0.05$; **$p < 0.01$ for values compared to control. From Ref. [29].

transfer of a glucuronic acid moiety from uridine diphosphoglucuronic acid (UDPGA) to a wide range of structurally diverse endogenous compounds and xenobiotics. The resulting glucuronide conjugates are more polar than the parent compound and are subsequently eliminated in the bile or urine. The liver represents one of the major sites of glucuronidation; however, UGTs are also expressed in extrahepatic tissues, including those in the gastrointestinal tract, the kidney, and the brain. Seventeen UGTs have been identified in humans so far and these have been assigned to two families, namely, UGT1 and UGT2, which are further divided on the basis of sequence homology into the subfamilies UGT1A, UGT2A, and UGT2B [30]. Apparent decreases or increases in the amount of glucuronide excreted into urine or bile will strongly affect blood concentration and drug efficacy that has been shown in several studies [30]. These apparent effects on glucuronidation could occur via direct inhibition or induction of the isoenzymes or by competition or inhibition of transport mechanism responsible for their excretion.

The multidrug resistance protein 2 (MRP2, ABCC2) is a major xenobiotic efflux pump on the canalicular membrane. ABCC2 plays a key role in the biliary excretion of organic anions, including bilirubin-diglucuronide, glutathione conjugates, sulfated bile salts, and numerous drugs, such as sulfopyrozone, indomethacin, penicillin, vinblastine, methotrexate, and telmisartan. Patients with Dubin–Johnson syndrome suffer from defective hepatic biliary excretion due to the absence of ABCC2. [31]. Many of these observations were made using one of the two rat strains with a hereditary deficiency in this exporter, namely, Esai rats, or transporter-deficient TR⁻ rats.

Data from Jäger and coworkers could demonstrate a clinically important interplay between UGT-catalyzed glucuronidation of the novel anticancer drug flavopiridol and ABCC2 [32]. Flavopiridol undergoes extensive metabolism in the rat liver to form two monoglucuronides M1 and M2, mainly excreted into bile [33]. Pronounced glucuronidation followed by biliary elimination could also be observed in cancer patients as indicated by high glucuronide levels in plasma and enterohepatic circulation [34]. Moreover, the main side effect, diarrhea, is also linked to biliary retention of flavopiridol glucuronides [35]. Another toxicity of note during flavopiridol treatment is the induction of reversible conjugated hyperbilirubinemia, which was observed in up to 22% of patients [36]. As conjugated bilirubin excretion into bile is also mediated with high affinity by the ATP-dependent transporter ABCC2, our hypothesis was that flavopiridol glucuronides may also be actively transported across the canalicular membrane into bile via ABCC2. This hypothesis is supported by previous studies from our lab showing a clear preference to ABCC2 for the biliary excretion of glucuronides from a structurally similar flavonoid, genistein [37]. Addition of genistein to an unconjugated bilirubin-containing rat liver perfusion medium also gradually decreased biliary excretion of bilirubin conjugates by 76% due to a competition for this canalicular anion carrier [37]. Flavopiridol conjugates may therefore also act as competitive inhibitors of this transporter by modifying the hepatic disposition of bilirubin glucuronides. A possible interaction on the level of the enzymatic pathway between flavopiridol and bilirubin, causing hyperbilirubinemia, can be excluded as bilirubin is glucuronidated selectively by the UDP-glucuronosyltransferase UGT1A1, whereas UGT1A9 is the major UGT involved in hepatic flavopiridol conjugation [38].

Because of the clinical importance of biliary flavopiridol elimination, Jäger *et al.*, studied whether the excretion of flavopiridol and its glucuronides in the isolated perfused liver is dependent on ABCC2 [32]. For this approach, the release of flavopiridol and flavopiridol glucuronides into bile and the perfusion medium was monitored in mutant TR$^-$ rats lacking a functional ABCC2 at the canalicular membrane [39, 40]. ABCC2-competent Wistar rats acted as controls. In addition, they investigated whether flavopiridol can influence the hepatic excretion of the ABCC2 substrate bilirubin in control rats. Using an isolated perfused rat liver model of ABCC2-deficient TR$^-$ rats, they found that the biliary excretion of the metabolites M1 and M2 was reduced to 4.3 and 5.4%, respectively, compared to Wistar rats (Figure 12.5). This inability of excretion indicates that M1 and M2 are almost exclusively eliminated into bile by ABCC2 in control rats.

However, excretion of unconjugated flavopiridol is decreased only by 48% in TR$^-$ rats (see Figure 12.5), suggesting that besides CMOAT other transporters, for example, ABCG2, which was found to be responsible for flavopiridol excretion in ABCG2-transfected mammary tumor cells [41], might also be involved in the biliary elimination of flavopiridol. This is in accordance with recent literature data indeed showing a high expression of this transport protein in the canalicular membrane of human liver cells [42]. In parallel, efflux of M1 and M2 into the effluent perfusate of TR$^-$ rats increased by 1.5- and 4.2-fold indicating that the basolateral release of flavopiridol glucuronides into the perfusion medium might be mediated through

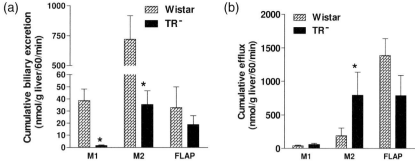

Figure 12.5 Cumulative secretion of flavopiridol, M1, and M2 into bile (a) and effluent perfusate (b) of Wistar and TR⁻ rats. *$p < 0.05$ significantly different from control. From Ref. [32].

other ABC transporters. Candidates for this transport are ABCC1 and ABCC3. Particularly, ABCC3, responsible for the transport of glucuronides in sinusoidal space, might be important, as Western blot analysis of total liver membrane from Wistar rat and TR⁻ rat indeed revealed that the expression of ABCC3 protein, but not that of ABCC1, increases up to fivefold in TR⁻ rat compared to control [43]. Pronounced induction of ABCC3 in liver of TR⁻ rat was also confirmed by immunofluorescence using a polyclonal antibody against ABCC3 [44]. Upregulation of ABCC3 may therefore compensate for the ABCC2 deficiency in mutant TR⁻ rats [45]. ABCC3 might also prevent enhanced the hepatocellular accumulation of flavopiridol glucuronides under clinical conditions where a defective biliary transport is present. In contrast to the enhanced efflux of metabolites, unconjugated flavopiridol secretion into perfusate of TR⁻ rats was reduced by 43.2% excluding ABCC3 as a candidate for flavopiridol efflux. Therefore, another not yet identified efflux pump for unconjugated flavopiridol may exist.

To test whether flavopiridol glucuronides compete with ABCC2-specific substrates, Jäger *et al.* also studied the biliary secretion of the organic anions bilirubin and BSP, and its modulation by flavopiridol in Wistar rats. Jäger and coworkers found that 30 µM flavopiridol reversibly inhibited ABCC2-mediated biliary elimination of bilirubin glucuronides and BSP-reduced glutathion to about the same degree by 54 and 51%. After withdrawal of flavopiridol from the perfusion medium, the biliary excretion of glucuronidated bilirubin and conjugated BSP rapidly recovered within about 10 min to reach levels before flavopiridol administration (see Figure 12.6).

The observed inhibition of conjugated bilirubin by flavopiridol is in accordance with a previous clinical phase I study, showing increased conjugated serum bilirubin during flavopiridol therapy, which might be caused by the reduced excretion of conjugated bilirubin through the canalicular transporter in the presence of flavopiridol glucuronides [36]. The dramatic reduction in the biliary excretion of flavopiridol glucuronides in TR⁻ rats and the inhibition of bilirubin and BSP excretion by flavopiridol strongly suggest a predominant role of ABCC2 in flavopiridol glucuronides excretion. This can explain the decrease in conjugated bilirubin excretion

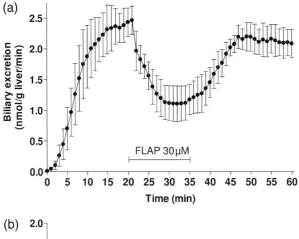

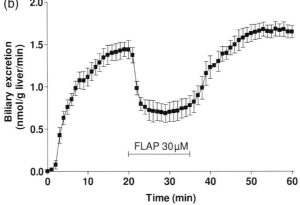

Figure 12.6 Effect of 30 μM flavopiridol on the biliary excretion of 5 μM bilirubin (a) and 1 μM bromsulfthalein (b) in the isolated perfused rat liver of Wistar rats. After achieving a constant biliary excretion of bilirubin and bromsulfthalein ($t = 20$ min), 30 μM flavopiridol was applied for 15 min. From Ref. [32].

during flavopiridol perfusion and may also apply to human liver. Therefore, conjugated serum bilirubin should be monitored under flavopiridol therapy.

The potential for an interaction between ABCC2 and UGTs might involve various drugs that are primarily glucuronidated and further eliminated into bile. A clinically important example is the interaction of mycophenolate mofetil, a potent immunosuppressant, and cyclosporin [46]. Mycophenolate mofetil is almost completely absorbed from the gut and is rapidly de-esterified into the active drug, mycophenolic acid. Mycophenolic acid is then converted by the UGT enzyme family in the liver to the inactive conjugate 7-hydroxy-mycophenolic acid glucuronide that is almost exclusively excreted into bile via ABCC2 [46]. Several studies have reported a lower exposure to mycophenolic acid in patients receiving mycophenolic acid in combination with the calcineurin inhibitor cyclosporin [46, 47]. These clinical findings were confirmed by an animal study by van Gelder and coworkers who treated Lewis rats with mycophenolate mofetil plus cyclosporin and with mycophenolate mofetil plus placebo [48]. Rats in the cyclosporin-plus-mycophenolate mofetil group showed significantly increased mycophenolic acid glucuronide but decreased mycophenolic acid plasma levels, ruling out any inhibition of UGTs by cyclosporin but strongly indicating that cyclosporin interferes with the biliary excretion of mycophenolic acid. Using ABCC2-deficient TR⁻ rats, Hesselink et al. could indeed show that cyclosporin-mediated inhibition of the biliary excretion of mycophenolic acid glucuronide is mediated by ABCC2 [49].

Evidence for a role of ABCC2 was also recently demonstrated for irinotecan, a semisynthetic camptothecin anticancer drug [50]. Using ABCC2-deficient Esai rats, the biliary excretion not only of irinotecan, the active metabolite SN-38, but also of the inactive SN-38 glucuronide was lower in the mutant rat strain compared to control Sprague-Dawley rats [50]. In addition, uptake by hepatic canalicular membrane vesicles isolated from the rats carrying the ABCC2 mutation was also lower than in those prepared from the other strain.

Recently, polymorphisms in *ABCC2* have been found to influence irinotecan disposition in patients. In a study of 64 patients, Innocenti et al. recently reported that the 3972TT genotype was associated with higher AUCs of irinotecan and SN-38 compared to patients with TC or CC genotypes [51]. Genetic variants of ABCC2 will therefore influence irinotecan toxicity and tumor response.

12.4
Biopharmaceutical Classification System

Amidon and coworkers recognized that aqueous solubility and gastrointestinal permeability are the fundamental parameters controlling the rate and extent of oral drug absorption from the intestine [52]. A generally accepted practical definition of the extent and rate at which a drug is delivered from a pharmaceutical form and becomes available in general circulation is the term bioavailability [53, 54]. Two oral dosage forms are considered to be bioequivalent if both rate and extent of absorption are the same. In the 1960s, several case studies were published demonstrating that any difference in the bioequivalence of drugs could result in either undermedication or intoxication [55]. Therefore, regulatory agencies such as the FDA in the United States or EMEA in the EU ask for bioequivalence studies when the pharmaceutical dosage form of clinically used drugs is changed [53, 54]. This is definitely important for low-soluble drugs and for compounds that show low bioavailability. However, for drugs that are readily soluble in the gastrointestinal fluid and highly permeable across the gastrointestinal membrane (routinely measured in permeability studies using human-derived colon adenocarcinoma (Caco-2) cells, changes in the formulation of drugs should hardly affect the dissolution behavior of tablets or capsules and the *in vivo* absorption. Drugs with these properties are therefore considered to be bioequivalent [56].

Based on solubility and permeability, the Biopharmaceutics Classification System (BCS) was established [52] that categorizes drugs into four classes to predict *in vivo*

	High permeability	
High permeability	**Class 1** High solubility High permeability Rapid dissolution	**Class 2** Low solubility High permeability
Low permeability	**Class 3** High solubility Low permeability	**Class 4** Low solubility Low permeability

Figure 12.7 The biopharmaceutics classification as defined by the FDA. After Ref. [52].

pharmacokinetics of drugs from *in vitro* measurements of permeability and solubility (see Figure 12.7). The BCS test may also help define which *in vitro* tests are most predictable for *in vivo* bioavailability [57]. The FDA has recently published a guideline on how to obtain biowaivers for high-soluble and high-permeable class 1 drugs [58]. Also, the EMEA is working with the concept of biowaivers for class 1 compounds; however, contrary to the FDA guidelines, additional requirements were defined (e.g., bioavailability should be higher than 90% and there should be an absence of metabolism) [53]. On the basis of the BCS classification of the top-selling drugs in the United States, Great Britain, Spain, and Japan, it is suggested that a minimum of 25–30% of the drug products on these markets are BCS class 1 and candidates for waiver of *in vivo* bioequivalence testing [59]. Waivers for highly soluble, low permeable class 3 drugs have also been scientifically recommended on the basis of dissolution studies alone [59, 60], although not implemented by the FDA as small changes in the formulation may significantly modify bioavailability of class 3 compounds.

Recently, Wu and Benet proposed a Biopharmaceutics Drug Disposition Classification System (BDDCS) based upon solubility and extent of metabolism [61]. According to the authors, the major route of elimination may be the more appropriate criterion for BCS classification than the currently used permeability criterion. Based on the BDDCS classification, a drug substance is considered to be "highly permeable" when the extent of the intestinal absorption (parent drug plus metabolites) in humans is determined to be $\geq 90\%$ of an administered dose in comparison to an intravenous reference dose. Table 12.3 shows the list of compounds according to the four BCS classes selected from literature [52, 54, 57, 60, 62–75]. Examining the drug substances listed in the four BCS classes in Table 12.3, it becomes obvious for Wu and Benet [61] that class 1 and class 2 compounds are eliminated primarily via metabolism, whereas class 3 and class 4 compounds are primarily eliminated unchanged into the urine and bile (Figure 12.8).

As demonstrated in Figure 12.9, Wu and Benet [61] also hypothesize that any effect of transporters will be minimal for class 1 compounds because of its high solubility and permeability properties leading to high concentrations in the gut mucosa, thereby saturating uptake and efflux transporters. Although *in vitro*

Table 12.3 Biopharmaceutical Classification System substrates [61]: high permeability.

High solubility: class 1	Low solubility: class 2
Abacavir	Amiodarone[I]
Acetaminophen	Atorvastatin[S,I]
Acyclovir[a]	Azithromycin[S,I]
Amiloride[S,I]	Carbamazepine[S,I]
Amitryptyline[S,I]	**Carvedilol**
Antipyrine	Chlorpromazin[I]
Atropine	**Cisapride**[S]
Buspirone[b]	*Ciprofloxacin*[S]
Caffeine	**Cyclosporin**[S,I]
Captopril	**Danazol**
Chloroquine[S,I]	**Dapsone**
Chlorpheniramine	Diclofenac
Cyclophosphamide	Diflunisal
Desipramine	Digoxin[S]
Diazepam	*Erythromycin*[S,I]
Diltiazem[S,I]	Flurbiprofen
Diphenhydramine	**Glipizide**
Disopyramide	Glyburide[S,I]
Doxepin	Griseofulvin
Doxycycline	Ibuprofen
Enalapril	**Indinavir**[S]
Ephedrine	Indomethacin
Ergonovine	**Itraconazole**[S,I]
Ethambutol	**Ketoconazole**[I]
Ketorolac	**Lansoprazole**[I]
Ketoprofen	**Lovastatin**[S,I]
Labetolol	*Mebendazole*
Levodopa[S]	Naproxen
Levofloxacin[S]	Nelfinavir[S,I]
Lidocaine[I]	*Nifedipine*[S]
Lomefloxacin	Ofloxacin
Meperidine	Oxaprozin
Metoprolol	Phenazopyridine
Metronidazole	Phenytoin[S]
Midazolam[S,I]	Piroxicam
Minocycline	Raloxifene[S]
Misoprostol	**Ritonavir**[S,I]
Nifedipine[S]	**Saquinavir**[S,I]
Phenobarbital	**Sirolimus**[S]
Phenylalanine	Spironolactone[I]
Prednisolone	**Tacrolimus**[S,I]
Primaquine[S]	Talinolol[S]
Promazine	**Tamoxifen**[I]
Propranolol[I]	**Terfenadine**[I]
Quinidine[S,I]	Warfarin

(Continued)

Table 12.3 (Continued)

High solubility: class 1	Low solubility: class 2
Rosiglitazone	
Salicylic acid	
Theophylline	
Valproic acid	
Verapamil[I]	
Zidovudine	

[a]The compounds listed in *italic* are those falling in more than one category by different authors, which could be a result of the definition of the experimental conditions.
[b]The compounds listed in bold are primarily CYP3A substrates where metabolism accounts for more than 70% of the elimination; superscript I and/or S indicate ABCB1 inhibitors and/or substrate, respectively.

cellular systems showed that many class 1 compounds are substrates for various transport proteins, for example, midazolam [76] and nifedipine [77] that are substrates for ABCB1, transporter effects on patients after peroral application should be negligible.

	High solubility	Low solubility
High permeability	**Class 1** Metabolism	**Class 2** Metabolism
Low permeability	**Class 3** Renal and/or biliary elimination of unchanged drug	**Class 4** Renal and/or biliary elimination of unchanged drug

Figure 12.8 Predominant routes of drug elimination for drug substances by BCS classification. After Ref. [61].

	High solubility	Low solubility
High permeability	**Class 1** Transporter effects minimal	**Class 2** Efflux transporter effects predominate
Low permeability	**Class 3** Absortive transporter effects predominate	**Class 4** Absorptive and efflux transporter effects could be important

Figure 12.9 Transporter effects on drug disposition by BCS classification. After Ref. [61].

However, transporter effects are important for class 2 compounds. As these highly lipid-soluble drugs rapidly penetrate the enterocytes, intestinal uptake transporters hardly play any role in their absorption [61, 72]. Due to the low solubility of these compounds, the concentration of class 2 drugs in the intestinal mucosa will be low, and there will be little opportunity to saturate apical efflux transporters and intestinal enzymes. As a result, the rate of absorption and the extent of oral bioavailability of class 2 compounds will strongly depend on efflux transporters and drug metabolizing enzymes such as CYP3A4 and UGTs. Thus, induction or inhibition in the expression of intestinal efflux transporters will also change intestinal metabolism of drugs that are substrates for the intestinal metabolic enzymes [61].

Indeed, Wu and Benet recognized a large number of class 2 compounds in Table 12.3 as being substrates for CYP3A (see Table 12.4) as well as substrates or

Table 12.4 Biopharmaceutical Classification System substrates [61]: low permeability.

High solubility: class 3	Low solubility: class 4
Acyclovir[a]	Amphotericin B
Amiloride[S,I]	Chlorthalidone
Amoxicillin[S,I]	Chlorothiazide
Atenolol	Colistin
Atropine	Ciprofloxacin[S]
Bisphosphonates	*Furosemide*
Bidisomide	*Hydrochlorothiazide*
Captopril	Mebendazole
Cefazolin	*Methotrexate*
Cetirizine	Neomycin
Cimetidine[S]	
Ciprofloxacin[S]	
Cloxacillin	
Dicloxacillin[S]	
Erythromycin[S,I]	
Famotidine	
Fexofenadine[S]	
Folinic acid	
Furosemide	
Ganciclovir	
Hydrochlorothiazide	
Lisinopril	
Metformin	
Methotrexate	
Nadolol	
Pravastatin[S]	
Penicillin	
Ranitidine[S]	
Tetracycline	
Trimethoprim[S]	
Valsartan	
Zalcitabine	

[a] The compounds listed in *italic* are those falling in more than one category by different authors, which could be a result of the definition of the experimental conditions.

inhibitors of the efflux transporter ABCB1 [61]. Oral dosing of class 2 compounds will therefore lead to significant interactions due to their potential for inhibition of intestinal enzymes (e.g., CYP3A and UGTs) as well as apical efflux transporters (e.g., ABCB1, ABCC2, ABCG2). Through this concomitant inhibition of the intestinal enzymes and the apical efflux transporters, systemic drug concentrations are synergistically increased. It is, therefore, not surprising that drugs removed from the market at the FDA's recommendation due to drug–drug interactions are predominately orally administered drugs that are substrates for both CYP3A and ABCB1 [78].

Class 3 compounds are well available in the gut lumen due to their good solubility. However, based on the poor permeability, uptake transporters such as OATPs will be necessary to increase the poor permeability of these compounds. Efflux proteins may also play a major role in the absorption of class 3 drugs counteracting the increased intestinal permeability via uptake transporters [61].

Based on their low permeability and low solubility characteristics, it might be expected that class 4 compounds would hardly be effective drugs. However, a considerable number of class 4 compounds may be misclassified in terms of *in vivo* characteristics, as solubility in aqueous solutions may not reflect solubility in gut content. For example, the FDA publication [72] and others have suggested that solubility measurements in surfactant-containing solution may be a more appropriate basis for the solubility criteria. Oral bioavailability for true class 4 compounds is already minimal so that any transporter effect could be relevant, as a small increase in bioavailability (e.g., from 2 to 4%) would make a significant difference [61].

The BDDCS is therefore an innovative suggestion with many implications for metabolism and drug disposition. In general, a significant correlation between nonpolarity of an uncharged drug molecule, metabolism, and membrane permeability is expected. However, comparisons or correlations between the BDDCS and BCS will have some limitations since they are based on different processes at the molecular level [72]. While BDDCS is based on transport and enzyme binding, the BCS classification is based on passive membrane permeation transport. Furthermore, contrary to BCS, which is based on lipophilicity considerations alone, BDDCS uses *in vitro* cell systems expressing drug metabolizing enzymes and drug transporters therefore better predicting *in vivo* bioavailability [56]. Despite these differences, there is a substantial agreement on the two approaches in the classification of drugs. However, a more detailed examination of the importance and reliability of key parameters, particularly for drug compounds that are classified differently by the two approaches, might be beneficial [72].

References

1 Christians, U., Schmitz, V., and Haschke, M. (2005) Functional interactions between P-glycoprotein and CYP3A in drug metabolism. *Expert Opinion on Drug Metabolism & Toxicology*, **1**, 641–654.

2 Westlind-Johnsson, A., Malbebo, S., Johansson, A., Otter, C., Andersson, T.B., Johansson, I., Edwards, R.J., Bobbis, A.R., and Ingelman-Sundberg, M. (2003) Comparative analysis of CYP3A

expression in human liver suggests only a minor role for CYP3A5 in drug metabolism. *Drug Metabolism and Disposition*, **31**, 755–761.

3 Schuetz, J.D., Beach, D.L., and Guzelian, P.S. (1994) Selective expression of cytochrome P450 CYP3A mRNAs in embryonic and adult human liver. *Pharmacogenetics*, **4**, 11–20.

4 Thiebaut, F., Tsuruo, T., Hamada, H., Gottesman, M.M., Pastan, I., and Willingham, M.C. (1987) Cellular localization of the multidrug-resistance gene product P-glycoprotein in normal human tissues. *Proceedings of the National Academy of Sciences of the United States of America*, **84**, 7735–7738.

5 Zhang, Y. and Bent, L.Z. (2001) The gut as a barrier to drug absorption. *Clinical Pharmacokinetics*, **40**, 159–168.

6 Mouly, S. and Paine, M.F. (2003) P-glycoprotein increases from proximal to distal regions of human small intestine. *Pharmaceutical Research*, **20**, 1595–1599.

7 Zhang, Q.Y., Dunbar, D., Ostrowska, A., Zeisloft, S., Yang, J., and Kaminsky, L.S. (1999) Characterization of human small intestinal cytochromes P450. *Drug Metabolism and Disposition*, **27**, 804–809.

8 Lown, K.S., Kolars, J.C., Thummel, K.E., Barnett, J.L., Kunze, K.L., Wrighton, S.A., Merion, R.M., and Watkin, P.B. (1994) Interpatient heterogeneitiy in expression of CYP3A4 and CYP3A5 in small bowel. Lack of prediction by the erythromycin breast test. *Drug Metabolism and Disposition*, **22**, 947–955.

9 Lown, K.S., Mayo, R.R., Leichtmann, A.B., Hsiao, H.L., Turgeon, D.K., Schmiedlin-Ren, P., Rossi, S.J., Brown, M.B., Guo, W., Benet, L.Z., and Watkins, P.B. (1997) Role of intestinal P-glycoprotein (mdr1) in interpatient variation in the oral bioavailability of cyclosporine A. *Clinical Pharmacology and Therapeutics*, **62**, 248–260.

10 Wacher, V.J., Silverman, J.A., Zhang, Y., and Benet, L.Z. (1998) Role of P-glycoprotein and cytochrome P450 3A in limiting oral absorption of peptides and peptidomimetics. *Journal of Pharmaceutical Sciences*, **87**, 1322–1330.

11 Benet, L.Z., Cummings, C.L., and Wu, C.Y. (2003) Transporter–enzyme interaction: implications for predicting drug–drug interactions from *in vitro* data. *Current Drug Metabolism*, **4**, 393–398.

12 Christian, U., Schmitz, V., and Haschke, M. (2005) Functional interactions between P-glycoprotein and CYP3A in drug metabolism. *Expert Opinion on Drug Metabolism & Toxicology*, **1**, 641–654.

13 Cummings, C.L., Jacobsen, W., and Benet, L.Z. (2002) Unmasking the dynamic interplay between intestinal P-glycoprotein and CYP3A4. *The Journal of Pharmacology and Experimental Therapeutics*, **300**, 1036–1045.

14 Cummings, C.L., Jacobsen, W., Christians, U., and Benet, L.Z. (2004) CYP3A4 transfected Caco-2 cells as a tool for understanding biochemical absorption barriers: studies with sirolimus and midazolam. *The Journal of Pharmacology and Experimental Therapeutics*, **308**, 143–155.

15 Wu, C.Y. and Benet, L.Z. (2003) Disposition of tacrolimus in isolated perfused rat liver: influence of troleandomycin, cyclosporine and GG918. *Drug Metabolism and Disposition*, **31**, 1292–1295.

16 Synold, T.W., Dussault, I., and Forman, B.M. (2001) The orphan nuclear receptor SXR co-ordinately regulates drug metabolism and efflux. *Nature Medicine*, **7**, 584–590.

17 Pal, D. and Mitra, A.K. (2006) MDR- and CYP3A4-mediated drug–herbal interactions. *Life Sciences*, **78**, 2131–2145.

18 Dresser, G.K., Schwarz, U.I., Wilkinson, G.R., and Kim, R.B. (2003) Coordinate induction of both cytochrome P4503A and MDR1 by John's wort in healthy subjects. *Clinical Pharmacology & Therapeutics*, **73**, 41–50.

19 Lin, J.H. and Yamazaki, M. (2003) Role of P-glycoprotein in pharmacokinetics. *Clinical Pharmacokinetics*, **42**, 59–98.

20 Hagenbuch, B. and Meier, P.J. (2004) Organic anion transporting polypeptides of the OATP/SLC21 family: phylogenetic classification as OATP/SLCO superfamily, new nomenclature and molecular/functional properties. *Pflugers Archives*, **447**, 653–665.

21 König, J., Seithel, A., Gradhand, U., and Fromm, M.F. (2006) Pharmacogenomics of human OATP transporters. *Naunyn-Schmiedeberg's Archives of Pharmacology*, **372**, 432–443.

22 Mück, W., Mai, I., Ochmann, K., Rohde, G., Unger, S., Johne, A., Bauer, S., Budde, K., Roots, I., Neumayer, H.H., and Kuhlmann, J. (1999) Increase in cerivastatin systemic exposure after single and multiple dosing in cyclosporine-treated kidney transplant recipients. *Clinical Pharmacology & Therapeutics*, **65**, 251–261.

23 Lin, J.H. (2007) Transporter mediated drug interactions: clinical implications and *in vitro* assessment. *Expert Opinion on Drug Metabolism & Toxicology*, **3**, 81–92.

24 Shitara, Y., Hirano, M., Adachi, Y., Itoh, T., Sato, H., and Sugiyama, Y. (2004) *In vitro* and *in vivo* correlation of the inhibitory effect of cyclosporin A on the transporter-mediated hepatic uptake of cerivastatin in rats. *Drug Metabolism and Disposition*, **32**, 1468–1475.

25 Lennernas, H. (2003) Clinical pharmacokinetics of atorvastatin. *Clinical Pharmacokinetics*, **42**, 1141–1160.

26 Kantola, T., Kivisto, K.T., and Neuvone, P.J. (1998) Effect of itroconazole on the pharmacokinetics of atorvastatin. *Clinical Pharmacology & Therapeutics*, **64**, 58–65.

27 Siedlik, P.H., Olson, S.C., Yang, B.B., and Stern, R.H. (1999) Erythromycin coadministration increases plasma atorvastatin concentrations. *Journal of Clinical Pharmacology*, **39**, 501–504.

28 Lau, Y.Y., Okochi, H., Huang, Y., and Benet, L.Z. (2006) Pharmacokinetics of atorvastatin and its hydroxyl metabolites in rats and the effects of concomitant rifampicin single doses: relevance of first-pass effect from hepatic transporters, and intestinal and hepatic metabolism. *Drug Metabolism and Disposition*, **34**, 1175–1181.

29 Lau, Y.Y., Wu, C.Y., Okochi, H., and Benet, L.Z. (2004) *Ex situ* inhibition of hepatic uptake and efflux significantly changes metabolism: hepatic enzyme–transporter interplay. *The Journal of Pharmacology and Experimental Therapeutics*, **308**, 1040–1045.

30 Kiang, T.K., Ensom, M.H., and Chang, T.K. (2005) UDP-glucuronosyltransferase drug–drug interaction. *Pharmacology & Therapeutics*, **106**, 97–132.

31 Tsujii, H., Konig, J., Rost, D., Stockel, B., Leuscher, U., and Keppler, D. (1999) Exon–intron organization of the human multidrug-resistance protein 2 (MRP2) gene mutated in Dubin–Johnson syndrome. *Gastroenterology*, **117**, 653–660.

32 Jäger, W., Gehring, E., Hagenauer, B., Aust, S., Senderowicz, A., and Thalhammer, T. (2003) Biliary excretion of flavopiridol and its glucuronides in the isolated perfused rat liver: role of multidrug resistance protein 2 (Mrp2). *Life Sciences*, **17**, 2841–2854.

33 Jäger, W., Zembsch, B., Wolschann, P., Pittenauer, E., Senderowicz, A.M., Sausville, E.A., Sedlacek, H.H., Graf, J., and Thalhammer, T. (1998) Metabolism of the anticancer drug flavopiridol, a new inhibitor of cyclin dependent kinases, in rat liver. *Life Sciences*, **62**, 1861–1873.

34 Lush, R., Stinson, S., Senderovicz, A.M., Hill, K., Feuer, J., Headlee, D., Figg, W.D., and Sausville, E.A. (1997) Flavopiridol pharmacokinetics suggests enterohepatic circulation. *Clinical Pharmacology & Therapeutics*, **61**, 145.

35 Innocenti, F., Stadler, W.M., Iyer, L., Ramirez, J., Volkes, E.E., and Ratain, M.J. (2000) Flavopiridol metabolism in cancer patients is associated with the occurrence of diarrhea. *Clinical Cancer Research*, **6**, 3400–3405.

36 Senderowicz, A., Headlee, D., Stinson, S.F., Lush, R.M., Kalil, N., Villalba, L., Hill,

K., Steinberg, S.M., Figg, W.D., Tompkins, A., Arbruck, S.G., and Sausville, E.A. (1998) Phase I trial of continuous infusion of flavopiridol, a novel cyclin-dependent kinase inhibitor, in patients with refractory neoplasms. *Journal of Clinical Oncology*, **16**, 2986–2999.

37 Jäger, W., Winter, O., Halper, B., Salamon, A., Satori, M., Gajdzik, L., Hamilton, G., Theyer, G., Graf, J., and Thalhammer, T. (1997) Modulation of liver canalicular transport processes by the tyrosine kinase inhibitor genistein: implication of genistein metabolites in rats. *Hepatology*, **26**, 1467–1476.

38 Hagenauer, B., Salamon, A., Thalhammer, T., Kunert, O., Haslinger, E., Klinger, P., Senderowicz, A.M., Sausville, E.A., and Jäger, W. (2001) *In vitro* glucuronidation of cyclin-dependent kinase inhibitor flavopiridol by rat and human liver microsomes: involvement of UDP-glucuronosyltransferases 1A1 and 1A9. *Drug Metabolism and Disposition*, **29**, 407–414.

39 Jansen, P.L., Peters, W.H., and Lamers, W.H. (1985) Hereditary chronic conjugated hyperbilirubinemia in mutant rats caused by defective hepatic anion transport. *Hepatology*, **4**, 573–579.

40 Jansen, P., Groothuis, G.M.M., Peters, W.H.M., and Meijer, D.K.F. (1987) Selective hepatobiliary transport defect for organic anions and neutral steroids in mutant rats with hereditary-conjugated hyperbilirubinemia. *Hepatology*, **7**, 71–76.

41 Robey, R.W., Medina-Pérez, W.Y., Nishiyama, K., Lahusen, T., Miyake, K., Litman, T., Senderowicz, A.M., Ross, D.D., and Bates, S.E. (2001) Overexpression of the ATP-binding cassette half-transporter, ABCG2 (MXR/BCRP/ABCP1), in flavopiridol-resistant human breast cancer cells. *Clinical Cancer Research*, **1**, 145–152.

42 Maliepaard, M., Scheffer, G.L., Faneyte, I.F., van Gastelen, M.A., Pijnenborg, A.C., Schinkel, A.H., van de Vijver, M.J., Scheper, R.J., and Schellens, J.H. (2001) Subcellular localization and distribution of the breast cancer resistance protein transporter in normal human tissues. *Cancer Research*, **61**, 3458–3464.

43 Cao, J.S., Stieger, B., Meier, P.J., and Vore, M. (2002) Expression of rat hepatic multidrug resistance-associated proteins and organic anion transporters in pregnancy. *American Journal of Physiology. Gastrointestinal and Liver Physiology*, **283**, 757–766.

44 Donner, M.G. and Keppler, D. (2001) Up-regulation of basolateral multidrug resistance protein 3 (MRP3) in cholestatic rat liver. *Hepatology*, **34**, 351–359.

45 Soroka, C.J., Lee, J.M., Azzaroli, F., and Boyer, J.L. (2001) Cellular localization and up-regulation of multidrug resistance-associated protein 3 in hepatocytes and cholangiocytes during obstructive cholestasis in rat liver. *Hepatology*, **33**, 783–791.

46 Picard, N., Premaud, A., Rousseau, A., Le Meur, Y., and Marquet, P. (2006) A comparison of the effect of cyclosporine and sirolimus on the pharmacokinetics of mycophenolate in renal transplant patients. *British Journal of Clinical Pharmacology*, **62**, 477–484.

47 Cattaneo, D., Merlini, S., Zenoni, S., Balzelli, S., Gotti, E., Remuzzi, G., and Perico, N. (2005) Influence of co-medication with sirolimus or cyclosporine on mycophenolic acid pharmacokinetics in kidney transplantation. *American Journal of Transplantation*, **5**, 2937–2944.

48 van Gelder, T., Klupp, J., Barten, M.J., Christian, U., and Morris, R.E. (2001) Comparison of the effects of tacrolimus and cyclosporine on the pharmacokinetics of mycophenolic acid. *Therapeutic Drug Monitoring*, **23**, 119–128.

49 Hesselink, D.A., van Hest, R.M., Mathot, R.A.A., Bonthuis, F., Weimar, W., Bruin, R.W.F., and van Gelder, T. (2005) Cyclosporine interactions with mycophenolic acid by the multidrug resistance-associated protein 2. *American Journal of Transplantation*, **5**, 987–994.

50 Chu, X.Y., Kato, Y., Niinuma, K., Sudo, K.I., Hakusui, H., and Sugiyama, Y. (1997) Multispecific organic anion transporter is responsible for the biliary excretion of the camptothecin derivative irinotecan and its metabolites in rats. *The Journal of Pharmacology and Experimental Therapeutics*, **281**, 304–314.

51 Innocenti, F., Undevia, S.D., Chen, P.X., Das, S., Ramirez, J., Dolan, M.E., Relling, M.V., Kroetz, D.L., and Ratain, M.J. (2004) Pharmacogenetic analysis of interindividual irinotecan (CPT-11) pharmacokinetics (PK) variability: evidence for a functional variant of ABCC2. *Journal of Clinical Oncology*, **22**, 2010.

52 Amidon, G.L., Lennernas, H., Shah, V.P., and Crison, J.R. (1995) A theoretical basis for a biopharmaceutics drug classification: the correlation of *in vitro* drug product dissolution and *in vivo* bioavailability. *Pharmaceutical Research*, **12**, 413–420.

53 European Agency for the Evaluation of Medicinal Products, Committee for Proprietary Medicinal Products (2004) Note for Guidance on the Investigation of Bioavailability and Bioequivalence [online]. Available at http://www.emea.eu.int/pdf/human/ewp/140198en.pdf (accessed 12 September 2004).

54 US Department of Health and Human Services (2004) Guidance for Industry: Bioavailability and Bioequivalence Studies for Orally Administered Drug Products: General Considerations [online]. Available at http://www.fda.gov/cder/guidance/4964dft.pdf (accessed 12 September 2004).

55 Wijnand, H.P. (1994) Bioequivalence assessment of drug formulations: non-parametric versus parametric analysis. Ph.D. thesis. University of Leiden, Leiden.

56 Faasen, F. and Vromans, H. (2004) Biowaivers for oral immediate-release products: implications of linear pharmacokinetics. *Clinical Pharmacokinetics*, **43**, 1117–1126.

57 Lennernas, H. (1997) Human jejunal effective permeability and its correlation with preclinical drug absorption models. *Journal of Pharmacy and Pharmacology*, **49**, 627–638.

58 US Department of Health and Human Services (2004) Guidance for Industry: Waiver of *In Vivo* Bioavailability and Bioequivalence Studies for Immediate-Release Solid Oral Dosage Forms Based on a Biopharmaceutics Classification System [online]. Available at http://www.fda.gov/cder/guidance/3618nl.pdf (accessed 12 September 2004).

59 Takagi, T., Ramachandran, C., Bermejo, M., Yamashita, S., Yu, L.X., and Amidon, G.L. (2006) A provisional biopharmaceutical classification of top 200 oral drug products in the united states, Great Britain, Spain and Japan. *Molecular Pharmaceutics*, **6**, 631–643.

60 Kanfer, I. (2002) Report on the International Workshop on the Biopharmaceutics Classification System (BCS): scientific and regulatory aspects in practice. *Journal of Pharmacy & Pharmaceutical Sciences*, **5**, 1–4.

61 Wu, C.Y. and Benet, L.Z. (2005) Predicting drug disposition via application of BCS: transport/absorption/elimination interplay and development of a biopharmaceutics drug disposition classification system. *Pharmaceutical Research*, **22**, 11–23.

62 van de Waterbeemd, H. (1998) The fundamental variables of the Biopharmaceutics Classification System (BCS): a commentary. *European Journal of Pharmaceutical Sciences*, **7**, 1–3.

63 Blume, H.H. and Schug, B.S. (1999) The Biopharmaceutics Classification System (BCS): class III drugs – better candidates for BA/BE waiver? *European Journal of Pharmaceutical Sciences*, **9**, 117–121.

64 Fleisher, D., Li, C., Zhou, Y., Pao, L.H., and Karim, A. (1999) Drug, meal and formulation interactions influencing drug absorption after oral administration. Clinical implications. *Clinical Pharmacokinetics*, **36**, 233–254.

65 Lobenberg, R. and Amidon, G.L. (2000) Modern bioavailability, bioequivalence and

Biopharmaceutics Classification System. New scientific approaches to international regulatory standards. *European Journal of Pharmaceutics and Biopharmaceutics*, **50**, 3–12.

66 Avdeef, A. (2001) Physicochemical profiling (solubility, permeability and charge state). *Current Topics in Medicinal Chemistry*, **1**, 277–351.

67 Rege, B.D., Yu, L.X., Hussain, A.S., and Polli, J.E. (2001) Effect of common excipients on Caco-2 transport of low-permeability drugs. *Journal of Pharmaceutical Sciences*, **90**, 1776–1786.

68 Tannergren, C.P., Langguth, P., and Hoffmann, K.J. (2001) Compound mixtures in Caco-2 cell permeability screens as a means to increase screening capacity. *Die Pharmazie*, **56**, 337–342.

69 Lennernas, H., Knutson, L., Knutson, T., Hussain, A., Lesko, L., Salmonson, T., and Amidon, G.L. (2002) The effect of amiloride on the *in vivo* effective permeability of amoxicillin in human jejunum: experience from a regional perfusion technique. *European Journal of Pharmaceutical Sciences*, **15**, 271–277.

70 Martinez, M.N. and Amidon, G.L. (2002) A mechanistic approach to understanding the factors affecting drug absorption: a review of fundamentals. *Journal of Clinical Pharmacology*, **42**, 620–643.

71 Taub, M.E., Kristensen, L., and Frokjaer, S. (2002) Optimized conditions for MDCK permeability and turbidimetric solubility studies using compounds representative of BCS classes I–IV. *European Journal of Pharmaceutical Sciences*, **15**, 331–340.

72 Yu, L.X., Amidon, G.L., Polli, J.E., Zhao, H., Mehta, M.U., Conner, D.P., Shah, V.P., Lesko, L.J., Chen, M.L., Lee, V.H., and Hussain, A.S. (2002) Biopharmaceutics classification system: the scientific basis for biowaiver extensions. *Pharmaceutical Research*, **19**, 921–925.

73 Bergstrom, C.A., Strafford, M., Lazorova, L., Avdeef, A., Luthman, K., and Artursson, P. (2003) Absorption classification of oral drugs based on molecular surface properties. *Journal of Medicinal Chemistry*, **46**, 558–570.

74 Tannergreen, C., Knutson, T., Knutson, L., and Lennernas, H. (2003) The effect of ketoconazole on the *in vivo* intestinal permeability of fexofenadine using a regional perfusion technique. *British Journal of Clinical Pharmacology*, **55**, 182–190.

75 Lindenberg, M., Kopp, S., and Dressman, J.B. (2004) Classification of orally administered drugs on the World Health Organization Model List of Essential Medicines according to the biopharmaceutical classification system. *European Journal of Pharmaceutics and Biopharmaceutics*, **58**, 265–278.

76 Tolle-Sander, S., Rautio, J., Wring, S., Polli, J.W., and Polli, J.E. (2003) Midazolam exhibits characteristics of a highly permeable P-glycoprotein substrate. *Pharmaceutical Research*, **20**, 757–764.

77 Watanabe, N. and Benet, L.Z. (2004) The effect of the interplay between CYP3A4 and P-gp on the metabolism of saquinavir and nifedipine in CYP3A4-transfected Caco-2 cells. Pharmaceutical Sciences World Congress, June 2004, Kyoto, Japan, Abstract NE-II-026.

78 Huang, S.M. and Lesko, L.J. (2004) Drug–drug, drug–dietary supplement and drug–citrus fruit and other food interactions: what have we learned? *Journal of Clinical Pharmacology*, **44**, 559–569.

13
ABC Transporters – From Targets to Antitargets?
Gerhard F. Ecker and Peter Chiba

13.1
Introduction

At the 2006 ABC transporter meeting in Innsbruck, the community celebrated the 30th anniversary of the discovery of P-glycoprotein (P-gp, ABCB1) [1]. The multispecific nature of this drug efflux transporter and its potential role in clinical drug resistance initiated development of inhibitors that would engage P-gp and thus re-establish sensitivity to standard therapeutic regimens. This concept was considered to solve the problem of drug resistance, one of the fundamental challenges in the treatment of cancer [2]. Since the identification of the P-gp inhibitory potential of verapamil [3], however, more than 25 years have passed and still no P-gp inhibitor has entered the market. Furthermore, since the discovery of P-gp in 1976 [4], additional 47 human ABC transporters have been identified of which several have been related to either human disease or drug resistance [5]. Nevertheless, none of them is currently targeted by a marketed drug, which underscores the special role of ABC transporters with respect to their druggability. In the past decade, considerable progress has been made in unraveling the physiological function of P-gp and other ABC transporters. Results clearly demonstrated a multiple involvement of several members of the ABC transporter family in drug uptake, disposition, and elimination [6] making them antitargets rather than classical targets suited for drug therapy. This chapter will highlight the conceptual changes from the design of inhibitors for ABC transporter to predicting potential substrates.

13.2
ABC Transporters as Targets

In 2002, Hopkins and Groom published a study in which they estimated the number of druggable proteins in the human genome to be around 3000 [7]. Among the top protein classes listed, almost 50% belong to G-protein-coupled receptors,

Transporters as Drug Carriers: Structure, Function, Substrates. Edited by Gerhard Ecker and Peter Chiba
Copyright © 2009 WILEY-VCH Verlag GmbH & Co. KGaA, Weinheim
ISBN: 978-3-527-31661-8

serine/threonine and tyrosin kinases, zinc metallopeptidases, serine proteases, nuclear hormone receptors, and phosphodiesterases. According to their definition, druggability means that the activity of these proteins can be modulated by drug-like compounds. However, the ability of a protein to bind drug-like compounds does not necessarily make the protein a drug target. Proteins considered as drug targets clearly need to be linked to a disease. Taking this into account, the number of druggable targets was estimated at 600–1500. According to this definition, several ABC transporters would represent versatile targets. They are involved in physiological and pathophysiological processes such as drug resistance, steroid transport, bile acid transport, and brain uptake. Most important, the involvement of ABCB1, ABCC1, and ABCG2 in cancer drug resistance clearly identified them as potential targets for treatment of multiple drug resistance in cancer chemotherapy [8]. The following section will outline in more detail the attempts to identify and develop drugs targeting ABC transporter.

13.2.1
P-Glycoprotein (ABCB1)

Almost 50% of all cancers are either intrinsically resistant or rapidly develop resistance during treatment with antitumor agents. Resistance to chemotherapy has been investigated since the late 1950s, and reduced intracellular accumulation was identified as one of the basic underlying mechanisms in the mid-1970s. In 1976, the pioneering work of Juliano and Ling linked the phenotype of multidrug resistance to overexpression of a single protein termed "permeability glycoprotein" (P-glycoprotein) [4]. Thus, the phenomenon of resistance to a broad panel of structurally and functionally diverse compounds could be connected to the function of a distinct protein. This protein was immediately considered a new and promising target for overcoming multidrug resistance in tumor therapy. Five years later, verapamil was discovered as the first lead compound that resensitizes vinca alkaloid-resistant P388 leukemia cells to vincristine and vinblastine [3]. The verapamil-induced restoration of cytotoxicity, also observed in case of anthracyclines, was subsequently extended to two other classes of calcium channel blockers, benzothiazepines and 1,4-dihydropyridines. Other therapeutically used drugs such as phenothiazines, quinine, tamoxifen, and cyclosporin A were also identified to inhibit P-gp. Because of their inherent pharmacological activity, these drugs were referred to as first-generation P-gp inhibitors. Early clinical studies with these compounds failed to demonstrate a significant effect when used in combination with vinblastine. In case of verapamil, severe cardiotoxicity was observed at doses significantly lower than those required to inhibit P-gp function *in vivo*. Also, the inherent pharmacological activity led to dose-limiting side effects, preventing the achievement of sufficiently high plasma concentrations (Table 13.1) [9].

Attempts to apply the concept of chiral switching, that is, using dexverapamil or dexniguldipine (the respective distomers with respect to cardiovascular activity) unfortunately failed and both compounds had to be withdrawn from clinical studies due to their severe side effects. The second-generation P-gp inhibitors were designed

Table 13.1 Achievable *in vivo* and optimal *in vitro* concentrations of first-generation MDR modulators.

Compound	Achievable *in vivo* concentration	Optimal *in vitro* concentration
Quinidine	4.5–5.6 µM	3.3–9.9 µM
Trifluoperazine	130 ng/ml	1–6 µg/ml
Tamoxifen	6 µM	~10 µM
Toremifene	10–15 µM	~15 µM
Cyclosporin	2.5–8.5 µg/ml	6 µg/ml
Verapamil	1–2 µM	6–10 µM

to get rid of the inherent pharmacological activity that limited the use of the first-generation compounds. Valspodar (PSC833) represents a nonimmunosuppressive cyclosporin analogue and biricodar (VX-710), a derivative of the macrocyclic antibiotic FK-506 (Table 13.2 and Figure 13.1)

However, these compounds interfered with the metabolism of anticancer drugs at the level of cytochrome P450 3A4, resulting in prolonged half-life and increased plasma levels of the anticancer drug used in clinical coadministration protocols. Increased response rates due to higher dose levels led to initial enthusiasm, but were shown to be due to higher AUC values in controlled pharmacokinetic studies. Dose reduction protocols were hard to design because of the large interindividual variation in the metabolism of anticancer drugs. The significant overlap of substrate profiles of P-glycoprotein and CYP P450 3A4 is now well appreciated along with the fact that these two proteins seem to complement each other in metabolizing and eliminating drugs [10].

Third-generation modulators of P-gp, such as tariquidar, zosuquidar, and elacridar have been developed to avoid the interference at the level of CYP P450 3A4.

Table 13.2 Compounds undergoing clinical investigation as MDR modulator.

Drug	Phase	Objective
Elacridar	I	Malignant neoplastic disease, solid tumors
SN-22995	I	Solid tumors
Biricodar	II	Prostate, lung, ovarian, and breast carcinomas
Ethacrynic acid	II	Malignant neoplastic disease, carcinoma
Irofulven	II	Recurrent ovarian epithelial and peritoneal cancer
Laniquidar	II	Metastatic breast cancer
ONT-093	II	Metastatic breast cancer
Timcodar	II	Malignant neoplastic disease
Zosuquidar	II	Advanced solid tumors
MS-209	III	Breast cancer, advanced solid tumors
Tariquidar	III	Ovarian cancer
Valspodar	III	Acute myeloid leukaemia

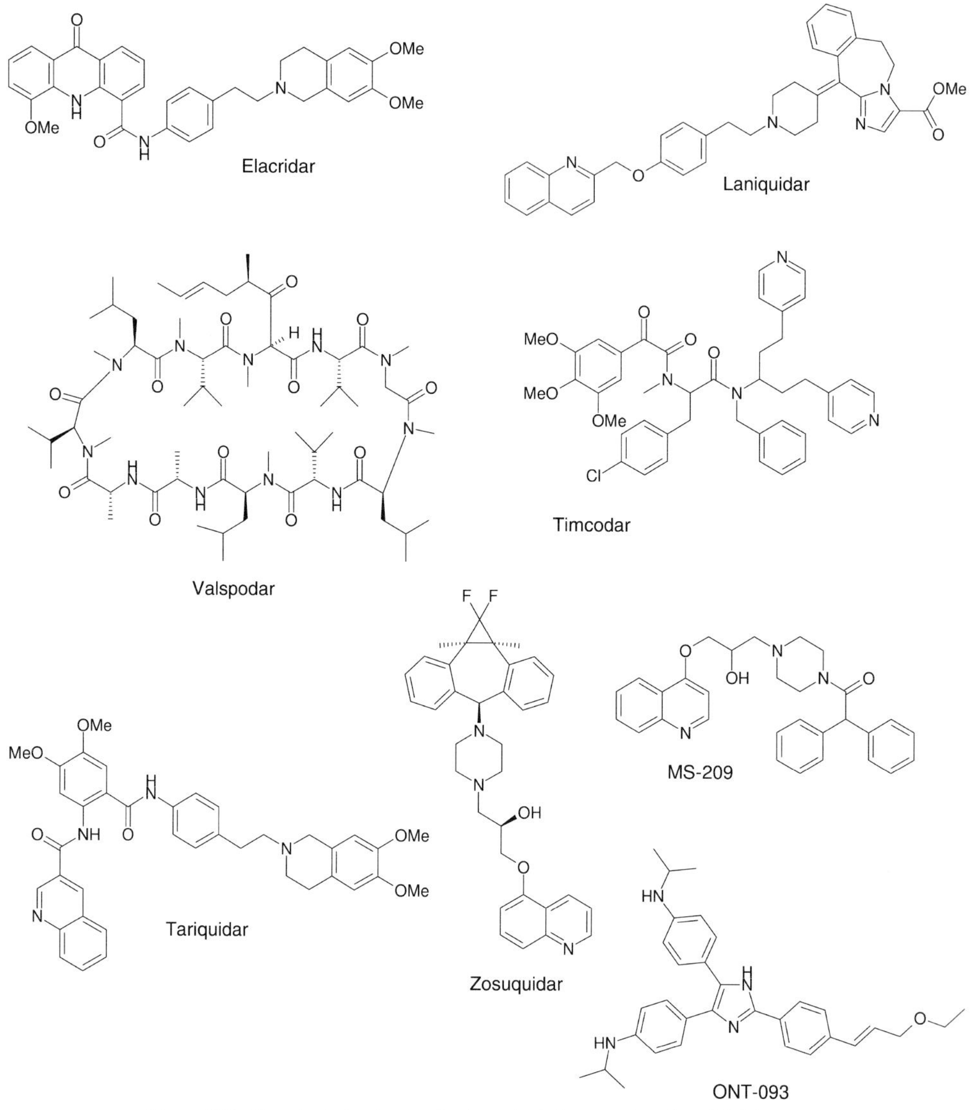

Figure 13.1 Chemical structures of selected third-generation P-gp inhibitors.

These drugs, however, have also not lived up to the expectations, and most studies fell short of demonstrating beneficial effects in a clinical setting. Although several clinical studies based on the concept of P-gp inhibition are ongoing, concerns about the general applicability of this concept remain. The drug development process from lead identification to drug approval on average takes about 12 years. This time has been exceeded in the case of P-gp by more than twofold given the fact that more than 25 years elapsed since the discovery of the first P-gp inhibitor. Although the reasons for this might be manifold, the therapeutic concept of coadministration of anticancer drugs and P-gp inhibitors might not meet the high expectations spurred by early *in vitro* experiments.

13.2.2
Other ABC Transporter as Drug Targets

In the last decade, the MDR-related proteins ABCC1 (MRP1) and ABCC2 (MRP2), the breast cancer resistance protein ABCG2 (BCRP), and the sister of P-gp ABCB11 (SPGP, BSEP) have also been considered drug targets and inhibitors have been developed [11]. More ABC proteins that have been shown to be capable of transporting drugs are ABCC3 (MRP3), ABCC4 (MRP4), ABCC5 (MRP5), and ABCA2 [12]. These transporters have attracted increasing interest as drug targets recently.

13.2.2.1 ABCG2 (Breast Cancer Resistance Protein, MXR)
ABCG2 is expressed mainly in the small intestine, placenta, liver, and at the blood–brain barrier (BBB) and transports mitoxantrone, methotrexate, camptothecins (topotecan, irinotecan), anthracyclines, etoposide, and flavonoids [13, 14]. The latter have also served as lead structures for the development of inhibitors. Zhang *et al.* selected a panel of 25 flavonoids covering 5 different structural subclasses in order to identify structural features important for ABCG2 inhibitory activity. Results showed that the presence of a 2,3-double bond in ring C, ring B attached at position 2, hydroxylation at position 5, lack of an OH group at position 3, and hydrophobic substituents at positions 6, 7, 8, or 4′ are prerequisites for strong interaction with ABCG2 [15]. Boumendjel *et al.* linked piperazines and phenylalkylamines to benzopyranones in order to obtain new inhibitors of ABCG2 [16]. The most active compounds shared several structural features, such as an alkylpiperazine moiety or methoxyphenylalkylamino groups with the highly active ABCG2 inhibitors imatinib (STI 571) and the natural product fumitremorgin C (FTC). The latter served as starting point for the synthesis of a series of 42 structurally analogous indolyl diketopiperazines [17]. Results obtained for the class of propafenones indicate that ABCG2 is more tolerant to structural modification than ABCB1. Selectivity is therefore determined mainly by the distinct QSAR pattern with respect to ABCB1 rather than by the specific interaction with ABCG2 [18].

13.2.2.2 ABCC1 and ABCC2 (Multidrug Resistance Proteins 1 and 2)
Multidrug resistance protein 1 (MRP1, ABCC1) is a high-affinity transporter of leukotriene C4. In addition, it confers resistance against vinca alkaloids, anthracyclines, epipodophyllotoxins, mitoxanthrone, and methotrexate, but not against taxanes and bisantrene [19]. In contrast to ABCB1, ABCC1 mainly functions as a (co)transporter of amphipathic organic anions. It transports hydrophobic drugs that are conjugated with or complexed to the anionic tripeptide glutathione (GSH), to glucuronic acid, or to sulfate [20]. As is the case for ABCB1 and ABCG2, a lot of structurally and functionally diverse inhibitors have been identified also for ABCC1. These have been discussed in a recent review [21] and comprise verapamil, flavonoids, raloxifene, isoxazoles, quinazolinones, quinolines, pyrrolopyrimidines,

and peptides. ABCC2 (cMOAT) has also been characterized as an organic anion transporter with a broad range of substrates such as methotrexate and drugs conjugated to glutathione [22].

Finally, it has to be noted that several ABC transporters may definitely be considered as versatile targets, especially in the field of multiple drug resistance in tumor therapy. However, so far the final proof of concept that this concept also works in patients is still missing. This is mainly due to the physiological function of these proteins that makes it difficult to block them at the systemic level. Thus, tissue-selective inhibitors and/or more advanced drug targeting strategies are needed to achieve further progress in this area.

13.3
P-Glycoprotein – An Antitarget?

With our increasing knowledge on the molecular basis of side effects and toxicity, it became evident that there are several distinct proteins that are responsible for severe side effects. They are frequently termed antitargets or off-pharmacologies. The paradigm protein in this rapidly developing field is the hERG potassium channel. Interaction with the hERG channel has been associated with severe and lethal cardiac arrhythmias, and prediction of hERG ligands is one of the priorities in drug safety profiling. Proteins are also considered antitargets when they are involved in drug–drug interactions, as is the case for the cytochrome P450 enzyme family. When "targeting" off-pharmacologies, the focus shifts from the design of inhibitors to the design of "nonligands." Thus, the major challenge is to establish models for the prediction of substrate properties with the ultimate goal to avoid interaction with these proteins. In case of P-glycoprotein and other ABC transporters, the main focus of interest in the scientific community seems to shift from target-related to antitarget-related concepts. This is mainly due to the increasing knowledge about the physiological role of these transporters and their multiple involvement in processes related to ADME.

13.3.1
Gastrointestinal Absorption

Besides its expression in the kidney, the liver, and at the blood–brain barrier, P-glycoprotein is constitutively expressed in the intestine. There it plays an important role in limiting the intestinal absorption of a wide variety of orally administered drugs. One well-known example is the quinidine–digoxin interaction, where the P-gp inhibitor quinidine increases the digoxin absorption rate by 30%, the peak plasma concentration by 81%, and the plasma AUC by 77% [23]. Since digoxin is not metabolized by cytochrome P450 enzymes and can be administered both orally and intravenously, it has become a well-established model substrate for determining P-gp transporter activity *in vivo*. The importance of drug transporters for uptake and disposition is now widely accepted, and Benet and coworkers recently suggested a Biopharmaceutics Classification System (BCS) that allows prediction of *in vivo* pharmacokinetic performance of drug candidates based on measurements of their permeability (determined as the extent of oral absorption) and solubility [24]. Later on, this classification system was modified in order to allow prediction of overall drug disposition, including routes of drug elimination and the effects of efflux and absorptive transporters on oral drug absorption [25]. In short, compounds with low water solubility acting as substrates of P-glycoprotein have a high likelihood of low bioavailability. For a more detailed description the reader is referred to Chapter 12.

13.3.2
Brain Uptake

The blood–brain barrier separates the brain and central nervous system (CNS) from the bloodstream. Therefore, in CNS drug development, it is of vital importance that the compounds are able to cross the BBB. Conversely, compounds designed for non-CNS targets should not cross the BBB to avoid unwanted side effects. In principle, drug uptake into the brain is influenced by five different main factors: passive diffusion, paracellular transport, carrier-mediated transport and receptor-mediated transcytosis for transporting compounds into the brain (active influx), and multidrug transport pumps for actively protecting the brain from unwanted chemicals (active efflux). In case of transport into the brain, numerous systems have been discovered, including transport proteins for amino acids, monocarboxylic acids, organic cations, hexoses, nucleotides, and peptides. With respect to active efflux, the important role of ABC pumps (ATP binding cassette) such as P-gp is increasingly recognized. *In vitro* studies demonstrated that the uptake of vincristine was reduced in primary cultured bovine capillary endothelial cells expressing P-gp at the luminal side and that this decreased accumulation was due to active efflux. Steady-state uptake was significantly increased in the presence of the P-gp blocking agent verapamil [26]. In addition, *mdr1a* double-knockout mice show hypersensitivity to a range of drugs known to be transported by P-gp [27]. Undoubtedly, P-gp is an important impediment to the entry of hydrophobic drugs into the brain. Recently, breast cancer resistance protein (ABCG2) has also been reported as playing a role in the brain uptake of a variety of compounds.

13.4
Predicting Substrate Properties for P-Glycoprotein

As already outlined above, P-glycoprotein is constitutively expressed in several organs, such as kidney, liver, intestine, and also at the blood–brain barrier. P-gp substrates therefore show poor oral absorption, enhanced renal and biliary excretion, and usually do not enter the brain [28]. This spurred the development of medium- and high-throughput systems addressing the P-gp substrate properties of compounds of interest. These systems mostly rely on transport studies through a monolayer of P-gp expressing Caco-2 [29] or MDCK cells [30]. In parallel, *in silico* methods have also been developed that span the whole range of classification algorithms including decision trees, discriminant analysis, support vector machines, and self-organizing maps.

13.4.1
Data Sets

Data sets used in P-gp substrate studies are rather small and sometimes also inconsistent. After analyzing six publications dealing with P-gp substrate/nonsubstrate classification, we identified 50 compounds (out of 326) as being classified differently in the literature. Especially on the molecular level, the classification into substrates and inhibitors is rather blurred. On the cellular level, compounds for which a net transport is observed are considered substrates and compounds blocking transport of model substrates are classified as inhibitors. On a biochemical level, substrates are frequently classified by evaluating their effect on P-gp-associated ATPase activity. Substrates are considered to stimulate ATPase activity in a biphasic manner, while inhibitors normally show a monophasic inhibition. However, some inhibitors have been shown to stimulate basal ATP activity and thus might be transport substrates as well [31]. Due to high lipophilicity, these compounds rapidly rediffuse into the membrane and thus might block the pump by keeping it engaged. Thus, a proper annotation of substrates is not always clear, and different results may be obtained depending on the experimental system used. Furthermore, much larger data sets are needed in order to carefully address these details and to expand the chemical space for development of *in silico* models. Recently, the group of Gottesman published a comprehensive study analyzing data from the NCI60 screen [32]. mRNA levels of all 48 human ABC transporters in 60 human tumor cell lines of the NCI60 anticancer drug screening panel were evaluated and correlated with cellular toxicity values of 1400 selected compounds. An inverse correlation between transporter mRNA levels and compound toxicity was considered to indicate that a compound is a substrate for the respective transporter. Potentially compromising is the measurement of mRNA levels, not protein expression, and the problem of setting the right threshold for the correlation coefficient for annotating a compound as substrate. Undoubtedly, this is by far the largest consistent data set available and studies from our group indicate that it might be successfully used as a basis for P-gp substrate prediction models (see below).

13.4.2
Classification Models

As large data sets for binding affinity of substrates are not available, almost all models rely on a binary substrate/nonsubstrate classification. Seelig analyzed a data set of 100 compounds with respect to substructures related to H-bonding and suggested a general recognition pattern for P-gp substrates [33]. In this classification, a P-gp substrate is characterized by two H-bond acceptor groups with a spatial distance of 2.5 and 4.6 Å, respectively. The latter might also contain a third H-bond acceptor group, whereby the outer two groups are separated by 4.6 Å. In an analogy to Lipinski's rule of five, Didziapetris et al. introduced the "rule of fours." Their analysis is based on a set of 220 compounds and the following filter rules: compounds with a number of N and O atoms ≥ 8, molecular weight > 400, and acid $pK_a < 4$ are likely to be P-gp substrates, whereas compounds with a number of N and O atoms ≤ 4, molecular weight < 400, and base $pK_a < 8$ are likely to be nonsubstrates [34]. Even less complex is the so-called Gombar–Polli rule: Compounds with molecular E-state values (MolES) 110 are predominantly substrates and those with MolES < 49 are nonsubstrates. However, only 30% of the compounds analyzed comply with these two thresholds, all others have values between 49 and 110.

Cabrera et al. pursued a topological substructural approach for the prediction of P-gp substrates. A linear discriminant model classified 163 compounds with an accuracy of 81% based on standard bond distance, polarizability, and the Gasteiger–Marsilli atomic charge [35]. Furthermore, the predictive potential of this TOPS-MODE approach was demonstrated for a set of 6-fluoroquinolones not included in the training set.

To overcome the problem of a priori descriptor selection, which is inherent to all QSAR attempts, Tropsha introduced the combinatorial QSAR approach [36]. In this setting, several sets of descriptors are used with several different methods in a combinatorial way as outlined in detail below. The authors used the experimental data set of Penzotti et al. and calculated molecular connectivity indices, atom pair descriptors, VolSurf descriptors, and MOE descriptors. These input matrices were then analyzed with k-nearest neighbor classification, decision tree, binary QSAR, and support vector machines, respectively [37]. The best model obtained used VolSurf descriptors and a support vector machine-based classifier, showing an overall accuracy of 94% for the training set of 94% and of 0.81 for the test set. VolSurf descriptors were also applied by Crivori et al. They developed both a model discriminating between substrates and nonsubstrates and a model that classified P-gp substrates with poor inhibitory activity and inhibitors showing no evidence of significant net transport [38]. Using GRIND descriptors, the partial least squares discriminant (PLSD) analysis allowed identification of key pharmacophoric features for substrates and inhibitors. The main descriptors for P-gp substrate properties were related to H-bonding properties. GRIND descriptors were also successfully applied by Cianchetta et al., who derived a hypothesis for P-gp substrate recognition [39]. Using a set of 129 compounds, the authors created a pharmacophore hypothesis that contains the following recognition elements: two hydrophobic groups at a distance of 16.5 Å, two H-bond acceptor groups at a distance of 11.5 Å apart, and a size of the molecule of 21.5 Å (between the two edges of the molecule).

On the basis of the Gottesman data set and a set of 259 compounds compiled from the literature, we explored the performance of several classification methods combined with different descriptor sets. These include simple ADME-type descriptors (log P, number of rotatable bonds, number of H-bond donors, and acceptors), VSA descriptors as described by Labute [40], and 2D autocorrelation vectors. The latter have already been successfully applied for the prediction of P-gp inhibitors [41].

When comparing binary QSAR and support vector machines, the latter gave more robust models with total accuracies in the range of 80%. In general, the prediction of nonsubstrates is better than those for substrates [42].

13.4.3
Pharmacophore Models

Pharmacophore models have been shown to be valuable tools for *in silico* screening of compound libraries. For P-gp substrates, several pharmacophore models have been derived and validated. Penzotti *et al.* used an ensemble of 100 two-, three-, and four-point pharmacophore models that discriminated between P-gp substrates and nonsubstrates [43]. The ensemble model correctly classified 50–60% of substrates and 80% of nonsubstrates. Also, in this case, prediction accuracy for nonsubstrates is generally higher than for substrates. Recently, the group of Ekins published a series of pharmacophore models for rapid identification of P-gp substrates and inhibitors [44]. They used a combination of their previously generated CATALYST models for substrates and inhibitors to establish one additional inhibitor model. All three models were used for *in silico* screening of an in-house database of 600 frequently prescribed drugs. A selected subset of predicted positives was subjected to pharmacological evaluation and supported the validity and applicability of the model.

13.4.4
Nonlinear Methods

As already outlined for P-gp inhibitors (Chapter 3), artificial neural networks are an excellent tool for classifying actives and inactives. Recently, Xue *et al.* reported the application of supported vector machines (SVMs) for the prediction of P-gp substrates [45]. Using a set of 201 compounds comprising 116 substrates and 85 nonsubstrates and a set of 159 molecular descriptors, the SVM yielded prediction accuracies of 81% for substrates and of 79% for nonsubstrates. Yang and coworkers developed a self-organizing map to separate P-gp substrates from inhibitors on the basis of a set of molecular connectivity indices and electrotopological state descriptors. The average accuracy of classification obtained was 82.3%. Comparison with feedforward backpropagation neural networks showed the superiority of the SOM method [46].

13.5
Conclusions

Although P-glycoprotein is known since more than 30 years and its prominent role in tumor multidrug resistance identifies this protein as a clear target, up to now no P-gp inhibitor has reached the market. Thus, there are increasing concerns about the druggability of P-glycoprotein and ABC transporter in general. In the past decade, the focus of interest thus shifted toward the role of ABC transporters for drug safety and drug/drug interactions. This makes these protein antitargets. Several pharmaceutical companies established high-throughput screening systems for measuring P-gp substrate properties of their compound libraries. In parallel, *in silico* methods have also been developed that on average show a classification accuracy of around 80%. However, data sets used are too small to ensure broad applicability of these models. Furthermore, till March 2009, all attempts had to rely on ligand-based approaches only. In March 2009 Aller *et al.* published the first structure of P-glycoprotein (mouse P-gp) in a reasonable resolution of 3.8 A [47] (see also the article by Ford *et al.* in this volume). This protein structure definitely will aid in the understanding of the molecular principles underlying the ligand-polyspecificity of these transporters and might encourage structure-based design approaches for prediction of P-gp substrates.

Acknowledgments

We gratefully acknowledge financial support from the Austrian Science Fund (grant # F3502) and from the Austrian Research Promotion Agency (grant # B1-812074).

References

1 Gottesman, M.M. and Ling, V. (2006) The molecular basis of multidrug resistance in cancer: the early years of P-glycoprotein research. *FEBS Letters*, **580**, 998–1009.

2 Szakacs, G., Paterson, J.K., Ludwig, J.A., Booth-Genthe, C., and Gottesman, M.M. (2006) Targeting multidrug resistance in cancer. *Nature Reviews. Drug Discovery*, **5**, 219–234.

3 Tsuruo, T., Iida, H., Tsukagoshi, S., and Sakurai, Y. (1981) Overcoming of vincristine resistance in P388 leukemia *in vivo* and *in vitro* through enhanced cytotoxicity on vincristine and vinblastine by verapamil. *Cancer Research*, **41**, 1967–1972.

4 Juliano, R.L. and Ling, V. (1976) A surface glycoprotein modulating drug permeability in Chinese hamster ovary cell mutants. *Biochimica et Biophysica Acta*, **455**, 152–162.

5 Dean, M., Hamon, Y., and Chimini, G. (2001) The human ATP-binding cassette (ABC) transporter superfamily. *Journal of Lipid Research*, **42**, 1007–1017.

6 Szakács, G., Váradi, A., Ozvegy-Laczka, C., and Sarkadi, B. (2008) The role of ABC transporters in drug absorption, distribution, metabolism, excretion and toxicity (ADME-Tox). *Drug Discovery Today*, **13**, 379–393.

7 Hopkins, A.L. and Groom, C.R. (2002) The druggable genome. *Nature Reviews. Drug Discovery*, **1**, 727–730.

8 Gottesman, M.M., Fojo, T., and Bates, S.E. (2002) Multidrug resistance in cancer: role of ATP-dependent transporters. *Nature Reviews. Cancer*, **2**, 48–58.

9 Raderer, M. and Scheithauer, W. (1993) Clinical trials of agents that reverse multidrug resistance. A literature review. *Cancer*, **72**, 3553–3563.

10 Kivistö, K.T., Niemi, M., and Fromm, M.F. (2004) Functional interaction of intestinal CYP3A4 and P-glycoprotein. *Fundamental & Clinical Pharmacology*, **18**, 621–626.

11 Chiba, P. and Ecker, G.F. (2004) Inhibitors of ABC-type drug efflux pumps – an overview on the actual patent situation.

Expert Opinion on Therapeutic Patents, **14**, 499–508.

12 Gottesman, M.M., Fojo, T., and Bates, S.E. (2002) Multidrug resistance in cancer: role of ATP-dependent transporters. *Nature Reviews. Cancer*, **2**, 48–58.

13 van Herwaarden, A.E. and Schinkel, A.H. (2006) The function of breast cancer resistance protein in epithelial barriers, stem cells and milk secretion of drugs and xenotoxins. *Trends in Pharmacological Sciences*, **27**, 10–16.

14 Mao, Q. and Unadkat, J.D. (2005) Role of breast cancer resistance protein (ABCG2) in drug transport. *The AAPS Journal*, **7**, E118–E133.

15 Zhang, S., Yang, X., Coburn, R.A., and Morris, M.E. (2005) Structure–activity relationships and quantitative structure–activity relationships for the flavonoid-mediated inhibition of breast cancer resistance protein. *Biochemical Pharmacology*, **70**, 627–639.

16 Boumendjel, A., Nicolle, E., Moraux, T., Gerby, B., Blanc, M., Ronot, X., and Boutonnat, J. (2005) Piperazinobenzopyranones and phenylalkylaminobenzopyranones: potent inhibitors of breast cancer resistance protein (ABCG2). *Journal of Medicinal Chemistry*, **48**, 7275–7281.

17 van Loevezijn, A., Allen, J.D., Schinkel, A.H., and Koomen, G.J. (2001) Inhibition of BCRP-mediated drug efflux by fumitremorgin-type indolyl diketopiperazines. *Bioorganic & Medicinal Chemistry Letters*, **11**, 29–32.

18 Cramer, J., Kopp, S., Bates, S.E., Chiba, P., and Ecker, G.F. (2007) Multispecificity of drug transporters: probing inhibitor selectivity for the human drug efflux transporters ABCB1 and ABCG2. *ChemMedChem*, **2**, 1783–1788.

19 Schinkel, A.H. and Jonker, J.W. (2003) Mammalian drug efflux transporters of the ATP binding cassette (ABC) family: an overview. *Advanced Drug Delivery Reviews*, **55**, 3–29.

20 Deeley, R.G. and Cole, S.P.C. (2006) Substrate recognition and transport by multidrug resistance protein 1 (ABCC1). *FEBS Letters*, **580**, 1103–1111.

21 Boumendjel, A., Baubichon-Cortay, H., Trompier, D., Perrotton, T., and Di Pietro, A. (2005) Anticancer multidrug resistance mediated by MRP1: recent advances in the discovery of reversal agents. *Medicinal Research Reviews*, **25**, 453–472.

22 Faber, K.N., Muller, M., and Jansen, P.L. (2003) Drug transport proteins in the liver. *Advanced Drug Delivery Reviews*, **55**, 107–124.

23 Fromm, M.F., Kim, R.B., Stein, C.M., Wilkinson, G.R., and Roden, D.M. (1999) Inhibition of P-glycoprotein-mediated drug transport: a unifying mechanism to explain the interaction between digoxin and quinidine. *Circulation*, **99**, 552–557.

24 Wu, C.Y. and Benet, L.Z. (2005) Predicting drug disposition via application of BCS: transport/absorption/elimination interplay and development of a biopharmaceutics drug disposition classification system. *Pharmaceutical Research*, **22**, 11–23.

25 Custodio, J.M., Wu, C.Y., and Benet, L.Z. (2008) Predicting drug disposition, absorption/elimination/transporter interplay and the role of food on drug absorption. *Advanced Drug Delivery Reviews*, **60**, 717–733.

26 Tsuji, A., Terasaki, T., Takabatake, Y., Tenda, Y., Tamai, I., Yamashima, T., Moritani, S., Tsuruo, T., and Yamashita, J. (1992) P-glycoprotein as the drug efflux pump in primary cultured bovine brain capillary endothelial cells. *Life Sciences*, **51**, 1427.

27 Borst, P. and Schinkel, A.H. (1998) P-glycoprotein, a guardian of the brain, in *Introduction to the Blood–Brain Barrier* (ed. W.M. Partridge), Cambridge University Press, Cambridge, pp. 198–206.

28 Chan, L.M., Lowes, S., and Hirst, B.H. (2004) The ABCs of drug transport in intestine and liver: efflux proteins limiting

drug absorption and bioavailability. *European Journal of Pharmaceutical Sciences*, **21**, 25–51.

29 Delie, F. and Rubas, W.A. (1997) A human colonic cell line sharing similarities with enterocytes as a model to examine oral absorption: advantages and limitations of the Caco-2 model. *Critical Reviews in Therapeutic Drug Carrier Systems*, **14**, 221–286.

30 Irvine, J.D., Takahashi, L., Lockhart, K., Cheong, J., Tolan, J.W., Selick, H.E., and Grove, J.R. (1999) MDCK (Madin-Darby canine kidney) cells: a tool for membrane permeability screening. *Journal of Pharmaceutical Sciences*, **88**, 28–33.

31 Schmid, D., Ecker, G., Richter, E., Hitzler, M., and Chiba, P. (1999) Structure–activity relationship studies of propafenone analogs based on P-glycoprotein ATPase activity measurements. *Biochemical Pharmacology*, **58**, 1447–1456.

32 Szakács, G., Annereau, J.P., Lababidi, S., Shankavaram, U., Arciello, A., Bussey, K.J., Reinhold, W., Guo, Y., Kruh, G.D., Reimers, M., Weinstein, J.N., and Gottesman, M.M. (2004) Predicting drug sensitivity and resistance: profiling ABC transporter genes in cancer cells. *Cancer Cell*, **6**, 129–137.

33 Seelig, A. (1998) A general pattern for substrate recognition by P-glycoprotein. *European Journal of Biochemistry*, **251**, 252–261.

34 Didziapetris, R., Japertas, P., Avdeef, A., and Petrauskas, A. (2003) Classification analysis of P-glycoprotein substrate specificity. *Journal of Drug Targeting*, **11**, 391–406.

35 Cabrera, M.A., Gonzalez, I., Fernandez, C., Navarro, C., and Bermejo, M. (2006) A topological substructural approach for the prediction of P-glycoprotein substrates. *Journal of Pharmaceutical Sciences*, **95**, 589–606.

36 Kovatcheva, A., Golbraikh, A., Oloff, S., Xiao, Y.D., Zheng, W., Wolschann, P., Buchbauer, G., and Tropsha, A. (2004) A combinatorial QSAR of Ambergris fragrance compounds. *Journal of Chemical Information and Computer Sciences*, **44**, 582–595.

37 De Cerqueira Lima, P., Golbraikh, A., Oloff, S., Xiao, Y., and Tropsha, A. (2006) Combinatorial QSAR modeling of P-glycoprotein substrates. *Journal of Chemical Information and Modeling*, **46**, 1245–1254.

38 Crivori, P., Reinach, B., Pezzetta, D., and Poggesi, I. (2006) Computational models for identifying potential P-glycoprotein substrates and inhibitors. *Molecular Pharmaceutics*, **3**, 33–44.

39 Cianchetta, G., Singleton, R.W., Zhang, M., Wildgoose, M., Giesing, D., Fravolini, A., Cruciani, G., and Vaz, R.J. (2005) A pharmacophore hypothesis for P-glycoprotein substrate recognition using GRIND-based 3D-QSAR. *Journal of Medicinal Chemistry*, **48**, 2927–2935.

40 Labute, P. (2000) A widely applicable set of descriptors. *Journal of Molecular Graphics & Modelling*, **18**, 464–477.

41 Kaiser, D., Terfloth, L., Kopp, S., de Laet, R., Chiba, P., Ecker, G.F., and Gasteiger, J. (2007) Self-organising maps for identification of new inhibitors of P-glycoprotein. *Journal of Medicinal Chemistry*, **50**, 1698–1702.

42 Zdrazil, B., Prakasvudhisarn, C., and Ecker, G.F. (2007) NCI60 screening data – a versatile tool for *in silico* models predicting substrate properties for ABC-transporter. Poster presentation at the Autumn 2007 National Meeting and Exhibition of the American Chemical Society, Boston, MA.

43 Penzotti, J.E., Lamb, M.L., Evensen, E., and Grootenhuis, P.D.J. (2002) A computational ensemble pharmacophore model for identifying substrates of P-glycoprotein. *Journal of Medicinal Chemistry*, **45**, 1737–1740.

44 Chang, C., Bahadduri, P.M., Polli, J.E., Swaan, P.W., and Ekins, S. (2006) Rapid identification of P-glycoprotein substrates and inhibitors. *Drug Metabolism and Disposition*, **58**, 1431–1450.

45 Xue, Y., Yap, C.W., Sun, L.Z., Cao, Z.W., Wang, J.F., and Chen, Y.Z. (2004) Prediction of P-glycoprotein substrates by a support vector machine approach. *Journal of Chemical Information and Computer Sciences*, **44**, 1497–1505.

46 Wang, Y.H., Li, Y., Yang, S.L., and Yang, L. (2005) Classification of substrates and inhibitors of P-glycoprotein using unsupervised machine learning approach. *Journal of Chemical Information and Modeling*, **45**, 750–757.

47 Aller, S.G., Yu, J., Ward, A., Weng, Y., Chittaboina, S., Zhuo, R., Harrell, P.M., Trinh, Y.T., Zhang, Q., Urbatsch, I.L., Chang, G. (2009) Structure of P-glycoprotein reveals a molecular basis for poly-specific drug binding. *Science*, **323**, 1718–1722.

Part Five:
A Systems View of Drug Transport

14
A Systems Biology View of Drug Transporters
Sean Ekins and Dana L. Abramovitz

14.1
Introduction

The human genome contains approximately 900 transporter genes, if not more, that have not yet been identified, which encode proteins responsible for transporting a diverse array of molecules across the membrane as described elsewhere in this volume and in other reviews [1]. Transporters can be classified into distinct superfamilies such as the solute carrier class (SLC) containing over 30 families and 200 members http://www.bioparadigms.org/slc/menu.asp [2] while the ATP binding cassette (ABC) family contains 7 families and over 48 members, including the widely studied P-glycoprotein (P-gp) and MRP subfamilies [3, 4]. Such transporters play an important role in clinical pharmacology as many drugs specifically target them either intentionally or unintentionally due to their overlapping molecular pharmacophores for the biological target/s and transporter/s. Drugs share transport pathways with nutrients, and transporters play a role in oral absorption, drug bioavailability, drug resistance, excretion, and ultimately pharmacokinetics and pharmacodynamics [5]. The importance of drug transport in hepatocytes is widely known both for basolateral uptake of hydrophobic compounds and efflux of these compounds or metabolites into the bile [6]. Similarly, intestinal membrane transport of drugs and nutrients is critical for their absorption [1, 7]. Human tumor cells also express various transporters, and these can alter drug sensitivity and resistance [8, 9].

Polymorphisms in drug transporters may also be a key factor in drug interactions and lack of effectiveness. Phenotypic variability is caused mainly by single-nucleotide polymorphisms (SNPs) resulting in lower protein activity, incorrect folding, or rapid degradation via proteosomes [10]. P-glycoprotein, which is expressed in many tissues, has numerous SNPs, one of which (C3435T) affects the expression level in the duodenum and therefore can affect absorption of molecules that are substrates for this transporter [11]. Nine SNPs were found in the human proton-dependent dipeptide transporter (hPEPT1) that can affect the absorption of molecules in the intestine, while only one displays a reduced transport capacity [12]. Recently, four

SNPs were identified in the Japanese population for the organic cation transporter 1 (OCT1), and when functionally characterized *in vitro*, the uptake of cations was reduced significantly for some of these mutations, indicating that this would likely contribute to interindividual variations in metabolism of drugs that are transported via OCT1 [13]. The sodium-dependent carnitine cotransporter OCTN2 can also possess mutations that result in primary carnitine deficiency, thus impacting fatty acid oxidation that is characterized by many clinical manifestations [14].

There are numerous examples of the importance of drug transporters to the clinical development of drugs. For instance, the major metabolite of the insulin sensitizer troglitazone (withdrawn due to hepatotoxicity) is a sulfated species that is suspected of being responsible for the observed toxicity. Sulfated troglitazone has a higher affinity for the organic anion transporting polypeptide OATP1B1 (also known as OATP-C, LST-1, OATP2, and *SLC21A6*) and possibly a lower affinity for OATP8 expressed on the hepatocyte basolateral membrane [15]. This metabolite could accumulate in hepatocytes and inhibit the bile salt export pump and ABC family member, MRP2. The OATPs are key membrane-bound transporters expressed in many organs including intestine, liver, lung, choroid plexus, and blood–brain barrier [16]. This family of transporters is capable of mediating the sodium-independent transport of a diverse array of molecules such as steroid conjugates, organic anions, and xenobiotics by coupling uptake with efflux of bicarbonate [17], glutathione, or its conjugates [18]. OATP1B1 represents the most studied human OATP to date [19]. The inhibition of hepatic uptake of other compounds by this transporter may be important for reported drug–drug interactions [20], for cerivastatin with cyclosporin A [21], and for cerivastatin with gemfibrozil [22]. Single-nucleotide polymorphisms have been shown for OATP1B1 [23] identified in European-Americans, African-Americans [23], and Japanese [24], and these dramatically impact the transport of ligands such as pravastatin [25, 26], estrone-3-sulfate [23, 24], rifampin [27], and estradiol 17β-D-glucuronide [23]. These polymorphisms may result in the accumulation of drug metabolites and in turn elicit idiosyncratic toxicity.

Another clinically relevant example is the MRP2-mediated transport of HIV-protease inhibitors saquinavir, ritonavir, and indinavir *in vitro*, which other drugs, such as probenecid and sulfinpyrazone, are able to enhance. The transport by MRP2 suggests that these compounds will have decreased bioavailability due to increased clearance while other drugs could aggravate this situation by further enhancing transport [28]. Analogous to this is the rifampicin-mediated induction of MRP2 and P-gp in healthy subjects that was found to significantly decrease the AUC of coadministered drugs and was also correlated with intestinal expression of these transporters. MRP2 is also inducible by cisplatin, 2-AAF, and phenobarbital [29], indicating that multiple mechanisms may be involved in its regulation.

As there are a large number of transporters, understanding their role and regulation in individual tissues upon treatment with xenobiotics definitely presents a challenge. A molecule may represent a substrate and/or inhibitor for one or many transporters or other proteins while at the same time acting as a regulator for these or other proteins. The development of systems biology methods that capture the majority of the biological processes assists us in piecing together small-scale biology studies and at the same time enables a platform for the evaluation of higher throughput biology for visualizing and interpreting the network of interactions that may occur *in vitro* or *in vivo*. These tools may serve as a means to better understand the role or biological function and provide a systems view of transporters [30] alongside all other proteins. This has value in drug discovery for optimizing uptake, as well as understanding bioavailability and toxicology data. The following sections will discuss further this systems view of transporters.

14.2
Regulation of Transporters

Nuclear hormone receptors play a key role in the regulation of transporters. For example, the pregnane X receptor (PXR) is a transcriptional regulator of the enzyme human MDR1 (P-gp), MRPs, and OATP [31], as well as many other genes involved in the transport, metabolism [32–35], and biosynthesis of bile acids [36]. Additional receptors such as the constitutive androstane receptor (CAR), farnesoid X receptor (FXR), liver X receptor (LXR), and other nuclear receptors and transcriptional factors take part in a complex network of interactions. Elucidation of the regulatory networks, which control the expression of efflux transporters and uptake transporters such as OATP [37], is of considerable interest to researchers in this area. For example, the bile salt export pump in human hepatocytes was shown to be regulated by FXR and not by LXR as it was inducible by 22(R)-hydroxycholesterol and appeared to have different ligand binding determinants in the receptor from chenodeoxycholic acid [38]. FXR has also been shown to regulate the organic solute transporters α and β in human adrenal gland, kidney, and intestine [39]. Although the exact physiological function of these transporters has not yet been defined, there may be a role in bile acid resorption. These above examples represent a very small sampling of the hundreds to thousands of smaller scale biology studies that have looked at the role of genes regulating individual transporters.

In species commonly used for *in vivo* toxicology studies such as the rat, orthologues of the transporters such as oatp2 are expressed and are inducible with PXR ligands such as PCN [40]. This is a useful knowledge because the advent of high-content and high-throughput genomics, proteomics/microarray technologies enables one to dose a rat with a xenobiotic and assess thousands of genes/proteins simultaneously in a particular tissue. These then allow one to look at the effects of a compound on the regulation of enzymes and transporters that could in turn influence clearance, excretion, or uptake. For instance, animals dosed with known nephrotoxins demonstrate upregulation of the Na–K–Cl transporter [41]. Some transporters may be differentially targeted by drugs in different tissues (e.g., the CNS), but these may also be expressed elsewhere, representing a site for off-target toxic effects. Well-known examples are P-gp, expressed at the blood–brain barrier and intestine, impacting the efficacy and bioavailability of drugs, and the serotonin transporter, expressed in the lungs and brain, where substrates such as fenfluramine can result in primary pulmonary hypertension as they accumulate in lung cells [42].

14.3
Network Analysis of Transporters and Ligands

We are now gaining a deeper insight into protein and ligand interactions using computationally generated networks that can be derived from high-throughput data [43] with applications for understanding adverse events [44], the identification and validation of drug targets [45], and complex metabolic interactions [46]. Using commercially available pathway databases and network building tools enables network comparisons and visualization to be undertaken. Current biological knowledge can be captured on static maps of interactions or by building custom interaction networks using many different algorithms. For example, pathway tools and various resources have also been applied to modeling the networks of nuclear hormone receptors and their connections with other genes and small molecules using a manually compiled database such as MetaDrug [47] or MetaCore™ (described further below) [48]. The transcriptional regulation of many proteins involved in drug metabolism, transport, and elimination, such as drug transporters, CYPs, and phase II enzymes, are regulated by these nuclear hormone receptors that may also impact cell growth, proliferation, and oxidative stress [49, 50]. Networks created by small molecules interacting with many nuclear receptors produce a very complex picture of interactions [47]. A molecule thought to bind with only one nuclear receptor may also interact with many others such that there is an overlap between signaling pathways. Other research groups have used a natural language processing method, CCNet, to show the genes regulated by the nuclear hormone receptor FXR [51] including transporters. Automated methods can facilitate a more complete understanding of the transcriptional factors [49, 52, 53] although they rely heavily on the quality of the content of the underlying database. Mestres *et al.* generated a database [54] of over 2000 molecules and used Shannon entropy descriptors and Euclidean distance with Cytoscape to map the direct interactions between ligands and nuclear receptors over a predefined cut-off [55]. A ligand-based approach to nuclear receptor profiling was also described [55]. Another potentially valuable approach is to use such network tools to visualize the results of quantitative structure–activity models for predicting molecules binding to enzymes, transporters, receptors, and ion channels [56, 57]. Alternatively, network approaches can be used to interpret high-throughput data providing a unique approach to predicting potential off-target effects in the area of systems pharmacology. For example, a data set using percentage inhibition data for two compounds screened against many different biological assays [58] could be visualized on a network [59].

There is a growing body of literature on absorption, distribution, metabolism, excretion, and toxicology (ADME/TOX) that is captured in some of the network building software and databases [56]. Limited academic efforts have captured data for drug transporters in the human membrane transporter database [60], TP-search transporter database (http://www.ilab.rise.waseda.ac.jp/tp-search/), and drug interaction database (http://www.druginteractioninfo.org/DatabaseInfo.aspx), while several other academic efforts, such as for the nuclear hormone receptors [61], the ADME-AP database [62], and PharmaGKB [34] are also available with data that could be linked to future network building software.

(a)
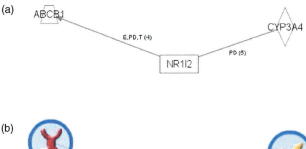

(b)

Figure 14.1 Examples of networks showing how two different tools A (IPA) and B (MetaCore) can be used to describe the direct interactions between PXR (NR1I2), CYP3A4, and P-gp (ABCB1, MDR1). The edges are highlighted to show the type of interactions annotated in each database (E, expression; PD, protein–DNA interaction; T or TR, transcription).

Important pathway/network building tools that could potentially be applied to transporter data include Ingenuity Pathways Analysis (IPA) (http://www.ingenuity.com/), PathArt (http://www.jubilantbiosys.com/pd.htm), Pathway Assist (http://www.ariadnegenomics.com/products/pathway.html) [63], and several other databases deposited at the Pathway Resource List (http://cbio.mskcc.org/prl). The majority of these products have unique underlying proprietary pathway databases, and to date there have been little comparison of the different methods. These tools enable the connection of protein nodes via edges representing their interactions. These tools have been used in a limited manner with respect to drug transporters and these efforts will be briefly discussed here.

Figure 14.1 provides a straightforward example of networks showing how two different tools A (IPA) and B (MetaCore) can be used to describe the direct interactions between PXR (NR1I2), CYP3A4, and P-gp (ABCB1, MDR1). The edges are highlighted to show the type of interactions annotated in each database.

14.4
Transporter Network Examples

14.4.1
MetaCore

A manually annotated database called MetaCore has been developed and consists of high-confidence human protein–protein interactions from the original full-text papers (as well as disease-relevant information from OMIM, EntrezGene). MetaCore represents an integrated software platform for network generation, statistical analysis, and mapping of experimental high-throughput data on human networks [47, 64]. The interaction database can be used to visualize disease-related genes and compounds as nodes on networks connected by edges that represent the published interactions. The networks that can be built with MetaCore are condition specific (as defined by the high-throughput data sets uploaded) and are nonrandom, as determined by statistical methods for analysis of networks. The system architecture, network-generating algorithms, and the process of mapping experimental data on the networks have all been previously described [65, 66]. The MetaCore database has been used to show that the ABCA1 transporter, which mediates the first step of cholesterol transport, appears on three manually curated pathway maps [59]. One can also use the maps to interface and access the underlying information about the transporter, including the genes/splice variants with known SNPs. For example, information related to mutations in ABCA1 gene responsible for causing Tangier disease, which results in severe HDL deficiency, cholesterol accumulation in macrophages, and attendant atherosclerosis, is available within the software along with many other examples. This transporter also represents a drug target for upregulation, modulation of cholesterol metabolism, and prevention of cardiovascular disease [67] that has been shown to be inhibited *in vitro* by the sulfonylurea glybenclamide [68, 69]. A custom network was previously constructed around ABCA1 using the network construction tool [59]. The ABCA1 network showed the transporter as linked directly to 25 other objects such as APOE1 and LXR. Neighboring genes have their own SNPs that could be key determinants of interactions between drug transport and endogenous ligand transport in health or disease. This type of visualization may be helpful for identifying putative pathways around a particular transporter or compound of interest.

The regulation of some transporters may be affected during extrahepatic cholestasis, bile duct ligation, bile salt-induced cholestatic hepatitis [19, 70, 71], and primary sclerosing cholangitis [72]. MetaCore has been used to visualize OATP1B1, its ligands, regulatory factors, and signaling molecules as some of the literature for OATP1B1 human substrates has been annotated in this database [59]. A network around OATP1B1 was generated with the autoexpand algorithm in the software. The gene details were viewed upon querying the database and links were provided to other public databases. For example, OATP1B1 is shown in MetaCore to be regulated by the liver-enriched transcription factor hepatocyte nuclear factor 1α (HNF1α) that binds to the promoter region of this transporter [73]. It would be useful if this database could also capture other related knowledge for the following. Experimentally determined site-directed mutagenesis of this binding site results in inactivation, suggesting the critical nature of the interaction with HNF1α. Bile acids such as CDCA have been shown to transcriptionally repress HNF1α *in vitro* inhibiting the transactivating effect of HNF4α on HNF1α [74]. After screening many rat and human uptake transporters *in vitro*, OATP1B1 was also shown to modulate the PXR response by controlling rifampin retention in the cell and therefore affecting the induction of CYP3A4 and other gene products such as P-gp [27]. The regulation of multiple genes by FXR is shown in Figure 14.2 including transporters such as the organic solute transporters β, as described above [39].

14.4 Transporter Network Examples | 371

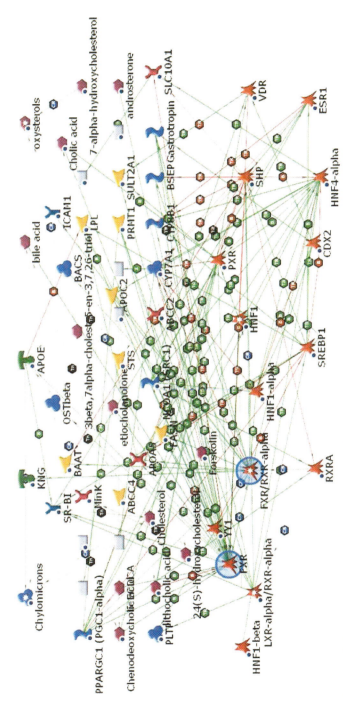

Figure 14.2 An autoexpand network generated around FXR illustrating the links to transporters and enzymes regulated by this nuclear receptor. Small molecules are shown as purple hexagons. The network was built using MetaCore version 4.

14.4.2
Ingenuity Pathways Analysis

Ingenuity Pathways Analysis is an application that uses content structured in the Ingenuity Pathways Knowledge Base (IPKB) to facilitate the analysis and interpretation of experimental data in a biological context of molecular relationships, compounds, functions, diseases, and pathways. The proprietary IPKB database consists of a multibranched ontology that contains millions of experimental findings manually extracted from the full text of peer-reviewed literature. These findings, which are supported by a figure, graph, or table in the original article, describe molecular interactions centered on mammalian biology, with a focus on human, mouse, and rat systems. Content in the IPKB is also supplemented with curated relationships parsed from Medline Abstracts and high-confidence relationships from OMIM, GeneOntology, and ToxNet databases. The molecular relationships contained in the IPKB are used to form a large interaction network that defines the direct and indirect relationships between the molecules (genes, gene products, and compounds). The biological events that define these relationships include functional relationships such as expression, activation, inhibition, and phosphorylation, as well as physical interactions such as protein–protein, protein–DNA, protein–chemical, and chemical–chemical interactions. As part of a data analysis, the IPA algorithm generates networks based on relationships in this larger interaction network that are focused around a scientist's particular genes of interest. In addition to the networks, detailed information about the biology around the genes, gene products, and chemicals is also provided. This includes the association of these molecules with biological functions, diseases, and well-understood canonical pathways. Information about mutations and allelic variations, cell and tissue expression, drug and clinical candidates, and modulation and regulation that is associated with the genes, gene products, or compounds of interest can also be found through the use of IPA. Thus, IPA can be used to build a network around a particular transporter, as there are over 1000 mammalian transporters represented in IPA, to identify genes and chemicals that it has relationships with and to view their associated biological functions. It can then be used to understand the effects of mutations and allele changes, to determine the impact of drugs, and to visualize the relationships with well-understood biological pathways to use the content from the peer-reviewed literature to better understand the roles of transporters in interesting areas of biology. We are aware of an example using expression of the transporter aquaporin-4 (AQP-4) that has been studied using RT-PCR and found to be upregulated in human hippocampi from patients with temporal lobe epilepsy. Microarray expression analysis followed by IPA analysis of the genes expressed above a 1.5-fold cut-off was used to produce a network containing six upregulated and five downregulated genes including dystrophin that may be important for AQP-4 anchoring and distribution in astrocytes [75].

One example of using IPA to visualize the FXR-mediated regulation of ABC genes (Figure 14.3) shows the overlap in the regulation of several transporters by multiple nuclear receptors and transcriptional factors. Networks can be used to highlight the direct interaction network around a transporter such as P-gp (ABCB1,

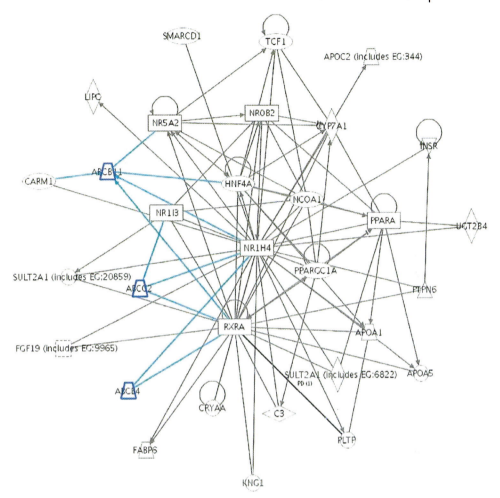

Figure 14.3 Direct interaction network generated around FXR (NR1H4) illustrating the links to transporters ABCB4, ABCC2, and ABCB1. The network was built using IPA version 4.

Figure 14.4a), showing the regulatory factors controlling gene expression, and several drugs that have been/are in clinical trials as P-gp inhibitors for preventing drug resistance. These may help researchers to understand up- or downstream regulators and what are already considered therapeutic targets. This network can be expanded with the indirect interactions (Figure 14.4b) becoming more complex and incorporating many more small molecules that have a role in the biology around this central node.

The identification of a new ABC transporter termed the breast cancer resistance protein (BCRP, MXR, ABCG2) has revealed that increased expression of this transporter results in resistance to anticancer therapeutics. Conversely, one group has recently shown that in 12 out of 19 human tissue cancers (including colorectal

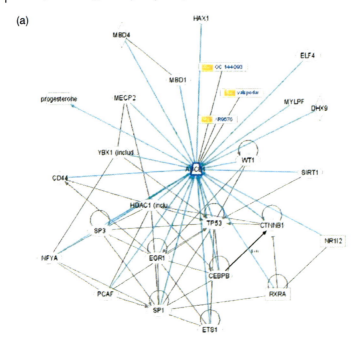

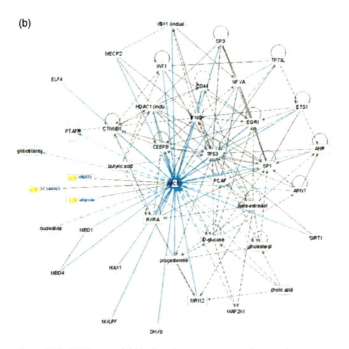

Figure 14.4 (a) Direct and (b) indirect interaction network around P-gp (ABCB1), showing the regulatory factors controlling gene expression, small molecules (metabolites), and several drugs in clinical trials as P-gp inhibitors for preventing drug resistance. Direct interactions: solid lines; indirect interactions: dashed lines. These networks were built using IPA version 5 (beta version).

and cervical cancers) decreased expression of mRNA and BCRP is observed compared to normal tissues [76]. It was suggested that this transporter is also expressed in the intestine indicating that this transporter could be important in limiting bioavailability and that Caco-2 cells express this at the level seen in the jejunum [77]. The functional role is likely protecting the organism from dietary genotoxins. Figure 14.5 shows a direct interaction network around P-gp (ABCB1) and BCRP (ABCC2), showing the regulatory factors controlling gene expression. Although neither transporter appears to share the same regulator, the estrogen receptor α (ESR1) and nuclear transcription factor Y α (NFYA) appear as nodes that connect both individual transporter networks. Estradiol has been reported to down-

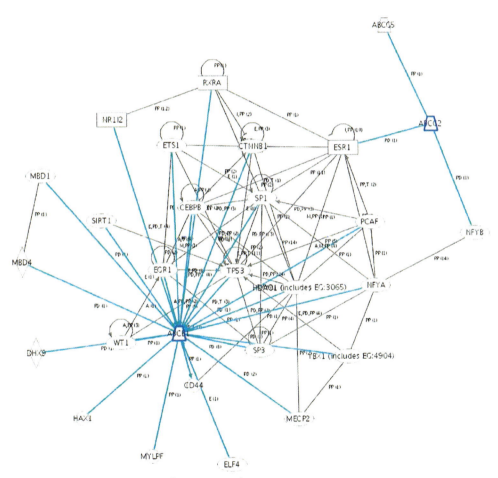

Figure 14.5 Direct interaction network (upstream and downstream) around P-gp (ABCB1) and BCRP (ABCC2), showing the regulatory factors controlling gene expression and protein–protein interactions. The network was built using IPA version 4.

regulate BCRP and P-gp in human breast cancer cells *in vitro* by post-transcriptional processes, representing a potential target for sensitizing cells to therapy [78]. The impact of modulating a particular receptor could be simulated on such a network by knocking it out by manually removing it.

14.5
Transporter Gene Expression Data

Focused use of microarrays to generate large gene expression data sets relevant to transporters has been rare. To date these microarrays have been limited in the number of transporters present on them [8]. They have been used in an attempt to correlate pharmacokinetic properties with gene expression, for example, for valacyclovir [79], as well as to understand the transporter expression profile in different tissues or cell lines upon dietary component or xenobiotic treatment [1]. The lack of transporters on many commercially available microarrays has prompted some groups to produce arrays with a heavier emphasis on transporters. These arrays have, for example, then been used to demonstrate the upregulation of ABC transporters and downregulation of GST-Pi in cell lines resistant to colchicines or 9-nitro-camptothecin [80]. The data from this study was previously used with MetaCore to show key genes that were significantly up- or downregulated and the similarities between them were assessed [59]. Only a few significantly changed genes were in common (IL-8, Fos, GST-Pi, calpactin, and ubiquitin hydrolase) across these compounds, so it is likely that a much larger common gene network is important for drug resistance via regulation of transporters and other proteins. Confirming this would require a much larger number of cell lines and drug treatments to produce a definitive drug resistance signature involving transporters, enzymes, and transcriptional regulators. Quantitative RT-PCR has also been performed to profile mRNA expression of 48 human ABC transporters in the NCI cancer cell lines that was then correlated with the results of over 1400 anticancer drugs tested in the same cells. Several correlations were found between transporter expression and response to such cytotoxic drugs that were further verified *in vitro* such as several new substrates for P-gp and MRP2 [8]. Inversely correlated molecules were suggested as potentially helpful for future efforts in defining the pharmacophore for P-gp substrates from diverse structural scaffolds. A recent low-density microarray for 38 ABC transporters (DualChip humanABC) was developed to investigate the expression of these genes in multidrug-resistant tumor cells. Three resistant cell lines were evaluated and compared with their parental lines showing in all cases the overexpression of numerous transporters, which was further confirmed with RT-PCR [9]. This group also generated a COMPARE analysis of the NCI standard agents database with the 31 ABC transporters with mRNA expression data. An $r > 0.4$ was used as a cut-off to identify substrates for each of these transporters. However, known substrates for some of the transporters were not picked up [9]. Several groups have taken a similar approach to studying the gene–drug relationship [81], which may be useful for identifying which molecules may be substrates for different transporters [1] and this in turn may be useful for data mining efforts [82]. It is likely that some compounds could be effluxed by multiple ABC transporters that have been less well characterized than P-gp and MRP1, and many of these may result in drug resistance. Hence, understanding the network of interactions a molecule may have with transporters, ion channels, regulatory proteins, and enzymes will be important for predicting drug resistance.

Several groups have looked at gene expression using commercially available gene chips with intestinal cells or Caco-2 cells to show site-specific localization of transporters [1, 77]. Changes in the profile of nine transporters in the intestine have been studied along the anterior–posterior and crypt–villus axes. It was noted that expression profiles changed along the intestine and most significantly between ileum and colon while the transporters maintained their crypt- or villus-specific localization in different intestine segments. The authors also indicated that transporter expression profiles in human intestine were similar to mouse. SLC5A8 was found to be expressed preferentially in the small intestine and may represent a new target for influencing drug uptake in the intestine [7].

14.6
Summary and Future Perspectives

From the bulk of publications and preceding sections in this and other chapters of this book, it is apparent that it may be useful to build an integrated computational approach that will (i) predict which transporters will have affinity for novel xenobiotic compounds, (ii) identify whether the role of one or more human transporters (and their polymorphisms) may be a risk factor for patients and drug developers, (iii) propose potential toxic effects, and (iv) enable the combination of predicted and experimental data (gene expression, *in vitro* biology, etc.) for drug transporters (Figure 14.6).

Such a tool could use any of the databases described above that are available and incorporate expanded content on human transporters, their regulation, and involvement in signaling and metabolic pathways in normal and disease states. A starting point would be to build a prototype database to capture drug–drug and drug–endobiotic interactions for key transporters with known xenobiotic ligands and their connections to transcriptional regulators. This could ultimately be expanded to cover all transporters in humans involved in xenobiotic transport for which data are available. The *in vitro* data in the literature will provide a basis for predictive quantitative (or binary) structure–activity relationship (QSAR) model building to enable assignment of transporter binding from molecular structure alone. Predictive computational models could be combined with the visualization of high-throughput data (gene expression) in the context of the whole system of interacting genes, as has been illustrated recently for enzymes [83]. Both xenobiotic and endobiotic transport data on humans could be linked in such a database and used for improving predictions of toxicity. To date, few commercially available predictive algorithm approaches have taken transporters into account. A systems biology approach that

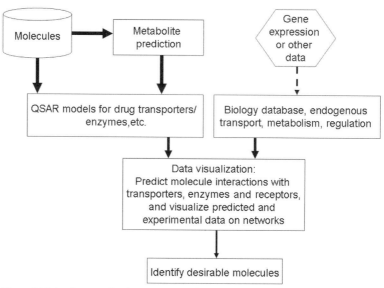

Figure 14.6 A schematic for the integration of predictive QSAR and systems biology methods to enable the prediction of the interactions of xenobiotics with transporters, enzymes, and the cell signaling and regulatory processes.

incorporates transporter information with regulation, signaling, and metabolic pathways may be the ultimate goal and this will require tools for visualizing networks and mapping of experimental data.

Expanding this data further to include transporters in other species, such as rat, would be useful. This is important as the rat is a widely used experimental model and the understanding of the function of its transporters is perhaps still in its infancy. Many similarities and differences in the substrate specificity of rat and human transporters are perhaps still to be discovered. As pharmaceutical companies are generating significant quantities of transporter data in animal models, this will be important to not only provide a collection of literature data but also enable them to add their own content to supplement this and to obtain a further understanding of drug–transporter interactions across species. The development of a method that would incorporate many of the features described could enable companies not only to visualize literature data and make predictions but also to store and visualize their proprietary empirical data on transporters. The extension of the proposed database from human to other species will be important for making extrapolations for metabolic, transporter, and toxicity findings in the future.

As microarray gene expression data are increasingly generated, the role of transporter regulation in toxicity of certain xenobiotics will become more apparent from either *in vivo* or *in vitro* studies. The visualization of gene networks involving transporters, their ligands, and regulatory factors will also be important for future toxicity prediction methods. By combining approaches to the prediction of interactions with various proteins and the visualization of xenobiotic–transporter (or other protein) interactions, we should be able to obtain a higher level of accuracy for the prediction of toxicity than is possible with individual computational methods.

We have summarized a small number of efforts in using systems biology approaches for understanding the regulation of transporters and drug–transporter relationships derived from correlations with transporter expression in different cells. Our analysis so far says that the available biology databases with integrated tools for network building and data analysis contain limited transporter content. These tools could learn from what is available for transporters in other databases such as those described earlier and a metabolism and drug–transport interaction database from the University of Washington (http://www.druginteractioninfo.org/) that includes Medline publications and describes *in vivo* and *in vitro* interactions, including some transporter abstracts. A dedicated effort to build a database of human membrane transporters and associated ligands and regulatory information would allow more scope in the areas queried. However, to our knowledge no current transporter database contains molecule structures, enables predictions to be made for a xenobiotic structure, or places the transporter in the context of surrounding genes while providing a means for visualizing the information. Ultimately, the development of a combined predictive approach will enable the identification of possible idiosyncratic drug reactions that are often not detected until the drug has been released on the market. Our understanding of some of the factors resulting in drug-induced toxicity has expanded to focus on the molecular mechanisms involved. In particular, there has been a considerable interest in hepatotoxicity mediated by drug–drug interactions (where transporters might be implicated) seen as a major cause of the failure of clinical candidates. The incorporation of knowledge for transporters alongside enzymes and other proteins will be important in this effort due to their key role in transporting endogenous and exogenous molecules.

Acknowledgments

SE gratefully acknowledges Mr. Clifford W. Mason and Dr. Peter W. Swaan (University of Maryland) for providing Figure 14.2.

References

1 Anderle, P., Huang, Y., and Sadee, W. (2004) Intestinal membrane transport of drugs and nutrients: genomics of membrane transporters using expression microarrays. *European Journal of Pharmaceutical Sciences*, **21**, 17–24.

2 Hediger, M.A., Romero, M.F., Peng, J.B., Rolfs, A., Takanaga, H., and Bruford, E.A. (2004) The ABCs of solute carriers: physiological, pathological and therapeutic implications of human membrane transport proteins: introduction. *Pflugers Archives*, **447** (5), 465–468.

3 Zhang, E.Y., Phelps, M.A., Cheng, C., Ekins, S., and Swaan, P.W. (2002) Modeling of active transport systems.

Advanced Drug Delivery Reviews, **54**, 329–354.

4 Zhang, E.Y., Knipp, G.T., Ekins, S., and Swaan, P.W. (2002) Structural biology and function of solute transporters: implications for identifying and designing substrates. *Drug Metabolism Reviews*, **34**, 709–750.

5 Chang, C. and Swaan, P.W. (2005) Computational approaches to modeling drug transporters. *European Journal of Pharmaceutical Sciences*, **27**, 411–424.

6 Pauli-Magnus, C. and Meier, P.J. (2003) Pharmacogenetics of hepatocellular transporters. *Pharmacogenetics*, **13**, 189–198.

7 Anderle, P., Sengstag, T., Mutch, D.M., Rumbo, M., Praz, V., Mansourian, R., Delorenzi, M., Williamson, G., and Roberts, M.-A. (2005) Changes in the transcriptional profile of transporters in the intestine along the anterior–posterior and crypt–villus axes. *BMC Genomics*, **6** (69), 1–17.

8 Szakacs, G., Annereau, J.P., Lababidi, S., Shankavaram, U., Arciello, A., Bussey, K.J., Reinhold, W., Guo, Y., Kruh, G.D., Reimers, M., Weinstein, J.N., and Gottesman, M.M. (2004) Predicting drug sensitivity and resistance: profiling ABC transporter genes in cancer cells. *Cancer Cell*, **6** (2), 129–137.

9 Gillet, J.P., Efferth, T., Steinbach, D., Hamels, J., de Logueville, F., and Bertholet, V. (2004) Microarray-based detection of multidrug resistance in human tumor cell by expression profiling of ATP-binding cassette transporter genes. *Cancer Research*, **64**, 8987–8993.

10 Weinshilboum, R. and Wang, L. (2004) Pharmacogenetics: inherited variation in amino acid sequence and altered protein quantity. *Clinical Pharmacology and Therapeutics*, **75** (4), 253–258.

11 Sakaeda, T., Nakamura, T., and Okumura, K. (2002) MDR1 genotype-related pharmacokinetics and pharmacodynamics. *Biological & Pharmaceutical Bulletin*, **25** (11), 1391–1400.

12 Gaucher, S.P., Taylor, S.W., Fahy, E., Zhang, B., Warnock, D.E., Ghosh, S.S., and Gibson, B.W. (2004) Expanded coverage of the human heart mitochondrial proteome using multidimensional liquid chromatography coupled with tandem mass spectrometry. *Journal of Proteome Research*, **3** (3), 495–505.

13 Sakata, T., Anzai, N., Shin, H.J., Noshiro, R., Hirata, T., Yokoyama, H., Kanai, Y., and Endou, H. (2004) Novel single nucleotide polymorphisms of organic cation transporter 1 (SLC22A1) affecting transport functions. *Biochemical and Biophysical Research Communications*, **313** (3), 789–793.

14 Lahjouji, K., Mitchell, G.A., and Qureshi, I.A. (2001) Carnitine transport by organic cation transporters and systemic carnitine deficiency. *Molecular Genetics and Metabolism*, **73** (4), 287–297.

15 Nozawa, T., Sugiura, S., Nakajima, M., Goto, A., Yokoi, T., Nezu, J., Tsuji, A., and Tamai, I. (2004) Involvement of organic anion transporting polypeptides in the transport of troglitazone sulfate: implications for understanding troglitazone hepatotoxicity. *Drug Metabolism and Disposition*, **32** (3), 291–294.

16 Tamai, I., Nezu, J., Uchino, H., Sai, Y., Oku, A., Shimane, M., and Tsuji, A. (2000) Molecular identification and characterization of novel members of the human organic anion transporter (OATP) family. *Biochemical and Biophysical Research Communications*, **273** (1), 251–260.

17 Satlin, L.M., Amin, V., and Wolkoff, A.W. (1997) Organic anion transporting polypeptide mediates organic anion/HCO_3^- exchange. *The Journal of Biological Chemistry*, **272** (42), 26340–26345.

18 Hagenbuch, B. and Meier, P.J. (2004) Organic anion transporting polypeptides of the OATP/SLC21 family: phylogenetic classification as OATP/SLCO superfamily, new nomenclature and molecular/functional properties. *Pflugers Archives*, **447** (5), 653–665.

19 Meier, P.J. and Stieger, B. (2000) Molecular mechanisms in bile formation. *News in Physiological Sciences*, **15**, 89–93.

20 Kim, R.B. (2003) Organic anion-transporting polypeptide (OATP) transporter family and drug disposition. *European Journal of Clinical Investigation*, **33** (Suppl 2), 1–5.

21 Shitara, Y., Itoh, T., Sato, H., Li, A.P., and Sugiyama, Y. (2003) Inhibition of transporter-mediated hepatic uptake as a mechanism for drug–drug interaction between cerivastatin and cyclosporin A. *The Journal of Pharmacology and Experimental Therapeutics*, **304** (2), 610–616.

22 Shitara, Y., Hirano, M., Sato, H., and Sugiyama, Y. (2004) Gemfibrozil and its glucuronide inhibit the OATP2(OATP1B1: SLC21A6)-mediated hepatic uptake and CYP2C8-mediated metabolism of cerivastatin – analysis of the mechanism of the clinically relevant drug–drug interaction between cerivastatin and gemfibrozil. *The Journal of Pharmacology and Experimental Therapeutics*, **311**, 228–236.

23 Tirona, R.G., Leake, B.F., Merino, G., and Kim, R.B. (2001) Polymorphisms in OATP-C: identification of multiple allelic variants associated with altered transport activity among European- and African-Americans. *The Journal of Biological Chemistry*, **276** (38), 35669–35675.

24 Nozawa, T., Nakajima, M., Tamai, I., Noda, K., Nezu, J., Sai, Y., Tsuji, A., and Yokoi, T. (2002) Genetic polymorphisms of human organic anion transporters OATP-C (SLC21A6) and OATP-B (SLC21A9): allele frequencies in the Japanese population and functional analysis. *The Journal of Pharmacology and Experimental Therapeutics*, **302** (2), 804–813.

25 Nishizato, Y., Ieiri, I., Suzuki, H., Kimura, M., Kawabata, K., Hirota, T., Takane, H., Irie, S., Kusuhara, H., Urasaki, Y., Urae, A., Higuchi, S., Otsubo, K., and Sugiyama, Y. (2003) Polymorphisms of OATP-C (SLC21A6) and OAT3 (SLC22A8) genes: consequences for pravastatin pharmacokinetics. *Clinical Pharmacology and Therapeutics*, **73** (6), 554–565.

26 Mwinyi, J., Johne, A., Bauer, S., Roots, I., and Gerloff, T. (2004) Evidence for inverse effects of OATP-C (SLC21A6) 5 and 1b haplotypes on pravastatin kinetics. *Clinical Pharmacology and Therapeutics*, **75** (5), 415–421.

27 Tirona, R.G., Leake, B.F., Wolkoff, A.W., and Kim, R.B. (2003) Human organic anion transporting polypeptide-C (SLC21A6) is a major determinant of the rifampin-mediated pregnane X receptor activation. *The Journal of Pharmacology and Experimental Therapeutics*, **304**, 223–228.

28 Huisman, M.T., Smit, J.W., Crommentuyn, K.M., Zelcer, N., Wiltshire, H.R., Beijnen, J.H., and Schinkel, A.H. (2002) Multidrug resistance protein 2 (MRP2) transports HIV protease inhibitors, and transport can be enhanced by other drugs. *AIDS*, **16** (17), 2295–2301.

29 Schrenk, D., Baus, P.R., Ermel, N., Klein, C., Vorderstemann, B., and Kauffmann, H.M. (2001) Up-regulation of transporters of the MRP family by drugs and toxins. *Toxicology Letters*, **120** (1–3), 51–57.

30 Yan, Q. (2005) Pharmacogenomics and systems biology of membrane transporters. *Molecular Biotechnology*, **29**, 75–87.

31 Luo, G., Lin, J., Fiske, W.D., Dai, R., Yang, T.J., Kim, S., Sinz, M., LeCluyse, E., Solon, E., Brennan, J.M., Benedek, I.H., Jolley, S., Gilbert, D., Wang, L., Lee, F.W., and Gan, L.S. (2003) Concurrent induction and mechanism-based inactivation of CYP3A4 by an L-valinamide derivative. *Drug Metabolism and Disposition*, **31** (9), 1170–1175.

32 Bertilsson, G., Heidrich, J., Svensson, K., Asman, M., Jendeberg, L., Sydow-Backman, M., Ohlsson, R., Postlind, H., Blomquist, P., and Berkenstam, A. (1998) Identification of a human nuclear receptor defines a new signaling pathway for CYP3A induction. *Proceedings of the*

National Academy of Sciences of the United States of America, **95** (21), 12208–12213.

33 Blumberg, B., Sabbagh, W., Jr., Juguilon, H., Bolado, J., Jr., van Meter, C.M., Ong, E.S., and Evans, R.M. (1998) SXR, a novel steroid and xenobiotic-sensing nuclear receptor. *Genes and Development*, **12** (20), 3195–3205.

34 Kliewer, S.A., Moore, J.T., Wade, L., Staudinger, J.L., Watson, M.A., Jones, S.A., McKee, D.D., Oliver, B.B., Willson, T.M., Zetterstrom, R.H., Perlmann, T., and Lehmann, J.M. (1998) An orphan nuclear receptor activated by pregnanes defines a novel steroid signalling pathway. *Cell*, **92**, 73–82.

35 Synold, T.W., Dussault, I., and Forman, B.M. (2001) The orphan nuclear receptor SXR coordinately regulates drug metabolism and efflux. *Nature Medicine*, **7**, 584–590.

36 Staudinger, J., Liu, Y., Madan, A., Habeebu, S., and Klaassen, C.D. (2001) Coordinate regulation of xenobiotic and bile acid homeostasis by pregnane X receptor. *Drug Metabolism and Disposition*, **29** (11), 1467–1472.

37 Yasukawa, T., Ogura, Y., Kimura, H., Sakurai, E., and Tabata, Y. (2006) Drug delivery from ocular implants. *Expert Opinion on Drug Delivery*, **3** (2), 261–273.

38 Deng, R., Yang, D., Yang, J., and Yan, B. (2006) Oxysterol 22(R)-hydroxycholesterol induces the expression of the bile salt export pump through nuclear receptor farnesoid X receptor but not liver X receptor. *The Journal of Pharmacology and Experimental Therapeutics*, **317**, 317–325.

39 Lee, H., Zhang, Y., Lee, F.Y., Nelson, S.F., Gonzalez, F.G., and Edwards, P.A. (2006) FXR regulates organic solute transporters alpha and beta in the adrenal gland, kidney, and intestine. *Journal of Lipid Research*, **47**, 201–214.

40 Guo, G.L., Choudhuri, S., and Klaassen, C.D. (2002) Induction profile of rat organic anion transporting polypeptide 2 (oatp2) by prototypical drug-metabolizing enzyme inducers that activate gene expression through ligand-activated transcription factor pathways. *The Journal of Pharmacology and Experimental Therapeutics*, **300** (1), 206–212.

41 Fleck, C., Sutter, L., Appenroth, D., Koch, B., Meinhold, T., Pitack, M., and Gasser, R. (2003) Use of gene chip technology for the characterisation of the regulation of renal transport processes and of nephrotoxicity in rats. *Experimental and Toxicologic Pathology*, **54** (5–6), 401–410.

42 Rothman, R.B., Baumann, M.H., Savage, J.E., Rauser, L., McBride, A., Hufeisen, S.J., and Roth, B.L. (2000) Evidence for possible involvement of 5-HT(2B) receptors in the cardiac valvulopathy associated with fenfluramine and other serotonergic medications. *Circulation*, **102** (23), 2836–2841.

43 Barabasi, A.-L. and Oltvai, Z.N. (2004) Network biology: understanding the cell's functional organization. *Nature Reviews. Genetics*, **5**, 101–113.

44 Hood, L. and Perlmutter, R.M. (2004) The impact of systems approaches on biological problems in drug discovery. *Nature Biotechnology*, **22** (10), 1215–1217.

45 Butcher, E.C., Berg, E.L., and Kunkel, E.J. (2004) Systems biology in drug discovery. *Nature Biotechnology*, **22** (10), 1253–1259.

46 Nicholson, J.K., Holmes, E., Lindon, J.C., and Wilson, I.D. (2004) The challenges of modeling mammalian biocomplexity. *Nature Biotechnology*, **22** (10), 1268–1274.

47 Ekins, S., Kirillov, E., Rakhmatulin, E., and Nikolskaya, T. (2005) A novel method for visualizing nuclear hormone receptor networks relevant to drug metabolism. *Drug Metabolism and Disposition*, **33**, 474–481.

48 Ekins, S., Bugrim, A., Nikolsky, Y., and Nikolskaya, T. (2005) Systems biology: applications in drug discovery, in *Drug Discovery Handbook* (ed. S. Gad), John Wiley & Sons, Inc., New York.

49 Ulrich, R.G. (2003) The toxicogenomics of nuclear receptor agonists. *Current Opinion in Chemical Biology*, **7** (4), 505–510.

50 Hartley, D.P., Dai, X., He, Y.D., Carlini, E.J., Wang, B., Huskey, S.E., Ulrich, R.G., Rushmore, T.H., Evers, R., and Evans, D.C. (2004) Activators of the rat pregnane X receptor differentially modulate hepatic and intestinal gene expression. *Molecular Pharmacology*, **65** (5), 1159–1171.

51 Apic, G., Ignjatovic, T., Boyer, S., and Russell, R.B. (2005) Illuminating drug discovery with biological pathways. *FEBS Letters*, **579** (8), 1872–1877.

52 Ekins, S., Mirny, L., and Schuetz, E.G. (2002) A ligand-based approach to understanding selectivity of nuclear hormone receptors PXR, CAR, FXR, LXRa and LXRb. *Pharmaceutical Research*, **19**, 1788–1800.

53 Plant, N. (2004) Interaction networks: coordinating responses to xenobiotic exposure. *Toxicology*, **202** (1–2), 21–32.

54 Cases, M., Garcia-Serna, R., Hettne, K., Weeber, M., van der Lei, J., Boyer, S., and Mestres, J. (2005) Chemical and biological profiling of an annotated compound library directed to the nuclear receptor family. *Current Topics in Medicinal Chemistry*, **5** (8), 763–772.

55 Mestres, J., Couce-Martin, L., Gregori-Puigjane, E., Cases, M., and Boyer, S. (2006) Ligand-based approach to *in silico* pharmacology: nuclear receptor profiling. *Journal of Chemical Information and Modeling*, **46**, 2725–2736.

56 Ekins, S., Nikolsky, Y., and Nikolskaya, T. (2005) Techniques: application of systems biology to absorption, distribution, metabolism, excretion, and toxicity. *Trends in Pharmacological Sciences*, **26**, 202–209.

57 Ekins, S., Andreyev, S., Ryabov, A., Kirilov, E., Rakhmatulin, E.A., Bugrim, A., and Nikolskaya, T. (2005) Computational prediction of human drug metabolism. *Expert Opinion on Drug Metabolism & Toxicology*, **1** (2), 303–324.

58 Fliri, A.F., Loging, W.T., Thadeio, P.F., and Volkmann, R.A. (2005) Biological spectra analysis: linking biological activity profiles to molecular structure. *Proceedings of the National Academy of Sciences of the United States of America*, **102** (2), 261–266.

59 Ekins, S. (2006) Systems-ADME/Tox: resources and network approaches. *Journal of Pharmacological and Toxicological Methods*, **53** (1), 38–66.

60 Yang, C. and Burt, H.M. (2006) Drug-eluting stents: factors governing local pharmacokinetics. *Advanced Drug Delivery Reviews*, **58** (3), 402–411.

61 Nakata, K., Yukawa, M., Komiyama, N., Nakano, T., and Kaminuma, T. (2002) A nuclear receptor database that maps pathways to diseases. *Genome Informatics*, **13**, 515–516.

62 Sun, L.Z., Ji, Z.L., Chen, X., Wang, J.F., and Chen, Y.Z. (2002) ADME-AP: a database of ADME associated proteins. *Bioinformatics*, **18**, 1699–1700.

63 Nikitin, A., Egorov, S., Daraselia, N., and Mazo, I. (2003) Pathway studio – the analysis and navigation of molecular networks. *Bioinformatics*, **19**, 2155–2157.

64 Ekins, S., Bugrim, A., Nikolsky, Y., and Nikolskaya, T. (2005) Systems biology: applications in drug discovery, in *Drug Discovery Handbook* (ed. S.C. Gad), John Wiley & Sons, Inc., New York, pp. 123–183.

65 Ekins, S., Bugrim, A., Brovold, L., Kirillov, E., Nikolsky, Y., Rakhmatulin, E.A., Sorokina, S., Ryabov, A., Serebryiskaya, T., Melnikov, A., Metz, J., and Nikolskaya, T. (2006) Algorithms for network analysis in systems-ADME/Tox using the MetaCore and MetaDrug platforms. *Xenobiotica*, **36** (10–11), 877–901.

66 Ekins, S., Nikolsky, Y., Bugrim, A., Kirillov, E., and Nikolskaya, T. (2006) Pathway mapping tools for analysis of high content data, in *High Content Screening: A Powerful Approach to Systems Cell Biology and Drug Discovery* (eds K. Giuliano, D.L. Taylor, and J. Haskin), Humana Press, Totowa, NJ.

67 Oram, J.F. and Lawn, R.M. (2001) ABCA1: the gatekeeper for eliminating excess tissue cholesterol. *Journal of Lipid Research*, **42** (8), 1173–1179.

68 Wang, N., Silver, D.L., Thiele, C., and Tall, A.R. (2001) ATP-binding cassette

69 Heitz-Mayfield, L.J. and Lang, N.P. (2004) Antimicrobial treatment of peri-implant diseases. *International Journal of Oral & Maxillofacial Implants*, **19** (Suppl), 128–139.

70 Rost, D., Herrmann, T., Sauer, P., Schmidts, H.L., Stieger, B., Meier, P.J., Stremmel, W., and Stiehl, A. (2003) Regulation of rat organic anion transporters in bile salt-induced cholestatic hepatitis: effect of ursodeoxycholate. *Hepatology*, **38** (1), 187–195.

71 Dumont, M., Jacquemin, E., D'Hont, C., Descout, C., Cresteil, D., Haouzi, D., Desrochers, M., Stieger, B., Hadchouel, M., and Erlinger, S. (1997) Expression of the liver Na^+-independent organic anion transporting polypeptide (oatp-1) in rats with bile duct ligation. *Journal of Hepatology*, **27** (6), 1051–1056.

72 Oswald, M., Kullak-Ublick, G.A., Paumgartner, G., and Beuers, U. (2001) Expression of hepatic transporters OATP-C and MRP2 in primary sclerosing cholangitis. *Liver*, **21** (4), 247–253.

73 Jung, D., Hagenbuch, B., Gresh, L., Pontoglio, M., Meier, P.J., and Kullak-Ublick, G.A. (2001) Characterization of the human OATP-C (SLC21A6) gene promoter and regulation of liver-specific OATP genes by hepatocyte nuclear factor 1 alpha. *The Journal of Biological Chemistry*, **276** (40), 37206–37214.

74 Jung, D. and Kullak-Ublick, G.A. (2003) Hepatocyte nuclear factor 1 alpha: a key mediator of the effect of bile acids on gene expression. *Hepatology*, **37** (3), 622–631.

75 Lee, T.S., Eid, T., Mane, S., Kim, J.H., Spencer, D.D., Ottersen, O.P., and de Lanerolle, N.C. (2004) Aquaporin-4 is increased in the sclerotic hippocampus in human temporal lobe epilepsy. *Acta Neuropathologica*, **108** (6), 493–502.

76 Gupta, N., Martin, P.M., Miyauchi, S., Ananth, S., Herdman, A.V., Martindale, R.G., Podolsky, R., and Ganapathy, V. (2006) Down-regulation of BCRP/ABCG2 in colorectal and cervical cancer. *Biochemical and Biophysical Research Communications*, **343** (2), 571–577.

77 Taipalensuu, J., Tornblom, H., Lindberg, G., Einarsson, C., Sjoqvist, F., Melhus, H., Garberg, P., Lundgren, B., and Artursson, P. (2001) Correlation of gene expression of ten drug efflux proteins of the ATP-binding cassette transporter family in normal human jejunum and in human intestinal epithelial Caco-2 cell monolayers. *The Journal of Pharmacology and Experimental Therapeutics*, **299**, 164–170.

78 Mutoh, K., Tsukahara, S., Mitsuhashi, J., Katayama, K., and Sugimoto, Y. (2006) Estrogen-mediated post transcriptional down-regulation of P-glycoprotein in MDR1-transduced human breast cancer cells. *Cancer Science*, **97** (11), 1198–1204.

79 Landowski, C.P., Sun, D., Foster, D.R., Menon, S.S., Barnett, J.L., Welage, L.S., Ramachandran, C., and Amidon, G.L. (2003) Gene expression in the human intestine and correlation with oral valacyclovir pharmacokinetic parameters. *The Journal of Pharmacology and Experimental Therapeutics*, **306** (2), 778–786.

80 Annereau, J.P., Szakacs, G., Tucker, C.J., Arciello, A., Cardarelli, C., Collins, J., Grissom, S., Zeeberg, B.R., Reinhold, W., Weinstein, J.N., Pommier, Y., Paules, R.S., and Gottesman, M.M. (2004) Analysis of ATP-binding cassette transporter expression in drug-selected cell lines by a microarray dedicated to multidrug resistance. *Molecular Pharmacology*, **66** (6), 1397–1405.

81 Huang, Y., Anderle, P., Bussey, K.J., Barbacioru, C., Shankavaram, U., Dai, Z., Reinhold, W.C., Papp, A., Weinstein, J.N., and Sadee, W. (2004) Membrane transporters and channels: role of the transportome in cancer chemosensitivity

and chemoresistance. *Cancer Research*, **64** (12), 4294–4301.

82 Ekins, S., Shimada, J., and Chang, C. (2006) Application of data mining approaches to drug delivery. *Advanced Drug Delivery Reviews*, **58** (12–13), 1409–1430.

83 Ekins, S., Andreyev, S., Ryabov, A., Kirillov, E., Rakhmatulin, E.A., Sorokina, S., Bugrim, A., and Nikolskaya, T. (2006) A combined approach to drug metabolism and toxicity assessment. *Drug Metabolism and Disposition*, **34**, 495–503.

15
Drug Transporters in Health and Disease
Barbara Bennani-Baïti and Christian R. Noe

15.1
Introduction

Regulation of drug transporters can occur at multiple levels. Expression of drug transporters is directed both by ubiquitously expressed and tissue-specific transcription factors and is further modulated by corepressors and coactivators that fine-tune the expression of a distinct drug transporter in a given context. Furthermore, epigenetic and posttranscriptional mechanisms play a role in drug transporter modulation. In the context of drug transporters and disease, several relatively unexplored regulatory pathways may also be involved. For instance, the targeting of efflux pumps to the cell membrane, a step that is as crucial to the function of transporters as their expression levels, is also tightly regulated. Once the drug transporter is adequately expressed and localized at the proper compartment, its concentration can be further modulated by pathways that involve regulated protein degradation.

Recent advances in medical research have mostly aimed at overcoming multidrug resistance (MDR) by decreasing the activity of drug transporters. This is, however, hampered by the fact that drug transporters are either ubiquitously expressed or least expressed in many different organs. Thus, blocking drug transporters is generally associated with a multitude of side effects.

It is essential, therefore, to explore the regulatory mechanisms of drug transporters and to understand the impact of disease and of medical regimen on the expression and proper localization of these transporters. These investigations would help uncover efficacious therapeutic strategies for the treatment of patients suffering from multidrug resistance. The well-studied ABC transporters serve as prototypes in the exploration of the complexity of these regulatory pathways. For instance, ABCB1 is constitutively expressed and is also highly inducible by environmental factors. *De novo* activation of the ABCB1 gene can be found in neoplasms derived from tissues not normally expressing the gene or alternatively in relapsed tumors, but not in the parental primary tumors [1]. ABCB1 expression can be further induced by heat shock, arsenite, partial hepatectomy, extracellular matrix components, growth

---

Transporters as Drug Carriers: Structure, Function, Substrates. Edited by Gerhard Ecker and Peter Chiba
Copyright © 2009 WILEY-VCH Verlag GmbH & Co. KGaA, Weinheim
ISBN: 978-3-527-31661-8

factors, sodium butyrate, protein kinase C agonists, and even its substrates and inhibitors [2]. Expression of ABCB1 can also change during tissue culture as exemplified by rat brain endothelial cells in culture [3]. Many aspects of ABC transporter regulation are still poorly understood and remain to be elucidated. Regulation of other transporters is described in more detail within the context of health and diseases in the relevant subchapters.

15.2
General Mechanisms of Drug Transporter Expression

Drug transporters may show high levels of basal constitutive transcription, or may be activated by external stimuli. The elements and mechanisms involved in constitutive expression (i.e., operative under normal growth conditions) under physiological conditions are described below.

To date, all human drug-related transporters examined lack a functional TATA box in their promoter sequences. Instead, their promoters feature a consensus CCAAT box and two GC box-like sequences as first documented in ABCB1 [4]. Figure 15.1 gives a schematic overview of the ABCB1 promoter elements that support its basal expression.

An initiator sequence (Inr) is the nucleation site of Pol II preinitiation complex and in the case of ABCB1, spans from nucleotides -6 to $+11$. Although Inr elements have not yet been functionally described in other drug transporters, consensus or near-consensus Inr sequences were identified in the promoters of MRP2 (GTACTTT) and BCRP (CCACTGC) genes [5]. The overlapping inverted CCAAT box (-82 to -73) and the first GC-rich element (-56 to -43) are binding sites for NF-Y and Sp families of transcription factors. The latter are ubiquitously expressed transcription factors that support basal expression of many genes, including that of MDR1. In addition, NF-Y has been shown to mediate the regulation of MDR1 by epigenetic modulation of histone acetylation states [5]. YB-1, a gene regulatory protein that normally interacts with single-stranded DNA and RNA, has been suggested to also mediate transcription via the ABCB1 inverted CCAAT box. However, YB-1 binds only single-stranded (and not double-stranded) ABCB1 inverted CCAAT box-containing DNA, and mutations in the CCAAT box do not affect this binding. It is more likely that transcription of the ABCB1 inverted CCAAT box is mediated via NF-Y. Nonetheless, YB-1 might contribute to ABCB1 expression, probably at an early step of transcription elongation, as nuclear localization of YB-1 was found to be associated with an increase in ABCB1 transcripts. Whether YB-1

Figure 15.1 Regulatory elements and transcription factors involved in basal transcription located within the promoter region of ABCB1. See text for details.

participates in the development of drug resistance due to environmental changes remains to be elucidated.

The transcription factor that binds to the second GC-rich region (-110 to -103) has not yet been identified. Immediately downstream of this GC-rich element is an inverted MED1 element (multiple start site downstream 1, iMED), an element involved in constitutive expression of ABCB1 in neuroblastoma and leukemia cell lines. It is not clear whether iMED also contributes to ABCB1 induction in multidrug-resistant cells.

Importantly, GC-rich elements also control constitutive expression of a number of other drug transporters. The MRP1 promoter contains a GC-rich domain (-91 to $+103$) that has been shown to interact with SP-1 and is essential for its basal transcription. MRP3 features multiple GC-rich SP-1 binding sites (-91 to -21). BCRP, another TATA-less promoter, harbors several putative SP-1 binding sites, about 300 bp upstream of the transcription start site, which confer basal expression. The MRP2 promoter, which lacks GC-rich sequences, possesses a putative CCAAT box that interacts *in vitro* with YB-1 rather than NF-Y. In summary, constitutive expression of ABC drug transporters is conferred mainly by SP-1 binding to GC-rich sequences and by NF-Y binding to an inverted CCAAT box [6, 7].

Physiological regulation of drug transporters involves the regulation of both their expression and their localization (e.g., apical versus basolateral membrane expression, cell surface expression versus cytoplasmic localization). Colocalization of the two transporters may also be necessary for transporter function. In some instances, control of transporter function occurs via other pumps and ion channels that create the driving force necessary for the transport. Thus, peptide uptake via PEPT1 requires an inwardly directed H^+ gradient, established by the Na^+/H^+ exchanger (NHE). PEPT1 needs to be in proximity to NHE, and there must be a mechanism that targets transporters in such a way that they form functional units within their membrane subcompartments. Adaptor proteins have recently been suggested to mediate this function. They contain the so-called PDZ-domains (PSD-95/ Discs-large/ZO-1) that bind to PDZ binding motifs located at C-termini of drug transporters. As expected, they seem to be essential for the proper localization of drug transporters. There are at least two different adaptor proteins designated PDZK-1 and PDZK-2. These feature multiple PDZ-domains for which different drug transporters exhibit different affinities. It was further found that the expression of both adaptor proteins and drug transporters is regulated by PPARα. It is important to note that species differ in their regulation of PDZK-1 and 2. Animal models, therefore, are not useful in predicting the impact of PDZK-1/2 on human drug transporter regulation [8–10].

15.3
Neurological Disorders

The brain is a unique compartment shielded from the peripheral circulation by the so-called blood–brain barrier (BBB) (see Chapter 10 for a more detailed review).

In conjuncture with the specialized scaffolding of the BBB, transporters ensure nutrient supply and protection from xenobiotic substances. Disruption of the BBB is almost always a cause or a symptom of disease. Furthermore, major changes in drug transporter expression may be the result of a disease (e.g., systemic inflammation, epilepsy) or may be causative of a disease (e.g., Parkinsonism). Importantly, drug transporters pose a major obstacle to the treatment of the diseased brain, as xenobiotics are often subjected to efflux before they reach their brain target. We survey in the following section the implication of drug transporters in epilepsy and describe their function in neurodegenerative diseases.

15.3.1
Epilepsy

Epilepsy is a pathological disorder characterized by the paroxysmal, repetitive occurrence of seizures. It is important to note that one seizure (from Latin, *sacire*; to take possession of) does not necessarily lead to epilepsy. As much as 5–10% of the general population experiences at least one seizure in life, whereas only 1–2% of the total population suffers from epilepsy [11, 12]. Seizures manifest themselves in a broad variety of clinical symptoms ranging from barely discernible experiential phenomena to dramatic convulsive activity. The underlying chronic process that leads to epilepsy is caused by abnormal, excessive hypersynchronous firing of central nervous system (CNS) neurons [12]. The most important clinical obstacle in treating epilepsy is the onset of refractory epilepsy (RE), a condition in which the patient is administered adequate levels of antiepileptic drugs (AEDs) but seizures remain uncontrolled. Following a normal treatment course, roughly 50% of patients respond to their first medication, another 25% of patients respond to either the second or the third treatment attempt, and about 6% of patients achieve seizure control with combination therapy. About 20–40% of epileptic patients suffer from refractory epilepsy and achieve only poor or no control over their seizures with drug therapy [13–15].

Out of this subgroup of patients, another 30% of patients are eligible for surgical removal of the epileptic foci and have a relatively good prognosis for seizure control. Yet, 70% of patients with refractory epilepsy (affecting about 0.2% of the total population) do not achieve seizure control and are significantly impaired in their daily lives (see Figure 15.2) [13].

It is not clear at diagnosis which patient will develop refractive epilepsy. This has recently been illustrated in a case study of two sisters who developed idiopathic generalized epilepsy in their teens. Following is an excerpt from this report "Electroencephalogram (EEG) findings demonstrated that they both had primary generalized spike and wave activity at the time of diagnosis. Despite apparently similar etiologies, and underlying genetics, these sisters had vastly different clinical outcomes. One sister has had only two generalized tonic–clonic convulsions in her entire life and is well controlled on carbamazepine, even though this is the wrong medication for her epilepsy syndrome. In contrast, the other sister has refractory juvenile myoclonic epilepsy (JME) and has failed multiple antiepileptic drugs,

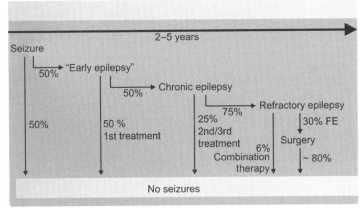

Figure 15.2 Treatment response in epilepsy. Only 50% of the people who experience a seizure will develop epilepsy. These patients enter an epilepsy stage termed "early epilepsy," and 50% of them achieve seizure control with their first AED. Another 50% develop chronic epilepsy. Out of these 50%, 25% respond to a second or third AED given as a monotherapy. Seventy-five percent of patients suffering from chronic epilepsy subsequently develop refractory epilepsy. Only 6% of patients with RE may achieve seizure control with combination therapy. Thirty percent of patients with RE are eligible for surgery and about 80% of these patients will become seizure free after surgical treatment. Seventy percent of patients with RE suffer from uncontrolled seizures with no treatment options (modified from Ref. [12]).

including valproic acid, lamotrigine, and topiramate" [16]. Why did these two sisters respond so differently to treatment and why did the treatment fail?

In order to answer the first question, one must address the physiopathology of refractory epilepsy. There are at least three possible reasons to AED treatment failure:

1. Increased systemic elimination of drugs (mediated by drug transporters and/or cytochrome P450, CYP).
2. Decreased absorption in the gut (this is questionable as in some patients, intravenous administration also results in subtherapeutic levels of AEDs [17]).
3. Increased elimination from the brain (substrate of endothelial efflux pumps).

Subtherapeutic AED plasma concentrations may in some RE patients persist independent of the route of administration [17]. These findings support the notion that increased systemic clearance, and not decreased absorption in the gut, contributes to RE.

Although it is unclear why this is so, it can be speculated that administration of AEDs increases the expression of drug transporters not only in the immediate environment of the epileptic foci but also in other organs, including the liver and kidneys. In conjunction with cytochrome P450 enzyme expression in the liver, this would lead to both degradation and excretion of AEDs. It is also plausible that endocrine factors released into the blood stream in the course of a seizure may be directed toward organs such as the BBB, the kidneys, and the liver, but not the luminal side of the gut.

Importantly, some AEDs influence the expression levels of both cytochrome P450 and drug transporters. For example, phenytoin treatment was shown to upregulate CYP3A isoforms in mouse hippocampus, which in turn mediate phenytoin degradation. In addition to phenytoin, carbamazepine, phenobarbital, and primidone induce many CYP and glucuronyl transferase enzymes. Because these drugs are substrates for the enzymes they induce, this leads to drastically reduced drug serum concentrations [18, 19]. It may be concluded, therefore, that the systemic subtherapeutic levels of AEDs observed in some patients are at least partially caused by direct AED-mediated upregulation of CYP. In this context, it would be of interest to test the coadministration of modulators of CYP expression with AEDs.

A main contributing factor to the onset of refractory epilepsy is thought to be the increased expression of drug efflux pumps. These include the major vault protein, MVP (a.k.a., lung resistance-related protein or LRP), BCRP, MRPs, ABCB1, and possibly RLIP67. Table 15.1 provides an overview of the expression of individual drug transporters detected by immunohistological staining of brain specimen in several different epileptic disorders.

ABCB1 is ubiquitously expressed and upregulated in all cell types across all seizure disorders. MRP1 expression was hitherto not detected in brain sites other than in

Table 15.1 Expression of drug transporters according to seizure disorder and location.

Pathology	Endothelium	Astrocytes (reactive)	Neurons	References
Hippocampal sclerosis	P-gp	P-gp	P-gp	[20–25]
	MRP2	MRP1	MRP1	
	BCRP	MRP2	MVP	
	MVP			
Focal cortical dysplasia	P-gp	P-gp	P-gp	[20, 22, 25–27]
	BCRP	MRP1	MRP1 (dysplastic neurons)	
	MVP		MVP (dysplastic neurons)	
Tuberous sclerosis	P-gp	P-gp	P-gp	[22, 25, 29]
	BCRP	MRP1	MRP1	
	?MVP			
Dysembryoplastic neuroepithelial tumor (DNET)	BCRP	P-gp	MVP	[22, 25, 29]
	MVP	MRP1		
Ganglioglioma	P-gp	MRP1	P-gp	[27, 29–31]
	BCRP	MVP	MRP1	
	MVP			
Rasmussen encephalitis	P-gp	P-gp		[3]

cerebrovascular endothelial cells [21, 25–27]. MRP2 is not very well investigated yet, but was found upregulated in endothelial cells and astrocytes in hippocampal tissues from sclerosis patients. BCRP (ABCG2) is exclusively expressed in cerebrovascular endothelial cells and significantly upregulated in epileptogenic brain tumors [29]. The implication of RLIP76 in epilepsy is controversial and therefore not listed in the table. RLIP76 was initially described to be highly expressed in epileptic tissue and to significantly contribute to the efflux of AEDs. According to this study, RLIP76 is exclusively localized to the luminal surface of endothelial cells in the brain, colocalizes with ABCB1 expression, and is significantly upregulated in epileptic disorders. The authors of this study did not differentiate between different epileptic disorders [32]. Others have not been able to confirm these findings and have questioned the importance of RLIP76 as an efflux pump in RE. The fact that different groups used different antibodies may explain the discrepancy in the findings. Taken together, however, all results underscore the importance of drug transporters in RE.

Table 15.2 lists antiepileptic drugs and their respective transporters. Our knowledge of AED/transporter specificities, however, is rather incomplete and further research is warranted to identify all transporters and respective pharmacological properties for all AEDs. This is of paramount importance for the effective treatment of various epileptic disorders and may help circumvent RE. Our current knowledge of drug/transporter specificities is further hampered by the use of different species in different studies. For example, MRP2, which ensures the transport of both phenytoin and carbamazepine in the rat, does not appear to do so in cells derived from dog or pig kidneys. The findings in the rat were substantiated *in vivo* by the use of transport-deficient (TR$^-$) Wistar rats, which lack MRP2-dependent transport and showed increased extracellular concentrations of phenytoin and carbamazepine in the brain [33, 34]. It is obvious from all studies that there are large species variabilities that account for differences in expression and substrate specificity. This underscores the need to develop *in vitro* human models for the study of AED/transporter properties [35, 36]. Interestingly, although highly expressed in epileptogenic tissues, BCRP does not seem to significantly contribute to AED transport. Phenobarbital, phenytoin, ethosuximide, primidone, valproate, carbamazepine, clonazepam, and lamotrigine have been found to be neither substrates nor inhibitors of BCRP in the BBB of epileptogenic brain tumors [37].

Importantly, Loscher *et al.* found that carbamazepine, felbamate, gabapentin, lamotrigine, phenobarbital, and topiramate are substrates of ABCB1 (P-gp) [38]. Crowe *et al.* also studied the transport of a variety of antiepileptic drugs including vigabatrin, gabapentin, phenobarbitone, lamotrigine, phenytoin, carbamazepine, and acetazolamide in colorectal tumor-derived Caco-2 cell monolayers. They found that only one antiepileptic, acetazolamide, is a weak ABCB1 substrate [39].

Although it has been established that drug transporters are upregulated in RE and that AEDs are substrates of the very drug transporters, there is to date only one study that links overexpression of drug transporters to reduced AED brain concentration of systemically administered AEDs in patients. In this elegant clinical trial, Rambeck and colleagues used microdialysis probes to measure

Table 15.2 AED substrates of drug transporters.

AED	Transporters	Species	References
Phenytoin	~~BCRP~~	Human	[37]
	P-gp	Rat, mouse	
	RLIP76	Human	
	MRP2	Rat (not mouse, human)	
Carbamazepine	BCRP	Human	[37]
	P-gp?	Human	
	RLIP76	Human	
	MRP2	Rat (not mouse, human)	
Valproat	~~BCRP~~	Human	[37]
	~~P-gp~~	Human	
	~~MRP1/2~~	Human	
Lamotrigine	~~BCRP~~	Human	[37]
	P-gp	Human	
	~~MRP2~~	Rat	
Topiramat	P-gp	Human	[33, 38, 40–42]
Ethosuximide	~~BCRP~~	Human	[37]
Felbamate	P-gp	Human	[33, 38, 40–42]
	~~MRP2~~	Rat	
Diazepam			
Clobazam			
Clonazepam	~~BCRP~~	Human	[37]
Phenobarbital	~~BCRP~~	Human	[37]
	P-gp	Human	
	~~MRP2~~	Human	
Primidone	~~BCRP~~	Human	[37]
Vigabatrin			
Tiagabin			
Gabapentin	P-gp	Human	[33, 38, 40–42]
Levetiracetam	P-gp	Mouse	[37]

Listed AEDs are not substrates for the corresponding crossed transporters (e.g., Phenytoin is a substrate for P-gp, RLIP76, and MRP2 in the indicated species, but not for BCRP, shown as ~~BCRP~~).

AED concentrations in the extracellular space of epileptogenic tissue, cerebrospinal fluid (CSF), and blood plasma of patients undergoing resective surgery. They documented that perfusates, collected 1 h prior to tissue excision, exhibited significantly decreased AED concentrations in the epileptogenic zone compared to CSF samples [42, 43].

15.3.1.1 Seizure Frequency and RE

Several hypotheses have been formulated on the involvement of drug transporters in the manifestation of RE (*vide supra*). As a rule of thumb, the more seizures have occurred before treatment initiation the more likely is the treatment to fail [15]. This suggests that the driving force of drug transporters' upregulation in RE is the seizures per se, rather than the exposure to AEDs alone.

Marson *et al.*, however, found no significant difference between deferred treatment (patients experienced multiple seizures before treatment onset) and immediate antiepileptic drug treatment [44]. Accordingly, patients from an underdeveloped region from northern Ecuador with a long history of seizures and no prior AED treatment showed RE incidences comparable to those who had experienced only one or two seizures prior to treatment [45].

Importantly, defined structural lesions (e.g., hippocampal sclerosis) are more prone to becoming drug resistant than more diffuse seizure disorders [15]. Mesial temporal sclerosis, characterized by hippocampal sclerosis, exhibits seizure clustering and displays the highest rate of RE. Furthermore, seizure clustering alone has been associated with a significantly increased incidence of RE [12, 16, 46]. Drug transporter upregulation, therefore, is unlikely to be triggered by "simple" seizure activity but rather to be specifically associated with certain forms of epilepsy that are presented to the physician only after having experienced several seizures.

In concordance with these findings, animal models showed that ABCB1 is transiently increased shortly after a seizure, and drops back to normal levels within 1–2 days [47, 48], indicating that an isolated seizure event cannot affect RE. It is, however, conceivable that multiple events occurring within a short time frame may lead to a reinforcement of the otherwise transient increase in efflux pump expression and may render the patient permanently refractory to treatment with anticonvulsants. Alternatively, a hitherto uninvestigated efflux pump may be upregulated as a result of prolonged and repetitive seizure activity.

15.3.1.2 The Role of Drug Transporters in RE Development

In 2002, Kwan and Brodie proposed the "P-gp positive seizure axis." This theory suggests that expression of ABCB1 is a progressive process that depends on intensity and time constancy of seizure injury. It was found in a cohort of 525 epileptic patients that the number of patients who develop RE directly correlates with the number and frequency of epileptic seizures before the onset of drug therapy.

ABCB1 gene promoter proximal sequences harbor AP-1 and NFκB DNA binding sites, and both AP-1 and NFκB are found at high levels in epileptic tissues. This suggests that these two transcription factors may be involved in ABCB1 upregulation in response to prolonged seizure activity [7, 17]. It was also found that administration of the more common AEDs leads to higher ABCB1 brain levels [28]. Taken together, these results point to the fact that both AED exposure and prolonged seizure activity contribute to the onset of RE.

Wadkins and Roepe found that ABCB1-expressing neurons exhibit significantly lower membrane potentials ($\Delta\psi_0 = -10$ to -20 mV; physiological membrane potential: $\Delta\psi_0$ of -60 mV) [49, 50]. Based on this observation, Lazarowski *et al.* put forth a hypothesis to explain the contribution of neuronal ABCB1 expression to RE development. They stipulate that the persistently low resting membrane potential in ABCB1-positive neurons would facilitate glutamergic signaling and would thus

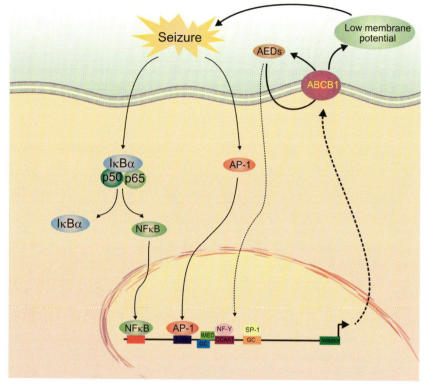

Figure 15.3 Epilepsy and ABCB1 expression. Recurrent seizures lead to the activation of NFκB and AP-1. Inactive NFκB is bound by IκBα and is normally present in the cytoplasm. Upon activation, IκBα is degraded and NFκB translocates to the nucleus where it binds to the promoter region of MDR1 and activates transcription of MDR1. Active AP-1, another transcription factor, also activates MDR1. In turn, increased MDR1 expression leads to a decrease in the neuronal membrane potential, which increases the cell's susceptibility to seizures resulting in a positive feedback loop.

render the patient more susceptible to seizures [17]. Figure 15.3 depicts physiopathological changes that contribute to the development of RE.

15.3.1.3 Overcoming RE

"Recently it was described that an 11-year-old boy who developed status epilepticus after a prolonged right-side simple partial motor seizure, which was unresponsive to long-term aggressive treatment with several AEDs [51]. The control of seizures was achieved at a plasma valproic acid level of 108 μg/ml, but electrical status epilepticus persisted, and the child remained comatose. On day 37, a treatment with verapamil (a calcium L-channel blocker) was started, and 1.5 h after the initiation of the infusion, the patient regained consciousness, breathed spontaneously, and the electrical status promptly disappeared. The authors suggested that verapamil, a known P-gp inhibitor acted by facilitating the brain penetration of AEDs simultaneously administered to the patient, however, because surgical treatment was not developed, brain-overexpression of P-gp can't be confirmed in this case. [17]"

Similar to this patient, Summers *et al.* describe the case of a 24-year-old woman suffering from intractable epilepsy, who also profited from verapamil coadministration [52]. These findings suggest that despite the fact that most AEDs seem to be poor ABCB1 substrates, verapamil can help increase AED concentration in epileptogenic tissues and thus can alleviate RE. It has to be noted, however, that verapamil is not a specific inhibitor of ABCB1 and may rather alter the properties of the BBB (i.e., tight junctions) by blocking Ca^{2+}-channels, thus enabling paracellular influx of AEDs into the brain. The exact mechanism of how verapamil enables AED treatment of RE remains to be elucidated.

Additional studies carried out in animal models of temporal lobe epilepsy (TLE) further support the use of selective ABCB1 inhibitors to reverse drug resistance in RE. Thus, tariquidar potentiates the effect of phenytoin and countervails resistance to phenobarbital in a rat model of TLE. Furthermore, cyclosporin A helps reverse resistance to phenytoin in a rat model of AED-resistant status epilepticus. Finally, verapamil counteracts resistance to oxacarbazepine in rats with pilocarpine-induced seizures [42].

These findings highlight the necessity of drug transporter research in the context of MDR. Patients suffering from RE who cannot receive surgical treatment have now a chance to overcome their resistance to AEDs with the use of highly selective drug transporter inhibitors. This area of research clearly needs further investigation before coadministration of inhibitors to select patients becomes part of routine treatment regimens.

15.3.2
MDR1 Expression and Neurodegenerative Disorders

Epidemiological data have unveiled a possible role of pesticide exposure in the etiology of Parkinson's disease (PD), the exact mechanism of which is hitherto unknown. Statistical analysis has implicated MDR1 in the development of pesticide-induced PD [53]. A mutation in the MDR1 gene that leads to decreased MDR1 expression at the BBB predisposes to the damaging effects of pesticides and possibly to those of other toxic xenobiotics transported by P-glycoprotein, thus further contributing to PD [54, 55]. Conversely, other mutations in the MDR1 gene protect people from developing PD through unknown mechanisms [56, 57].

Interestingly, Alzheimer's disease (AD) appears to be also linked to MDR1. MDR1 transports soluble Abeta40 and Abeta41 out of the brain, thus reducing the amyloid charge in the brain. A decrease in MDR1 expression, therefore, leads to an accumulation of amyloid proteins in the brain [58]. Furthermore, MDR1 is involved in protecting neuronal cells from apoptosis and it also shields the brain from noxious agents [59, 60]. It is therefore conceivable that inducers of MDR1 may slow down the progress of AD. Accordingly, rifampicin, a potent inducer of P-gp, leads to a notable improvement in the cognitive function of AD patients after a treatment period of 3 months [61].

Taken together, the presented findings support the notion that MDR1 is a target for the prevention or adjuvant therapy of cerebral amyloid angiopathy, AD, and possibly pesticide exposure-associated PD.

15.4
Inflammation

As many diseases have an inflammatory component (e.g., diabetes mellitus, Alzheimer's disease), are triggered by an underlying inflammatory process (e.g., rheumatoid arthritis, inflammatory bowel disease (IBD)), or are invoked by an infection, it is evident that inflammation plays an important role in a large number of disease states. Most of these diseases warrant medical treatment and some of the drugs administered are substrates of drug carriers that hamper their bioactivity. Unfortunately, most drug transporters have only been characterized under normal physiological conditions. Importantly, recent experimental findings, together with patient data, have shown that inflammation has a significant impact on the regulation of drug transporter expression and treatment response. This part of the chapter aims at elucidating the impact of inflammation on the expression and function of drug carriers and their role in treatment success.

15.4.1
Inflammation – The Acute-Phase Response

The immediate reaction of inflammation is called the acute-phase response (APR). It encompasses a complex series of physiological reactions that occur shortly after the onset of an infection, tissue damage, malignancy, or an inflammatory process. The APR is the body's tool to prevent ongoing tissue damage, to eliminate an infective organism, and to initiate the necessary repair processes to restore homeostasis. This process usually lasts between 24 and 48 h. A failure in APR termination within this time frame can lead to a chronic inflammatory disorder. A patient suffering from APR presents with fever, metabolic changes, leukocytosis, and vasodilation. During APR, expression and plasma concentration of acute-phase proteins (APPs) change. An APP is defined as any protein whose plasma concentration increases (positive acute-phase proteins such as fibrinogen, serum amyloid A, alpha1-acid glycoprotein, C-reactive protein, alpha1-antitrypsin) or decreases (negative acute-phase proteins such as transferrin, many of the P450 cytochromes, insulin growth factor I) by at least 25% during an inflammatory disorder [62]. The changes in APP plasma levels are triggered by inflammation-associated cytokines, such as interleukins IL-6 and IL-1β, tumor necrosis factor-α (TNF-α), interferon-γ (INF-γ), transforming growth factor-β (TGF-β), and possibly IL-8 [63–65]. Chronologically, the initial event (i.e., tissue damage) initiates proinflammatory cytokine release. Upon receptor binding of the cytokines, a signaling cascade culminating in the synthesis and release of APPs is triggered. As stated above, not all APPs are upregulated during the APR. It is hypothesized that the decrease and degradation of several of the negative APPs (e.g., albumin) serve as an amino acid source for the increase of positive APPs (e.g., C-reactive protein).

There are two main models used to investigate APR: (i) the lipopolysaccharide (LPS) endotoxin model and (ii) the turpentine model. The intraperitoneal or intravenous administration of endotoxins, cell-wall components of Gram-negative bacteria, triggers a pronounced systemic inflammatory response, inferring fever, hypotension, and tachycardia. Importantly, endotoxins from different bacterial strains lead to the release of different cytokines. As stated above, the most relevant cytokines to APR are IL-1β, TNF-α, IL-6, and Interferon-γ. The turpentine model comprises the subcutaneous or intramuscular injection of turpentine, a mixture of terpenes obtained by distillation of resin from pine trees. The word turpentine is derived from the Greek terebinthine, the name of the terebinth tree from whose sap the turpentine spirit was originally distilled. The application of turpentine results in a local aseptic inflammatory response, which yields a dermal abscess that triggers a systemic APR. In contrast to the LPS model, the turpentine reaction is based mainly on the release of IL-6, as IL-6$^{-/-}$ mice do not develop systemic APR following turpentine treatment. This example shows that the underlying molecular pattern of APR depends on the stimulus and the cytokines released thereupon. Further *in vivo* models that mimic inflammatory conditions include transgenic or chemically induced animal models of cholestasis, ulcerative cholitis, inflammatory bowel disease, arthritis, and chronic renal failure.

As inflammation is an integral part of many disorders that affect different organs, the following section is organized by organ systems. To date, most studies have investigated the impact of inflammation on hepatic gene expression. In addition, we will also discuss the impact of inflammation on the blood–brain barrier, the intestine, and the kidney.

15.4.2
BBB and Inflammation

The diseased brain (e.g., ischemia/reperfusion, seizure disorders, neurodegenerative diseases, meningitis, etc.) often exhibits an inflammatory component. As observed in other organs, expression of drug transporters may be induced or repressed by various stimuli. The most studied transporter in this context is ABCB1, which is abundantly expressed in most brain cells. Puig *et al.* reported an increased efficacy of several centrally acting ABCB1 substrates, including loperamide, morphine, and fentanyl in a mouse model of intestinal inflammation [66]. Endotoxin-treated mice showed increased intracranial accumulation of doxorubicin and concomitantly reduced levels of ABCB1 expression [67]. In rats, systemic or CNS inflammation, triggered by either intracranial or intraperitoneal injection of LPS, led to a marked decrease of ABCB1 expression and function [68, 69]. The same study showed that brain inflammation leads to a marked decrease in the drug transporter Oatp2 mRNA expression in the brain and liver of rats [68]. In sharp contrast to these findings, Zhao *et al.* reported that Shiga-like toxin-treated mice display an increase in ABCB1 expression and function [70]. Furthermore, Yu *et al.* recently reported that in the RBE4 rat brain endothelial cell line, TNF-α exposure mediates a significant increase in ABCB1 expression and function as evidenced by augmented H^3-vinblastine efflux. Microarray-based gene expression profiling by the same authors showed that other drug transporters in the RBE4 cell line are not affected by TNF-α exposure [71]. These studies confirm that different inflammatory stimuli modulate various drug transporters differently. For example, administration of LPS triggers a release of several cytokines including IL-6 and TNF-α. IL-6 has been postulated to mediate ABCB1 repression whereas TNF-α effects an increase in ABCB1 expression [72]. It can be concluded that IL-6 signaling prevails over that of TNF-α when inflammation is triggered by LPS (i.e., upon infection with Gram-negative bacteria). This example illustrates that further studies need to be conducted to closely mimic *in vivo* conditions of specific inflammatory states and to evaluate their impact on drug transporter expression and function.

TNF-α is released from virtually all brain parenchymal cells after trauma, hypoxia, epilepsy, neuro-AIDS, and inflammation [73]. Interestingly, TNF-α is not only specifically transported across the BBB but also modulates the functions of the specialized endothelial cells lining the BBB [74–76]. TNF-α transport across the BBB follows a circadian rhythm, and strikingly ABCB1 expression at the BBB also displays a circadian rhythm [71, 77]. Upon receptor binding, TNF-α probably affects ABCB1 expression by activating NFκB, which binds to the proximal promoter of ABCB1 and activates its transcription [78].

Taken together, these results underline the important role of inflammation at the BBB. As the BBB is a natural barrier that shields the brain from potentially noxious substances, inflammatory mediators such as TNF-α may serve as potentiators of BBB's barrier function. Unfortunately, the increased barrier function also shields the brain from treatment when the inflammatory signal originates from within the brain. These considerations lead us to suggest the coadministration of anti-inflammatory agents in brain disease states with an inflammatory component (e.g., neuro-AIDS).

15.4.3
Kidney and Inflammation

Although the vast majority of renal transport studies have focused on glucose, sodium, and urea transport, yet the kidney is a major player in absorption, distribution, metabolism, and excretion (ADME) that determines both the active and the passive renal elimination of xenobiotics. Relatively few studies examined the role of inflammation in the regulation of renal drug transporters. Furthermore, the little data available seems to be conflicting due to different inflammatory models and species employed. Renal excretion of rhodamine-123, an ABCB1 substrate, is significantly reduced in endotoxin-challenged rats. In agreement with this finding, these rats displayed reduced ABCB1 mRNA expression in their kidneys [79]. On the other hand, endotoxin-challenged mice exhibited a significant induction of ABCB1 kidney expression and a concomitant increase in kidney-mediated doxorubicin clearance [80]. Additional studies are clearly warranted to define the interplay of drug transporters in the kidney in the context of inflammation.

15.4.4
Liver and Inflammation

The liver is responsible for both metabolism and excretion thereby affecting the bioavailability of a given substance. Many drug transporters contribute to this function and are, to a certain extent, affected by inflammation. The efflux pumps whose upregulation in the liver has been thoroughly documented are depicted in Figure 15.4 along with their location and function in liver physiology [85]. Lack of expression of some drug transporters such as MDR3 (ABCB4) leads to the accumulation of bile and to cholangitis [81, 82]. As most of the findings on liver disease and drug transporters were derived from rodent models, it has to be pointed out that there are significant differences between human and rodent drug transporters' regulation, both at the transcriptional and the posttranscriptional levels. This is exemplified by the much milder phenotype observed in the hepatobiliary transporter Bsep knockout mice compared to patients lacking the very same drug transporter [83, 84]. Furthermore, when the inflammatory condition develops in rodent models, it manifests itself within hours to weeks, whereas it takes months to decades to develop in humans.

Hepatobiliary transporters are affected by both systemic inflammation (e.g., arising from an infection) and inflammation intrinsic to the liver (e.g., acute inflammatory cholestasis caused by drug or alcohol abuse). As described above, endotoxin or turpentine are used to trigger systemic inflammation in rodents. Other rodent models of cholangitis include ethinylestradiol (oral contraceptive-induced cholestasis/cholestasis of pregnancy), alpha-naphthylisocyanate (vanishing bile duct syndrome), and common bile duct ligation (extrahepatic biliary obstruction) [87, 88].

Despite the major differences in etiology, systemic inflammation and inflammation arising from the liver translate into almost the same changes in drug transporter expression patterns. Most of the observed changes are now attributed to altered

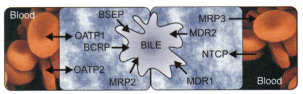

Figure 15.4 Hepatic drug transporters affected by inflammation. NTCP (Slc10a1), OATP1 (Slc21a1), and OATP2 (Slc21a5) are responsible for sinusoidal uptake of substrates into the liver. OATP1 (Slc21a1) and OATP2 (Slc21a5), together with MRP3 (ABCC3), are also involved in substrate efflux into the plasma. MDR1 (ABCB1), MDR2 (ABCB4), BSEP (ABCB11), MRP2 (ABCC2), and BCRP (ABCG2) mediate the efflux of substrates into the bile (modified from Refs [85, 86]).

activity of nuclear receptors. Interestingly, biliary compounds retained in the liver during cholestasis are nuclear receptor ligands. These include bile salts and bilirubin. Nuclear receptors involved include the constitutive androstane receptor (CAR), farnesoid X receptor (FXR), pregnane X receptor (PXR), vitamin D nuclear receptor (VDR), and possibly also the liver X receptor (LXR) [89–92]. There is an excellent review by Trauner *et al.* that covers the general principles of hepatobiliary transporter gene regulation [92]. After an inflammatory event, the accumulation of bile leads to activation of these nuclear receptors and ultimately to a change in the expression patterns of drug transporters.

Most experimental data were derived from rodent models in which drug transporter regulation occurs mostly at the transcriptional level. Posttranscriptional regulation (e.g., targeting and retrieval of transporters from membranes), however, seems to be important in humans also [93, 94].

Inflammatory conditions of the liver, in particular inflammatory hepatocellular cholestasis, are one of the most frequent causes of jaundice in the clinic. The major underlying denominator of this disorder is the inhibition of transporter expression and function by proinflammatory cytokines, which are either induced systemically or within the liver. Alcoholic hepatitis accounts for up to two-thirds of patients and is the most frequent trigger, followed by idiosyncratic drug reactions, sepsis or other extrahepatic bacterial infections, some variants of viral hepatitis, and total parenteral nutrition [95, 96].

Changes in drug transporter expression due to alcoholic hepatitis or idiosyncratic drug reactions are illustrated in Figure 15.5. Figure 15.6 summarizes the effects of endotoxins on hepatobiliary transporter expression, whereas Figure 15.7 portrays the differences between IL-6- and TNF-α-mediated regulation of drug transporter expression. Data presented are biased by the respective models employed and are only an approximation of patient's physiopathology. Data derived from animal models, however, conclusively indicate a lack of hepatobiliar drug transporter upregulation following inflammation. It has been speculated that some transport systems such as MRP3 and certain MDRs, which are either maintained at the same

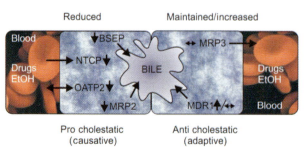

Figure 15.5 Hepatic drug transporter expression changes due to alcoholic hepatitis or idiosyncratic drug reactions. Drug- or ethanol-induced inflammation of the liver results in downregulation of NTCP, OATP2, BSEP, and MRP2. MRP3 levels remain unchanged, whereas MDR1 levels either remain constant or increase.

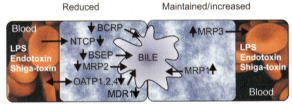

Figure 15.6 The effect of endotoxins on hepatobiliary drug transporter expression. Endotoxins mediate a decrease in BCRP, NTCP, BCEP, MDR1, MRP2, OATP1, OATP2, and OATP4. MRP3 transcription, however, remains unchanged whereas MRP1 expression levels increase following endotoxin exposure.

level of expression or upregulated, limit cholestatic liver injury. The exact mechanisms of this process remain to be uncovered [87, 94]. In fact, most drug transporters are downregulated in inflamed liver tissue. LPS-treated mice and rats display a decrease of expression of almost all drug transporters of the Slc and ABC families [97, 98]. This downregulation can be significantly attenuated following glucocorticoid administration. Further studies investigated the effect of IL-6 and TNF-α on drug transporter expression [85, 97–102], and the findings of these studies are summarized in Figure 15.7.

As in other organs, ABCB1 is the best investigated transporter in the liver. It was demonstrated that turpentine-induced APR leads to a 50–70% reduction of ABCB1 expression and function in rat liver tissue 48 h after treatment [103]. Similar results were observed in studies of endotoxin-triggered inflammation in which both constitutive and induced expression were affected in rodents [104]. Further experiments have shown that reduction of ABCB1 expression is also linked to a reduction in ABCB1 function. In two distinct experimental setups, Shiga-toxin II and endotoxin administration to rats prompted a substantial reduction in ABCB1 function (assessed by hepatobiliary doxorubicin and ^{99m}TC-sestamibi clearance), accompanied by a significant reduction in ABCB1 protein expression [69, 105]. Likewise, endotoxin-treated mice displayed decreased liver-mediated doxorubcin clearance as a result of

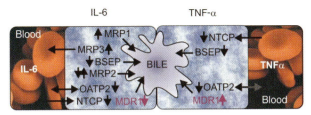

Figure 15.7 Differences between IL-6- and TNF-α-mediated regulation of hepatobiliary drug transporter expression. IL-6 induces hepatobiliary MRP1 and MRP3 expression and leads to downregulation of hepatobiliary MDR1, NTCP, OATP2, and BSEP. Data on its effect on MRP2 expression are contradictory. Similarly, TNF-α reduces hepatobiliary OATP2, NTCP, and BSEP expression. Importantly, TNF-α reduces MDR1 expression whereas IL-6 induces it (indicated with purple arrows).

reduced ABCB1 expression [80]. Minor differences were observed in various rodents used. These were associated with the model-, strain-, and species-dependent release of cytokines [68, 69, 98, 106, 107].

IL-6- and TNF-α-specific responses were studied in more detail, showing that IL-6 is responsible for ABCB1 downregulation, whereas TNF-α upregulates ABCB1 expression [108–110]. On the other hand, TNF-α participates in endotoxin-induced repression of ABCB1 via a yet unidentified mechanism (see Figure 15.7) [110].

Studies on human-derived materials also support the theory of a cytokine-mediated ABCB1 decrease. *In vitro* findings indicate that IFN-γ, TNF-α, IL-2, and leukoregulin boost chemosensitivity in colocarcinoma cells concomitant with a reduced ABCB1 gene expression. The only study carried out involving liver tissue found that IL-6 and IL-1β reduced ABCB1 activity and mRNA levels [111–113]. This study was, however, carried out on hepatoma cells in culture, and these results may or may not be valid in the inflamed but otherwise normal liver tissue. Studies on perfused human liver tissues or primary human hepatocytes in culture need to be conducted to further substantiate these data.

Similar experiments would be equally beneficial to uncover the hitherto less investigated interplay between inflammation and expression of other drug transporters in the liver. MRP2 expression is dramatically reduced upon inflammation, and both IL-6 and IL-1α contribute to its downregulation. Similar to ABCB1, both transcriptional and posttranscriptional mechanisms seem to modulate MRP2 expression and function [99, 107, 114–119]. Recent data emphasize that these posttranscriptional mechanisms may play an even more significant role in human drug transporter regulation than in rodents [120]. Figures 15.4–15.7 summarize the effect of inflammation on the expression of various drug transporters in the liver [85, 94, 99–102]. Almost all drug transporters are affected by inflammation and inflammatory mediators, BCRP (not shown in the table) was shown by one group to be downregulated in the liver in endotoxin-treated rats [85].

As most of these data were derived from animal models, it is reasonable to question the clinical impact of these findings. However, recent studies found that cancer patients receiving IFN-γ or TNF-α along with cytotoxic drugs display better chemotherapeutic response rates than those treated with anticancer drugs only [121, 122]. We envision that a better understanding of the effects of inflammation on the hepatobiliary expression and function of drug transporters would ultimately lead to better treatment options for patients suffering from systemic/liver inflammation or liver cirrhosis.

15.4.5
Intestine and Inflammation

The last major barrier in ADME to be discussed in this review is the intestinal surface, which limits drug adsorption and oral bioavailability. This subsection is of particular interest as there is a vast patient repertoire that has been traditionally treated with both anti-inflammatory drugs and substrates/inhibitors of drug transporters. We will first outline the general mechanisms of inflammation in the gastrointestinal (GI) tract and then discuss the physiopathological implications of inflammatory bowel disease.

APR of the GI tract is almost identical to the one observed in the liver [123, 124]. For instance, endotoxin-induced inflammation causes an increase in IL-6 and a subsequent decrease in ABCB1 expression in rat intestines [125]. Similar results were obtained from studies based on dextrane sulfate sodium (DSS)-induced colitis of both mice and rats [126, 127]. In agreement with hepatobiliary findings on the impact of inflammation on ABCB1 function, the mucosal-to-serosal absorption of ABCB1 substrates across the intestinal barrier was increased in the presence of inflammatory stimuli [127, 128]. Colon tissue samples derived from patients with ulcerative colitis show dramatically reduced levels of ABCB1 protein expression validating the data obtained in rodent models [129]. Studies on patients with various gastrointestinal inflammatory disorders further confirmed the link between cytokines and downregulation of several drug transporters [130–132]. Interestingly, tacrolimus and cyclosporin are also better absorbed in pediatric patients suffering from diarrhea [133]. Unfortunately, to date no similar study has been carried out in patients suffering from inflammatory bowel disease, but it stands to reason that cyclosporin may be better absorbed in these patients due to the nature of the underlying inflammatory disease. It can be speculated that IBD patients would benefit from downregulation of drug transporters affected by inflammatory signals, which would enable a better uptake of the known ABCB1 substrate cyclosporin. In addition, cyclosporin impairs ABCB1, thus exacerbating the effects of ABCB1 downregulation seen in IBD patients. However, both in IBD-afflicted patients and in rat models, noninflamed intestinal biopsies display higher ABCB1 expression, resulting in an overall increase of ABCB1 expression in the GI tract. This indicates a feedback mechanism that compensates for ABCB1 downregulation in inflamed intestinal tissues.

The effects of TNF-α and IL-6 were investigated in the human colorectal adenocarcinoma Caco-2 cells. Contrary to the findings in liver tissue, IL-6 mediates upregulation of ABCB1 in Caco-2 cells. In the same cell model, IL-1β and IFN-γ also upregulated ABCB1 mRNA expression. However, despite the increase in ABCB1 expression triggered by IFN-γ, there is no increase in ABCB1 function. This has been attributed to a concomitant redistribution of ABCB1 within the cell. ABCB1 activity is reduced after plasma from rats with an acute renal failure (presumably containing increased levels of both IL-6 and TNF-α) is added to Caco-2 cells [134–136]. The decrease in ABCB1 expression mediated by TNF-α is likely to supersede IL-6-mediated ABCB1 induction. More studies are necessary to find out why inflammatory mediators affect the hepatobiliary and the intestinal expression of ABCB1 differently. Unidentified molecular players linked to inflammation are likely to mediate these effects.

Interestingly, ABCB1 expression seems to be directly linked to the tumor suppressor gene APC (adenomatous polyposis coli). Mutations of APC occur in the majority of patients with sporadic and hereditary colorectal cancers. APC leads to accumulation of β-catenin, which is a coactivator for the transcription complex TCF/LEF

(T-cell factor/lymphoid enhancer factor). TCF/LEF in turn binds and activates the ABCB1 promoter [137].

Other transporters investigated in the context of inflammation of the GI tract include MRP2 and BCRP. Similar to the findings on ABCB1, BCRP expression is decreased in inflamed colonic and rectal mucosa samples from patients with ulcerative colitis but remains unchanged in duodenal biopsies from patients with obstructive cholestasis. Only MRP2 protein was downregulated in duodenal biopsies from patients with obstructive cholestasis. MRP2 was also shown to be downregulated in various rodent models of gastrointestinal inflammation. All findings on MRP2 in this context suggest that both transcriptional and posttranscriptional mechanisms mediate its downregulation. Contrary to MRP2, duodenal ABCB1 and BCRP expression levels remain unaffected in patients suffering from obstructive cholestasis, indicating that these two transporters are mostly regulated at the transcriptional level [85, 125, 129, 138]. However, transcription of all three transporters, namely, ABCB1, MRP2, and BCRP seems to be induced by PXR following exposure to drugs such as rifampin [139–142].

Finally, approximately 2 out of 10 000 Europeans 15 years or older suffer from inflammatory bowel disease, that is, ulcerative cholitis and Crohn's disease. Ulcerative cholitis occurs with a higher incidence than Crohn's disease [143]. The standard treatment for IBD includes glucocorticoids and other immunomodulators, and as many as 30% of Crohn's disease and 25% of ulcerative cholitis patients are glucocorticoid dependent. However, 50% of patients suffering from Crohn's disease and 20% of patients with ulcerative colitis experience bowel resection. IBD treatment failure can occur when either or all of the three mechanisms listed below come into effect:

1. Decreased cytoplasmatic drug concentration as a result of ABCB1-mediated efflux.
2. Impaired glucocorticoid signaling due to defects in the glucocorticoid receptor-mediated response.
3. Constitutive epithelial activation of proinflammatory mediators that inhibit glucocorticoid receptor transcriptional activity.

As outlined above, IBD patients display highly elevated levels of ABCB1 expression in the inflamed bowel epithelium. Moreover, specific pump inhibitors have been shown to significantly increase cortisol and cyclosporin levels within the intestinal cells [144]. Patients suffering from other autoimmune disorders that require steroid treatment (e.g., rheumatoid arthritis, lupus erythematosus) can also express higher levels of ABCB1 [145, 146]. Glucocorticoid failure in IBD is particularly interesting as glucocorticoids influence the inflammatory condition, which in turn is responsible for ABCB1 modulation. The interplay between glucocorticoids, inflammatory mediators, and multidrug resistance in IBD is depicted in Figure 15.8 [147].

This last example further illustrates the need for a better understanding of drug transporters in health and disease. Specific treatments are available for all of the physiopathological conditions outlined in this chapter, but drug transporters can and often do hamper the delivery of these drugs to the diseased tissue. It is vital to further

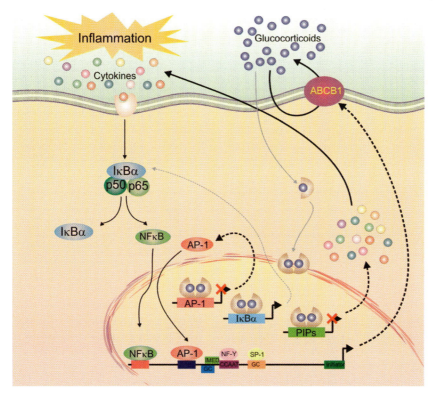

Figure 15.8 IBD and glucocorticoid resistance. Normally, glucocorticoids permeate through the cellular membrane into the cytoplasm where they bind to glucocorticoid receptors (gray arrows). These receptors dimerize and translocate into the nucleus, where they bind to the promoters of their target genes and either activate or inactive transcription. Glucocorticoids inhibit the transcription of AP-1 and of proinflammatory proteins (PIPs) but activate the transcription of IκBα. In turn, IκBα binds cytoplasmic NFκB and inhibits its activation. Glucocorticoid therapy, therefore, reduces MDR1 expression by inhibiting AP-1 and NFκB-mediated MDR1 transcription. In glucocorticoid-resistant IBD patients, MDR1 blocks glucocorticoid entry into the cell (black arrows). AP-1 and proinflammatory protein transcription is no longer repressed, resulting onto an MDR1 transcription positive feedback loop.

investigate the physiopathology of drug transporters in specific disease states for the design and implementation of effective therapies.

References

1 Gottesman, M.M. and Pastan, I. (1993) Biochemistry of multidrug resistance mediated by the multidrug transporter. *Annual Review of Biochemistry*, **62**, 385–427.

2 Gottesman, M.M. *et al.* (1995) Genetic analysis of the multidrug transporter. *Annual Review of Genetics*, **29**, 607–649.

3 Kwan, P. and Brodie, M.J. (2005) Potential role of drug transporters in the

pathogenesis of medically intractable epilepsy. *Epilepsia*, **46** (2), 224–235.

4 Ueda, K., Pastan, I., and Gottesman, M.M. (1987) Isolation and sequence of the promoter region of the human multidrug-resistance (P-glycoprotein) gene. *The Journal of Biological Chemistry*, **262** (36), 17432–17436.

5 Jin, S. and Scotto, K.W. (1998) Transcriptional regulation of the MDR1 gene by histone acetyltransferase and deacetylase is mediated by NF-Y. *Molecular and Cellular Biology*, **18** (7), 4377–4384.

6 Scotto, K.W. and Johnson, R.A. (2001) Transcription of the multidrug resistance gene MDR1: a therapeutic target. *Molecular Interventions*, **1** (2), 117–125.

7 Scotto, K.W. (2003) Transcriptional regulation of ABC drug transporters. *Oncogene*, **22** (47), 7496–7511.

8 Thwaites, D.T. et al. (2002) H/dipeptide absorption across the human intestinal epithelium is controlled indirectly via a functional Na/H exchanger. *Gastroenterology*, **122** (5), 1322–1333.

9 Kato, Y., Watanabe, C., and Tsuji, A. (2006) Regulation of drug transporters by PDZ adaptor proteins and nuclear receptors. *European Journal of Pharmaceutical Sciences*, **27** (5), 487–500.

10 Sugiura, T., Kato, Y., and Tsuji, A. (2006) Role of SLC xenobiotic transporters and their regulatory mechanisms PDZ proteins in drug delivery and disposition. *Journal of Controlled Release*, **116** (2), 238–246.

11 Commission on Classification and Terminology of the International League Against Epilepsy (1981) Proposal for revised clinical and electroencephalographic classification of epileptic seizures. *Epilepsia*, **22** (4), 489–501.

12 Kasper, D.L. (2005) *Harrison's Principles of Internal Medicine*, vol. **16**, McGraw-Hill, New York, p. 2607.

13 Dodel, R., Rosenow, F., and Hamer, H.M. (2007) The costs of epilepsy in Germany. *Pharmazie in Unserer Zeit*, **36** (4), 298–305.

14 Collaborative Group for the Study of Epilepsy (1992) Prognosis of epilepsy in newly referred patients: a multicenter prospective study of the effects of monotherapy on the long-term course of epilepsy. *Epilepsia*, **33** (1), 45–51.

15 Kwan, P. and Brodie, M.J. (2000) Early identification of refractory epilepsy. *The New England Journal of Medicine*, **342** (5), 314–319.

16 French, J.A. (2007) Refractory epilepsy: clinical overview. *Epilepsia*, **48** (Suppl 1), 3–7.

17 Lazarowski, A. et al. (2007) ABC transporters during epilepsy and mechanisms underlying multidrug resistance in refractory epilepsy. *Epilepsia*, **48** (Suppl 5), 140–149.

18 Gehlhaus, M. et al. (2007) Antiepileptic drugs affect neuronal androgen signaling via a cytochrome P450-dependent pathway. *The Journal of Pharmacology and Experimental Therapeutics*, **322** (2), 550–559.

19 Perucca, E. (2006) Clinically relevant drug interactions with antiepileptic drugs. *British Journal of Clinical Pharmacology*, **61** (3), 246–255.

20 Sisodiya, S.M. et al. (2006) Vascular colocalization of P-glycoprotein, multidrug-resistance associated protein 1, breast cancer resistance protein and major vault protein in human epileptogenic pathologies. *Neuropathology and Applied Neurobiology*, **32** (1), 51–63.

21 Aronica, E. et al. (2004) Expression and cellular distribution of multidrug resistance-related proteins in the hippocampus of patients with mesial temporal lobe epilepsy. *Epilepsia*, **45** (5), 441–451.

22 Sisodiya, S.M. et al. (2003) Major vault protein, a marker of drug resistance, is upregulated in refractory epilepsy. *Epilepsia*, **44** (11), 1388–1396.

23 Tishler, D.M. et al. (1995) MDR1 gene expression in brain of patients with medically intractable epilepsy. *Epilepsia*, **36** (1), 1–6.

24 Dombrowski, S.M. et al. (2001) Overexpression of multiple drug resistance genes in endothelial cells from patients with refractory epilepsy. *Epilepsia*, **42** (12), 1501–1506.

25 Sisodiya, S.M. et al. (2002) Drug resistance in epilepsy: expression of drug resistance proteins in common causes of refractory epilepsy. *Brain*, **125** (Pt 1), 22–31.

26 Sisodiya, S.M. et al. (2001) Multidrug-resistance protein 1 in focal cortical dysplasia. *Lancet*, **357** (9249), 42–43.

27 Aronica, E. et al. (2003) Expression and cellular distribution of multidrug transporter proteins in two major causes of medically intractable epilepsy: focal cortical dysplasia and glioneuronal tumors. *Neuroscience*, **118** (2), 417–429.

28 Wang, Y. et al. (2003) A kindling model of pharmacoresistant temporal lobe epilepsy in Sprague-Dawley rats induced by Coriaria lactone and its possible mechanism. *Epilepsia*, **44** (4), 475–488.

29 Aronica, E. et al. (2005) Localization of breast cancer resistance protein (BCRP) in microvessel endothelium of human control and epileptic brain. *Epilepsia*, **46** (6), 849–857.

30 Berger, W. et al. (2001) Overexpression of the human major vault protein in astrocytic brain tumor cells. *International Journal of Cancer*, **94** (3), 377–382.

31 Aronica, E. et al. (2003) Overexpression of the human major vault protein in gangliogliomas. *Epilepsia*, **44** (9), 1166–1175.

32 Awasthi, S. et al. (2005) RLIP76, a non-ABC transporter, and drug resistance in epilepsy. *BMC Neuroscience*, **6**, 61.

33 Baltes, S. et al. (2007) Differences in the transport of the antiepileptic drugs phenytoin, levetiracetam and carbamazepine by human and mouse P-glycoprotein. *Neuropharmacology*, **52** (2), 333–346.

34 Potschka, H., Fedrowitz, M., and Loscher, W. (2003) Multidrug resistance protein MRP2 contributes to blood–brain barrier function and restricts antiepileptic drug activity. *The Journal of Pharmacology and Experimental Therapeutics*, **306** (1), 124–131.

35 Cucullo, L. et al. (2005) Drug delivery and *in vitro* models of the blood–brain barrier. *Current Opinion in Drug Discovery & Development*, **8** (1), 89–99.

36 Cucullo, L., Oby, E., Hallene, K., Aumayr, B., Rapp, E., and Janigro, D. (2006) Artificial blood–brain barrier, in *Blood–Brain Barriers* (eds R. Dermietzel, D.C. Spray, and M. Nedergaard), Wiley-VCH Verlag GmbH, Weinheim, pp. 375–402.

37 Cerveny, L. et al. (2006) Lack of interactions between breast cancer resistance protein (bcrp/abcg2) and selected antiepileptic agents. *Epilepsia*, **47** (3), 461–468.

38 Loscher, W. and Potschka, H. (2005) Drug resistance in brain diseases and the role of drug efflux transporters. *Nature Reviews. Neuroscience*, **6** (8), 591–602.

39 Crowe, A. and Teoh, Y.K. (2006) Limited P-glycoprotein mediated efflux for anti-epileptic drugs. *Journal of Drug Targeting*, **14** (5), 291–300.

40 Potschka, H., Fedrowitz, M., and Loscher, W. (2001) P-glycoprotein and multidrug resistance-associated protein are involved in the regulation of extracellular levels of the major antiepileptic drug carbamazepine in the brain. *Neuroreport*, **12** (16), 3557–3560.

41 Baltes, S. et al. (2007) Valproic acid is not a substrate for P-glycoprotein or multidrug resistance proteins 1 and 2 in a number of *in vitro* and *in vivo* transport assays. *The Journal of Pharmacology and Experimental Therapeutics*, **320** (1), 331–343.

42 Loscher, W. (2007) Drug transporters in the epileptic brain. *Epilepsia*, **48** (Suppl 1), 8–13.

43 Rambeck, B. et al. (2006) Comparison of brain extracellular fluid, brain tissue, cerebrospinal fluid, and serum concentrations of antiepileptic drugs measured intraoperatively in patients with intractable epilepsy. *Epilepsia*, **47** (4), 681–694.

44 Marson, A. et al. (2005) Immediate *versus* deferred antiepileptic drug treatment for early epilepsy and single seizures: a randomised controlled trial. *Lancet*, **365** (9476), 2007–2013.

45 Placencia, M. et al. (1994) The characteristics of epilepsy in a largely untreated population in rural Ecuador. *Journal of Neurology, Neurosurgery, and Psychiatry*, **57** (3), 320–325.

46 Brandao, E.M. and de Manreza, M.L. (2007) Mesial temporal sclerosis in children. *Arquivos de Neuro-Psiquiatria*, **65** (4A), 947–950.

47 Volk, H.A. and Loscher, W. (2005) Multidrug resistance in epilepsy: rats with drug-resistant seizures exhibit enhanced brain expression of P-glycoprotein compared with rats with drug-responsive seizures. *Brain*, **128** (Pt 6), 1358–1368.

48 Volk, H.A., Potschka, H., and Loscher, W. (2004) Increased expression of the multidrug transporter P-glycoprotein in limbic brain regions after amygdala-kindled seizures in rats. *Epilepsy Research*, **58** (1), 67–79.

49 Roepe, P.D. (2000) What is the precise role of human MDR 1 protein in chemotherapeutic drug resistance? *Current Pharmaceutical Design*, **6** (3), 241–260.

50 Wadkins, R.M. and Roepe, P.D. (1997) Biophysical aspects of P-glycoprotein-mediated multidrug resistance. *International Review of Cytology*, **171**, 121–165.

51 Iannetti, P., Spalice, A., and Parisi, P. (2005) Calcium-channel blocker verapamil administration in prolonged and refractory status epilepticus. *Epilepsia*, **46** (6), 967–969.

52 Summers, M.A., Moore, J.L., and McAuley, J.W. (2004) Use of verapamil as a potential P-glycoprotein inhibitor in a patient with refractory epilepsy. *The Annals of Pharmacotherapy*, **38** (10), 1631–1634.

53 Dick, F.D. (2006) Parkinson's disease and pesticide exposures. *British Medical Bulletin*, **79–80**, 219–231.

54 Drozdzik, M. et al. (2003) Polymorphism in the P-glycoprotein drug transporter MDR1 gene: a possible link between environmental and genetic factors in Parkinson's disease. *Pharmacogenetics*, **13** (5), 259–263.

55 Furuno, T. et al. (2002) Expression polymorphism of the blood–brain barrier component P-glycoprotein (MDR1) in relation to Parkinson's disease. *Pharmacogenetics*, **12** (7), 529–534.

56 Tan, E.K. et al. (2004) Analysis of MDR1 haplotypes in Parkinson's disease in a white population. *Neuroscience Letters*, **372** (3), 240–244.

57 Tan, E.K. et al. (2005) Effect of MDR1 haplotype on risk of Parkinson disease. *Archives of Neurology*, **62** (3), 460–464.

58 Kuhnke, D. et al. (2007) MDR1-P-glycoprotein (ABCB1) mediates transport of Alzheimer's amyloid-beta peptides – implications for the mechanisms of Abeta clearance at the blood–brain barrier. *Brain Pathology*, **17** (4), 347–353.

59 Smyth, M.J. et al. (1998) The drug efflux protein, P-glycoprotein, additionally protects drug-resistant tumor cells from multiple forms of caspase-dependent apoptosis. *Proceedings of the National Academy of Sciences of the United States of America*, **95** (12), 7024–7029.

60 Tainton, K.M. et al. (2004) Mutational analysis of P-glycoprotein: suppression of caspase activation in the absence of ATP-dependent drug efflux. *Cell Death and Differentiation*, **11** (9), 1028–1037.

61 Loeb, M.B. et al. (2004) A randomized, controlled trial of doxycycline and rifampin for patients with Alzheimer's

disease. *Journal of the American Geriatrics Society*, **52** (3), 381–387.

62 Morley, J.J. and Kushner, I. (1982) Serum C-reactive protein levels in disease. *Annals of the New York Academy of Sciences*, **389**, 406–18.

63 Kushner, I. (1993) Regulation of the acute phase response by cytokines. *Perspectives in Biology and Medicine*, **36**, 611–622.

64 Wigmore, S.J., Fearon, K.C., Maingay, J.P., Lai, P.B. and Ross, J.A. (1997) Interleukin-8 can mediate acute-phase protein production by isolated human hepatocytes. *American Journal of Physiology*, **273** (4 Pt 1), E720–6.

65 Gabay, C. and Kushner, I., (1999). Acute-phase proteins and other systemic response to inflammation. *New England Journal of Medicine*, **340**, 448–454.

66 Puig, M.M. and Pol, O. (1998) Peripheral effects of opioids in a model of chronic intestinal inflammation in mice. *The Journal of Pharmacology and Experimental Therapeutics*, **287** (3), 1068–1075.

67 Zhao, Y.L. *et al.* (2002) Effect of endotoxin on doxorubicin transport across blood–brain barrier and P-glycoprotein function in mice. *European Journal of Pharmacology*, **445** (1–2), 115–123.

68 Goralski, K.B. *et al.* (2003) Downregulation of mdr1a expression in the brain and liver during CNS inflammation alters the *in vivo* disposition of digoxin. *British Journal of Pharmacology*, **139** (1), 35–48.

69 Wang, J.H. *et al.* (2005) Detection of P-glycoprotein activity in endotoxemic rats by 99mTc-sestamibi imaging. *Journal of Nuclear Medicine*, **46** (9), 1537–1545.

70 Zhao, Y.L. *et al.* (2002) Shiga-like toxin II modifies brain distribution of a P-glycoprotein substrate, doxorubicin, and P-glycoprotein expression in mice. *Brain Research*, **956** (2), 246–253.

71 Yu, C. *et al.* (2007) TNF activates P-glycoprotein in cerebral microvascular endothelial cells. *Cellular Physiology and Biochemistry*, **20** (6), 853–858.

72 Ronaldson, P.T. and Bendayan, R. (2006) HIV-1 viral envelope glycoprotein gp120 triggers an inflammatory response in cultured rat astrocytes and regulates the functional expression of P-glycoprotein. *Molecular Pharmacology*, **70** (3), 1087–1098.

73 Pan, W. *et al.* (1997) Tumor necrosis factor-alpha: a neuromodulator in the CNS. *Neuroscience & Biobehavioral Reviews*, **21** (5), 603–613.

74 Gutierrez, E.G., Banks, W.A., and Kastin, A.J. (1993) Murine tumor necrosis factor alpha is transported from blood to brain in the mouse. *Journal of Neuroimmunology*, **47** (2), 169–176.

75 Pan, W., Banks, W.A., and Kastin, A.J. (1997) Permeability of the blood–brain and blood–spinal cord barriers to interferons. *Journal of Neuroimmunology*, **76** (1–2), 105–111.

76 Pan, W. *et al.* (2006) Stroke upregulates TNFalpha transport across the blood–brain barrier. *Experimental Neurology*, **198** (1), 222–233.

77 Pan, W. *et al.* (2002) Selected contribution: circadian rhythm of tumor necrosis factor-alpha uptake into mouse spinal cord. *Journal of Applied Physiology*, **92** (3), 1357–1362; discussion 1356.

78 Zhou, G. and Kuo, M.T. (1997) NF-kappaB-mediated induction of mdr1b expression by insulin in rat hepatoma cells. *The Journal of Biological Chemistry*, **272** (24), 15174–15183.

79 Ando, H. *et al.* (2001) Effect of endotoxin on P-glycoprotein-mediated biliary and renal excretion of rhodamine-123 in rats. *Antimicrobial Agents and Chemotherapy*, **45** (12), 3462–3467.

80 Hartmann, G., Vassileva, V., and Piquette-Miller, M. (2005) Impact of endotoxin-induced changes in P-glycoprotein expression on disposition of doxorubicin in mice. *Drug Metabolism and Disposition*, **33** (6), 820–828.

81 Fickert, P. *et al.* (2002) Ursodeoxycholic acid aggravates bile infarcts in bile duct-ligated and Mdr2 knockout mice via

disruption of cholangioles. *Gastroenterology*, **123** (4), 1238–1251.

82 Fickert, P. *et al.* (2004) Regurgitation of bile acids from leaky bile ducts causes sclerosing cholangitis in Mdr2 (Abcb4) knockout mice. *Gastroenterology*, **127** (1), 261–274.

83 Wang, R. *et al.* (2001) Targeted inactivation of sister of P-glycoprotein gene (spgp) in mice results in nonprogressive but persistent intrahepatic cholestasis. *Proceedings of the National Academy of Sciences of the United States of America*, **98** (4), 2011–2016.

84 Jansen, P.L. *et al.* (1999) Hepatocanalicular bile salt export pump deficiency in patients with progressive familial intrahepatic cholestasis. *Gastroenterology*, **117** (6), 1370–1379.

85 Petrovic, V., Teng, S., and Piquette-Miller, M. (2007) Regulation of drug transporters during infection and inflammation. *Molecular Interventions*, **7** (2), 99–111.

86 Botta Orfila, T. and Surià Albà, R.M. (2005) Anàlisi de la Regió ENm008 Alfa Globina.

87 Lee, J. and Boyer, J.L. (2000) Molecular alterations in hepatocyte transport mechanisms in acquired cholestatic liver disorders. *Seminars in Liver Disease*, **20** (3), 373–384.

88 Trauner, M. *et al.* (2005) Molecular regulation of hepatobiliary transport systems: clinical implications for understanding and treating cholestasis. *Journal of Clinical Gastroenterology*, **39** (4 Suppl 2), S111–S124.

89 Trauner, M. and Boyer, J.L. (2003) Bile salt transporters: molecular characterization, function, and regulation. *Physiological Reviews*, **83** (2), 633–671.

90 Karpen, S.J. (2002) Nuclear receptor regulation of hepatic function. *Journal of Hepatology*, **36** (6), 832–850.

91 Chiang, J.Y. (2003) Bile acid regulation of hepatic physiology. III. Bile acids and nuclear receptors. *American Journal of Physiology. Gastrointestinal and Liver Physiology*, **284** (3), G349–G356.

92 Geier, A. *et al.* (2007) Principles of hepatic organic anion transporter regulation during cholestasis, inflammation and liver regeneration. *Biochimica et Biophysica Acta*, **1773** (3), 283–308.

93 Zollner, G. *et al.* (2003) Adaptive changes in hepatobiliary transporter expression in primary biliary cirrhosis. *Journal of Hepatology*, **38** (6), 717–727.

94 Zollner, G. *et al.* (2001) Hepatobiliary transporter expression in percutaneous liver biopsies of patients with cholestatic liver diseases. *Hepatology*, **33** (3), 633–646.

95 Trauner, M., Fickert, P., and Stauber, R.E. (1999) Inflammation-induced cholestasis. *Journal of Gastroenterology and Hepatology*, **14** (10), 946–959.

96 Whitehead, M.W., Hainsworth, I., and Kingham, J.G. (2001) The causes of obvious jaundice in South West Wales: perceptions *versus* reality. *Gut*, **48** (3), 409–413.

97 Teng, S. and Piquette-Miller, M. (2005) The involvement of the pregnane X receptor in hepatic gene regulation during inflammation in mice. *The Journal of Pharmacology and Experimental Therapeutics*, **312** (2), 841–848.

98 Cherrington, N.J. *et al.* (2004) Lipopolysaccharide-mediated regulation of hepatic transporter mRNA levels in rats. *Drug Metabolism and Disposition*, **32** (7), 734–741.

99 Siewert, E. *et al.* (2004) Interleukin-6 regulates hepatic transporters during acute-phase response. *Biochemical and Biophysical Research Communications*, **322** (1), 232–238.

100 Geier, A. *et al.* (2005) Cytokine-dependent regulation of hepatic organic anion transporter gene transactivators in mouse liver. *American Journal of Physiology. Gastrointestinal and Liver Physiology*, **289** (5), G831–G841.

101 Green, R.M., Beier, D., and Gollan, J.L. (1996) Regulation of hepatocyte bile salt transporters by endotoxin and

inflammatory cytokines in rodents. *Gastroenterology*, **111** (1), 193–198.

102 Kim, P.K. et al. (2000) Intraabdominal sepsis down-regulates transcription of sodium taurocholate cotransporter and multidrug resistance-associated protein in rats. *Shock*, **14** (2), 176–181.

103 Piquette-Miller, M. et al. (1998) Decreased expression and activity of P-glycoprotein in rat liver during acute inflammation. *Pharmaceutical Research*, **15** (5), 706–711.

104 Tang, W. et al. (2000) Endotoxin downregulates hepatic expression of P-glycoprotein and MRP2 in 2-acetylaminofluorene-treated rats. *Molecular Cell Biology Research Communications*, **4** (2), 90–97.

105 Hidemura, K. et al. (2003) Shiga-like toxin II impairs hepatobiliary transport of doxorubicin in rats by down-regulation of hepatic P glycoprotein and multidrug resistance-associated protein Mrp2. *Antimicrobial Agents and Chemotherapy*, **47** (5), 1636–1642.

106 Cherrington, N.J. et al. (2003) Induction of multidrug resistance protein 3 (mrp3) *in vivo* is independent of constitutive androstane receptor. *Drug Metabolism and Disposition*, **31** (11), 1315–1319.

107 Vos, T.A. et al. (1998) Up-regulation of the multidrug resistance genes, Mrp1 and Mdr1b, and down-regulation of the organic anion transporter, Mrp2, and the bile salt transporter, Spgp, in endotoxemic rat liver. *Hepatology*, **28** (6), 1637–1644.

108 Hartmann, G., Kim, H., and Piquette-Miller, M. (2001) Regulation of the hepatic multidrug resistance gene expression by endotoxin and inflammatory cytokines in mice. *International Immunopharmacology*, **1** (2), 189–199.

109 Sukhai, M. et al. (2000) Inflammation and interleukin-6 mediate reductions in the hepatic expression and transcription of the mdr1a and mdr1b genes. *Molecular Cell Biology Research Communications*, **4** (4), 248–256.

110 Liu, F. et al. (2008) Hypoxia modulates lipopolysaccharide induced TNF-alpha expression in murine macrophages. *Experimental Cell Research*, **314** (6), 1327–1336.

111 Walther, W. and Stein, U. (1994) Influence of cytokines on mdr1 expression in human colon carcinoma cell lines: increased cytotoxicity of MDR relevant drugs. *Journal of Cancer Research and Clinical Oncology*, **120** (8), 471–478.

112 Evans, C.H. and Baker, P.D. (1992) Decreased P-glycoprotein expression in multidrug-sensitive and -resistant human myeloma cells induced by the cytokine leukoregulin. *Cancer Research*, **52** (21), 5893–5899.

113 Stein, U., Walther, W., and Shoemaker, R.H. (1996) Reversal of multidrug resistance by transduction of cytokine genes into human colon carcinoma cells. *Journal of the National Cancer Institute*, **88** (19), 1383–1392.

114 Trauner, M. et al. (1997) The rat canalicular conjugate export pump (Mrp2) is down-regulated in intrahepatic and obstructive cholestasis. *Gastroenterology*, **113** (1), 255–264.

115 Nakamura, J. et al. (1999) Kupffer cell-mediated down regulation of rat hepatic CMOAT/MRP2 gene expression. *Biochemical and Biophysical Research Communications*, **255** (1), 143–149.

116 Roelofsen, H. et al. (1995) Impaired hepatocanalicular organic anion transport in endotoxemic rats. *The American Journal of Physiology*, **269** (3 Pt 1), G427–G434.

117 Hartmann, G., Cheung, A.K., and Piquette-Miller, M. (2002) Inflammatory cytokines, but not bile acids, regulate expression of murine hepatic anion transporters in endotoxemia. *The Journal of Pharmacology and Experimental Therapeutics*, **303** (1), 273–281.

118 Kubitz, R. et al. (1999) Regulation of the multidrug resistance protein 2 in the rat liver by lipopolysaccharide and dexamethasone. *Gastroenterology*, **116** (2), 401–410.

119 Roelofsen, H. et al. (1995) Redistribution of canalicular organic anion transport activity in isolated and cultured rat hepatocytes. *Hepatology*, **21** (6), 1649–1657.

120 Elferink, M.G. et al. (2004) LPS-induced downregulation of MRP2 and BSEP in human liver is due to a posttranscriptional process. *American Journal of Physiology. Gastrointestinal and Liver Physiology*, **287** (5), G1008–G1016.

121 Wadler, S. and Schwartz, E.L. (1992) Principles in the biomodulation of cytotoxic drugs by interferons. *Seminars in Oncology*, **19** (2 Suppl 3), 45–48.

122 Kreuser, E.D., Wadler, S., and Thiel, E. (1995) Biochemical modulation of cytotoxic drugs by cytokines: molecular mechanisms in experimental oncology. *Recent Results in Cancer Research*, **139**, 371–382.

123 Molmenti, E.P., Ziambaras, T., and Perlmutter, D.H. (1993) Evidence for an acute phase response in human intestinal epithelial cells. *The Journal of Biological Chemistry*, **268** (19), 14116–14124.

124 Wang, Q. et al. (1998) Endotoxemia and IL-1 beta stimulate mucosal IL-6 production in different parts of the gastrointestinal tract. *The Journal of Surgical Research*, **76** (1), 27–31.

125 Kalitsky-Szirtes, J. et al. (2004) Suppression of drug-metabolizing enzymes and efflux transporters in the intestine of endotoxin-treated rats. *Drug Metabolism and Disposition*, **32** (1), 20–27.

126 Iizasa, H. et al. (2003) Altered expression and function of P-glycoprotein in dextran sodium sulfate-induced colitis in mice. *Journal of Pharmaceutical Sciences*, **92** (3), 569–576.

127 Naud, J. et al. (2007) Down-regulation of intestinal drug transporters in chronic renal failure in rats. *The Journal of Pharmacology and Experimental Therapeutics*, **320** (3), 978–985.

128 Veau, C. et al. (2001) Effect of chronic renal failure on the expression and function of rat intestinal P-glycoprotein in drug excretion. *Nephrology, Dialysis, Transplantation*, **16** (8), 1607–1614.

129 Englund, G. et al. (2007) Efflux transporters in ulcerative colitis: decreased expression of BCRP (ABCG2) and Pgp (ABCB1). *Inflammatory Bowel Diseases*, **13** (3), 291–297.

130 Blokzijl, H. et al. (2007) Decreased P-glycoprotein (P-gp/MDR1) expression in inflamed human intestinal epithelium is independent of PXR protein levels. *Inflammatory Bowel Diseases*, **13** (6), 710–720.

131 Martinez, A. et al. (2007) Role of the PXR gene locus in inflammatory bowel diseases. *Inflammatory Bowel Diseases*, **13** (12), 1484–1487.

132 Dring, M.M. et al. (2006) The pregnane X receptor locus is associated with susceptibility to inflammatory bowel disease. *Gastroenterology*, **130** (2), 341–348; quiz 592.

133 Maezono, S. et al. (2005) Elevated blood concentrations of calcineurin inhibitors during diarrheal episode in pediatric liver transplant recipients: involvement of the suppression of intestinal cytochrome P450 3A and P-glycoprotein. *Pediatric Transplantation*, **9** (3), 315–323.

134 Bertilsson, P.M., Olsson, P., and Magnusson, K.E. (2001) Cytokines influence mRNA expression of cytochrome P450 3A4 and MDRI in intestinal cells. *Journal of Pharmaceutical Sciences*, **90** (5), 638–646.

135 Huang, Z.H. et al. (2000) Expression and function of P-glycoprotein in rats with glycerol-induced acute renal failure. *European Journal of Pharmacology*, **406** (3), 453–460.

136 Belliard, A.M. et al. (2004) Effect of tumor necrosis factor-alpha and interferon-gamma on intestinal P-glycoprotein expression, activity, and localization in Caco-2 cells. *Journal of Pharmaceutical Sciences*, **93** (6), 1524–1536.

137 Wong, N.A. and Pignatelli, M. (2002) Beta-catenin – a linchpin in colorectal

carcinogenesis? *The American Journal of Pathology*, **160** (2), 389–401.

138 Dietrich, C.G. *et al.* (2004) Consequences of bile duct obstruction on intestinal expression and function of multidrug resistance-associated protein 2. *Gastroenterology*, **126** (4), 1044–1053.

139 Albermann, N. *et al.* (2005) Expression of the drug transporters MDR1/ABCB1, MRP1/ABCC1, MRP2/ABCC2, BCRP/ABCG2, and PXR in peripheral blood mononuclear cells and their relationship with the expression in intestine and liver. *Biochemical Pharmacology*, **70** (6), 949–958.

140 Synold, T.W., Dussault, I., and Forman, B.M. (2001) The orphan nuclear receptor SXR coordinately regulates drug metabolism and efflux. *Nature Medicine*, **7** (5), 584–590.

141 Kauffmann, H.M. *et al.* (2002) Influence of redox-active compounds and PXR-activators on human MRP1 and MRP2 gene expression. *Toxicology*, **171** (2–3), 137–146.

142 Kast, H.R. *et al.* (2002) Regulation of multidrug resistance-associated protein 2 (ABCC2) by the nuclear receptors pregnane X receptor, farnesoid X-activated receptor, and constitutive androstane receptor. *The Journal of Biological Chemistry*, **277** (4), 2908–2915.

143 Shivananda, S. *et al.* (1996) Incidence of inflammatory bowel disease across Europe: is there a difference between north and south? Results of the European Collaborative Study on Inflammatory Bowel Disease (EC-IBD). *Gut*, **39** (5), 690–697.

144 Farrell, R.J. *et al.* (2002) P-glycoprotein-170 inhibition significantly reduces cortisol and cyclosporin efflux from human intestinal epithelial cells and T lymphocytes. *Alimentary Pharmacology & Therapeutics*, **16** (5), 1021–1031.

145 Diaz-Borjon, A. *et al.* (2000) Multidrug resistance-1 (MDR-1) in rheumatic autoimmune disorders. Part II. Increased P-glycoprotein activity in lymphocytes from systemic lupus erythematosus patients might affect steroid requirements for disease control. *Joint, Bone, Spine*, **67** (1), 40–48.

146 Maillefert, J.F., Jorgensen, C., and Sany, J. (1996) Multidrug resistance in rheumatoid arthritis. *The Journal of Rheumatology*, **23** (12), 2182.

147 Farrell, R.J. and Kelleher, D. (2003) Glucocorticoid resistance in inflammatory bowel disease. *The Journal of Endocrinology*, **178** (3), 339–346.

Index

a

abacavir 64
ABC (ATP binding cassette) transporter protein 3, 124, 160, 206, 219
– antitarget 349ff.
– canalicular membrane 301
– cellular membrane 162
– drug metabolizing enzyme 325ff.
– drug transporter of human fungal pathogen 171
– efflux transporter (ABCB1, MDR1) 278
– family 123, 238, 278
– gene of *Saccharomyces cerevisiae* 162f.
– pharmacophore 217
– physiological roles of drug 176
– polyspecificity 197
– quaternary structure 11
– structure and function 3
– *Saccharomyces cerevisiae* 164f.
– target 349
– tertiary structure 7
– topology 4
ABC pump Snq2 163ff.
ABC1 176
ABC2 176
ABCA1 369f.
ABCA2 206
ABCB1, *see also* P-gp (MDR1) 197, 240, 300ff., 325ff., 388ff., 397ff.
– CYP3A 325ff.
– endogenous substrate 302
– exogenous substrate 305
– inhibitor 310
– interaction network 372
– substrate 279ff., 355, 393
ABCB4 300ff.
– endogenous substrate 308
– interaction network 372

ABCB11 (BSEP) 206ff., 246, 300ff.
– endogenous substrate 308
ABCC1-ABCC5, *see* MRP
ABCG family 284
– ABCG1 284
– Abcg2 252, 311
– ABCG2 (BCRP, MXR) 206, 284, 300ff., 353, 373, 393
– ABCG5 284, 300ff.
– ABCG8 284, 300ff.
– endogenous substrate of ABCG2 302
– endogenous substrate of ABCG5/G8 308
– exogenous substrate of ABCG2 305
– inhibitor 207
– promoter of ABCG2 389
Abeta40 397
Abeta41 397
access protomer 133
accessory protein
– Gram-negative bacteria 134
acetazolamide 393
N-acetyl neuraminic acid 232
AcrA 130ff.
AcrAB–TolC tripartite efflux complex 129f., 143
AcrB 129ff., 210
AcrEF–TolC 143
acute lymphoid leukemia (ALL) 67
acute myeloid leukemia (AML) 67
acute-phase protein (APP) 398
acute-phase response (APR) 398
acyclovir 65
adefovir 65
adherens junction 270
ADME (absorption, distribution, metabolism, and excretion)-type descriptor 357
ADME-AP database 368

Index

ADMET (ADME/TOX, absorption, distribution, metabolism, excretion, and toxicology) 197, 368
Adp1 166
adsorption-mediated transcytosis (AMT) 270
ALDP family 163
alkaline phosphatase 265
alkoxy- and alkylaminoquinoline 143
alternate access and release mechanism 127
alternative occupancy model 129ff.
Alzheimer's disease (AD) 397
amino acid
– transporter 115, 274
2-amino-1-methyl-6-phenylimidazo[4,5-b]-pyridine (PhIP) 302ff.
aminoglycoside 137ff.
ampicillin 302ff.
angiopoetin-1 (ANG-1) 267
anion exchanger 64
anthracycline 219
antibiotic 85
anticancer drug 61ff.
– NT in cultured cell model 65
– transport 61
antiepileptic drug (AED) 385ff.
– substrate of drug transporter 390
antifungal resistance-associated transporter
– pathogenic fungi 173
antihistamine 85, 272
antiporter 235
antiviral drug 62ff.
– transport 63
AP-1 (activator protein-1) 395
APC (adenomatous polyposis coli) 404
apical membrane polarity 270
apolipoprotein A1 253, 265
apolipoprotein E 253
arsenite 302
arylpiperazine 143
Aspergillus fumigatus 160, 176
– AfuMDR1 176
– AfuMDR2 176
– AfuMDR4 176
ASRI (allosteric serotonin reuptake inhibitor) 115
astemizole 272
astrocyte 266
astrocyte end-feet 251f., 266
atorvastatin (ATV) 310, 330
ATP binding 7
ATP binding cassette, *see* ABC
ATP hydrolysis 7
– transport 7
ATP switch model 125
attention deficit hyperactivity disorder (ADHD) 115
Aus1 166, 175ff.
azidopine 22
azidothymidine (AZT) 62ff.
azole resistance 175

b

bacteria
– accessory protein from Gram-negative bacteria 134
bacterial multidrug transporter
– molecular and clinical aspect 121ff.
basal lamina 266
basal membrane polarity 270
basement membrane 266
basic fibroblast growth factor (bFGF) 267
BBB, *see* blood–brain barrier
BCRP, *see* breast cancer resistance protein
BEN^r 174
bestatin 217
bile acid 85
bile canalicular transporter 299ff.
bile salt 402
bile salt export pump (BSEP, ABCB11) 206ff., 246, 307
– ABCB11 208, 246, 307
bilirubin 402
binding protomer 133
biological membrane 231ff.
biopharmaceutical classification system (BCS) 337, 355
– substrate 339ff.
biopharmaceutics drug disposition classification system (BDDCS) 338ff.
biricodar (VX-710) 351
blood circulation 254
blood–brain barrier (BBB) 251, 263ff.
– enzyme 285
– inflammation 396
– localization of drug transporter 275
– mechanism of transport 267
– P-gp 278ff.
– strategy to overcome 286
– transport 263ff.
blood–CSF barrier (BCSFB) 263f.
BmrA (YvcC) 14
BmrR 133
Bpt1 167
brain drug delivery 286
brain endothelial cell 265f.
brain homeostasis 269
brain microvessel 251, 266

brain uptake 355
breast cancer resistance protein (Bcrp/BCRP, MXR, ABCG2) 206, 219, 249ff., 268, 284, 300ff., 353, 373, 392
– promoter 388
– substrate for BCRP transport 284
BtuC 9
BtuD 9
bZIP family 168ff.

c

calpactin 376
camptothecin-derived topoisomerase I inhibitor 219
canalicular membrane 301
canalicular transporter 301
– drug delivery 309
cancer patient
– correlation to NT function 67
Candida albicans 160
– Ca*MDR1* 174
Candida dubliniensis 176
– Cd*CDR1* 175
– Cd*CDR2* 175
– Cd*MDR1* 176
Candida glabrata 160
– Cg*AP1* 175
– Cg*CDR1* 174
– Cg*CDR2* 174
– CgPdr1 175
Candida krusei
– ABC transporter 176
Cap1 174
capecitabine 61
carbamazepine 393
carboxydichlorofluorescein 302
cardiac glycoside 85
carnitine cotransporter OCTN2, *see* zwitterion/cation transporter (OCTN)
CATALYST model 204, 216, 358
Ca*YCF1* 174
Cdr1 172ff.
CDR1 172
Cdr2 172ff.
CDR2 172
Cdr3 174
Cdr4 174
CDR5 174
CDR11 174
CEF regimen 310
ceftriaxone 302ff.
central nervous system (CNS) 252, 263ff.
– CNT 59
– neuron 390

cerebral capillary 251
cerebrospinal fluid 254
cerivastatin 101, 330
– cyclosporin A 366
– gemfibrozil 366
cetirizine 272
CFTR, *see* cystic fibrosis transmembrane conductance regulator
chemical cross-linking 22ff.
cholestasis 311
– intrahepatic cholestasis of pregnancy (ICP) 311
– obstetric 311
choroid plexus 251ff.
cib (concentrative, insensitive to NBTI, with broad substrate selectivity) 49f.
cidofovir 65
cif (concentrative, insensitive to NBTI, transporting the purine analogue formycin-B) 49f.
cimetidine 278
ciprofloxacin 138f.
cit (concentrative, insensitive to NBTI, transporting the pyrimidine thymidine) 49f.
citalopram 114
cladribine (2-CdA, 2-chlorodeoxyadenosine) 61f.
clarithromycin 274
claudin-3 271
claudin-5 271
clinical drug resistance 174
clinical study 67
clofarabine (2-chloro-2′-fluoro-deoxy-9-β-arabinosyladenine) 62
CoMFA 218
– input matrix 203
compartment of uncoupling receptor and ligand (CURL) 270
CoMSIA 218
concentrative nucleoside transporter (CNT) 51f.
– absorptive epithelia 57
– central nervous system 59
– CNT1 (SLC28A1) 49ff.
– CNT2 (SLC28A2) 49ff.
– CNT3 (SLC28A3) 49ff.
– drug transport 61
– immune system cell 58
– liver parenchymal cell 57
– SLC28 family 274
– tissue distribution and regulation 53ff.
constitutive androstane receptor (CAR) 303, 367, 402

CRC220 85
Crm1 171
Crohn's disease 404
Cryptococcus neoformans 176
– Cn*AFR1* 176
cultured cell model
– role of NT in sensitivity to nucleoside anticancer drug 65
cyclosporin 310, 330ff., 405
cyclosporine A (CsA) 101, 241, 350
– cerivastatin 366
– derivative 28
cysteine scanning mutagenesis 22ff.
cystic fibrosis transmembrane conductance regulator (CFTR) 12f., 160
cytarabine (ara-C, 1-β-D-arabinosilfuranosilcytosine) 61f.
cytochrome P450 (CYP) protein 304ff., 392
– CYP3A 325ff., 392
– CYP3A4 325ff., 351, 369
– CYP3A5 325ff.
cytokine 57
– inflammation-associated 398
cytotic mechanism 238
– transport of macromolecule 238
cytotoxicity 65

d
DAN/TIR family 170
database searching
– transporter pharmacophore 217
daunorubicin 219, 274
dehydrosilybin 219
deltrophinII 85
5′-deoxy-5-fluorouridine (5′-DFUR) 61ff.
dexamethasone 303
dexniguldipine 350
dexverapamil 350
dibromosulfophthalein (DBSP) 302
didanosine (2′,3′-dideoxyinosine, ddI) 62ff.
distance comparison (DISCO) 216
digoxin 85, 219, 241, 274, 332, 354
disease
– drug transporter 387ff.
– transporter 302
DMT (drug/metabolite transporter) superfamily 123
docetaxel 219
dopamine transporter (DAT) 115
doxorubicin 131f., 310
drug
– physiological roles of drug transporting ABC proteins 176
drug absorption

– intestinal 239
drug binding
– structural interpretation 25
drug delivery
– canalicular transporter 309
drug disposition 101
drug efflux
– *Candida glabrata* 175
drug efflux family 124
drug interaction
– OCT1 243
drug interaction database 368
drug metabolism
– gut mucosa 328
– hepatocyte 328
drug metabolizing enzyme 325ff.
– ABC transporter 325ff.
drug resistance
– clinical 174
– in yeast and fungal pathogen 159ff.
drug response element (DRE) 172
drug sensitivity
– NT function 67
drug target 113
– transporter 113
drug transport 65, 231ff.
– nucleoside- and nucleobase-derived 61
– responsiveness to treatment 65
drug transporter
– expression 388ff.
– health and disease 387ff.
– hepatic 402
– localization at the blood–brain barrier 275
– pharmacophore 215ff.
– physiological roles of drug transporting ABC proteins 176
– polymorphism 365
– refractory epilepsy (RE) 392
– systems biology 365ff.
drug–drug interaction (DDI) 101, 377f.
Dubin–Johnson syndrome 333

e
EAA loop 9f.
EAAT, *see* excitatory amino acid transporter
Ecm22 170
efflux 135ff.
– antibiotic resistance 137
– barrier 272
efflux pump 138
efflux pump inhibitor (EPI) 139ff.
efflux transport system 265
– gut mucosa 328
– hepatocyte 328

efflux resistance 137
– Gram-negative 139
– Gram-positive 137
ei (insensitivity to inhibition by nanomolar concentrations of the nucleoside analogue NBTI) 49f.
elacridar 28f., 219, 285, 310
elevator mechanism 132
EmhB 131
EmrD 126
– structure 127
EmrE 128
enalapril 85, 217
endothelial cell 251f., 264ff.
– brain 265f.
endothelin receptor antagonist BQ-123 85
endotoxin 403
enterohepatic hormone CCK-8 85
epilepsy 390ff.
– ABCB1 expression 395
epithelial cell (EC) 252
– CNT 57
equilibrative nucleoside transporter (ENT)
– drug transport 61
– ENT1 (SLC29A1) 50ff.
– ENT2 (SLC29A2) 50ff.
– ENT3 (SLC29A3) 50
– ENT4/PMAT (SLC29A4) 50
– SLC29 family 274
– tissue distribution and regulation 53ff.
es (sensitivity to inhibition by nanomolar concentrations of the nucleoside analogue NBTI) 49f.
Escherichia coli 126
escitalopram 114
estradiol 17β-D-glucuronide 366
estrogen receptor α (ESR1) 375
estrone-3-sulfate 366
etoposide 274
excitatory amino acid transporter (EAAT) 116, 274
export protein 238
extrusion protomer 133
ezetimibe 312

f

facilitated diffusion 235
FAM regimen 309
farnesoid X (activated) receptor (FXR) 303ff., 367, 402
– interaction network 372
FCR1 174
felbamate 393
felodipine 326

fexofenadine 85, 272
Fick's law 234
FK-506 351
flavopiridol 219, 334f.
– glucuronide 334f.
flippase 21, 177
Flr1 167
FLR1 174
Flu1 174
fludarabine (F-ara-A, 9-β-D-arabinosyl-2-fluoroadenine) 61f.
fluoroquinolone 137ff.
5-fluorouracil (5-FU) 61
fluoxetine 114
cis-flupentixol 22
fluvoxamine 114
Fos 376
Free-Wilson-type indicator variable 205
full transporter 4
fungal pathogen
– membrane transporters in pleiotropic drug resistance and stress response 159ff.
– antifungal resistance-associated transporter in pathogenic fungi 173

g

GABA 113
GABA transporter (GAT) 115
– GAT1 115
GABA uptake inhibitor 115
gabapentin 393
gancyclovir 65
gastrointestinal absorption 354
GBR12909 115
gemcitabine (dFdC, 2′,2′-difluorodeoxycytidine) 61f.
gemfibrozil 101
– cerivastatin 366
genetic algorithm similarity program (GASP) 204, 216
GF120918 (elacridar) 28f., 219, 285, 310
glial cell 266
glial-derived neurotrophic factor (GDNF) 266
glomerular filtration rate (GFR) 248
glucose transporter (GLUT) 116
– Glut1 252
– GLUT-1 265
– GLUT-2 239
– GLUT4 116
– GLUT-5 239
glutamate transporter
– Na$^+$-dependent 268
γ-glutamyltranspeptidase 265

glutathione sulfotransferases 304
glycine 113
glycine transporter-1 (GlyT1) 116
GP170, see P-gp
Gram-negative bacteria
– accessory protein 134
gramicidin D 278
GRID-alignment-independent descriptor (GRIND) 218
– descriptor 210, 357
GST-Pi 376
gut mucosa 328
– drug metabolism 328
– efflux transporter 328

h
H^+/peptide cotransporter 235
half transporter 4, 219
hepatic drug transporter 402
hepatic transport system
– localization in hepatocyte membrane 247
hepatobiliary transporter 402
hepatocyte 55ff.
– apical (canalicular) membrane 300
– drug metabolism 328
– efflux transporter 328
– transport system participating in drug transport in hepatocyte membrane 244
hepatocyte nuclear factor 1α (HNF1α) 370
hERG (human ether-a-go-go related gene) potassium channel 354
high-osmolarity glycerol (HOG) pathway 171
HIPHOP pharmacophore model 217
hippocampal sclerosis 392
HIV therapy
– drug 63ff.
HMG-CoA (3-hydroxy-3-methylglutaryl-coenzyme A) reductase inhibitor 85
human concentrative nucleoside transporter (hCNT)
– hCNT1 52, 67f.
– hCNT2 52, 68
– hCNT3 52, 67f.
human equilibrative nucleoside transporter (hENT)
– hENT1 51, 66ff.
– hENT2 51, 66ff.
– hENT3 51
– tissue distribution and regulation 54f.
human OATP family 97
– OATP1A2 97
– OATP1B1 98ff.
– OATP1B3 98
– OATP1C1 99
– OATP2A1 99
– OATP2B1 99
– OATP3A1 100
– OATP4A1 100
– OATP4C1 100
– OATP5A1 100
– OATP6A1 100
human peptide transporter PEPT1 217, 239, 389
human proton-dependent dipeptide transporter (hPEPT1) 365
Hxt9 168
Hxt11 168
hydrogen bond acceptor (HBA) 103, 200f.
hydrogen bond donor (HBD) 103
hydrophobicity coefficient 200

i
idiosyncratic drug reaction 402
ileal bile acid transporter (IBAT) 240
imatinib (STI 571) 353
immune system cell
– CNT 58
immuno-liposome 253
indinavir 219, 366
indocyanine green (ICG) 302
indolopyrimidine 219
induced best fit 134
inflammation 398ff.
– BBB 400
– intestine 405
– kidney 400
– liver 404
inflammatory bowel disease (IBD) 398, 406
ingenuity pathways analysis (IPA) 369ff.
ingenuity pathways knowledge base (IPKB) 372f.
interferon-γ (INF-γ) 399
interleukin
– IL-6 398ff.
– IL-8 376, 398
interstitial fluid (ISF) 264
intestine
– inflammation 405f.
intestinal peptide transport 240
intracytoplasmic loop (ICL) 10
intrahepatic cholestasis of pregnancy (ICP) 311
inverted MED1 (multiple start site downstream 1) element (iMED) 389
ion gradient-dependent transport 237
IPT1 177
irinotecan 337
itraconazole resistance 176

j

juvenile myoclonic epilepsy (JME)   391

k

kidney   246
kinetics
– carrier-mediated transport   236
Ko143   219

l

β-lactamase   141
Lactococcus lactis   124
lamivudine (2′,3′-dideoxy-3′-thiacytidine, 3tC)   62ff.
lamotrigine   393
laniquidar (R101933)   310
LAT1   274
LDL
– acetylated   265
LDL receptor   253
lectin   265
leukemia inhibitory factor (LIF)   267
levofloxacin   141
lincosamide   137
lipid A exporter/flippase MsbA   6ff.
lipid bilayer   231
lipid membrane   233
– transport   233
lipophilicity   202f.
lipopolysaccharide (LPS) endotoxin model   399
liver   244
– apical (canalicular) membrane of liver cell   300
liver parenchymal cell
– CNT   57
liver receptor homologue-1 (LRH-1)   309
liver X receptor (LXR)   367, 402
LmrA   15, 124
LmrCD   124
LmrP   127
lopinavir   219
loratadine   272
losartan   219
LY329146   219
LY335979   310
LY402913   219

m

MacA   136
macrolide   137
macrolide efflux transporter   136
macromolecule
– transport   238
macrophage   55
major vault protein (MVP)   392
MalF   10
MalG   10
maltose transporter MalFGK$_2$ 6ff.
MATE (multidrug and toxic extrusion) family   123
– MATE MDR transporter   127
maximum transport velocity   236
MC-002595   143
MC-004124   143
MC-207110   142ff.
MdfA   126f.
Mdl1   14
Mdr, *see* multidrug resistance
Mef pump   137
Mef transporter   137
membrane
– ABC protein   162
– abluminal   254
– apical (canalicular)   300
– luminal   254
membrane barrier
– pharmacokinetic-relevant   239
membrane carbohydrate   233
membrane lipid flippase   177
membrane protein   233
membrane transport   233
– protein-coupled   235
membrane transporter
– in yeast and fungal pathogen   159ff.
MepA   128
metabolic barrier   285
– enzyme at the BBB   285
metabolism and drug–transport interaction database   373
MetaCore™   368ff.
MetaDrug   368
methicillin-resistant *Staphylococcus aureus* (MRSA)   121, 139
methotrexate   85, 219, 302
MexA   135
MexAB–OprM efflux system   140ff.
MexCD–OprJ   140
MexEF–OprN   140
MexXY–OprM   140f.
MFP   135f.
MFS (major facilitator superfamily)   123, 168
– MFS permease   167ff.
– transporter   126, 176
Michaelis–Menten equation   236
midazolam   326, 340
minocycline   131f.
mitomycin   310

mitoxantrone 219
ModB 10
ModBC 10
MOE descriptor 357
molecular interaction field 203
monoamine transporter 114
monocarboxylate transporter (MCT) 274ff.
– blood–brain barrier 277
– Mct1 252
– MCT1 276
– MCT2 276
– MCT8 276
MOP (multidrug/oligosaccharidyl-lipid/polysaccharide) superfamily 123
moxifloxacin 139
MRP (multidrug resistance-associated/related protein) 235, 268, 282f., 367, 392
– endogenous substrate of ABCC2 302
– exogenous substrate of ABCC2 305
– interaction network for ABCC2 372
– MRP/CFTR subfamily 160ff.
– Mrp1/MRP1 (ABCC1) 17, 206f., 219, 246, 283, 335, 353
– MRP1 inhibitor 219
– MRP1 promoter 389
– Mrp2/MRP2 (ABCC2, cMOAT) 101, 206f., 242ff., 283, 300ff., 333ff., 353f., 366
– MRP2 promoter 389f.
– Mrp3/MRP3 (ABCC3) 206, 246, 303, 335, 403
– MRP4 (ABCC4) 206, 249, 283
– MRP5 (ABCC5) 206, 283
– Mrp6/MRP6 (ABCC6) 246, 283
– MRP7-MRP9 (ABCC10-ABCC12) 283
MRP/CFTR subfamily 160ff.
MsbA, see lipid A exporter/flippase MsbA
Msn2 171
multidrug resistance (MDR) protein, see also MRP 15, 123, 235, 403
– bacterial MDR efflux system 123
– MDR efflux pump 138
– Mdr1, see also P-gp 174, 240ff.
– MDR1, see also P-gp 278, 367
– MDR1 expression and neurodegenerative disorder 397
– mdr1a 278
– mdr1b 278
– mdr2 278
– MDR3 (ABCB4) 246, 278
– modulator 351
multidrug resistance (MDR) gene
– tissue distribution and physiological role 17
multidrug resistance transporter 122

multidrug resistance-associated/related protein, see MRP
multiple antibiotic resistance problem 121
multispecific organic anion transporter (cMOAT), see also MRP1 246, 334, 354
MXR, see breast cancer resistance protein
Mycobacterium tuberculosis 129
mycophenolate mofetil 336

n
N1 49
N2 49
N3 49
N4 49
N5 49
Na^+/L-carnitine cotransporter (OCTN2) 64
Na^+/dicarboxylate transporter 249
Na^+/glucose cotransporter 235
Na^+/H^+-exchanger (NHE) 389
Na^+/K^+-antiporter 235
Na^+/K^+-ATPase 235ff., 268
Na^+/phosphate cotransporter NPT1 249
Na^+/taurocholate cotransport system
– Ntcp 244
α-naphthylisothiocyanate 302
NBTI (nitrobenzylthioinosine) 49ff.
nelfinavir 219
network analysis
– transporter and ligand 368
neurological disorder 389
neurovascular unit 267
new chemical entity (NCE) 144f.
NFkB (nuclear factor 'kappa-light-chain-enhancer' of activated B-cells) 395
Nft1 168
Ngg1 171
nicardipine 219
Niemann-Pick C1-like 1 protein (NPC1L1) 312
nifedipine 340
NMDA receptor antagonist 116
nonlinear method 358
nonmammalian Oatp family 97
nonneurotransmitter transporter 116
NorA 138
noradrenaline 113
noradrenaline transporter (NAT) 114
norepinephrine 113
NorM transporter 128
NPC1 protein 129
nuclear hormone receptor 367
nuclear receptor 402
nuclear transcription factor Y α (NFYA) 375
nucleobase transporter 49ff.

nucleoside reverse transcriptase inhibitor (NRTI) 63
nucleoside transporter (NT) 49ff.
– drug sensitivity 67
– sensitivity to nucleoside anticancer drug 65
– transceptor 60
nucleotide binding domain (NBD) 4, 125, 160f.
– conserved motif 5
– NBD–NBD interaction 7

o

OAT, see organic anion transporter
Oatk1 249
Oatk2 249
Oatp/OATP, see organic anion transporting polypeptide
oc144093 (ONT-093) 310
OCT, see organic cation transporter
OCTN, see zwitterion/cation transporter
opioid receptor agonist 85
OprM 135
organic anion transporter (OAT) 64, 268
– human OAT 276
– Oat1 248ff.
– OAT1 (SLC22A6) 277
– Oat2 248ff.
– Oat3 248ff.
– OAT3 (SLC22A8) 277
– Oat4 248ff.
– Oat5 248
– rodent oat 276
organic anion transporting polypeptide (Oatp/OATP) 82, 268, 367
– cytochrome P450 3A 329f.
– function 83
– gene classification 82ff.
– human OATP 276f.
– nomenclature 82
– Oatp1 (Oatp1a1, Slc21a1) 244ff.
– OATP1A2 (SLC21A3) 277
– oatp1a4 (Slco21a5) 277
– oatp1a5 (Slco21a7) 277
– OATP1B1 (OATP-C, LST-1, OATP2, SLC21A6) 330, 366ff.
– OATP1B3 330
– oatp1c1 (Slco21a14) 277
– oatp2 277
– Oatp2 (Oatp1a4, Slc21a5) 245
– OATP2B1 330
– oatp3 277
– Oatp3 245
– Oatp4 (Oatp1b2, Slc21a10) 245
– oatp14 277
– OATP-A 277
– QSAR study 102
– rodent oatp 276f.
– structure 83
– substrate spectrum 85
– tissue distribution 83
organic cation transporter (OCT) 64, 244
– Oct1 250f.
– OCT1 243
– Oct2 250f.
– OCT2 243
– OCT3 243
organic ion transporter and transporting peptide 276
outer membrane factor (OMF) 134f.

p

P-glycoprotein (permeability glycoprotein, P-gp, Mdr1, mdr1a, GP170, ABCB1) 6ff., 217, 240, 249ff., 265, 278ff., 350ff.
– antitarget 354
– BBB 278ff.
– classification model 356
– comparison with Sav1866 19f.
– conformational change 17
– CYP3A 325ff.
– descriptor 357
– drug binding site 21
– drug resistance (MDR) 16, 278ff.
– in silico model 356
– inhibitor 27, 197, 282, 352
– pharmacophore model 358
– propafenone-type inhibitor 200ff.
– property for drug interaction 29
– substrate 392
– substrate for P-gp transport 279ff.
– substrate property 355
– transporter network 369ff.
paclitaxel (taxol) 242, 274
paracellular transport 270
Parkinson's disease (PD) 115, 397
paroxetine 114
partial least squares discriminant (PLSD) analysis 357
PathArt 369
Pathway Assist 369
Pathway Resource List 369
Pdh1 175
PDH1 174f.
PDR, see pleiotropic drug resistance
PDR family 163
PDR regulatory network
– modulator and regulator 169
– yeast 163ff.

PDZ-domains (PSD-95/Discs-large/ZO-1) 389
PDZK-1 389f.
PDZK-2 389f.
PEPT1 217, 239, 389
pericyte 251f., 266
permeability 355
– paracellular 272
permeability coefficient 272
pharmacophore 215
– MDR modulator-based pharmacophore model 218
– model 358
– modeling 203ff.
PharmaGKB 368
phenobarbital 393
phenothiazine 350
phenoxymethyl quinoxalinone II 219
PhIP, see 2-amino-1-methyl-6-phenylimidazo-[4,5-b]pyridine
phosphatidylcholine (PC) 232, 306
photolabeling 22ff.
physical barrier 270
plasma membrane
– canalicular 244
– sinusoidal (basolateral) 244
pleiotropic drug resistance (PDR) 160ff.
– Pdr1 168ff.
– Pdr3 168ff.
– Pdr5 163ff., 176
– PDR5 170ff.
– Pdr5p 13
– Pdr10 166ff.
– Pdr11 166ff.
– Pdr12 166ff.
– PDR12 175
– Pdr13 171
– PDR13/SSZ1 171
– Pdr15 168ff.
– PDR15 171
– regulatory network 163ff.
pleiotropic drug resistance element (PDRE) 168
PMAT (plasma membrane monoamine transporter) 50
polymorphism
– drug transporter 365
polyspecific transporter 122
polyspecificity/promiscuity 197
– structural basis 210
pravastatin 85, 302, 366
pregnane X receptor (PXR) 133, 303, 367, 402
probenecid 302
progressive familial intrahepatic cholestasis

– type 2 (PFIC2) 307
– type 3 (PFIC3) 306
propafenone-type inhibitor of P-gp 204ff.
prostaglandin transporter (PGT)
– Oatp2a1/Slc21a2 245
– PGT-2 96
prostanoid 85
protein-coupled membrane transport 235
proximal convoluted tubule (PCT) 57
proximal tubule 248ff.
prozac 114
PSC833 (valspodar) 28, 351
Pseudomonas aeruginosa 122
PXR, see pregnane X receptor

q
QacB transporter 126
QacR 133
quantitave structure–activity relationship (QSAR) study 197ff., 358, 377
– 3D-QSAR model 203, 216
– ABC transporter 197ff.
– combinatorial 209
– design inhibitors of P-glycoprotein (ABCB1) 197
– Oatp/OATP 102
– polyspecificity 197
quinidine 274, 354
quinine 350

r
rat concentrative nucleoside transporter (rCNT)
– rCNT1 53
– rCNT2 (SPNT) 52
rat organic anion transporter
– rOat2 (Slc2a7) 245
– rOat3 245
receptor-mediated transcytosis (RMT) 270
refractory epilepsy (RE) 390ff.
– drug transporter 395
– overcoming 395
– seizure frequency 394
resistome 122
9-cis-retinoic acid receptor α (RXRα) 304
RLIP76 (RALBP-1) 285, 393
rhodamine 25
ribavirin 62ff.
rifampicin (RIF) 241, 330ff.
rifampin 366
ritonavir 219, 274, 366
RND (resistance-nodulation-cell division) family 123
– inhibiting Gram-negative RND pump 141

RND MDR transporter   129
RND transporter inhibitor   142
rodent Oatp family   93
– Oatp1a1   93
– Oatp1a3-v1/v2   93
– oatp1a4 (Slco21a5)   277
– Oatp1a4   93
– oatp1a5 (Slco21a7)   277
– Oatp1a5   94
– Oatp1a6   94
– Oatp1b2   94
– oatp1c1 (Slco21a14)   277
– Oatp1c1   95
– oatp2   277
– Oatp2a1   95
– Oatp2b1   95
– oatp3   277
– Oatp3a1   95
– Oatp4a1   96
– Oatp4c1   96
– Oatp6b1/Oatp6c1   96
– oatp14   277
– rodent oatp   276f.

s

Saccharomyces cerevisiae   160
– ABC protein gene   162f.
saquinavir   219, 274, 366
Sav1866   19
– comparison with P-gp   19f.
– structure   125
selective serotonin reuptake inhibitor (SSRI)   114
serotonin   113
serotonin transporter (SERT)   114
sertraline   114
SGLT1   239
Shiga-toxin II   403
sialic acid   232
SIBAR descriptor   209
single-nucleotide polymorphisms (SNP)   365
sirolimus   326
sister of P-glycoprotein (SPGP)   206, 246
SLC (solute carrier family)   82, 244, 274
SLC transporter pharmacophore   216
SLC1 family   113f.
SLC2 family   113
– Slc2a7 (rOat2)   245
SLC5A8   377
SLC6 family   113
SLC21 family   82
SLC21A family   277
– Slc21a1 (Oatp1, Oatp1a1)   244ff.
– Slc21a2 (Oatp2a1)   245

– SLC21A3 (OATP1A2)   277
– Slc21a5 (Oatp2, Oatp1a4)   245
– SLC21A6 (OATP1B1, OATP-C, LST-1, OATP2)   330, 366
– Slc21a10 (Oatp4, Oatp1b2)   245
– Slco21a14 (oatp1c1)   277
SLC22 family   64, 243
SLC22A family   245, 276
– SLC29A3 (OCTN2)   250f., 278, 366
– SLC22A6 (OAT1)   277
– SLC22A8 (OAT3)   277
SLC28 family   274
– SLC28A1 (CNT1)   49
– SLC28A2 (CNT2)   49
– SLC28A3 (CNT3)   49
SLC29 family   274
– SLC29A1 (ENT1)   50
– SLC29A2 (ENT2)   50
– SLC29A3 (ENT3)   50
– SLC29A4 (ENT4/PMAT)   50
Slco21a
– Slco21a5 (oatp1a4)   277
– Slco21a7 (oatp1a5)   277
– Slco21a14 (oatp1c1)   277
small heterodimeric partner (SHP)   308
SMR (small multidrug resistance) family   123
– SMR MDR transporter   128
SN-38   337
Snq2   163ff.
SNQ2   170ff.
sodium–glucose cotransporter (SGLT)   116
solute carrier family, *see* SLC
SPECS compound library   205
SPNT/rCNT2 (sodium-dependent, purine-preferring nucleoside transporter)   52
St John's wort (*Hypericum perforatum*)   329
Staphylococcus aureus   128
stavudine (dideoxythymidine, d4T)   62ff.
Stb5   170
steroid-responsive Q1 element (SRE)   172
steroidhormone   85
Streptomyces   124
streptogramin   137
stress response
– in yeast and fungal pathogen   159ff.
stress response elements (STRE)   171
substrate binding protein (SBP)   4
substrate-induced fit   306
sulfasalazine   219
sulfinpyrazone   302
superbug   121, 137
supported vector machine (SVM)   358
SUR1   14
SXR, *see* pregnane X receptor

symporter 235
synercid 137
systems biology 365ff.
– drug transporter 365ff.

t

tacrolimus 406
tamoxifen 310, 350
Tangier disease 370
tariquidar (XR9576) 29f., 310
TAT-1 276
taurodeoxycholate (TUDC) 311f.
taxane (IDN 5109) 242
taxol 242, 274
TCF/LEF (T-cell factor/lymphoid enhancer factor) 404
temocaprilat 85
temporal lobe epilepsy (TLE) 397
testis-specific transporter (TST)
– TST-1 97
– TST-2 97
Tet transporter 137
tetracycline 137
thrombin inhibitor CRC220 85
thyroid hormone 85, 276
tiagabine 115
tight junction (TJ) 264ff.
time-share mechanism 129
tobramycin 141
TolC 130ff.
topiramate 393
transceptor 60
– NT 60
transcytosis 253
transendothelial electrical resistance (TEER) 265ff.
transforming growth factor-β (TGF-β) 266, 398
translocation
– energy 6
transmembrane domain (TMD) 4f., 125, 160f.
transmembrane α-helix (TMH) 125
transmembrane segment (TMS) 161
transport
– anticancer drug 61
– ATP hydrolysis 7
– blood–brain barrier 263ff.
– efflux barrier 272
– hepatic 247
– hepatocyte membrane 244
– ion gradient-dependent 237
– kinetics of carrier-mediated process 236
– lipid membrane 233

– macromolecule 238
– mechanism 5, 233, 267
– paracellular 270
transporter 113
– ABC transporter family 278
– amino acid 115
– antifungal resistance-associated transporter in pathogenic fungi 173
– bile canalicular 299ff.
– disease 302
– drug target 113
– hepatobiliary 402
– Na^+-dependent 268
– network example 369
– pharmacophore 217
– polyspecific 122
– regulation 367
– structure 113
transporter database 368
transporter gene expression data 376
Trichophyton rubrum 176
– *TruMDR1* 176
– *TruMDR2* 176
troxacitabine ((−)-2′-deoxy-3′-oxacitabine) 61f.
tubular membrane
– apical 249
tubule 250
– proximal 250
TUDC, *see* taurodeoxycholate
tumor necrosis factor-α (TNF-α) 398ff.
turpentine model 399

u

ubiquitin hydrolase 376
UDP-glucuronosyltransferase (UGT) 304, 332ff.
– ABCC2 332f.
ulcerative cholitis 404
Upc2 170ff.
uridine diphosphoglucuronic acid (UDPGA) 333
URAT1 64

v

vacuolar ABC transporter 167
valspodar (PSC833) 28, 351
VceC 135
verapamil 25, 241, 274
vesicular GABA transporter (VGAT) 115
vesicular monoamine transporter 2 (VMAT2) 115
Vibrio parahaemolyticus 128
vinblastine 22, 241, 274, 302, 350

vincristine   219, 350
Virchow–Robin space   266
vitamin B12 importer (BtuCD)   9
vitamin D receptor (VDR)   308, 402
Vmr1   168
VolSurf descriptor   357
VolSurf/GRIND descriptor   210
von Willebrand factor (vWF)   265
VX-710   351

w
Walker A region   5
Walker B region   5
War1   170
weak acid response element (WARE)   170

x
XR9576 (tariquidar)   29f., 310

y
Yap1   171
YAP1   175
Yap1 response element (YRE)   171
Yap2   171
Yap8   171
Ybt1   167
Ycf1   167
YCF1   174

yeast
– ABC protein   164f.
– cellular membrane containing ABC protein   162
– PDR regulatory network   163ff.
YEF3/Rli family   163
Yor1   167ff.
YOR1   170ff.
YRM1 (yeast reveromycin resistance modulator)   170
Yrr1   170
YRR1   170
YvcC (BmrA)   14

z
zafirlukast   219
zalcitabine (2′,3′-dideoxycytidine, ddC)   62ff.
zidovudine   63
zimelidine   114
ZO-1   271
ZO-2   271
ZO-3   271
zosuquidar (LY335979)   310
ZRT1   175
zwitterion/cation transporter (OCTN)   64
– OCTN1   250f.
– OCTN2 (SLC22A3)   250f., 278, 366